Ceramic Materials
for Electronics

MATERIALS ENGINEERING

Additional Volumes in Preparation

Ceramic Materials for Electronics
Third Edition, Revised and Expanded

edited by
Relva C. Buchanan
University of Cincinnati
Cincinnati, Ohio, U.S.A.

CRC Press
Taylor & Francis Group
Boca Raton London New York

CRC Press is an imprint of the
Taylor & Francis Group, an **informa** business

CRC Press
Taylor & Francis Group
6000 Broken Sound Parkway NW, Suite 300
Boca Raton, FL 33487-2742

First issued in paperback 2019

© 2004 by Taylor & Francis Group, LLC
CRC Press is an imprint of Taylor & Francis Group, an Informa business

No claim to original U.S. Government works

ISBN-13: 978-0-8247-4028-3 (hbk)
ISBN-13: 978-0-367-39413-4 (pbk)

Visit the Taylor & Francis Web site at
http://www.taylorandfrancis.com

and the CRC Press Web site at
http://www.crcpress.com

Library of Congress Cataloging-in-Publication Data
A catalog record for this book is available from the Library of Congress.

Preface

The scope and breadth of ceramic materials usage in electrical circuitry, related to electronic packaging and allied electroceramics applications, have expanded significantly during the decade or more since the second edition was published. The field has naturally evolved, with the heavy emphasis on bulk electroceramic materials gradually shifting toward an expanding interest in ceramic thin films, functional ceramic devices, and sensors. All nine original chapters of the second edition, therefore, have been expanded and updated to reflect this new focus, and a new chapter—specifically dealing with ceramic thin films (composition, technology, and applications), an area of emerging importance for device and memory storage applications—has been added. The chapter on multilayer ceramics technology has been substantially refocused and expanded to reflect the widespread and growing use of LTCC (low temperature cofired ceramic) components in microwave and wireless communications packaging technology. Likewise, the closely related chapter on ceramic thick film technology has been rewritten and updated, with greater emphasis placed on ink chemistry and formulations, especially for low temperature firing and base-metal usage.

The overall philosophy of the book remains in which each chapter initially presents one or more basic sections which reviews and discusses the underlying scientific principles for the particular electroceramics category, followed by a description of composition, structure, processing, characterization, device or component fabrication and utilization. The intended audience also remains largely the same. The book is intended for upper level undergraduate students in materials science and engineering or related disciplines, as well as for practicing engineers and scientists with broad interest in electroceramics material properties and their utilization.

Presentation of the chapters has also undergone some rearrangement, with the earlier chapters covering more broadly the areas of basic ceramics science, defect structure and relevant physical properties, with the later chapters more closely allied to component design, processing and device utilization, where much of this materials background and principles would be assumed.

The other major electroceramic material categories, dealing with insulators, highly conducting ceramics, capacitors, piezoelectric and electro-optic materials, ferrites, ceramic sensors and varistors, have all substantially evolved, in varying degrees, into commodity items. These material categories, nevertheless, are of active technological interest because of the expanding demand for component miniaturization, enhanced electrical performance, and economics of scale. This has led to the development of new materials formulations, more precise control and understanding of basic properties, and to innovations in processing, testing, and packaging protocols. The revised chapters reflect and develop these trends, with the incorporation of new, revised, and updated text material, figures, tables, and references.

The first chapter on insulators presents a revised and updated review of dielectric, electrical conduction and defect properties in ceramic materials, including oxide materials, non-oxide carbide and nitride materials, glasses and porcelains, related to composition, processing, thermo-mechanical properties, and uses. Chapter 2 addresses fundamental issues relating to composition, bonding, defect structure and charge transport in highly conducting ceramic materials, including oxide conductors for electrolyte, thick film, fuel cell and battery applications, superconductors, fast ion conductors, including glasses, and mixed ionic/electronic conductors. The chapter on ceramic capacitors has been substantially revised with inclusion of new material that reflects the impressive innovations in materials development and processing technology, leading to capacitive devices with micron thick layers, high volumetric capacitance, base metal electrodes, and wireless communications use. Chapter 4 presents a comprehensive and in-depth review of piezoelectric and electro-optic phenomena in ceramic materials (single crystal, polycrystalline, electro-optic), relating to their composition, structural aspects, processing, tensor properties, as well as the widespread applications of these piezoelectric and electro-optic ceramics as useful components and devices. Several appendices and many basic references provide substantial detail regarding the more fundamental aspects of these materials.

In the expanded chapter on ferrite ceramics, the range of ferrite compositions including thin films are discussed, with emphasis being placed on processing control and on the expanding range of component applications for these materials, including transformer cores and microwave designs for such components as circulators and isolators, which are widely employed in microwave cellular and satellite communication systems. In the revised chapter, the material on positive temperature coefficient (PTC) thermistors, negative temperature coefficient (NTC) thermistors, gas sensors, humidity sensors, capacitive pressure sensors, and detectors for "uncooled" thermal imaging applications have been updated, with new sections added on pyroelectric-ferroelectric sensors for infrared imaging, combustible and toxic gas sensors, humidity sensors, and pressure sensors. The chapter on ZnO ceramic varistors reviews the current mechanistic understanding of the highly nonlinear current-voltage characteristic, at the single grain–grain junction, also with the complexities introduced by variability in the grain structure. These varistor devices are now preferred for protecting electronic, electrical and power distribution and transmission circuits from destructive voltage and switching surges.

The overall result of these revisions is a fairly detailed and comprehensive treatment on electroceramic materials, with particular attention to their compositions, defect states, unique properties and processing and, most importantly, the fundamental underlying physical principles that control these properties and their circuit utilization. The scope of this utilization reflects state-of-the-art practice and innovation, with respect to useful products and components, both current and projected. The book contributes to a broader awareness and understanding of electroceramic materials, their composition, properties, and electronic uses. It constitutes a valuable resource for those working in the field of electroceramics and component device packaging.

Finally, I wish to acknowledge the invaluable assistance of Mr. Rajesh Surana, graduate research associate in the department of chemical and materials engineering at the University of Cincinnati, during the various phases of the manuscript preparation, referencing, and proofreading of the text. Thanks are also due to other graduate research associates, including Mr. Sherjang Singh and Chiyat-Ben Yau, for their assistance with referencing and text material.

Relva C. Buchanan

Contents

Contributors

Ahmed Amin Scientist, Transducer Materials Branch, Naval Sea Systems Command, Newport, Rhode Island, U.S.A.

John Ballato Professor, Department of Ceramic and Materials Engineering, Clemson University, Clemson, South Carolina, U.S.A.

William Borland Research Fellow, DuPont Electronic Technologies, Research Triangle Park, North Carolina, U.S.A.

Relva C. Buchanan Professor, Department of Chemical and Materials Engineering, University of Cincinnati, Cincinnati, Ohio, U.S.A.

Oliver Dernovsek Corporate Technology, Siemens AG, Munich, Germany

Gene H. Haertling Department of Ceramic and Materials Engineering, Clemson University, Clemson, South Carolina, U.S.A.

A. Kingon Professor, Materials Science and Engineering Department, North Carolina State University, Raleigh, North Carolina, U.S.A.

Bernard M. Kulwicki Former Manager, Research and Development, Texas Instruments Incorporated, Attleboro, Massachusetts, U.S.A.

Lionel M. Levinson Vartek Associates LLC, Schenectady, New York, U.S.A.

Rong-Fuh Louh Professor, Department of Materials Science, Feng Chia University, Taichung, Taiwan

Stanley J. Lukasiewicz Distinguished Member Technical Staff, Texas Instruments Incorporated, Attleboro, Massachusetts, U.S.A.

P. Muralt Department of Ceramic Engineering, Swiss Federal Institute of Technology, Lausanne, Switzerland

Thomas G. Reynolds III* Niceville, Florida, U.S.A.

Robert W. Schwartz Professor, Department of Ceramic Engineering, University of Missouri-Rolla, Rolla, Missouri, U.S.A.

N. Setter Professor, Department of Ceramic Engineering, Swiss Federal Institute of Technology, Lausanne, Switzerland

Savithri Subramanyam Senior Member Technical Staff, Advanced Analysis and Computational Development, Texas Instruments Incorporated, Attleboro, Massachusetts, U.S.A.

Harry L. Tuller Professor, Massachusetts Institute of Technology, Cambridge, Massachusetts, U.S.A.

R. Waser Professor, Institut f. Werkstoffe der Elektrotech, Aachen Univerisity of Technology, Aachen, Germany

Wolfram Wersing Corporate Technology, Siemens AG, Munich, Germany

---

* Formerly Mr. Reynolds was Director, Research and Development, Murata Electronics North America Inc., Smyrna, Georgia, U.S.A.

Ceramic Materials for Electronics

1

Ceramic Insulators

Relva C. Buchanan
University of Cincinnati
Cincinnati, Ohio, U.S.A.

1. INTRODUCTION

The primary function of insulators in electrical circuits is to provide physical separation between conductors and to regulate or prevent current flow between them. Important other functions include providing mechanical support, heat dissipation, thermal shock resistance, and a chemically stable environment for the conductors. Ceramic materials that provide primarily these functions are classified as ceramic insulators. They include most glasses, porcelains, oxide and nitride materials, as well as mica. The main advantage of ceramics as insulators is their capability for high-temperature operation without hazardous degradation in chemical, mechanical, or dielectric properties. In particular, the ceramic insulators must satisfy use specifications where high electrical resistance and dielectric strength, as well as relatively low dielectric constant and loss, are essential requirements [1–3].

Insulators of the type described above belong to the class of materials known as linear dielectrics. These are materials in which the electric displacement (D) increases in direct proportion to the electric field (E), where the proportionality constant is the relative dielectric constant (ε_r), a characteristic material property. The relationship is given in Eq. (1):

$$D = \varepsilon_0 E_a = \varepsilon_0\, \varepsilon_r E \qquad (1)$$

1

The relative dielectric constant ε_r is a measure of the ability of the material to store charge, relative to the permittivity or dielectric constant of vacuum, given as $\varepsilon_o = 8.85 \times 10^{-12}$ F/m [1].

In addition to the dielectric constant (ε_r), three other material properties are important in determining the insulating characteristics of the material. These are the electrical (volume) resistivity (ρ), the dissipation factor (tan δ), and the dielectric strength (DS). The electrical resistivity is simply a measure of the resistance that a unit cube of the material offers to current flow in a given (dc) field. The dielectric strength is a measure of the maximal voltage gradient that can be impressed across the dielectric without physical degradation of its insulating properties, leading to breakdown. Although critical to the insulating use of the dielectric, it is not a true material property since it is significantly influenced by measurement conditions, sample size, and flaws in the ceramic. In an ac field, the electrical conductivity and resistivity can be related to the dielectric constant and dissipation factor (which measures the energy loss per cycle, usually in the form of heat) of the material. This relationship is given by:

$$\sigma = \frac{1}{\rho} = \omega\varepsilon_0\varepsilon_r \tan \delta$$

(2)

where σ is conductivity $(\Omega\text{-m})^{-1}$; ρ is resistivity (Ω-m); ω is frequency ($2\pi f$); ε_0 is permittivity of vacuum (8.85×10^{-14} F/cm); ε_r is dielectric constant; tan δ is dissipation factor; and ε_r tan δ is dielectric loss factor is ε_r''. Equation (2) also gives the frequency dependence of the resistivity (ρ) for a lossy resistance in parallel with the capacitance.

For general use, ceramic materials that satisfy the following property criteria at 25°C are usually classified as good insulators [4–9]:

Dielectric constant (ε_r) ≤ 30
Electrical resistivity (ρ) $\geq 10^{12}$ (Ω-cm)
Dissipation factor (tan δ) ≤ 0.001
Dielectric strength (DS) ≥ 5.0 kV/mm
Dielectric loss factor (ε_r'') ≤ 0.03

Table 1 lists typical dielectric property values for some commonly used ceramic insulators, with the frequency and temperature conditions specified as appropriate. These dielectric properties are important in specifying use conditions for the insulators. Excluded from Table 1 are such classes of ceramic materials as ferroelectrics, ferrites, and sensor types such as positive temperature coefficients of resistance (PTCRs), varistors, and ceramic thermistors, which typically exhibit a nonlinear response to changes in applied field or temperature, and for which insulating characteristics, though important, are often of secondary consideration. SiO_2 has the lowest dielectric loss properties of the inorganic materials. It is

Table 1 Dielectric Properties of Ceramic Insulators

Material	Properties at 1 MHz (room temp.)				Dielectric strength (kV/mm)	Resistivity at 25°C (Ω-cm)
	Tan δ	Dielectric constant	Loss factor			
Porcelain ($R_2O \cdot Al_2O_3 \cdot SiO_2$)	0.008–0.020	5.0–6.5	0.04–0.13		20	10^{14}
Zircon ($ZrO_2 \cdot SiO_2$)	0.001	8.0–9.6	0.008–0.0096		6.3–11.5	$>10^{14}$
Steatite ($MgO \cdot SiO_2$)	0.0008–0.0035	6.0	0.005–0.02		15	10^{17}
Forsterite ($2MgO \cdot SiO_2$)	0.0005–0.001	5.8–6.7	0.003–0.007		7.9–11.9	10^{17}
Cordierite ($2MgO \cdot 2Al_2O_3 \cdot 5SiO_2$)	0.003–0.005	4.1–5.3	0.012–0.025		10	10^{16}
Alumina (Al_2O_3 90–99.9%)	0.0003–0.002	8.8–10.1	0.003–0.02		17	10^{16}
Spinel ($MgO \cdot Al_2O_3$)	0.0004	7.5	0.003		11.9	10^{14}
Mullite ($3Al_2O_3 \cdot 2SiO_2$)	0.005	6.2–6.8	0.03–0.034		7.8	10^{14}
Magnesia (MgO)	0.0001	8.9	0.0089		8.5–11.0	$>10^{14}$
Beryllia (BeO 96–99%)	0.0001–0.001	6.0	0.006–0.06		13	$>10^{16}$
Zirconia (ZrO_2)	0.01	12.0	0.12		~5.0	10^9
Thoria (ThO_2)	0.0003	13.5	0.004		~5.3	10^{10}
Hafnia (HfO_2)	0.01	12	0.12		—	10^8
Ceria (CeO_2)	0.0007	15	0.011		—	10^9
Spodumene ($Li_2O \cdot Al_2O_3 \cdot SiO_2$)	0.005	6.5–7.5	0.03–0.04		—	10^{11}
Boron nitride (BN)	0.001	4.2	0.004		35.6–55.4	10^{14}
Silicon nitride (Si_3N_4)	0.0001	6.1	0.0006		20	10^{13-14}
Pyroceram	0.0017–0.013	5.5–6.3	0.01–0.07		9.9–11.9	10^{12}
Glass-bonded mica	0.0015–0.003	6.4–9.2	0.011–0.023		10.6–23.7	10^{14}
Mica	0.0002	5.4–8.7	0.001–0.002		39.5–79.1	10^{16}
Glass ($Na_2O \cdot CaO \cdot SiO_2$)	0.0005–0.01	4.0–8.0	0.002–0.08		7.8–13.2	10^{12}
Quartz (SiO_2)	0.0003	3.8–5.4	0.0015		15–25.0	10^{14-18}
Pb-Al silicate	0.001	8.2–15	0.008–0.015		8.9–16.0	10^{13}
Aluminum nitride (AlN)	0.0001	8.8–8.9	0.001		15	10^{13}
Silicon	—	11.9	—		—	—

commonly used in insulating fibers, fiberoptics, and as piezoelectric resonators, and in the development of electrical porcelains ($R_2O \cdot Al_2O_3 \cdot SiO_2$). The silica materials have high dielectric strengths with low loss and are therefore suitable for high-voltage use. Other properties that can significantly influence the choice of dielectric for insulation purposes are the mechanical and thermal properties of the material. Typical values for these properties are listed in Table 2 for the ceramic insulators shown in Table 1. Of these, the thermal conductivity, thermal expansion coefficient, and rupture strength exercise the most direct influence on the use of the ceramic as an insulator.

II. DIELECTRIC PROPERTIES

A. Dielectric Constant and Loss

From Eq. (1) and the capacitive cell illustrated in Figure 1a, expressions for the relative dielectric constant ε_r, total charge Q (coulombs), and capacitance C (farads) for the cell can be developed as follows [5–8];

$$\varepsilon_r = \frac{D}{\varepsilon_0 E} = \frac{Q/A}{\varepsilon_0 V/d} \tag{3}$$

Therefore,

$$Q = \varepsilon_0 \varepsilon_r \frac{A}{d} V = CV \tag{4}$$

where

$$C = \varepsilon_0 \varepsilon_r \frac{A}{d} \tag{5}$$

$$C_0 = \varepsilon_0 \frac{A}{d} \tag{6}$$

and

$$\varepsilon_r = \frac{C}{C_0} = \frac{\varepsilon'}{\varepsilon_0} \tag{7}$$

In the expressions above, A represents the area of the capacitive cell, d its thickness, C_0 and C the air and material capacitance, respectively, V the voltage impressed across the cell, and ε' the material permittivity (F/m). Thus, ε_r represents the ratio of the permittivities or the charge stored in the capacitive cell relative to air or vacuum as the dielectric.

Table 2 Thermomechanical Properties of Ceramic Insulators

Material	Specific gravity	Thermal conductivity at 25°C (cal/s·°C·cm)	Thermal coeff. of expansion 25–300°C (10^{-6}/°C)	Tensile strength (MPa)	MOR transv. strength (MPa)	Compressive strength (MPa)	Thermal shock resistance
Porcelain ($R_2O \cdot Al_2O_3 \cdot SiO_2$)	2.4	0.006	6.0	48	83	352	Fair
Zircon ($ZrO_2 \cdot SiO_2$)	3.7	0.012	4.3–4.8	96	172	524	Good
Steatite ($MgO \cdot SiO_2$)	2.8	0.006	6.9–7.8	100	125	650	Moderate
Forsterite ($2MgO \cdot SiO_2$)	2.8	0.006–0.01	10	76	140	550	Poor
Cordierite ($2MgO \cdot 2Al_2O_3 \cdot 5SiO_2$)	2.2–2.9	0.005–0.007	2.2–2.4	65	105	400	Excellent
Alumina (Al_2O_3 90–99.9%)	3.85–3.9	0.06–0.07	8.0	260	445	3400	Good
Spinel ($MgO \cdot Al_2O_3$)	2.8	0.018	6.6	95	103	1710	Fair
Mullite ($3Al_2O_3 \cdot 2SiO_2$)	2.6–3.2	0.01	4.3–5.0	90	150	1200	Fair
Magnesia	3.3–3.5	0.04–0.009	10–13	90	138	950	Fair
Beryllia (BeO)	2.8–2.95	0.4–0.7	7–9	120	248	1600	Good
Zirconia (ZrO_2)	5.43–5.56	0.02–0.05	4.3–8.3	148	186	940	Poor
Thoria (ThO_2)	9.7	0.033	5.3–9.0	115	131	1524	Poor
Hafnia (HfO_2)	9.0	0.004	6.5	90	110	1386	Poor
Ceria (CeO_2)	7.0	0.029	10.0	88	110	1386	Poor
Spodumene ($Li_2O \cdot Al_2O_3 \cdot SiO_2$)	2.4	0.012	2.0	30	55	900	Good
Boron nitride (BN)	2.2–2.3	0.04–0.07	4.5	25	52	250	Good
Silicon nitride (Si_3N_4)	3.2–3.4	0.03–0.07	2.5–3.5	410	610	2000	Excellent
Pyroceram	2.4–2.6	0.004–0.009	0.2–0.4	64	248	—	Good
Glass-bonded mica	2.6–3.8	0.0012	10–14.5	69	117	214	Fair
Mica	2.6–3.8	0.0008–0.002	18–27			221	Fair
Glass ($Na_2O \cdot CaO \cdot SiO_2$)	2.0–8.0	0.002–0.004	0.5–1.0	<34	120	697	
Quartz (SiO_2)	2.2	0.003	0.3–0.4	55	—	1130	Excellent
Aluminum nitride (AlN)	2.61–2.93	0.048–0.072	4.03–6.09	441	—	—	
Silicon carbide (SiC)	3.1–3.2	0.01	2.6–4.3	85	490	1380	Excellent

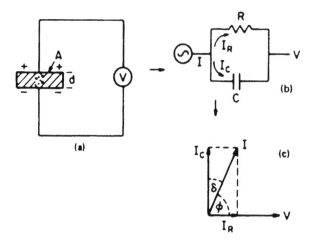

Figure 1 Equivalent circuit diagrams: (a) capacitive cell; (b) charging and loss current; (c) loss tangent for a typical dielectric.

For the case of V being sinusoidal, Eq. (4) may be written as:

$$Q = CV_0 e^{i\omega t} \tag{8}$$

Therefore,

$$I = \frac{dQ}{dt} = i\omega CV = i\omega C_0 \varepsilon_r V \tag{9}$$

where I represents the current flow on discharge of the capacitive cell in time t. However, for a real dielectric the current I has vector components I_c and I_R, as illustrated in Figure 1 for the condition of a lossy dielectric, represented by the circuit analog of a resistance in parallel with a capacitor. The current I_c represents a (wattless) capacitive current proportional to the charge stored in the capacitor. It is frequency dependent and leads the voltage by 90°. The current I_R is an ac conduction current in phase with the voltage V, which represents the energy loss or power dissipated in the dielectric. This condition can be represented by a complex permittivity or dielectric constant, in order to deal with the loss current, as follows:

$$I = i\omega C_0(\varepsilon_r - i\varepsilon_r'')V \tag{9a}$$

$$= i\omega C_0 \varepsilon_r V + \omega C_0 \varepsilon_r'' V \tag{9b}$$

$$= I_c + I_R \tag{9c}$$

From the magnitude of these currents, therefore, one can define a dissipation factor, tan δ, as

$$\tan \delta = \left| \frac{I_R}{I_C} \right| = \frac{\omega C_0 \varepsilon_0 \varepsilon_r'' V}{\omega C_0 \varepsilon_0 \varepsilon_r V} = \frac{\varepsilon_r''}{\varepsilon_r} \tag{10}$$

These vector relationships, illustrated in Figure 1, show the phase angle ϕ as (90 $- \delta^\phi$) and the loss tangent, δ representing the deviation from ideality or the loss in the dielectric.

From Eqs. (2) and (10), the dielectric loss factor, ε_r'', is shown to be the product of the material parameters ε_r and tan δ. A quality factor Q, defined as

$$Q = \frac{1}{\tan \delta} = \frac{\text{average energy stored per cycle}}{\text{energy dissipated per cycle}} \tag{11}$$

is often used as a design parameter for high-frequency dielectric use.

B. Polarization in Dielectric Solids

As indicated, the dielectric constant and dissipation factor are true material properties and can, therefore, be related to the composition and structure of the material. The macroscopic behavior can be described in terms of three vector quantities: dielectric displacement D, electric field E, and polarization P:

$$D = \varepsilon_0 \varepsilon_r E = \varepsilon_0 E + P \tag{12}$$

Therefore,

$$P = \varepsilon_0 (\varepsilon_r - 1)E = \varepsilon_0 x E \tag{13}$$

where x is defined as the electric susceptibility and is the proportionality factor that relates polarizability to field in a linear dielectric [5–8]. Polarization can be further described in terms of a bound charge density that opposes the surface charge density σ_s created by the applied field, i.e.:

$$P = \sigma \left(1 - \frac{1}{\varepsilon_r} \right) \tag{14}$$

P can also be considered in terms of a volume charge density, which is proportional to the concentration of dipoles per unit volume, N, and the local field E' in the dielectric:

$$P = \alpha N E' \tag{15}$$

with

$$E' = \frac{\varepsilon_r + 2}{3} E \tag{16}$$

where α is defined as the polarizability and is a true material parameter with components reflecting the different polarization mechanisms in the solid. From Eqs. (13) and (14) expressions can be developed (Clausius–Mosotti) relating the dielectric constant to the various polarizability mechanisms:

$$\frac{\varepsilon_r - 1}{\varepsilon_r + 2} = \frac{N_a}{3\varepsilon_0} = \frac{1}{3\varepsilon_0} \frac{N_0 \rho_d}{M} \alpha = \frac{1}{3\varepsilon_0} \sum N_i \alpha_i \tag{17}$$

where N_0 is Avogadro's number, and ρ, and M are the density and molecular weight of the solid, respectively. The components of the polarizability are:

$$\alpha = \alpha_e + \alpha_i + \alpha_0 + \alpha_s \tag{18}$$

representing susceptibilities associated with electronic, ionic, orientation, and space charge polarization, respectively.

Electronic polarization (P_e) is due to shifts or displacement of electron clouds in the dielectric field away from their equilibrium positions, resulting in a net dipolar response. It occurs in all solids up to optical frequencies of about 10^{16} Hz. Ionic polarization (P_i) results from similar ionic displacement in the field and occurs up to the infrared region of 10^{10}–10^{13} Hz. The polarization ($P_e + P_i$) = P_∞ is mostly composition dependent, instantaneous, or nearly frequency independent for most dielectrics and only marginally affected by temperature. In contrast, orientation polarization is both frequency (time) dependent and temperature dependent, since it represents dipole orientation and ion jump polarization. The parameter $P_s = P_o + P_\infty$ thus represents the total polarization from all mechanisms, including the static or low-frequency polarization.

According to Debye, P_o is time dependent, approaching its final value in an exponential manner described by:

$$P_o(t) = P_o(1 - e^{-t/\tau}) \tag{19}$$

where τ is the relaxation time [period needed to diminish the polarization charge to 36.8% ($1/e$) of its original value]. Solution of Eq. (19) in terms of complex quantities gives the Debye equations:

$$\varepsilon_r = \varepsilon_{r_\infty} + \frac{\varepsilon_{r_s} - \varepsilon_{r_\infty}}{1 + \omega^2 \tau^2} \tag{20a}$$

$$\varepsilon_r'' = \varepsilon_{r\infty} + \frac{(\varepsilon_{r_s} - \varepsilon_{r_\infty})\omega\tau}{1 + \omega^2 \tau^2} \tag{20b}$$

$$\tan \delta = \frac{\varepsilon_r''}{\varepsilon_r'} = \frac{(\varepsilon_{r_s} - \varepsilon_{r_\infty})\omega\tau}{\varepsilon_{r_s} - \varepsilon_{r_\infty}\omega^2\tau^2} \qquad (20c)$$

where ε_{r_s} and ε_{r_∞} represent the dielectric constants for the equivalent polarizations. The relaxation behavior representing these parameters for the ideal Debye solid (single relaxation mode) are illustrated in Figure 2. For ε_r and ε_r'', maximal relaxation and loss occur at $\omega\tau = 1$ and slightly higher for the tan δ peak [5–7].

The resonance losses for ionic and electronic oscillations occur at infrared and ultraviolet frequencies and are associated with absorption of the incident radiation at or near their natural vibration frequencies. Conversely, the relaxation losses are associated with dipolar orientation, ion jump, or electron hopping in such solids as SiO_2, Na silicate glasses, and semiconducting oxides, respectively. These relaxations can occur over a wide frequency range, as indicated. Space charge polarization losses typically occur at low frequency (<1 kHz) and are associated with ion migration and electrode contact losses or with the presence of grain boundary or inhomogeneous phases in the dielectric. Space charge losses in the latter case can occur at much higher ($\sim 10^5$ Hz) frequencies.

Figure 3 shows the frequency dispersion or change in dielectric properties with frequency and the corresponding circuit analogs for a general solid with several relaxation modes. The dielectric constant shows the expected decrease with frequency as successive polarization modes become damped out, and corresponding peaks occur in the loss spectra at the different relaxation and resonance frequencies [7–9]. Tables 3–5 define the relevant dielectric parameters and summarize the useful relationships between them for dielectric applications [1–3]. For tan $\delta > 0.032$, however, the series equivalent circuit gives too high a value for ε_r and ε_r'' and the parallel equivalent circuit should be used. That is, a correction should be made (using the equations above) if very accurate results are desired.

Contributions to the overall dielectric constant from different polarization mechanisms are related to the composition, frequency, and temperature of the dielectric. For predominantly electronic solids, such as diamond and molecular solids like polyethylene, the dielectric constant is closely related to the refractive index (i.e., $\varepsilon_r \sim n^2$). Otherwise, the electronic polarizability is given by [9–13]:

$$\alpha_e = 4\pi\varepsilon_0 r_0^3 = \frac{\varepsilon_0}{N_0} R_\infty \qquad (21)$$

where

$$R_\infty = \frac{(n^2-1)M}{(n^2+1)\rho_d} \qquad (22)$$

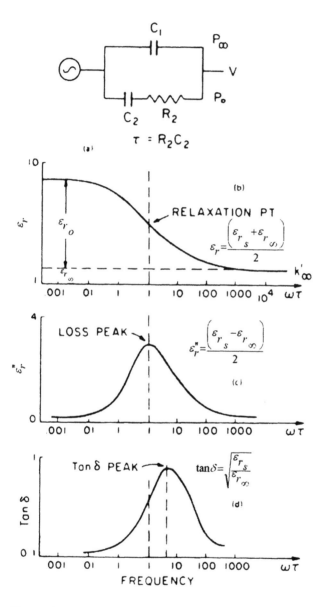

Figure 2 Schematic of (a) dielectric circuit for a Debye solid, (b) dielectric constant (polarization), (c) dielectric loss factor, and (d) dissipation factor, as a function of frequency for a solid with single relaxation mode.

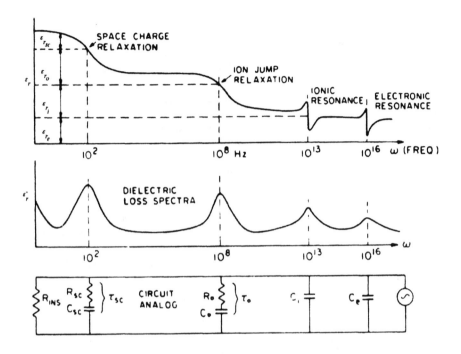

Figure 3 Frequency dispersion behavior for a general solid with multiple relaxation and resonance modes, with analog of equivalent circuits representing loss mechanisms.

Here r_0 represents the radius of the atom or ion, n the refractive index, and R_∞ the molar refractivity, values for which are available for different ions. In general, α_e increases with ion concentration, ion size, and ion polarizability; hence, the use of such ions as Ba^{2+}, Pb^{2+}, La^{3+}, and K^+ in optical glasses in order to achieve a higher dielectric constant and refractive index. For predominantly ionic solids such as MgO, Al_2O_3, and NaCl, the ionic polarizability α_i can be expressed analytically as:

$$\alpha_i = \frac{(ez)^2}{2\omega_0^2 r_0^3}\left(\frac{1}{m_1} - \frac{1}{m_2}\right) \tag{23}$$

where ω_0 is the ionic resonance frequency ($\sim 10^{13}$ Hz) at the interionic separation distance r_0, ez the charge on the ion, and m_1 and m_2 the mass of the cation and anion, respectively. Ion size and separation are seen to have a significant effect on the polarization, which in these solids generally exceeds the electronic polarization, and thus contribute significantly to the dielectric constant.

Table 3 Definitions of Dielectric Parameters

Unit	Parameter	Definition
Farad	C_0	Capacitance of cell with air/vacuum dielectric
Farad	C	Capacitance of cell with solid dielectric
Coulomb	Q	Total charge on capacitor
C/s (A)	I	dQ/dt = current due to charge flow
F/m	ε_0	Dielectric constant of vacuum (permittivity)
F/m	ε_-	Dielectric constant of solid
	ε_r	e/ε_0 = relative dielectric constant
	ε''	Loss permittivity of solid
	ε_r''	$\varepsilon''/\varepsilon_0$ = relative loss factor
$(\Omega\text{-m})^{-1}$	σ	Dielectric conductivity
	$\tan\delta$	D = dissipation factor = $\dfrac{\text{energy dissipated/cycle}}{\text{energy stored in solid}}$
	Q	Quality factor = $1/\tan\delta$
(Ω)	Z	Circuit impedance = ratio of the voltage to the total Current flowing in a circuit: $Z^*(\omega) = \dfrac{V*(\omega t)}{I*(\omega t)}$
s	τ	Time constant
	ε_0	8.85×10^{-12} F/m; 8.85×10^{-14} F/cm
	C'	Real capacitance
	C''	Imaginary part (loss part) of capacitance

Source: Data from Refs. 5–7.

For ceramic insulators with mobile ions, the orientation polarization can be considered equivalent to an ion jump polarization, α_j, where

$$\alpha_0 = \frac{p^2}{3kT} \text{ and } \alpha_j = \frac{(ezd)^2}{3kT} \tag{24}$$

and $p = ezd$ is the dipole moment associated with the jump of an ion of charge ez through a distance d. As a function of frequency, the real and imaginary loss parts of a complex α_j may be given as:

$$\alpha_j' = \frac{(ezd)^2}{3kT} \frac{1}{1+\omega^2\tau^2} \tag{25}$$

$$\alpha_j'' = \frac{(ezd)^2}{3kT} \frac{\omega\tau}{1+\omega^2\tau^2} \tag{26}$$

Table 4   Relationships Between Dielectric Parameters

Capacitance		Complex quantities
$C_0 = \varepsilon_0 \dfrac{A}{d}$	also	$\varepsilon^* = \varepsilon' - i\varepsilon''$
$C = \varepsilon' \dfrac{A}{d}$		$\varepsilon^* = \varepsilon_0 \varepsilon_r^*$
$\dfrac{C}{C_0} = \dfrac{\varepsilon'}{\varepsilon_0} = \varepsilon_r$		$\therefore \varepsilon_r^* = \varepsilon_r - i\varepsilon_r''$
	and	$\varepsilon'' = \varepsilon_0 \varepsilon_r''$
And		
$\varepsilon = \varepsilon_0 \varepsilon_r$		$C'' = \dfrac{C_s}{1+\omega^2\tau^2}; \; C'' = \dfrac{C_s\omega\tau}{1+\omega^2\tau^2}$
Time const./dissipation factor		*Charge current*
$\tan\delta = \omega R_s C_s = \dfrac{1}{\omega R_p C_p}$		$Q = CV; V = V_0 e^{i\omega\tau}$
$\tau = RC$ (series)		$I = dQ/dt = i\omega CV$

Source: Data from Refs. 5–7.

$$\tau = \frac{1}{2v} = \tau_0 e^{\mu/kT} \tag{27}$$

and

$$n = n_0 v = n_0 e^{-\mu/kT} \tag{28}$$

where τ is the relaxation time, v the jump frequency, n the number of successful ion jumps per second, and μ the activation energy for ion jump ($\sim$0.4–0.9 eV ceramics). Since for the ceramics τ decreases with temperature, the relaxations move to a higher temperature with increasing frequency. However, since n (number of successful ion jumps per unit time) also increases constantly with temperature, a relaxation peak is not found for materials such as porcelains and glasses. Instead, the dielectric constant and loss increase continuously with temperature but decrease with frequency, as illustrated for a lead-borosilicate glass in Figures 4a and 4b, where the behavior observed is due in part to the distribution of relaxation times arising from the irregular structure of the glass. Figure 5 illustrates similar behavior for a steatite porcelain, also as a function of temperature and frequency.

Table 5 Series and Parallel Dielectric Circuit Equations[a]

	Z	$|Z|$	Tan δ	τ
(a) R_S C_S		$\dfrac{1}{\omega C_s}\left(1+\omega^2 C_s^2 R_s^2\right)^{1/2}$ $=\dfrac{1}{\omega C_s}\left(1+\tan^2\delta\right)^{1/2}$ $=\dfrac{1}{\omega C_s}\left(1+\omega^2\tau^2\right)^{1/2}$	$\omega R_s C_s$	$R_s C_s$
(b) R_p C_p	$\dfrac{R_p}{1+i\omega R_p C_p}$	$\dfrac{R_p}{\left(1+\omega^2 R_p^2 C_p^2\right)^{1/2}}$ $=\dfrac{R_p}{\left(1+1/\tan^2\delta\right)^{1/2}}$	$\dfrac{1}{\omega R_p C_p}$	$\dfrac{1}{\omega^2 R_p C_p}$

(c) $R_p = R_s\left(1+\dfrac{1}{\tan^2\delta}\right)$

$C_p = \dfrac{C_s}{1+\tan^2\delta}$

$\varepsilon'_p = \dfrac{\varepsilon'_s}{1+\tan^2\delta}$; $\varepsilon_{r_p} = \dfrac{\varepsilon_{r_s}}{\left(1+\tan^2\delta\right)}$

[a] Series circuits are used mainly for low-loss materials (tan δ < 0.032), which constitute most dielectrics.

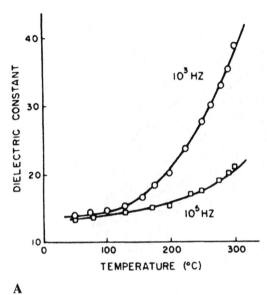

A

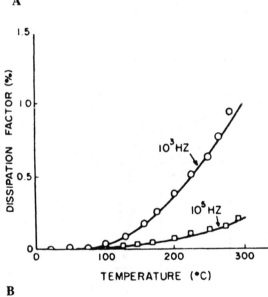

B

Figure 4 Plot of (a) dielectric constant and (b) dissipation factor, for a lead-alumina-borosilicate glass, showing change with frequency and temperature

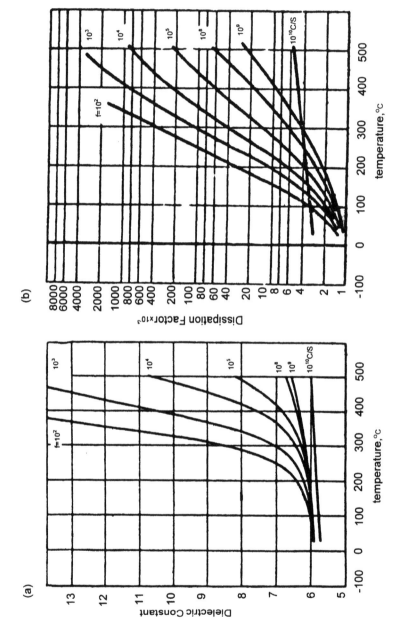

Figure 5 Plot of (a) dielectric constant and (b) dissipation factor, for a steatite porcelain ceramic, showing change with frequency and temperature. *Source:* Ref. 9.

III. DIELECTRIC STRENGTH

All dielectrics when placed in an electric field will lose their insulating properties if the field exceeds a certain critical value. This phenomenon is called dielectric breakdown, and the corresponding electric field is referred to as the dielectric breakdown strength [8–13]. Dielectric strength, therefore, may be defined as the maximal potential gradient to which a material can be subjected without loss of insulation, i.e.,

$$DS = \left(\frac{dV}{dx} \right)_{max} = \frac{V_B}{d} \tag{29}$$

where DS is the dielectric strength in kV/mm, V_B the breakdown voltage, and d the thickness of the sample. Figure 6 illustrates the current–voltage characteristics up to breakdown for a typical dielectric.

Dielectric breakdown occurs in gases due to corona discharge in nonuniform fields or to photoionization and collision of electrons with gaseous atoms, leading to further ionization, increased conduction, and dielectric breakdown. In liquids, breakdown is due primarily to dielectric heating and increased ionization processes. This process is greatly accelerated by the presence of impurities or high concentrations of mobile ions (Na^+, Li^+) in the liquid. For solids, the main breakdown mechanisms can be described as intrinsic, thermal, and ionization breakdown. A possible fourth mechanism, electrochemical degradation or breakdown, is associated with ion migration under a (dc) field gradient at elevated temperatures.

A. Intrinsic Breakdown

At high electric fields electrons may be ejected from the electrode material or may even be generated from the valence band. The electrons are then accelerated through the crystal under the influence of the high field, and some undergo electron–phonon interactions, generating heat and being absorbed in the process. Other electrons collide with ions or atoms in the solid, knocking out other electrons and elevating them to the conduction band. These newly generated electrons are in turn accelerated by the field generating other electrons, resulting finally in an "avalanche" of conducting electrons concentrated over a narrow area, causing breakdown generally through a channel. Electronic or intrinsic breakdown of this type can occur in all solids but is rarely observed, since the very high fields required ($\sim 10^6$–10^7 V/cm) usually trigger other breakdown mechanisms [10–12].

Temperature has a very pronounced effect on the electron–phonon interactions and hence on intrinsic breakdown. At low temperatures ($<< 25°C$) the

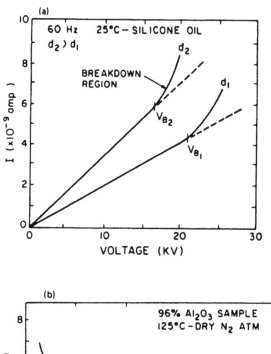

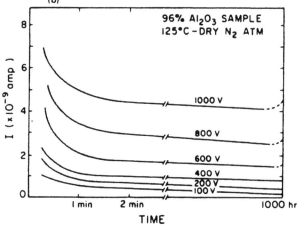

Figure 6 Breakdown voltage (V_B) as determined from current–voltage plot of a dielectric of different thickness; (b) demonstration of electrochemical breakdown from current–voltage–time characteristics. (Dashed lines show onset of breakdown.)

phonon interactions are weak, so that electrons gain more energy from the field and are able to generate many free electrons. Hence, breakdown voltages are relatively low. Intrinsic breakdown voltage peaks at room temperature, where phonon interactions become important, but decreases again above room temperature, since thermal vibrations can now generate free electrons as well. In glasses and amorphous materials the irregular structure makes the change in mean free path for electrons fairly gradual with temperature. Hence, the decrease in strength with temperature is more gradual and such factors as composition, pulse duration, and field strength become more important. Intrinsic breakdown is characterized by short breakdown times, about 10^{-8}–10^{-6} s, as well as by increased conductivity with increase in field strength.

B. Thermal Breakdown

Thermal breakdown is the most common dielectric failure mode in ceramic insulators, with dielectric strengths typically 2 orders of magnitude lower than for intrinsic breakdown. The causes for failure are generally attributed to local heating, due to a high but uneven field, and the resulting conduction losses. As the local temperature increases to the point of melting or even evaporation, the thermal gradients formed cause cracks to develop around the local hot spot. The combination of ionized gases and increased conduction associated with this condition is generally the cause of dielectric breakdown. The energy generated by an ac field can be given as:

$$W = \omega\varepsilon_0\varepsilon_r V^2 \tan\delta \sim \omega V^2 C \tan\delta \tag{30}$$

where W (watts/cycle) is the energy generated, V the voltage, and C the capacitance of the sample (farads). The heat dissipated from the sample can be approximated by

$$q = Ah(T_s - T_A) \tag{31}$$

where q represents the heat dissipated (W, cal/s), A the surface area of the sample (cm^2), h the surface heat transfer coefficient (cal/s-°C-cm^2), and T_s and T_A the surface and ambient temperatures, respectively. To avoid heat accumulation and breakdown, the heat dissipated must equal the heat generated. As is evident from the expressions above, heat generation is dependent both on frequency and voltage and on the material parameters ε_r and $\tan\delta$. These factors can, therefore, be manipulated to avoid thermal breakdown, as can the size, shape, and heat transfer environment of the ceramic part. Equation (31), for example, can be used in the design of discrete capacitors, where the heat dissipated is dependent on the size (surface area), temperature gradient, and surface heat transfer coefficient ($h \sim$ 0.0002–0.002 cal/s-°C-cm^2 for low to moderate airflow conditions).

C. Ionization Breakdown

Ionization breakdown in inhomogeneous dielectric solids occurs mainly through the mechanism of partial gas discharges, resulting from the presence of pores or cracks in the ceramic. Ionization of gases within the pores can then take place, due to the high local electric field across the gas phase [12]. This leads to local heating and the generation of a cascade of ionized charges, which transmit heat to the surrounding material. The resulting temperature gradients and development of thermal stresses leads to high dielectric losses and high conduction locally. The stresses can generate cracks, leading to further ionization and breakdown via a thermoionization process, similar to that described for thermal breakdown. The presence of porosity in an insulator, therefore, will degrade the dielectric strength, particularly in humid environments and at elevated temperatures [11–14].

D. Electromechanical Breakdown

Electrochemical breakdown is caused usually by ion migration (Na^+, Ag^+, OH) under a continuously applied dc (or low ac) field at moderately elevated temperatures (50–200°C). This can lead to interelectrode shorting of closely spaced electrodes, to ion buildup, or to depletion at one or the other electrodes, causing high local fields and electrode corrosion. Since these factors are all time dependent, this condition can lead to thermal breakdown after prolonged exposure. For most ceramic insulators, especially glasses, an initial absorption current is followed by a more gradual approach to equilibrium, reflecting rapid initial ion migration and subsequent slowing down by the uneven potential barriers in the structure. A steady-state condition is then maintained for several hours, culminating, if breakdown occurs, in a gradual increase in current flow. To evaluate electrochemical breakdown, dc fields >3000 V/cm at 100–300°C in dry ambient (N_2) for periods of time (1000 h) may be applied to the dielectric. The resulting current–voltage–time curves are shown in Figure 6b for an alumina ceramic substrate material. Figure 7 shows similar resistivity–temperature–electric field data for rutile ceramic [9–13]. The point of departure from steady-state condition can be taken as the electrochemical breakdown voltage for the given conditions, as illustrated in Figure 7 by the dashed line for rutile. Such information is frequently important for microelectronic circuit applications such as Ag conducting lines on an Al_2O_3 substrate, where a potential difference may exist between the lines. This situation is aggravated by the presence of high humidity; hence, encapsulation of such circuits is often necessary.

E. Factors Affecting Dielectric Strength

Factors that can affect dielectric strength include: composition, microstructural features (porosity, cracks, flaws, second phases), and measurement parameters

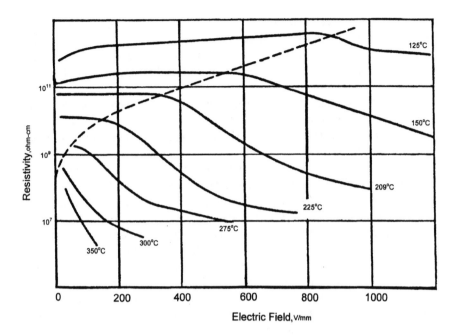

Figure 7   Plot showing change in resistivity of rutile porcelain, as a function of temperature and applied voltage. (Dashed lines show onset of breakdown.) *Source*: Ref. 9.

such as electrode configuration, specimen thickness, time, temperature, frequency, humidity, and heat transfer conditions [11–13,18].

Composition effects relate to the amorphous or crystalline nature of the
material and to the presence of mobile ions in the structure. Increased alkali
concentration not only leads to higher bulk conduction but also renders the surface
more hygroscopic. The combined effects of surface porosity, surface alkali, and
moisture may lead to high surface currents and breakdown; thus, porcelain bodies
intended for high-voltage applications are typically glazed to minimize these
effects. Microstructural features, such as cracks and porosity, can lead to ionization of entrapped gases at high fields, particularly under high-humidity conditions.
This in turn can lead to high current flow, local heating, cracking, and thermal
breakdown as described. Grain size and second phases will significantly affect
breakdown, with the smaller grains in general giving higher dielectric strengths,
reflecting the reduced conduction and lower dielectric loss in the solid.

Dielectric strength is perhaps most sensitive to changes in measurement
conditions. Electrode geometry is especially important since the breakdown voltage tends to decrease with electrode area. Because of the high fields involved,

point contacts can precipitate breakdown through field injection of electrons into the dielectric. To minimize this effect and provide environmental control, measurements are usually carried out with the sample immersed in a dielectric-grade silicone oil. This treatment also reduces the surface heat transfer and humidity effects.

Dielectric strength values are very sensitive to specimen thickness, which must be specified if results are to be meaningful. This is because flaw density and microstructural defects increase with the thickness (volume) of the specimen under test, leading to lower measured dielectric strengths as sample thickness is increased. Dielectric strength evaluations are typically carried out at room temperature and 60 Hz. The measured dielectric strengths are higher for shorter pulse or time duration of the applied voltage, due to the decrease in conduction and polarization losses in the dielectric at higher frequencies. For particular applications, therefore, dielectric strength evaluations should be carried out within the temperature and frequency range of interest. The effect of higher temperatures is to decrease the dielectric strength due to higher conduction in the solid. Dielectric strength values for ceramic insulators typically range from a low of 2×10^4 V/cm for some porcelains to 1×10^7 V/cm for some micas. The single-phase oxides, such as Al_2O_3, MgO, and BeO, are intermediate at 1.0–2.0×10^5 V/cm, with equivalent values for fused silica and the alkali-free insulating glasses; high alkali glasses may range as low as 5×10^4 V/cm. Dielectric strengths for a variety of ceramic insulators are given in Table 1.

IV. THERMAL AND MECHANICAL PROPERTIES

A. Thermal Properties of Materials

The most important properties are the thermal conductivity, the heat capacity, and the temperature coefficient of expansion [14–18].

1. Thermal Conductivity

Thermal conductivity is a relative measure of the ability of a material to dissipate heat. For ceramic materials, this encompasses a wide range of values (0.002–0.6 cal/s-°C-cm), as indicated in Table 2. It is determined mainly by the composition, structure, defect state, and temperature of the material. For equivalent conditions, highest thermal conduction is obtained for dense, single-phase oxides such as BeO, MgO, and Al_2O_3. The presence of complexing oxides ($MgO \cdot Al_2O_3$, $\cdot MgO \cdot SiO_2$) and microstructural features such as porosity, cracks, and second phases generally degrades the thermal conduction. Glasses, in particular, show low thermal conduction, which varies little with composition. Figure 8 compares the

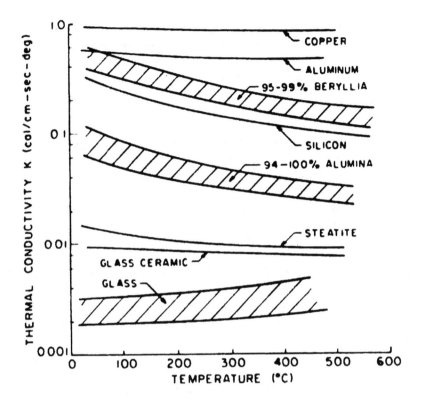

Figure 8 Change in thermal conductivity with temperature for some common materials. (From Ref. 18.)

change in thermal conductivity with temperature for some common materials [5,11,16,17]. The thermal conductivity of a material is defined as

$$q = -k\frac{dT}{dx} \tag{32}$$

where k is thermal conductivity in W/(m·°C), q is heat flux in W/cm^2, and dT/dx is the temperature gradient in °C/m at steady state. The negative sign indicates that heat flows from the areas of higher temperature to the areas of lower temperature.

Two mechanisms contribute to the thermal conductivity: the movement of free electrons and lattice vibrations, or phonons. When a material is locally heated, the kinetic energy of free electrons in the vicinity of the heat source increases, causing the electrons to migrate to cooler areas. These electrons undergo collisions with other atoms, losing their kinetic energy in the process. The net result is that heat is drawn away from the source toward the cooler areas. In similar fashion, increased lattice vibrations, as a result of higher temperature, transmit phonons which also carry energy away from the source. The thermal conductivity of the material is thus the sum of the contributions of these two mechanisms,

$$k = k_p + k_e \qquad (33)$$

where k_p is contribution due to phonons and k_e is contribution due to electrons.

In ceramics, heat flow is primarily due to phonon generation with crystalline materials, such as alumina and beryllia which are more efficient heat conductors than amorphous structures like glass. Impurities or other structural defects in the ceramics tend to lower the thermal conductivity by causing the electrons to undergo more collisions, which lowers the mobility and reduces the ability of electrons to transport heat away from the source. As the ambient temperature increases the number of collisions also increases and the thermal conductivity of most materials drops. In many practical cases this temperature dependence is critical, as ceramic materials are frequently the limiting factor for heat dissipation in microelectronic package design [16–18].

Heat flow due to convection is proportional to the surface area and the temperature differential between the surface and surrounding fluid as shown in the equation:

$$q = hA(T_s - T_f) \qquad (34)$$

where A is surface area (m^2), T_s is surface temperature (°C) T_f is fluid temperature (°C), and h is heat transfer coefficient [W/(m^2°C)]. In practice, the heat transfer coefficient, h, is quite difficult to obtain except in the most elementary cases.

Energy may be transmitted between two media not in direct contact by radiation. While conduction and convection require bulk matter to accomplish heat transfer, radiation is electromagnetic in nature and may transmit heat in a vacuum. The radiation equation is

$$q = Cf_a f_e (T_s^4 - T_r^4)S \qquad (35)$$

where C is constant, f_a is an arrangement factor, f_e is emissivity factor, S = surface area (m^2), T_s is source temperature, (°C), and T_r is receiver temperature (°C). The radiation equation is easily analyzed only for a point source, as the emissivity and arrangement factors are nonlinear and functions of temperature.

2. Thermal Resistance

According to Eq. (32), the existence of a temperature differential across a solid material, or between materials in intimate thermal contact, will lead to heat flow within the material(s) in the direction of the thermal gradient, depending on the area, effective thermal conductivity, and path length or thickness d. The rate of heat flow through a homogeneous slab material under steady-state conditions is given by the expression:

$$Q = A.\Delta T/R \tag{36a}$$

where Q = the total heat flow (Watts); A is the surface area normal to the heat flow (m^2); ΔT = the temperature difference between the warm and cool sides of the slab (°C); and R is the thermal resistance per unit area of the material ((m^2.°C)/W).

The thermal resistance R may be further defined as the temperature difference across the material which is needed to produce one unit of heat flow per unit of area. It is related to the thermal conductivity k by the relationship:

$$R = d/k \tag{36b}$$

where d represents the thickness of the slab. The concept of thermal resistance, while not a true material property in view of the thickness dependence is, nevertheless, a very useful design parameter, since the total resistance of a thermal pathway will include all of the resistances of the individual materials. For a composite structure consisting of n number of layers of different materials, for example, the *thermal resistances* can be added, such that the same area will conduct less heat energy for a given temperature difference. This total resistance can be given as:

$$R_t = \Sigma d_n/k_n \tag{36c}$$

where the resistance of the *nth* layer is: $R_n = d_n/k_n$

When the material layers are placed in parallel, their *thermal conductances* can be added and the total energy flow is increased for a given temperature difference. Hence, standard mixing rule concepts may be used to estimate the thermal resistance of a composite structure.

Thermal resistance data may also be given in terms of "R" values, which is the resistance to heat flow from one square meter of the material under a temperature difference of one degree. Therefore, a ceramic material with an R-value of 3.0 ((m^2.°C)/W), the area of insulation in square meters, multiplied by the temperature difference in degrees Centigrade and divided by 3.0, will give the heat flow in Watts.

3. Thermal Coefficient of Expansion

The thermal expansion coefficient (TCE) of a material is a measure of the characteristic change in its dimension (length or volume) for each degree change in temperature, as the material is (uniformly) heated or cooled. Thus, the linear thermal coefficient of expansion, α, can be written as:

$$\alpha = \frac{\Delta l}{l_0 \, \Delta T} \tag{37}$$

where Δl represents the change in length over a temperature span ΔT, starting from an initial length, l_0. It is a strongly temperature-dependent property, but for most ceramic insulators a near-linear range can be defined between approximately 25°C and 300°C, within which the expansion coefficient varies little. Over this range of interest the TCE of most materials is linear and may be expressed as a single number. Table 2 lists linear expansion coefficient values for a wide range of ceramic insulator materials. Thermal expansion mismatch is a major design consideration in the selection of suitable insulator materials for electrical component use.

Thermal expansion coefficient is a directional or anisotropic factor because of its strong dependence on crystal structure, bond strength, and density. Consequently, changes in volume thermal expansion with temperature closely parallel changes in specific heat capacity. For glasses, the open structure and low density typically lead to low thermal expansion coefficients, as with fused silica ($\alpha \sim 5.0 \times 10^{-7}\,°C^{-1}$). With the addition of modifier ions this value may approach $100 \times 10^{-7}\,°C^{-1}$ for a NaCa silicate glass, as a result of increased density and overall decrease in bond strength caused by the modifier ions. For strongly anisotropic materials (aluminum titanate, cordierite, spodumene), this may result in negative directional expansion and overall low-volume coefficient of expansion. Polycrystalline samples of these materials will, in consequence, show excessive grain boundary stresses and be inherently low in strength. However, thermal shock resistance may be improved [9,15,18].

B. Mechanical/Strength Properties

The strength properties of ceramic materials are usually given in terms of the elastic modulus and modulus of rupture or fracture strength, rather than in terms of the tensile strength which, because of the brittle nature of the ceramics, is generally low. In contrast, ceramics are usually strong in compression, but compressive strength is less useful as a design criteria, since it is much less sensitive to microstructural defects in the material [11,18–20].

The modulus of elasticity or Young's modulus E is a measure of elongation per unit length or strain which is developed, in the elastic range, when the material specimen is subjected to a uniform tensile stress, as given by Hooke's law:

$$\sigma = E\varepsilon \tag{38}$$

where σ is tensile stress (psi or N/m^2), E is elastic modulus (psi or N/m^2), and ε is strain (in./in. or m/m) or the net elongation per unit length. The elastic modulus is strongly dependent on composition and on the nature of the bonding and bond strength in the material. Thus, strongly bonded solids, such as many covalent (diamond, SiC, Si_3N_4) and ionic (Al_2O_3, MgO, $MgAL_2O_4$) compounds, will have high elastic modulii (>200 GPa), in contrast to more weakly bonded solids such as CsCl and many polymers.

The rupture modulus or fracture strength σ_f is a measure of the capability of the ceramic material to withstand the thermal and mechanical stresses to which it might be subjected during use. To measure the modulus of rupture, the sample is placed between two supports while load is applied at the center of the span until fracture occurs. It is also related to the composition, structure, and bonding in the material, but microstructural features such as pores, cracks, flaws, second phases, grain size, and grain boundary stresses are usually dominant. These factors are rarely quantifiable and therefore scatter in measured data tends to be large. Hence, test specimen uniformity and size are important factors, owing to the sensitivity to defects such as microcracks and voids. As with thermal conduction, dense single-phase oxides (and nitrides: AlN, Si_3N_4, BN) tend to have high modulus of rupture, whereas complex mixed oxides and glasses have moderate to low values. The practical range of values for ceramic insulators is 40–800 MPa, but higher values can be achieved with special processing.

Compressive strength is determined usually by placing the sample under compressive load and measuring the corresponding strain. By convention, the strain under compression is considered negative. Compared to metals, ceramics usually have greater compressive strengths. For example, the compressive strength of alumina at 85% purity is about 290 ksi (290,000 psi) at room temperature, increasing to 380 ksi at 99% purity. By contrast, tempered and hardened high-strength alloy steels have room temperature compressive strength ranging from 275 to 318 ksi. The tensile strength of ceramics is typically only about 10% of the compressive strength. For example, the strength of alumina ranges from 17 to 35 ksi. For applications in which tensile strength is important, thicker specimens may be utilized or the specimen design could be changed so as to take advantage of its compressive properties.

The properties discussed above often interact, especially where use conditions result in development of thermal stresses, due to the existence of thermal or expansion gradients in the material. Such conditions may occur where localized hot spots develop on a substrate due to direct heating or to the dissipation of heat from a resistor or active device. A combination of high specific heat and high thermal conductivity would minimize the rise in temperature and the thermal stresses developed. If not relieved, such stresses can result in mechanical and dielectric degradation of the system. Similar considerations arise where there is thermal expansion mismatch across interfaces, as in the use of enamel or glaze coatings and passivation layers on metal or ceramic substrates.

For ceramic insulators, possible destructive effects on heating are controlled by the thermal diffusivity, expansion coefficient, elastic modulus, and tensile strength of the material. These properties may be conveniently grouped as components of the thermal shock resistance, defined as the maximal quenching or heating rate to which the material can be subjected without undergoing physical damage. A practical measure of this quantity is the maximal temperature to which the ceramic can be heated and then quenched without physical damage resulting. Under these conditions, the thermal shock resistance represents simply a temperature rate $(T_o - T_a)/t$ change, which can be described by a semiquantitative expression relating the temperature change from T_0 to T_a (°C) in time t in the material to its thermomechanical properties, as follows [11,17,18]:

$$\frac{\Delta T}{t} = \frac{T - T_0}{t} \sim \frac{\sigma_T}{E\alpha_v}\left(\frac{\kappa}{\rho_d C_p}\right)^{1/2} S$$

(39)

where $\Delta T/t$ is the thermal shock resistance (°C/s), t the cooling time (s), σ_T and E the tensile strength and elastic modulus (MPa), α_v the volume expansion coefficient (°C^{-1}), and $\kappa/\rho_d C_p$ the thermal diffusivity (cm^2/s) of the material. Elements of the thermal diffusivity are the thermal conductivity, κ (cal-cm/s-°C-cm^2); the density, ρ_d (g/cm^3); and the specific heat capacity, C_p (cal/g − °C).

Typical values and ranges for most of these properties are given in Table 2, which, however, describes the thermal shock resistance only in qualitative terms. This is because it is not a true material property, as its magnitude depends strongly on size and shape (S) of the test specimens as well as on transient factors such as thermal diffusivity, which may be difficult to define under test conditions. Thermal shock resistance is nonetheless an important design parameter in choosing ceramic insulators for applications, where temperature changes may be severe, such as for heater and resistor core and thick-film substrate applications.

Of the properties contributing to the thermal shock resistance, the modulus ratio (σ_T/E) and the thermal diffusivity tend to vary only over narrow ranges. The thermal diffusivity range for most ceramic insulators is 0.1–0.3 cm^2/, reflecting the fact that both κ and C_p tend to increase or decrease in response to the same materials characteristics. This is generally true also for the fracture strength and elastic modulus, which are closely related to chemical bonding and structure in the different solids. The elastic moduli for ceramic insulators fall in the narrow range of 34–414 GPa, reflecting the very strong ionic or covalent bonding. Fracture strengths for ceramic insulators are much lower and more variable, since they are dependent on processing, test conditions, and flaws in the ceramic. Even so, high elastic moduli can usually be correlated with high fracture strengths, so that variability in the ratio of the two is small.

The dominant property controlling thermal shock resistance, therefore, is the thermal expansion coefficient, which for ceramic insulators typically ranges from −0.4 to 145 × 10^{-7} cm/cm − °C (Table 2). The volume thermal expansion α_v in Eq. (39) is approximately three times the linear expansion coefficient. Figure

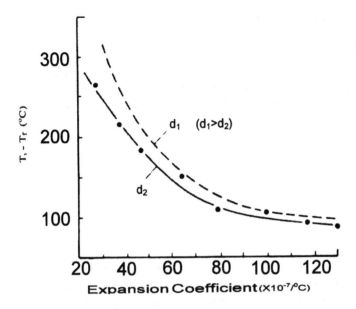

Figure 9 Change in thermal shock resistance for glass samples of thickness d_1 and d_2 $(d_2 > d_1)$ as a function of thermal expansion coefficient. (From Ref. 89.)

9 illustrates the sharp decrease in thermal shock resistance that is observed for glasses of different thickness $(d_1 > d_2)$ as the thermal expansion coefficient is increased. When measured under equivalent test conditions, the ΔT values range from about 50°C for high-alkali and lead (soft) glasses, to about 150°C for (hard) borosilicate glasses, to about 1000°C for a fused silica glass, at equivalent cooling rates and thickness. This range of thermal shock resistance values typically hold also for the nonvitreous ceramics in Table 2, where high thermal conduction, high fracture strength, and low to moderate thermal expansion coefficients generally yield good to excellent thermal shock values. Examples include BeO, cordierite, Al_2O_3, zircon, and spodumene, in contrast to such high-expansion ceramics as forsterite and zirconia, which exhibit generally poor thermal shock characteristics.

Aside from considerations of thermal shock, thermal stresses can also develop in glass coatings at steady state due to thermal expansion mismatch with the substrate or due to temperature gradients developed across the glass. The tensile stresses developed (proportional to α, E, and ΔT as well as the coating thickness) must be significantly lower than the rated tensile strength of the glass to prevent failure.

V. TRANSPORT PROPERTIES

A. Electrical Conduction

Electrical conduction is that property which relates current flux in a material to the electrical field impressed across it. The current flux or current density can also be defined in terms of both measurement and material parameters, as follows [10–12,14,21]:

$$j_c = \sigma \frac{dV}{dx} = \sigma E \tag{40a}$$

$$j_c = \frac{I}{A} \tag{40b}$$

$$j_c = nq\mu E \tag{40c}$$

where j_c (A/m^2) is the current density, $\sigma(\Omega\text{-m})^{-1}$ the conductivity, dV/dx the voltage gradient or field E, μ (m^2/V-s) the mobility, n the concentration of charge carriers, and $q = ez$ (coulombs) the charge on the carriers ($e = 1.6 \times 10^{-19}$ C and $z = $ valence or charge on the carriers). From Eq. (10), expressions for the conductivity can be developed in terms of measured parameters such as the resistance R:

$$\sigma = \frac{1}{R} \frac{d}{A} \tag{41a}$$

and

$$\frac{1}{\sigma} = \rho = R \frac{A}{d} \tag{41b}$$

or,

$$\sigma = nq\mu \tag{41c}$$

in terms of the material parameters, where ρ is the electrical resistivity (Ω-m), R (ohms) the sample resistance, A its area (m^2), and d the thickness (m). For more than one type of charge carrier being present, the resultant conductivity is given as the sum of component conductivities (σ_i) as follows:

$$\sigma = \sum_i n_i (ez)_i \mu_i = \sum_i \sigma_i \tag{42a}$$

and

$$\sigma_i = \sigma \, t_i \tag{42b}$$

where t_i ($0 \leq t_i \leq 1$) is the transference number, representing the fraction of the total current or conductivity carried by the mobile charge carriers, i, which may be electronic (electrons or holes) or ionic (cations and/or anions). Depending on which charge carriers dominate, the solid may be classified as primarily an electronic (n or p type) or ionic conductor. However, mixed conduction does occur, and this conduction is then represented as:

$$\sigma = \sigma_{elect} + \sigma_{ionic} \tag{43}$$

In such solids the ionic component tends to dominate at higher temperatures, typically above 200°C. Direct determination of the transference number is difficult to perform experimentally. In most cases, the ionic component is determined by use of a coulometric type of cell or, where appropriate, through a gas-phase reaction and use of the Nernst relationship, e.g., in measuring the oxygen ion component in ZrO_2 [11,22–24]. Table 6 gives data on transference number for a few ionic and electronic compounds.

Where the number of charge carriers taking part in the conduction process is essentially constant with temperature [Eq. (41c)], the conduction process may be described as extrinsic or impurity controlled. This condition arises from the presence of mobile impurities in the solid, or from selective doping, e.g., the addition of $BaCl_2$ (Figure 10) to KCl or P to Si. The extrinsic

Table 6 Transfer Numbers of Several Materials

Compound	T_c (cation)	t_a (anion)	t_a (electron)	T_p (hole)	Temp. (°C)
NaCl	1.00	—	—	—	400
	0.98	0.02	—	—	500
	0.91	0.09	—	—	600
	0.88	0.12	—	—	620
CuCl	1.00	—	—	—	250–400
NaF	1.00	—	—	—	500
	0.94	0.06	—	—	585
	0.86	0.14	—	—	625
KBr	0.5	0.5	—	—	605
$BaCl_2$	—	1.0	—	—	400–700
CaF_2	—	1.0	—	—	200–1000
ZrO_2	—	1.0	—	—	800–1000
x-Ag_2S	—	—	1.0	—	220
Cu_2O	—	—	—	1.0	800–1000
FeO	—	—	—	1.0	1000

Source: Ref. 22

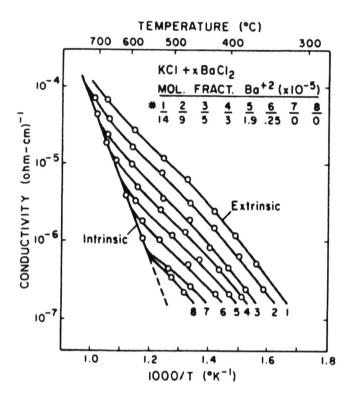

Figure 10 Plot of log conductivity versus reciprocal absolute temperature for Ba-doped KCl showing intrinsic and extrinsic conduction with increasing Ba^{2+} concentration.

conduction is typified by higher conductivity and lower activation energy, since energy is needed only for charge carrier migration. In contrast, for intrinsic conduction the fraction of charge carriers (n) contributing to the conduction increases with temperature by an activated process. The higher activation energy for this process is therefore indicated, as illustrated in Figure 11.

The mobility term is defined as $\mu = \bar{v}/E$, where $\bar{v}$ is the average drift velocity of the carriers under the influence of the field E. For semiconductor materials such as Si and GaAs, the mobility at all but the lowest temperatures (where impurity scattering predominates) is determined by lattice or phonon scattering of the charge carriers and decreases with temperature. This relationship may be given as

$$\mu \sim const \times T^{-3/2} \tag{44}$$

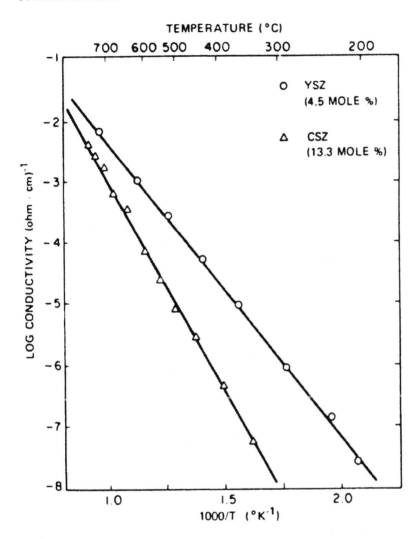

Figure 11 Comparison of conductivity versus reciprocal absolute temperature behavior for calcia (CSZ, 13.3 mol %)–and yttria (YSZ, 4.5 mol %)–stabilized zirconia ceramics.

and for low-temperature impurity scattering

$$\mu \sim \text{const} \times T^{3/2} \tag{45}$$

For these semiconductor materials, the mobility and conductivity are related by the Hall coefficient R_H, as follows:

$$\mu = R_H \sigma \tag{46}$$

where

$$R_H = \pm \frac{1}{ne} \tag{47}$$

and

$$\sigma = \sigma_o e^{-Eg/2kT} \text{ (intrinsic)} \tag{48}$$

or

$$\sigma = \sigma_o \, e^{-Ex/kT} \text{ (extrinsic)} \tag{49}$$

where E_g and E_x represent the activation energies for intrinsic (band gap) and extrinsic (impurity level) conduction, respectively; k is the Boltzmann constant; and T is the absolute temperature.

For an ionic solid, mobility is related to the diffusion coefficient D (cm^2/s) by the Einstein relationship

$$\mu = \frac{ez}{kT} D \tag{50}$$

Diffusion and conductivity are related by the Nernst–Einstein equation:

$$\sigma = \frac{n(ez)^2 D}{kT} \tag{51}$$

Both diffusion and the number of defects generated N are activated processes whereby

$$N = ne^{-w/2kT} \tag{52}$$

and

$$D = D_o e^{-\mu/kT} \tag{53}$$

then

$$\sigma = \sigma_o e^{-E/kT} \tag{54}$$

and

$$E = \frac{w}{2} + \mu \tag{55}$$

Here w and μ are the activation energies for defect generation and migration, respectively. For extrinsic conduction $w = 0$ and $E = \mu$; that is, the ionic mobility becomes the controlling factor in the conduction. It should be noted that for ionic vacancy diffusion, the product $N_v D_v = n_i D_i$, where N_v and D_v represent the vacancy concentration and diffusion coefficient at a given temperature, with n_i and D_i representing these parameters for the diffusing ionic species. Since N_v is typically several orders of magnitudes lower than n_i, it follows that vacancy diffusion rates are significantly higher for the ionic species.

B. Defects

Diffusion processes in ionic solids require the presence of point defects in the crystal lattice. In consequence, the conductivity is closely related to the existence of these ionic defects, which are of four kinds: (1) cation vacancies, (2) cation interstitials, (3) anion vacancies, and (4) anion interstitials [11,14,21,23–25]. Table 7 lists the Kröger–Vink nomenclature for the different types of ionic defects.

Defects may be generated in a solid through (1) entropy changes (Schottky, Frenkel); (2) chemical substitution; (3) oxidation or reduction; and (4) energetic radiation [21–24]. The thermodynamic Schottky and Frenkel defects are inherent defects, with one or the other type being always present in an ionic solid at all temperatures above absolute zero. The formation of these defects increases the configurational entropy of the solid, thereby decreasing the overall free energy. Relationships for Schottky defect formation may be given as:

$$M_M + X_x \rightarrow V''_M + V''_x \tag{56}$$

Table 7 Kröger–Vink Nomenclature for Ionic-Type Defects Referenced to MO lattice

Symbol	Definition[a]
M_M	Cation on regular lattice site (Ni_{Ni})
X_X	Anion on regular lattice site (O_o)
V_M	Cation vacancy
V''_M	Effective charge on cation vacancy
M_i	Cation on interstitial site
M_i^{**}	Effective charge on interstitial cation
X_i	Anion on interstitial site
X''_i	Effective charge on interstitial anion
V_x	Anion vacancy
V_x^{**}	Effective charge on anion vacancy

[a] Values for W_s, W_f, and μ in kcal/mole. 23.0 kcal/mole = 1.0 eV.
Source: Ref. 22.

and

$$n_s = ne^{-w_s/2kT} \tag{57}$$

where n_s is the concentration of Schottky defects and w_s is the activation energy for defect formation. Schottky defects are typically formed in close-packed (face-centered cubic, hexagonal close packed) structures such as the alkali halides, MO oxides, and corundum (Al_2O_3) since such structures cannot readily accommodate interstitial ions. The cation and anion vacancies are considered to form separately and migrate to the surface of the crystal, thereby increasing the volume and decreasing the density. Relationships for Frenkel defects may be written as:

$$M_M \rightarrow M_i^{\cdot\cdot} + V''_M \tag{58}$$

or

$$X_x \rightarrow X''_i + V_x^{\cdot\cdot} \tag{59}$$

and

$$n = (nn^*)^{1/2}e^{-w_f/2kT} \tag{60}$$

where n_f is the concentration of Frenkel defects, n^* the number of interstitial sites, and w_f the activation energy for defect formation. Frenkel defects typically form in more open structures (CaF_2), where there is a size and charge disparity between the cations and anions. This includes such solids as BaF_2, ZrO_2, UO_2, and the Ag and Cu halides. Frenkel solids are often electronic conductors at lower temperatures, due to the prevalence of covalent bonding, e.g., in the Cu halides shown in Table 8.

Chemical substitution can result in any of the four defects indicated being formed in the crystal lattice, as follows:

Table 8 Schottky and Frenkel Defect Compounds [11,25]

	Schottky			Frenkel				
		μ	μ			μ		
Compound	w_s	V_M	V_X	Compound	w_f	M_i or X_i	V_M or V_x	
LiCl	49	9.5	—	AgCl	37		3.2	8.5
NaCl	48	20	26	AgBr	28		2.6	5.5
CsBr	46	13	6	CaF_2	65	38	16	
MgO	90–150	69	62	ZrO_2	≈95	25	—	
Al_2O_3	470	—	≈58	UO_2	77	26–30	—	

1. Cation vacancy:

$$x(CaCl_2) \xrightarrow{NaCl} x(Ca_{Na}^{\cdot} + 2Cl_{Cl} + V_{Na}')$$ (61)

2. Anion vacancy:

$$x(CaO) \xrightarrow{ZrO_2} x(Ca_{Zr}'' + O_o + V_o^{\cdot\cdot})$$ (62)

3. Cation interstitial:

$$x(AlNa) \xrightarrow{SiO_2} x(Al_{Si}' + Na_i^{\cdot})$$ (63)

4. Anion interstitial:

$$x(NdF_3) \xrightarrow{CaF_2} x(Nd_{Ca}^{\cdot} + 2F_F + F_i')$$ (64)

Reaction (63) represents the classic weathering of silica to form a feldspar mineral. Constraints on the substitution are that the impacted ions should be within $\pm 15\%$ in ionic size, of not too great disparity in valence state, and of similar coordination structure. Any other condition leads to very limited substitution of the ion in the host lattice. Chemical substitution may also be used to induce valence changes in the material, as follows:

$$x(La_2O_3) \xrightarrow{BaTiO_3} x(2La_{Ba}^{\cdot +} 3O_o + 2Ti_{Ti}')$$ (65)

where, as indicated, a reduction of Ti^{4+} ions to the Ti^{3+} state takes place to offset the increased charge on the A site in $BaTiO_3$. The juxtaposition of Ti^{3+} and Ti^{4+} ions on equivalent sites facilitates electron transfer or hopping from the Ti^{3+} to the Ti^{4+} ions with increased n-type conduction, depending on the Ti^{3+} concentration. This mechanism defines another type of point defect, the polaron, which is an electron or hole trapped on a particular atom, ion, or vacancy. The electron or hole and its associated polarization field may be considered as a particle since it can propagate throughout the crystal by hopping from one ion to another. If the dimension of the polarization field is small compared to the lattice dimension of the crystal, this entity is described as a small polaron [25–28].

C. Defects in Nonstoichiometric Oxides

Point defects that are formed in oxides as a result of equilibrium in a gas-phase ambient may occur in two different ways: (1) by oxygen deficiency (excess metal) or (2) by metal deficiency (excess oxygen) with respect to a stoichiometric composition, as follows.

1. Oxygen-Deficient Oxides

For such an oxide, the overall nonstoichiometric reaction can be written as

$$MO \rightarrow MO_{1-x} + \frac{x}{2}O_2 \tag{66}$$

The oxygen vacancy is formed when an oxygen ion on a normal lattice site is transformed to the gaseous state [11,21–25]:

$$O_0 \rightarrow \frac{1}{2}O_2(g) + V_0^{\cdot\cdot} + 2e' \tag{67}$$

Since the vacancy has two trapped electrons, it can act as a donor and become singly or doubly ionized. This would lead to n-type conduction at high temperature in such fixed-valence oxides as Al_2O_3, MgO, and CaO. The free electrons may also become associated with variable-valence cations and the following reactions can occur:

$$M_M + V_0 \rightarrow V_0^{\cdot} + M_M' \tag{68}$$

$$M_M' + V_0^{\cdot} \rightarrow V_0^{\cdot\cdot} + M_M'' \tag{69}$$

Equation (68) is the more likely reaction and leads to n-type hopping conduction in such oxides as TiO_2, Fe_2O_3, CeO_2, and Nb_2O_5.

In the case of excess metal compounds, e.g., ZnO, the metal atoms may occupy interstitial sites, as follows:

$$M_M + O_0 \rightarrow M_i^{\cdot\cdot} + \frac{1}{2}O_2(g) + 2e'' \tag{70}$$

The neutral M_i atom can subsequently be ionized:

$$M_i \rightarrow M_i^{\cdot} + e' \tag{71}$$

$$M_i^{\cdot} \rightarrow M_i^{\cdot\cdot} + e' \tag{72}$$

Again, the singly ionized state is the more likely and leads to n-type conduction in the ZnO.

2. Metal-Deficient Oxides

In an MO oxide, cation vacancies can be formed through the reaction of ambient oxygen with the oxide:

$$\frac{1}{2}O_2(g) \rightarrow V_M'' + O_O + 2h^\bullet \tag{73}$$

The holes present in the vacancy may be excited and transferred to other parts of the crystal:

$$V_M'' \rightarrow V_M' + h^\bullet \tag{74}$$

This condition leads to p-type conduction in such oxides as MnO, NiO, FeO, and CoO. As a valence defect, Eq. (70) can be written as:

$$M_M + V_M'' \rightarrow V_M' + M_M^\bullet \tag{75}$$

With excess oxygen, neutral interstitial oxygen atoms may be formed thus:

$$\frac{1}{2}O_2(g) \rightarrow O_i \tag{76}$$

but this condition is not common in oxide lattices.

The relationship of oxygen partial pressure to conduction in metal-deficient oxides can be illustrated as follows:

$$2M_M + \frac{1}{2}O_2(g) \rightarrow V_M'' + 2M_M^\bullet + O_O \tag{77}$$

The equilibrium constant, K_T, for this reaction at temperature T is:

$$K_T = \frac{[M_M^\bullet]^2[V_M''][O_O]}{(P_{O_2})^{1/2}[M_M]^2} \tag{78}$$

Where $[O_o]$ and $[M_M] \sim 1$ and $V_M''' = 1/2\,[M_M^\bullet]$. Since σ, the conductivity, is proportional to $[M_M^\bullet]$ (e.g., $[Ni^{3+}]$), we can derive a relationship:

$$\sigma = KP_{O_2}^{1/6} \tag{79}$$

Thus, if the P_{O_2} is increased, the conductivity also increases. In oxygen-deficient oxides it can be shown, similarly, that the effect of increase in P_{O_2} is just the opposite, i.e.,

$$\sigma = KP_{O_2}^{-1/6} \tag{80}$$

Depending on the oxygen partial pressure, therefore, n- or p-type conduction may occur at very low and high P_{O_2}, respectively. For P_{O_2} ranges where the oxygen and metal vacancy concentrations are of similar magnitude, the oxide is essentially stoichiometric and ionic conduction predominates. Over a wide range of oxygen partial pressures, all three types of conduction modes are exhibited by such oxides as ZrO_2.

D. Band Conduction

Since ceramic insulators are typically large (≥ 4 eV) bandgap materials, with ionic conduction largely dominant, an energy-band model for conduction does not normally apply, except in limited cases where there is aliovalent doping of the ceramic [26–34]. For the transition metal oxides, the energy bands that are important for conduction are the 3d or inner orbital electron levels rather than the outermost valence (4s) electron levels. In an oxide where the 3d levels are only partially filled, therefore, the conductivity will strongly reflect the extent of overlap of the 3d orbitals [29–31]. If the wave function overlap in the oxide is sufficient to develop a narrow energy band, electron migration within this band can lead to significantly high conduction, bordering on metallic, which has been characterized as narrow-band conduction. Certain FCC transition metal oxides such as TiO, MnO, and VO exhibit this type of behavior, as do some higher transition metal oxides such as RuO_2 [25–27]. In other transition metal oxides such as TiO_2/Ti_2O_3, the $3d$ wave function overlap is too small for the $3d$ band to develop. Instead, the $3d$ electrons are localized on the individual ions and can move from ion to ion only by a hopping or polaron conduction process. Examples of materials exhibiting such conduction are Li-doped NiO, n-doped BaTiO3 [32,33], and Fe_3O_4.

In ionic host lattices, the presence of localized electrons or holes as charge carriers can cause polarization of the lattice, resulting in the formation of a new charged entity, consisting of the electronic carrier plus its polarization field, referred to as a polaron. If the polarization field around the electron is large compared to the unit cells in the lattice, the discrete dipolar state can be replaced by a polarization continuum, in which there is only weak coupling between the charge carrier and the polarization field. This is referred to as a large polaron state and results in conductivity similar to a quasi-free electron state. An intermediate polaron state is one in which the polaron radius is comparable to that of the unit cell and is typical of some perovskite materials [29–31]. In contrast, when the polaron radius is smaller than the unit cell dimension (small polaron state), the mobility becomes strongly coupled to the lattice, which must move along with the electronic carrier. This process is loosely referred to as a "hopping" mechanism.

Charge transport by hopping conduction occurs via the impurity states located within the band gap. The nature of the localized states depends on the presence and concentration of defects such as vacancies, interstitial atoms, and their ionized states. At high defect concentration the localized states overlap, with conduction occurring via thermally activated hopping between states. The activation energy for conduction is typically ≤ 1 eV, which allows conduction to occur at room temperature even in these wide band gap materials. This is because the discrete energy levels can become localized states of lower energy within the band gap. Polaron motion is then possible as a result of small overlap of the wave functions of neighboring ions. Electron mobility will increase according to the degree of overlap of the partially filled local states and not through the conduction band [27,30,34]. Mobility values for the small polaron are typically less than 1 $cm^2/V \cdot s$ but can be much lower.

E. Diffusion

As indicated, lattice diffusion in ionic solids is made possible by the presence of point defects in the lattice [12,21,23–25]. Transport of ionic species via these defects occurs primarily through a vacancy transport mechanism, but various interstitial migration modes also exist as described below:

Vacancy migration takes place when an ion on a normal lattice site jumps into an equivalent vacant lattice site. This moves the vacancy to the site exited by the ion, so that the vacancy migrates in a direction opposite to that of the ion. It should be noted that the vacancy carries a charge opposite to that of the ion. The controlling factor in this migration is the diffusion coefficient, D, which can be given as:

$$D = \alpha\lambda^2 v_0 \, (\exp - u/kT) \tag{81}$$

Where α is the number of equivalent jump sites for the vacancy; λ the average jump distance; v_0 the natural vibration frequency for the ion ($\sim 10^{-13}$/s); u the activation energy for the jump; k the Boltzman constant; and T the temperature (K). The available sites and the jump distance are both related to ion size, crystal structure, and unit cell dimension, as is the activation energy u, but with the added influence of ion charge. The vibration frequency v_0 increases slightly with ion size and elastic modulus of the solid, while the overall jump frequency [$v = v_0 \, (\exp - u/kT)$] is strongly temperature dependent. The dominant factors in vacancy migration, therefore, are crystal structure, ion size, and temperature. As indicated, it is the dominant mechanism in Schottky solids such as NaCl and MgO, where interstitial migration would be difficult, but also in a wide variety of other ceramic structures, including ZrO_2, CaF_2, Al_2O_3, and many others.

Interstitial migration occurs when an ion on an interstitial site moves to a neighboring interstitial site. The diffusion process is described as interstitial migration. The movement is responsive to all the elements of the diffusion process described above, except that α now represents the number of equivalent interstitial sites that are adjacent. The jump of the interstitial atom involves a distortion of the lattice, and this mechanism is probable when the interstitial atom or ion is smaller than those on the normal lattice sites. It is the dominant mechanism in glasses, however, where the open, amorphous structure permits ready migration of alkali and other ions via this mechanism.

Interstitialey migration can occur in a solid when the lattice distortion associated with the interstitial migration becomes too large to make this movement possible. In the interstitialcy mechanism, the interstitial ion pushes one of its nearest-neighbor ions located on a normal lattice site into another interstitial position and itself occupies the lattice site of the displaced ion. In the interstitialcy mechanism one may distinguish between two types of atom movements. If the atom on the normal site is pushed in the same direction as that of the interstitial atom, the jump is termed collinear. If the atom is pushed to one of the other neighboring sites so that the jump direction is different from that of the interstitial ion, the jump is termed noncollinear. This mechanism is important in Frenkel solids such as the Ag halides, where the direct migration of the large Ag^+ ion interstitially would cause severe lattice distortions.

Divacancy migration occurs when cation and anion vacancies combine and migrate in tandem through the solid. The migration can be considered as dissociative if it temporarily breaks up and reforms or nondissociative if no break-up occurs. The latter mechanism requires less energy for propagation. The divacancy mechanism can account for material transport without accompanying charge transport since the charges on the vacancies offset each other. Since no net charge is transported through this mechanism, it is of only limited interest for insulation purposes, but this mechanism can be important in processes such as sintering.

Diffusion effects on conductivity. Discrepancies between the measured and calculated diffusion coefficients using the Nernst–Einstein relationship, discussed previously, can be attributed in the diffusion mechanisms. The Nernst–Einstein relationship must, therefore, be modified as follows [11,21,24,26]:

$$\sigma_i = \sigma t_i = \frac{n_i (ez_i)^2 D_i}{kT} B \tag{82}$$

where the mobility parameter, B, is related to the various diffusion mechanisms as follows: $B = 1.27$ (vacancy); $B = 1.0$ (interstitial); $B = 1.38$ (interstialcy–collinear); $B = 3.0$ (interstialcy–noncollinear); and $B \ll 1.0$ (divacancy). These empirical factors are intended to normalize differences between the diffusion factors related to number of available jump sites and jump distances. For glasses, where interstitial migration dominates, there is good agreement between the Nernst–Einstein equation and measured data [11,26].

VII. INSULATOR MATERIALS

A. Ceramic Insulators

Insulators are materials that offer effective resistance to current flow in an electric field due to the very low concentration of mobile charge carriers. Important ceramic insulators such as SiO_2, Al_2O_3, mullite ($3Al_2O_3 \cdot 2SiO_2$), BeO, AlN, boron nitride (BN), and Si_3N_4 have resistivities of about 10^{14} ($\Omega - cm$). These high values are the result of a large energy gap between a filled valence band and the next available energy level, where the promotion of an electron into a higher state is energetically unfavorable [20,35–37]. The conductivity of these ceramics, therefore, is significantly influenced by both ionic and electronic defects. In insulating oxides, ionic defects arise from the presence of impurities of different valence from the host cation. An aluminum ion impurity substituting on a magnesium site in the MgO host lattice creates Mg vacancies. These vacancies facilitate ionic migration of Mg^{2+} ions under the influence of an electric field. This is a high-temperature process, however, and room temperature conductivities are very low.

Electronic conduction may arise in oxide materials from the natural loss of oxygen, which typically occur in oxides on heating to high temperatures. Electrons trapped at the vacancy can become partially or fully ionized, leading to weak n-type electronic conduction. Again, the conductivity is low. Conduction in materials such as AlN and SiC occur mainly through the presence of impurities having different valence states, leading to n- or p-type conduction. Similar conduction mechanisms can be expected for most nitride materials, including Si_3N_4 and BN. Depending on the level of impurity doping, SiC, which has only a moderate band gap ($E_g = 2.8$–3.2 eV), can become semiconducting and has been developed for device use. SiC is also widely employed as heating elements for furnace applications. Electrically insulating SiC can also be fabricated using BeO dopant additions. This is an important material for laser heat sink applications because of its high thermal conducting and electrical insulating properties.

Of the oxide insulators in the system ($MgO \cdot SiO_2$), neither MgO nor spinel ($MgO \cdot Al_2O_3$) is widely used, although both are excellent insulators, in part because of the high cost and difficulties in fabricating the materials. However, SiO_2 is widely used as a piezoelectric transducer in single-crystal form, and it also finds usage for delay lines and high-temperature insulation. Other oxide materials listed in Tables 1 and 2 (zirconia, thoria, hafnia, and ceria) have mainly specialized uses except for zirconia, which has substantial insulating, oxygen sensor, fuel cell, and heater applications. For these applications, the zirconia must be "stabilized" with CaO or MgO oxides (5–15 mol %) or with Y_2O_3 (2–6 mol %) to prevent the destructive monoclinic/tetragonal phase transformation, which occurs on heating or cooling. Zirconia also exhibits excellent strength, toughness, and wear resistance [37].

SiO$_2$ has the lowest dielectric loss properties of any inorganic material [36]. It is commonly used in insulating fibers and in the development of electrical porcelains ($R_2O \cdot Al_2O_3 \cdot SiO_2$). These materials have high dielectric strengths with low loss and are therefore suitable for high-voltage applications, such as transmission line insulators, high voltage circuit breakers, and cutouts. Mullite, $3Al_2O_3 \cdot 2SiO_2$, MgO, and steatite, $MgO \cdot SiO_2$, are extensively used for high-temperature electrical insulation and for high-frequency insulation because of their low loss characteristics. For electrical insulating applications and heat sinks, Al_2O_3, AlN, SiC, and Si_3N_4 are the most commonly used materials. SiC and Si_3N_4 are also industrially valuable as high-temperature heat exchangers because of high thermal conductivity and electrical insulating behavior, high hardness, durability, and excellent high-temperature, corrosion, and thermal shock resistance. Films of these materials, including diamond, have been developed with properties similar to those of the bulk materials. The conduction processes in the films mainly result from impurity and electrode injection effects, which degrades the intrinsic high resistivity and dielectric properties of the materials.

B. Oxide Insulator Materials

1. Alumina (Al_2O_3)

Of the oxide materials, Al_2O_3 is the most widely used, particularly as a substrate material for thick film and microelectronics circuitry, because of its outstanding thermal, mechanical, and dielectric insulating, properties as well as its relative chemical inertness. It is also easy to fabricate and can be sintered to high tolerances, surface texture and finish [38–42]. The high thermal conduction of alumina (second only to BeO as oxides) is a major asset for heat dissipation and thermal shock resistance in dense circuit applications. Its wide use in insulators, bushings, and circuit breakers reflects its high dielectric strength (10–17 kV/mm) and resistivity ($>10^{14}$ Ω-cm). Similarly, its high strength and refractoriness permit alumina to be used in spark-plug insulators, power resistor cores, and missile nose cones. The low dielectric constant and loss factor make alumina a suitable material also for microwave windows and electron-tube spacers and envelopes. The ceramic compositions for such applications typically range from 94 to 99 wt % Al_2O_3. Flux additives of talc, clay, CaO, and MgO aid in sintering and grain growth control, but these additives can also supply impurity ions [40,41].

Electrical conduction mechanisms in alumina has been much studied [40,41]. The high-purity polycrystalline ceramic is typically an n-type electronic conductor below ~450°C while this n-type electronic conduction might be expected to predominate up to ~1300°C, ionic conduction may contribute appreciably to the overall conduction, depending on the presence, type, and concentration of impurity ions.

The temperature and oxygen partial pressure dependence of the electrical conductivity have been determined for the single-crystal and polycrystalline α-alumina, over the range of 1300–1750°C and at oxygen partial pressures of 10^0–10^{-10} atm. The band gap was determined to be ~5.0 eV, with n-type, intermediate, and p-type conduction being present within the different ranges of oxygen partial pressures [42]. The activation energies for conduction were in the range of 2.48–2.97 eV below 1627°C. However, no simple or single conduction process can fully explain the electrical conductivity of α-Al_2O_3 over the wide range of compositions, processes, temperatures, and oxygen partial pressures normally encountered.

The dielectric properties, especially the dielectric strength, are affected by defects such as porosity, lattice distortions, grain boundaries, interfaces with a secondary phase, and impurities such as Si, Ti, Mg and Ca. Substitution of Si^{4+} or Ti^{4+} on Al^{3+} sites creates donor levels at the impurity site and acceptor levels at compensating cation vacancy sites. Conversely, the substitution of Mg^{2+} or Ca^{2+} for Al^{3+} creates acceptor levels at the impurity sites and donor levels at the compensating oxygen ion vacancies. These donor and acceptor ions contribute to charge migration, which can degrade both the dielectric and loss characteristics of the alumina. SiO_2, however, has only limited solubility in Al_2O_3; thus, any excess SiO_2 will form a glassy phase at the grain boundaries leading to ionic conduction. Minor additions of ~500 ppm of SiO_2 to Al_2O_3 have a relatively small influence on the dielectric breakdown strength (14.3 versus 15.2 kV/mm for 99.99% pure polycrystalline Al_2O_3), compared to the presence of MgO which decreases the dielectric strength from 15 to 13 [40,41]. However, the dielectric breakdown strength of polycrystalline Al_2O_3 with grain size between 1 and 3 μm (prepared from powders of 99.99% purity) is greater than the corresponding single-crystal value. This is because the diffusion along grain boundaries avoids the localization and concentration of charges, which is the chief cause of breakdown in dielectrics. For lower purity levels the dielectric strength decreases with increase in grain size, but this grain size dependence can be reduced by adding zirconia stabilized by Y_2O_3. The zirconia is present as an intergranular phase and the modified grain boundary enhances the diffusion of charges and hence improves the breakdown strength [20,39].

2. Beryllia (BeO)

Beryllia finds extensive use in applications requiring significant heat dissipation, such as substrates for high-performance thick-film circuits. For these applications, it is a strong competitor of alumina, but its toxicity in powder form and fabrication costs limit its more general use. It is mainly employed, therefore, in critical structural and insulating applications such as electron and traveling-wave tubes and microwave windows. At room temperature the thermal conductivity of BeO is half that of copper but it rapidly decreases with increase in temperature [43–44]. However, in spite of the strong temperature dependence of the thermal conduction in berrylia, the value at 800°C is still greater than the room temperature thermal conductivity of alumina, as seen in Table 9. The high strength and high thermal conductivity give BeO good thermal shock resistance. Moreover it is chemically stable in air, vacuum, hydrogen, carbon monoxide, argon, and nitrogen at tempera-

Table 9 Irradiated and Nonirradiated Conditions

Material	Unirradiated			Irradiated		
	Relative permittivity, ε_r	Loss tangent, $\tan \delta$	Thermal conductivity, λ (W/m^{-1}°C)	Relative permittivity, ε_r	Loss tangent, $\tan \delta$	Thermal conductivity, λ (W/m^{-1}°C)
Alumina (98–99% Al$_2$O$_3$)	9.5	5.0×10^{-4}	34.0 at 20°C 15.0 at 400°C 10.0 at 800°C	9.6	1.3×10^{-4}	11.0
Alumina (>99% Al$_2$O$_3$)	10.0	1.8×10^{-4}	34.0 at 20°C 15.0 at 400°C 10.0 at 800°C	10.2	2.1×10^{-4}	11.0
BeO	6.8	2.5×10^{-4}	210 at 20°C 125 at 400°C 40 at 800°C	6.8	3.0×10^{-4}	100.0*

Source: Ref. 43.

tures up to 1700°C. BeO is the most stable of all the oxides in contact with graphite. BeO has lower dielectric constant than alumina, but the temperature dependence is similar to that of Al_2O_3. It is resistant to corrosion by liquid alkali metals but reacts with water vapor above 1650°C to form volatile $Be(OH)_2$. A radiation resistant use for beryllia is as ceramic vacuum windows in cyclotron type reactors, where the material typically is exposed to high neutron fluence of up to $10^{22}n/m^2$. For such applications berrylia is preferred because the lower thermal conductivity of alumina would introduce excessive themal stresses if used for this purpose. Table 9 compares the thermal and electrical insulating properties of BeO and two grades of Al_2O_3 under such use conditions. The thermal and mechanical properties of the two oxide materials are again compared in Table 10, with reference to to the metals Al and Ti, emphasizing again the very high thermal conductivity and mechanical strength values for BeO.

3. Magnesia (MgO)

Magnesia has a high thermal expansion coefficient but relatively low mechanical strength compared to alumina, and this limits its widespread use due to poor thermal shock resistance. High cost is another factor limiting its low use as an insulator, as is its tendency to hydrate in powder form. However, it is a better insulator than Al_2O_3, particularly at high temperature. Also, its combination of high thermal conductivity and high electrical resistivity makes it suitable for use as insulating thermocouple tubes and for heating core elements. MgO has been shown to be a mixed electronic/ionic conductor, the relative importance of the ionic and electronic conduction being a function of temperature, oxygen partial pressure, and impurity content [45,46].

Extensive studies of these aspects have generally found that the total conductivity decreases with decreasing oxygen partial pressure at high oxygen pres-

Table 10 Thermal and Mechanical Properties of BeO and Al_2O_3 in Comparison with Ti and Al Metals

Material	Young's modulus E (MPa)	Possion's ratio, ν	Thermal conductivity, λ (W/m^{-1}°C)	Thermal coeff. of expansion (10^{-6}/°C)
BeO	340,000	0.3	210	8.4
Al_2O_3	350,000	0.3	34	8.0
Ti	99,000	0.35	17	8.8
Al	100,000	0.3	240	24

Source: Ref. 43.

sures, then goes through a minimum at low or high P_{O_2} and at higher temperatures ($> 1300°C$), while ionic conductivity is favored at low temperatures and at intermediate oxygen pressures. The minimum in conductivity reflects a roughly equal balance between magnesium vacancies, which predominate at high P_{O_2}, and oxygen vacancies, which predominate at low P_{O_2}. The conductivity in the impure oxides has been attributed to Mg vacancies. MgO is essentially similar to Al_2O_3 and BeO in its conduction behavior.

4. Zirconia (ZrO_2)

Zirconia is difficult to produce as a dense, crack-free product because of the disruptive volume change ($\sim 10\%$) that occurs during the tetragonal to monoclinic phase transformation at $1000°C$. This phase change can be eliminated by stabilizing the cubic form of the zirconia with solid solutions of CaO, MgO, Y_2O_3, Yb_2O_3, Nd_2O_3, or SC_2O_3. Stabilized ZrO_2 has a relatively poor thermal resistance because of its high thermal expansion coefficient ($\sim 11 \times 10^{-6}$ cm/cm·°C, about 1.5 times that of alumina) and its relatively low thermal conductivity [0.004 cal/ (s) (cm^2) (°C/cm), about one-fourth that of alumina]. The incorporation of the stabilizing ions into the zirconia lattice leads to enhanced oxygen ion mobility and to a significant increase in the electrical conductivity with temperature. This property has led to the use of yttria- and calcia-stabilized zirconia as the electrolyte materials for high-temperature fuel cell operations [46–55].

As indicated, the addition of stabilizers such as CaO, MgO, and Y_2O_3 results in formation of the temperature-stable cubic phase and also affects the conductivity [48–53]. The magnitude and type of conductivity in the ZrO_2 depends much on the structure, since the material has three distinct phase modifications, as follows:

(a.) *Cubic Stabilized ZrO_2.* The electrical conductivity for cubic $Ca_{0.15}Zr_{.85}O_{1.85}$ has been found to be independent of the oxygen partial pressure, P_{O_2}, indicating that the conductivity is wholly ionic [50,51]. Furthermore, the conductivity over the temperature range 700–1725°C can be represented by the empirical expression below, which shows its strong temperature dependence and relatively low activation energy for conduction.

$$\sigma = 1.5 \times 10^3 \, \exp\left(\frac{-1.26}{kT}\right)$$

(83)

From a calculation of the electrical conductivity resulting from oxygen ion diffusion it was concluded that the conductivity is completely ionic in calcia-stabilized ZrO_2 and that the transference number for oxygen ions is unity [50,51]. Similar observations can be made regarding yttria-stabilized zirconia, which shows even higher ionic conduction with a maximum at approximately 9.0 mol % Y_2O_3. The transport mechanism has been confirmed to be oxygen ion conduction via oxygen vacancies, the concentration of which is increased when the Y^{3+} or Ca^{2+} ions replace Zr^{4+} ions. Figure 11 compares the conductivity as a function of reciprocal absolute temperature for calcia-(CSZ:13.3 mol % CaO) and yttria-(YSZ: 4.5 mol % Y_2O_3) stabilized zirconia specimens. As expected, the YSZ sample shows a higher conductivity and lower activation energy for the temperature range of interest. Both samples had average grain sizes of ~1 μm and were sintered at 1300–1350°C for 4 h [52–54]. The development of *n*-type electronic conduction by relatively small concentrations of donor ions (vanadium, tantalum) in the presence of calcium, an acceptor ion, has been observed [53], but the mechanism by which this occurs is unclear. CeO_2 doping of the ZrO_2 system leads also to enhanced conductivity and lower temperatures [55].

(*b*). *Monoclinic ZrO₂.* The electrical conductivity in monoclinic ZrO_2 has been studied as a function of partial pressure of oxygen, P_{O_2}, at 1000°C using an ac technique [48]. The conductivity was found to go through a minimum at P_{O_2} ~10^{-16} atm, with *n*- and *p*-type conduction being observed at the lower and higher oxygen pressures, respectively, with ionic transport number being less than 0.01. The oxygen pressure dependence of the *p*-type conduction was interpreted to be proportional to $P_{O_2}^{1/5}$. The monoclinic ZrO_2 was determined to be stoichiometric at P_{O_2} ~10^{-16} atm, with the *p*-type conduction and a $P_{O_2}^{1/5}$ pressure dependence, attributed to the predominance of fully charged zirconium vacancies. Monoclinic ZrO_2 under these conditions is, therefore, a metal-deficient rather than an oxygen-deficient oxide conductor. The hole mobility at 1000°C was found to be 1.4×10^{-6} cm²/V-s, which indicates a thermally activated hopping mechanism for charge transport, since this mobility is too small for hole conduction to occur via any simple band structure model. The defect structure of monoclinic ZrO_2 has also been studied [48] by measuring the transference number and electrical conductivity as a function of P_{O_2} and temperature. The data suggest the existence of doubly ionized oxygen vacancies at low pressures (P_{O_2} < 10^{-19} atm) and singly ionized oxygen interstitials at P_{O_2} pressures >10^{-19} atm, consistent with an anti-Frenkel disorder. The monoclinic zirconia was found to be primarily an ionic conductor below ~700°C and an electronic conductor at 700–1000°C for $10^{-22} \leq P_{O_2} \leq 1$ atm.

(*c*) *Tetragonal ZrO₂.* Tetragonal ZrO_2 has been found to be a mixed electronic and ionic conductor, with a significant ionic contribution except at very high temperatures or very low oxygen partial pressures [47,49]. The *n*-type electronic conductivity, observed at high temperatures (1400°C) and low oxygen partial pressures (P_{O_2} < 10^{-13} atm), has been interpreted as arising from completely ionized oxygen vacancies. At the monoclinic-to-tetragonal transition (~1150°C) there is an abrupt isothermal change, with the tetragonal form showing the lower conductivity.

C. Spinels

The spinel crystal structure (AB_2O_4) is based on the cubic close packing of oxygen ions in which the cations are situated on both the tetrahedral A and octahedral B sites [56–63]. Based on their electrical conductivities, the spinels can be divided into four groups: (1) aluminates ($M^{2+}Al_2^{3+}O_4$) having low conductivities; (2) chromites ($M^{2+}Cr_2^{3+}O_4$) moderate; (3) ferrites ($M^{2+}Fe_2^{3+}O_4$) fair conductivities; and (4) magnetites ($Fe^{2+}Fe_2^{3+}O_4$) with fairly high electrical conductivity. In the aluminates the method of conduction is controlled by the divalent cation, with magnesium and zinc aluminates being electron conductors whereas the cobalt and nickel aluminates show hole conduction. With the exception of ferrous compounds, the conductivity of the chromites is controlled by the trivalent cation. The chromites are all hole conductors [59,61] whereas ferrites are electron conductors. Magnetite is an intrinsic semiconductor due to the presence of ions of the same metal (Fe^{2+}, Fe^{3+}) but with differing valence, on equivalent octahedral lattice sites. Most spinels obey the Arrhenius-type resistivity–temperature relationship to some degree. In particular, the aluminates of Co, Ni, and Zn, zinc and cobalt chromites, nickel and zinc ferrites, and magnetite conform to the law over the temperature range 650–1000°C. Table 11 compares electrical conductivity data for various spinel types.

Table 11 Comparison of Electrical Properties for Spinel Ceramics

Spinel type	E (eV)	Temp. range (°C)	σ at 900°C (mhos)	Method of conduction
Co aluminate	1.55	700–1020	8.55×10^{-6}	Hole
Mg aluminate	1.14	740–1010	1.33×10^{-6}	Electron
Ni aluminate	1.82	780–1020	2.75×10^{-6}	Hole
Zn aluminate	0.86	640–980	5.38×10^{-6}	Electron
Co chromite	1.05	500–1010	2.49×10^{-3}	Hole
Mg chromite	0.95	535–1000	9.71×10^{-4}	Hole
Zn chromite	1.27	800–1020	1.03×10^{-4}	Hole
Zn chromite	1.09	480–800		Hole
Mg ferrite	1.16	610–830	3.98×10^{-2}	Electron
Ni ferrite	1.10	500–1000	1.95×10^{-2}	Electron
Ni ferrite	0.77	225–500		Electron
Zn ferrite	0.96	500–1020	1.28×10^{-2}	Electron
Zn ferrite	0.83	290–500		Electron
Ferrous chromite	0.76	655–1000	1.83×10^{-3}	Hole
Magnetite	0.106	90–400	1.85×10^{-1} @ 300°C	Intrinsic semiconductor

In the magnetic spinel ferrites, cations such as Fe^{3+} or Mn^{2+} can occupy both A and B sites, giving an inverse $[(AB)BO_4]$ structure and higher net magnetization. In the spinel lattice, the normal cations (e.g., Ni, Co, Li, Mg, Zn, Cd) show a definite preference for either the tetrahedral or the octahedral sites. This cation concentration and site preference essentially determines the properties of the ferrite. Ferrites need to have high electrical resistivities in order to eliminate eddy current and dielectric losses, as well as to allow for full penetration of electromagnetic fields throughout the solid [57,58,60]. High resistivity is obtained when a cation has only one valence state in the lattice. In the processing of these materials, therefore, high sintering temperatures and reducing atmospheres must be avoided because these conditions can produce variable valence states on some cations.

The range of published resistivities for spinel and garnet ferrite materials is wide, from about 10^{-4} to 10^9 Ω-cm at room temperature [58]. The low resistivity is associated with the presence of both Fe^{2+} and Fe^{3+} ions on the octahedral lattice sites. In general, a condition for conduction in the ferrite structure is the presence of ions having multiple valence states on like crystallographic sites. Thus, the concentration of ions in Fe_3O_4 can be controlled by solid solution in which Fe^{2+} or Fe^{3+} ions are diluted by other ions that do not participate in the electron exchange.

There are three main commercial classes of ferrite spinels, namely, nickel-zinc ferrite, $(NiZn)Fe_2O_4$; manganese-zinc ferrite, $(MnZn)Fe_2O_4$; and magnesium-manganese ferrites, $(MnMg)Fe_2O_4$. The electrical conduction is by small-polaron mechanism. The electrical resistivity primarily determines the utility of these materials in the high megahertz or microwave frequency ranges. Low-frequency use requires a trade-off between high permeability and high resistivity. Nickel-zinc ferrites typically show an increase in permeability and a departure from stoichiometry in the iron-rich direction. Decreased resistivity results, therefore, when the formation of divalent iron becomes more probable. Manganese-zinc and magnesium-manganese zinc ferrites are typically used in low-frequency devices such as pulse transformers and memory core devices. For higher frequency use, the high dc resistivity needed for full magnetic penetration and low eddy current losses can be obtained with an Fe-deficient oxide powder. However, a more complex processing situation arises because of the three possible valence states of manganese cation and the site preference for each. Sintering temperature and cooling rate can also impact the magnesium site locations. Increasing the amount of manganese increases the lattice constant, making it easier for the Mg ions to occupy both cation sites, such that divalent and trivalent ions are present on both sites, with adverse impact on the permeability, resistivity, and utility of the material [58,59]. Processing factors such as binder content [63], calcination, and sintering can also significantly impact the electrical properties of the ferrite [57].

Solid solutions of Fe_3O_4 and $MgCr_2O_4$, hercynite ($FeAl_2O_4$), and ferrous chromite ($FeCr_2O_4$) have been prepared. Semiconductor materials of this type, with controlled temperature coefficient of resistivity, typically incorporate $MgAl_2O_4$, $MgCr_2O_4$, and titanium zinc oxide (Zn_2TiO_4) as the nonconducting component. Because the electrical conductivity of materials made in this way have a negative temperature coefficient (NTC), they are typically utilized in temperature sensing and thermistor applications. Copper-nickel manganites, $Mn_{3-x-y}Ni_yCu_xO_4$ spinels, are of technological importance for use in these low-resistance thermistors [61]. Through adjustment in stoichiometry, the electrical conductivity can increase to 0.1 $(\Omega\text{-cm})^{-1}$ at room temperature. In nickel-manganites the conduction is primarily through electron hopping from Mn^{3+} to Mn^{4+} cations on the B sites. In the copper-nickel-manganites the situation is quite complex since there is the presence of Ni^{2+}, Mn^{2+}, Mn^{3+}, and Mn^{4+} ions as well; hence there is uncertainty over distribution of copper cations and their ionic states.

D. Complex (Perovskite) Oxides

$BaTiO_3$ has an intrinsically high resistivity ($\geq 10^{10}$ Ω-cm) when prepared under oxidizing conditions. However, the material can become much less resistive through controlled doping of the A lattice sites with aliovalent cations such as La^{3+}, Y^{3+}, or Nd^{3+}, or of the B sites with Nb^{5+}, Ta^{5+}, and like cations. At low dopant levels, typically <1.0 mol %, Ti^{3+} ions are formed in the $BaTiO_3$ lattice, leading to local distortion of the TiO_6 octahedra, a change in the polarization field, and a splitting of the Ti ($3d$) orbitals [64–74]. This condition leads to n-type polaron conduction in the material, due to the hopping migration of the donor electrons on the Ti^{3+} ions. The mode of conduction overall reflects the combined effects of the local cell distortion caused by the Ti^{3+} valence state; the lattice distortion caused by the electron hopping; and the polarization field overlap of the localized Ti^{4+} and Ti^{3+} discrete energy levels [29–31]. Heat treatment of the $BaTiO_3$ under less than oxidizing conditions also produces Ti^{3+} states, with the conductivity of the material being directly related to the concentration of these states. Strontium titanate ($SrTiO_3$) also becomes an *n*-type semiconductor when additional electrons are created by donor doping of the Ti lattice sites or by heat treatment in a reducing atmosphere with resultant loss of oxygen. Electron mobility in the doped $SrTiO_3$ is about 6 $cm^2/V.s$.

At higher dopant levels (~0.3–1.0 mol %), the mechanism changes from electronic to ionic compensation, reflecting significant cation vacancy formation, leading to a significant increase in resistivity and a lowering of the Ti ($3d$) energy level associated with the Ti^{3+} state [67–71]. A complex defect structure can result from the presence of different charge compensation mechanisms as the dopant level is increased, with the result that the low resistance values are obtained only within a narrow dopant concentration range of $\sim(0.1$–$0.3)$ mol %.

Under controlled heat treatment, a thin, insulating grain boundary layer can be formed in the polycrystalline ceramic. In this grain boundary region oxygen is adsorbed during annealing, reducing the Ti^{3+} concentration [68,69]. Oxygen vacancies may also diffuse from the interior to the grain boundary region where they act as electron traps. A space-charge or barrier layer is thereby formed that repels electrons moving toward the grain boundary. The near-grain boundary region in doped and annealed $BaTiO_3$ shows regions of high strain, high defect segregation, and nonuniform domain structure. With the annealed material, a martensitic-type structure change is believed to occur within the narrow grain boundary region which, when coupled to the sudden release of stress, is considered to be the driving force for the observed abrupt change in grain boundary potential and resistivity [73,74]. At temperatures below the critical ferroelectric phase transition, spontaneous polarization charges neutralize the grain boundary charges in coherent regions along the grain boundaries, creating thereby low-resistance pathways [67,68] with significant overall lowering of the resistivity. The PTCR behavior in n-doped $BaTiO_3$, capitalizes on this effect. In the polar state the material exhibits low resistivity, which then changes abruptly by several orders of magnitude over a narrow range near the phase transition temperature. It is observed as a large increase in resistance (typically several orders of magnitude), near the phase transition temperature, over a narrow range of dopant concentration. The PTCR effect and room temperature resistivities are both highly dependent on dopant concentration and type. Materials with this type of behavior are used for current limiting, temperature control, resistive heating, motor starters, and many sensor applications [67].

Barrier layer capacitors having high dielectric constants and low loss can be produced by increasing the grain boundary thickness through the use of additional dopants, such that the grain boundary barrier layer becomes impassable to electrons, by creating a large concentration of acceptor states that nullify the effects of the spontaneous polarization. This traps the conducting electrons by creating a space-charge layer, which repels like charges. Several additional dopants, e.g., Sr, Zr, Pb, Si, Cu, and Bi, to $BaTiO_3$-based systems are used in the manufacture of both capacitor and PTCR sensor devices in order to adjust the switching temperature and other operating characteristics [67,68].

E. Carbides and Nitrides

I. Silicon Carbide (SiC)

Most of the highly refractory carbides and nitrides oxidize easily and must be provided with a protective atmosphere for use above 1000°C. Silicon carbide, on the other hand, is highly resistive to oxidation due to formation of a surface-protective SiO_2 layer, with maximal working temperatures of about 1700°C in oxidizing atmosphere and 2200°C in neutral.

Silicon carbide is an important electrical ceramic material that is used mainly for its resistive properties and high-temperature stability [75–81]. A significant use is as heating rods for furnaces that must operate at temperatures close to 1500°C. The heating rods are typically made by pressing of the granular SiC with temporary binder into rods, which are then fired at ~2000°C or higher. Electrical resistivity for these rods is typically 0.2 Ω-cm at 25°C, decreasing to about 0.1 Ω-cm at 1000°C. This behavior can be understood in terms of its electrical conduction mechanism. Starting from ambient temperatures, the number of conduction electrons increases with temperature up to ~750°C as more and more electrons are released by thermal excitation from impurity trap levels. Between 750°C and 1500°C, thermal vibrations decrease the mobility of the conduction electrons, causing a net increase in the resistance. Above 1500°C, intrinsic conduction becomes significant and the resistivity falls abruptly with increasing temperature. Silicon carbide can be doped with boron to provide acceptor levels within the band gap ($E_c \sim 0.3$ eV), thus making it a p-type conductor. Conversely, nitrogen can be added to provide donor levels below the conduction band ($E_d \sim 0.07$ eV), leading to n-type conduction.

Nonlinear electrical grade SiC is used for lightening arrestors and in voltage-limiting resistor applications. SiC-fiber reinforced glass has also been proposed for intelligent sensor applications. The silicon carbide fiber reportedly can detect damage to the composite due to mechanical loading or detect temperature change through changes in electrical resistance [79,80].

2. Silicon Nitride (Si_3N_4)

Silicon nitride has been developed primarily for its structural applications, but it has important potential for microelectronics use, owing to its superior dielectric properties and high thermal conduction [75–77]. It also exhibits high electrical resistivity ($>10^{12}$ Ω-cm), high strength ($>25,000$ psi at 1200°C), excellent thermal shock resistance, and chemical inertness. Applications as dielectric films and as substrates are also widespread.

3. Boron Nitride (BN)

Boron nitride, in its hexagonal modification, has a similar structure to graphite and can readily be machined. It possesses excellent dielectric and thermal shock properties and is a useful insulating material, in both bulk and film form, though difficult to bond to. End-use products include microelectronic substrates, integrated circuit packages, and planar diffusion sources [20,77].

4. Aluminum Nitride (AlN)

Aluminum nitride, in both thin-film and bulk form, is an actively developed material for substrates and microelectronics use. Advantages are its very high thermal conduction (close to BeO), low thermal expansion coefficient (close to Si), low density, and excellent dielectric insulating properties. Against these advantages must be placed its high sintering temperature ($\sim 1800°C$), sensitivity to impurities (which degrade its heat dissipation properties) high-temperature corrosion under humid and oxidizing conditions and relatively high cost. Also, techniques for reliably bonding to the substrate must still be optimized. Despite these challenges, AIN is projected to have an increasing impact on the microelectronics substrate, packaging, and device component market [20,76,77,81].

F. Glasses

Glasses are widely used as electrical insulators because of their generally excellent dielectric properties, but factors such as thermal expansion coefficient, glass transition temperature, and softening point temperature play an important role in the selection of compositions for particular applications. Glasses are generally favored because of their wide availability, low cost, and ease of fabrication. Traditional uses include ceramic glazes, wire and porcelain coatings, lamp and tube envelopes, electrical bushings, standoff insulators, fuse bodies, and fiber insulation. Microelectronic applications include use as substrates, delay lines, device passivation, capacitor dielectric, thick-film resistor components, dielectric film layers, and solder glass sealing.

1. Structure and Composition

Glass is characterized by an amorphous or random structure and wide variability in composition. Normal oxide glasses may be classified as silicates, borates, phosphates, or germanates, according to which the glass-forming oxides (SiO_2, B_2O_3, P_2O_5, or GeO_2) make up the structure [82,95,96]. This structure is normally modified by selected additions of such intermediate oxides as Al_2O_3, Bi_2O_3, and PbO, which at relatively high concentrations may contribute to glass formation, or by alkali (Li, Na, K, Rb, Cs) and alkaline earth (Mg, Ca, Sr, Ba) oxides. The alkali oxides in particular have the effect of loosening the structure, thereby lowering the softening point and providing increased ion mobility [82–105].

The basic ingredients of oxide glasses may be classified as network formers (SiO_2, B_2O_3, P_2O_5), modifiers (Na_2O, K_2O, CaO, MgO), and intermediates (Al_2O_3, PbO). During fusion of the glass mixture a structural readjustment takes place such that the very strong Si–O–Si bonds are broken and the bridging oxygens are converted to nonbridging oxygens. Such a reaction can be represented as:

$$-\overset{|}{\underset{|}{Si}}-O-\overset{|}{\underset{|}{Si}}- + Na_2O \rightarrow -\overset{|}{\underset{|}{Si}}-O-Na^+-O-\overset{|}{\underset{|}{Si}}- \tag{84}$$

where the alkali ions or alkali earth ions become electrostatically attached to the nonbridging oxygens and can migrate freely. As more alkalis are added to the glass, fewer and fewer bridging oxygens remain. The network "coherence" decreases and it becomes easier for the ions to move [82,88]. A coherence parameter Y, which describes these structural changes for silicate glasses, can be given as:

$$Y = 8 - 2R \tag{85}$$

For coherent glasses, the parameter Y typically has values between 3.2 and 4 (3.2 $\leq Y \leq 4$), representing the number of nonbridging oxygen ions per Si^{4+} polyhedra. The parameter R is given as:

$$R = \frac{2-x}{1-x} \tag{86}$$

where x is the mole fraction of modifier ion (Li^+, Na^+, K^+). R may also be related to the average number of oxygen ions (or other glass-forming) per Si^{4+} ion.

Of the glass systems above, borate glasses, although lower melting and of higher electrical resistivity and lower loss, are rarely used because their resistance to moisture is poor. However, they form an important component of low-melting glasses. The phosphate glasses are also low melting and provide better electrical insulation for many device applications, because they are usually free of alkali ions. However, moisture resistance, and thus corrosion resistance, is marginal for many applications. The silicate-based glasses, in consequence, are the most widely used. Other glass types, such as oxynitrides, chalcogenides, and halides [87–89], have been developed for specialized uses, including amorphous oxynitride films for microelectronic applications.

2. Properties

Of the factors that critically influence the choice of glass insulator, electrical resistivity, thermal expansion, temperature–viscosity behavior, and chemical durability are most important. The latter property relates to the resistance of the glass to weathering and chemical attack. This is usually specified in terms of a weight change per unit area of glass surface per unit time (at 25°C or 95°C) when immersed in an acid or alkaline aqueous solution at a given pH value [95,98].

The thermal expansion behavior reflects the change in volume or length over a temperature span below the glass transition temperature, T_g. Important points on the heating curve from just below T_g to melting T_m are strongly correlated to the viscosity characteristics of the particular glass. Figure 12 relates the thermal expansion behavior and some of these viscosity points to the change in temperature. In Figure 12, T_g is the glass transition temperature, which point represents a structural change from a glassy to a supercooled liquid state to a

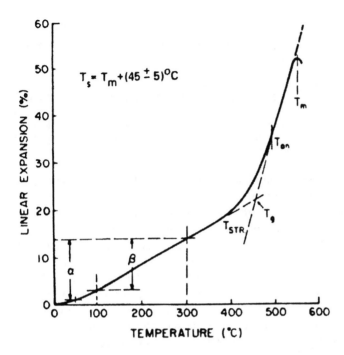

Figure 12 Percent linear expansion versus temperature of lead-alumina-borosilicate glass, showing relationship to the viscosity points in Figure 1, $T_s = T_m + (45 \pm 5)°C$.

glassy state on cooling, making T_g sensitive to the cooling rate. T_s represents the softening point of the glass, which can be simply related to T_d, the deformation point where the glass will no longer support a load. At the annealing temperature, T_{an}, strain should be effectively relieved in about one 1 h, whereas about 6 h is required at the lower limit of the annealing range, given as T_{str}. In terms of composition, high alkali content lowers both T_s and T_g, increases the thermal expansion coefficient and degrades the chemical durability. Partial replacement of the alkali content with RO -or R_2O_3-type modifiers reverses this trend. For most glasses, the dielectric constant at room temperature is linearly related to the density, which in turn can be related to the composition of the glasses, as follows [88,89]:

$$\rho_d = \sum_i f_i v_i \sim \sum_i f_i w_i \tag{87a}$$

$$\varepsilon_r \sim 2.2\rho_d \tag{87b}$$

where ρ_d is the density of the glass, f_i a density factor for each oxide, v_i the specific volume fraction, w_i the weight fraction, and ε_r the dielectric constant. Typical values of f_i for different oxides are as follows:

Oxide	SiO_2	Al_2O_3	B_2O_3	Na_2O	K_2O	CaO	MgO	PbO	ZnO	BaO
f_i	2.2	2.75	1.9	2.6	3.2	3.3	3.4	10.0	5.9	7.0

Table 12 gives the composition ranges for a number of commonly used glasses, with corresponding properties of these glasses listed in Tables 13 and 14. Soda-lime-silicate glasses are the most widely used for such purposes as power and telephone line insulators, fuses, lamp envelopes, and some substrates. Lead-alkali glasses are lower melting and lower in alkali but of higher PbO content. Typical uses are for lamp stems or as graded and pinched seals. The potash-lead glasses are typically made into ribbons for capacitor use. Borosilicate glasses may be used as photomask or substrate glasses, as power line insulation, and for glass-metal sealing with Kovar, W, and Mo metals. They are typically low-alkali, low-expansion, high-temperature glasses. Aluminosilicate glasses are used for electronic component mask substrates for high-pressure mercury lamps (high ultraviolet transmission) and for Cu, Fe, and Ni seal glasses. Fused silica finds electrical uses in delay lines as well as for device passivation and similar applications.

Table 12 Compositions of Typical Commercial Glasses

	wt %					
Oxide	Soda-lime	Lead-alkali	Potash-lead	Borosilicate	Aluminosilicate	Fused silica
SiO_2	71–73	56–63	34–37	70–81	54–55	99+
Na_2O	16–18	5–8.6	—	0–4	—	0–0.5
K_2O	0.8–1.0	6–7	6–8	0–1.0	—	0.02
CaO	9–12	—	—	—	13–15	—
MgO	1–3	—	—	—	—	—
BaO	—	—	—	—	3–4	—
PbO	—	21–30	57–60	0–1.5	—	—
B_2O_3	—	—	—	13–28	6–8	—
Al_2O_3	0.5–2.0	0.8–1.0	—	0–1.2	20–22	—

Table 13 Properties of Commercial Glasses

Property	Soda-lime	Alkali-zinc-borosilicate	Alkali-borosilicate	Lime-aluminosilicate
Str. point (°C)	472	506	520	670
Anneal. point (°C)	512	539	565	710
Soft. point (°C)	696	696	720	910
Thermal expansion (ppm/°C)	9.2	7.2	3.25	4.6
Density (g/cm^3)	2.47	2.51	2.23	2.63
Refractive index n_D	1.51	1.53	1.47	1.52
Vol. rest. (Ω-cm)	$10^{5.6}$	$10^{7.9}$	$10^{13.8}$	$10^{11.2}$
Dielectric constant (ε_r)	6.5	6.7	4.6	6.4
Young's modulus (psi)	10	10.8	9.1	12.4
Chem. durability, 5% HCl (mg/cm^2 at 100°C/24 h)	0.02	0.03	0.005	0.4

Source: Ref. 89.

Table 14 Properties of Commercial Glasses

Property	Fused silica	Alkali-lead-borosilicate	Aluminoborosilicate
Str. point (°C)	990	—	613
Anneal. point (°C)	1050	—	650
Soft. point (°C)	1580	725	820
Thermal expn. (ppm/°C)	0.56	5.5–6.1	4.5
Density (g/cm^3)	2.2	—	2.76
Refractive index, n_D	1.458	1.58	1.53
Vol. rest. (Ω-cm)	$10^{11.2}$	10^{12}	$10^{12.4}$
Dielectric constant (tan δ)	3.9	8.9	5.8
Young's modulus (psi)	10.5	—	9.8
Chem. durability, 5% HCl (mg/cm^2 at 100°C/24 h)	0.17	—	0.28

Among the important electronic application for glasses is their use as substrates for photomasks or for hybrid circuits or as porcelain enamel–coated substrates, which find use in some lower cost thick-film circuitry; here the glass coating is often a lead-borosilicate glass composition. The properties needed are atomically smooth surfaces, chemical inertness, high volume and surface resistivity, flatness, dimensional stability, good thermal shock resistance, and low cost. Freedom from volume defects and good transparency are also needed for photomask applications and low-alkali aluminoborosilicate glasses are typically used, though soda-lime-silicate glasses may be used for less critical applications. These aluminoborosilicate formulations are high-melting glasses with low to moderate expansion coefficients. A somewhat unique application for high-lead glasses reduced in hydrogen to yield a semiconducting surface has been their use in channel electron multipliers.

3. Glass Film Applications

Glass insulating films are widely used for the passivation of silicon and other integrated circuit devices, as crossovers or isolation layers in multilayer circuits, and for encapsulation and/or hermetic sealing of integrated and hybrid circuit packages. These various applications require very different glass compositions (PSG, SiO_2) and would also include both single- and mixed-oxide amorphous films (SiO, SiO_2, Al_2O_3, ZnO-SiO_2), and nitride-type films (e.g., Si_3N_4, AlN) prepared by vapor deposition (CVD, MOCVD) techniques [99,100].

The glass passivation layers are required to provide good dielectric isolation, mechanical and scratch protection, act as an alkali impurity and moisture barrier, and provide low surface recombination velocity and device stability at

Table 15 Typical Properties of Dielectric Films

Dielectric	Density (g/cm^3)	Melting point (°C)	Coefficient of thermal expansion (°C^{-1})	Dielectric constant	Index of refraction
SiO	2.1	—	—	3.3–5.2	1.55–2.0
SiO$_2$	2.2	~1600	0.55×10^{-6}	3.9	1.46
PSG	—	~900	0.87×10^{-6}	3.7–4.0	1.42
Si$_3$N$_4$	3.1	~1900 (pressure)	3.2×10^{-6}	6.1	2.1
Al$_2$O$_3$	3.6	~2020	7.58×10^{-6}	7.7	1.6
AlN	3.2	~2400	6.1×10^{-6}	9.1	2.16
TiO$_2$	2.6	~1850	8.22×10^{-6}	20–50	2.0
Ge$_3$N$_4$	5.0–5.4	—	—	6.5	2.06

Source: Data from Refs. 99 and 100.

elevated temperature under bias or operating conditions. These requirements are for the most part met by SiO_2 and phosphosilicate glass films, sometimes in conjunction with Si_3N_4 films as an alkali diffusion barrier. Typically, the films are deposited by low-pressure or regular CVD or by radio-frequency (RF) sputtering techniques. Some fusible phosphosilicate glass films are also prepared, as are some low-melting lead-alumina-borosilicate glass films for special applications. Properties of these films are listed in Table 15.

4. *Sealing* Glasses

In contrast to the passivation of films, the glasses used primarily for thick-film dielectric insulation or for solder glass sealing are the more conventional lead-borosilicate type. Typical composition and property ranges are given in Tables 16 and 17 for the dielectric and sealing glass compositions, respectively. The dielectric glasses are used for such applications as thick-film crossover insulator, ac plasma display coatings, insulator glazes, diode encapsulation, ribbon and thick-film capacitors, and as a component of some thick-film resistor pastes. Requirements for low softening points (T_s) and glass transition temperature (T_g), high dielectric constants (ε_r), and moderate thermal expansion coefficients (α) can be accommodated within the PbO-B_2O_3-SiO_2 composition ranges shown, with the modifier oxides (Al_2O_3, CaO, MgO, Na_2O, K_2O) being used to shift desired properties into specific directions.

The wide variety of applications indicated require controlled changes in one or more components. In some cases, statistical methods can be employed in

Table 16   Dielectric Glass Compositions (wt %) and Property Ranges

Component	Range	Mean	Standard deviation
PbO	50–74	5.9.90	5.270
B_2O_3	10–25	18.10	3.54
SiO_2	8–26	13.30	3.10
Al_2O_3	0–5	1.78	1.385
CaO	0–6	3.08	1.452
MgO	0–4	2.00	0.749
Na_2O	0–5	1.84	1.049
T_g (°C)	425–510	471.9	18.306
T_s (°C)	470–565	526.0	19.466
α (10^{-7} °C^{-1})	75–101	84.56	5.149
ε_r	12–17	14.3	0.958

Table 17 Sealing Glass Compositions (wt %) and Property Ranges

PbO	B$_2$O$_3$	SiO$_2$	ZnO	BaO	Al$_2$O$_3$	Bi$_2$O$_3$	CuO
68–75	9–12	2–4	10–12	0–3	0–5	0–4	0–3
T_s		T_g		T_{seal}		$\alpha_{0-300°C}$	
380–410°C		340–470°C		445–475°C		80–90 × 10^{-7} °C^{-1}	

an attempt to make these changes in a predictable manner, so that the properties can be tailored to specific applications.

For example, linear predictive equations of the form:

$$P_i = r_1w_1 + r_2w_2 + \cdots r_7w_7 + I_p = \Sigma r_iw_i + I_p \qquad (88)$$

can be developed using multiple correlation/regression analysis, where P_i is the property of interest (e.g. T_s), r_i the unstandardized regression coefficient for component i, w_i the weight percent associated with that component, and I_p an intercept value. Both r_i and I_p can be obtained from computer analysis. Analysis of generated data from more than 50 glasses within the composition/property ranges shown in Table 16 yielded coefficients for the softening point T_s (at the 99% confidence level), as follows:

$$T_s(°C) = -4.67PbO - 1.14B_2O_3 + 0SiO_2 - 1.32Al_2O_3 + \\ 1.08CaO + 5.5MgO - 9.83Na_2O + 832.4 \qquad (89a)$$

with an estimated $T_s = 526°C$. Similar analyses can be carried out for other properties for which data are available. The magnitude and sign of the regression coefficients, r_i, can be directly correlated to the effectiveness of the different oxide components in raising or lowering a particular property. These trends are illustrated in Table 18 for glasses in the PbO-B$_2$O$_3$-SiO$_2$ base system.

Table 18 Effects of Oxide Concentration on Glass Properties in Lead-Borosilicate Glasses

Property	Oxide decrease	Oxide increase
T_g (°C)	PbO, Na$_2$O	MgO
T_s (°C)	PbO, B$_2$O$_3$, Na$_2$O, Al$_2$O$_3$	CaO, MgO
α (10^{-7} °C^{-1})	B$_2$O$_3$, SiO$_2$, Al$_2$O$_3$, B$_2$O$_3$, Al$_2$O$_3$	PbO, CaO, Na$_2$O, PbO, MgO

Solder glasses are used for sealing various combinations of glass/metal/ceramic structures in a variety of configurations. This includes the bonding of magnetic (ferrite) heads for data recording (where the glass serves also as a gap material), hermetic seals for hybrid and integrated circuit packages, as well as some encapsulation functions. Typically, these glasses must be low melting, and since the integrity of the bonded structures is of prime importance, expansion coefficients must be matched to minimize stresses. This is illustrated in Figure 13, which compares the stress condition for two glass films on a metal substrate on cooling to 25°C. The net compressive stress component for the lower expansion glass A is the desired condition.

Control of the thermal expansion coefficient is influenced greatly by the ZnO, Bi_2O_3, and CuO concentrations, since these oxides have the desirable effect of lowering softening points without a significant increase in thermal expansion, in contrast to the alkali oxides [83,89]. Some sealing glass compositions may be almost wholly crystalline after being formed, affording greater strength and reheat capability for the sealed structure. These compositions are subject to the same type of composition/property analysis as that discussed for the dielectric glasses. Table 19 gives composition ranges and properties for typical crystallizable solder glass compositions.

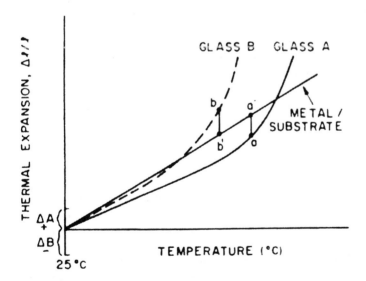

Figure 13 Thermal expansion versus temperature for different glasses (A and B) on metal substrates, showing residual stress components (ΔA, compressive; δB, tensile) on cooling.

Table 19 Crystallizable Solder Glasses (wt %)

Component	Range	Typical
PbO	75–82	75.0
B_2O_3	6.5–12	8.0
ZnO	7–14	10.0
SiO_2	1.5–3.0	2.5
Al_2O_3	0–3.0	1.0
CuO	0–3.0	2.5
Sb_2O_3	0–1.0	0.5
Na_2O	0–0.5	—
BaO	0–4.0	—
$\alpha(10^{-7}/°C)$ (0–300°C)	80–105	85.0
T_s, °C	375–395	385.0
T_{seal}, °C	400–450	430 ± 10

5. Electrical Conduction in Glasses

Electrical conduction in glasses is due mainly to the migration of mobile ions, such as Li^+, Na^+, K^+, OH^-, and H^+, under the influence of an applied field. At temperatures above 200°C, divalent ions such as Ca^{2+}, Mg^{2+}, and O^{2-} also contribute to the conduction process, although their mobility is significantly lower than that of the alkali ions [83,92,93,104]. Conduction in glass is an activated process, thus, the number of ions contributing to conduction increases with both temperature and applied field. The temperature/resistivity dependence is given as:

$$\rho = \frac{6kT}{\lambda \upsilon \alpha (ez)n} e^{E/kT} = \rho_0 e^{E/kT}$$

(91)

where ρ_0 defines the pre-exponential terms, v is the natural vibration frequency (10^{12}–10^{13} s^{-1}), λ (8–9 Å) the average jump distance, α (~3) the number of adjacent jump sites, n the concentration of charge carriers, and E the activation energy.

As the mole fraction and concentration (n) of modifier ions increase, the effect is to loosen the structure, resulting in freer migration of the conducting ions and a lowering of the activation energy, E. Although α does not vary much with modifier content, the jump distance λ increases because the network expands with increasing concentration of mobile ions. However, both λ and α are significantly affected by the presence of two or more mobile ionic species in the struc-

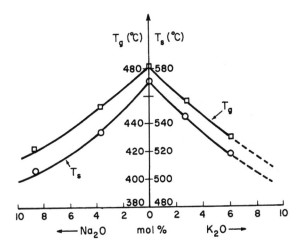

Figure 14 Plot showing change in T_g with increase in Na and K ion concentration for a base lead borosilicate glass.

ture. This condition leads to a blocking effect and mutual decrease in the mobility of the cations in the glass structure, the so-called mixed-alkali effect [85,96].

In Figures 14–16, the effects of different molar amounts of Na and K modifiers added to a lead-alkali-borosilicate glass are shown with respect to the temperature (T_g, T_s, see Fig. 14), dielectric constant (ε_r, see Fig. 15), as well as on the resistivity and activation energy properties of the glass (Fig. 16). These properties are seen to change significantly as either alkali content is increased, with the effect of Na being more pronounced.

Studies relating the electrical conductivity in calcium silicate glasses to the Na_2O content, in the temperature range of 250–500°C, found that the conductivity was insensitive to Na_2O content but increased with CaO content [77,78]. Correspondingly, the activation energy decreased from 1.4 to 1.25 eV as the CaO content changed from 40 to 55 mol %. Table 20 lists activation energy values as a function of modifier ion concentration for a number of alkali and alkali-earth-modified silicate glasses. The effect of various modifier oxides (MnO, ZnO, B_2O_3, Fe_2O_3, BaO, PbO, TiO_2, K_2O) on the electrical resistivity of glasses in the soda-lime-silica system has also been studied [84,86]. The specific resistance at 400°C was found to increase, in the above order, from 0.06 to 5.0×10^5 Ω-cm. However, a decrease in resistance was produced with Na_2O, CaO, and Al_2O_3 modifiers. These complex effects on the conductivity must be related to the detailed structure of glass, including coordination environment of the ions, and not simply to the random distribution and concentration of mobile ions in the glass.

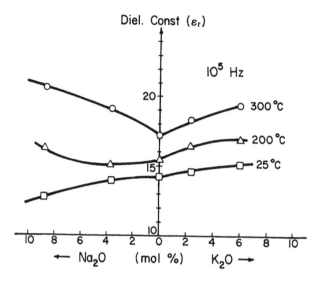

Figure 15 Plot showing change in dielectric constant with increase in Na and K ion as a function of frequency and temperature for lead borosilicate glass in Figure 14.

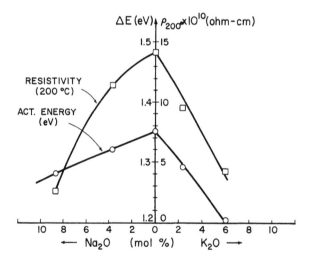

Figure 16 Plot showing change in activation energy and resistivity for alkali (Na and K)–modified lead borosilicate.

Table 20 Activation Energies of Alkali and Alkali Earth Glasses

Glass comp. (mol %)	Activation energy (eV)
$15K_2O-85SiO_2$	0.83
$40Na_2O-60SiO_2$	0.51
$40K_2O-60SiO_2$	0.63
$20Na_2O-20K_2O-60SiO_2$	1.03
$20K_2O-20Li_2O-60SiO_2$	1.05
$10Na_2O-20PbO-70SiO_2$	1.04
$30Na_2O-20PbO-50SiO_2$	0.82
$30Na_2O-20BaO-50SiO_2$	0.86
$20Na_2O-30BaO-50SiO_2$	1.10
$10Na_2O-10CaO-80B_2O_3$	1.17
$10Na_2O-30MgO-60B_2O_3$	1.13
$40CaO-60SiO_2$	1.10
$45CaO-55SiO_2$	1.35
$50CaO-50SiO_2$	1.30
$55CaO-45SiO_2$	1.25

Source: Ref. 84.

For most oxide glasses, the temperature dependence of the electrical resistivity over a fairly wide range (but below the glass transition temperature) is adequately described by the Rasch–Hinrichsen equation [97,103]:

$$\log \rho = A + \frac{B}{T} \tag{89}$$

or

$$\rho = A'e^{B/T} \tag{90}$$

where for most glasses $A = -0.45$ to 1.5, $B = 3000-6000 \ K^{-1}$, $\rho = 10^{10}-10^{15}$ (Ω-cm) at 25°C, and $E = 0.2 \times 10^{-3} B$ (eV) (E is the activation energy for conduction). Figure 17 shows a plot of log resistivity versus reciprocal temperature for various types of commercial glasses, where, as indicated, the presence of mobile cations generally leads to much lower resistivity values. The Rasch–Hinrichsen relationship is useful but gives no direct information related to the structure of the glass or the conduction mechanisms.

6. Conduction Mechanisms

Even though the glass structure can be described as open because of its random nature, there is no well-defined channel for conduction in glasses as occur in

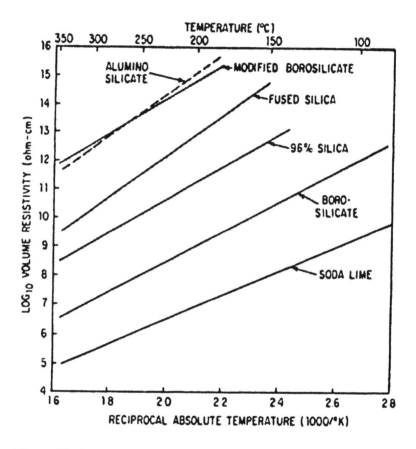

Figure 17 Plot of volume resistivity versus reciprocal absolute temperature for some common commercial glasses, illustrating wide variation in values.

crystalline ceramics. With respect to a basic conduction mechanism, there is also no general consensus that universally explains all aspects of ionic transport in glasses [94]. However, application of percolation theory within the well-known random-energy model leads to the most consistent explanation for both ac and dc conduction effects in glasses. Explanations for the observed strong dependence of ionic conductivity on composition have mainly been based on changes in the activation energy for electrical conduction [99]. For alkali migration in oxide glasses this consists of two parts: the bonding energy between the mobile cation and its charge compensating center; and the elastic strain energy associated with the distortion of the glass network as the ion moves from one site to another. For glasses with a high concentration of monovalent ions as charge carriers, ionic transport is determined not only by the interactions between the ions and the glass network but also between the ions themselves [100]. These effects strongly influence the structure and the geometry of the transport pathways, leading to a lowering of the activation energy and increased conductivity.

The current carriers are considered to be located in potential wells and undergo thermal vibrations of magnitude proportional to the temperature. In the absence of an applied field the ions make random jumps, according to the probability of surmounting the surrounding energy barriers, but migrate preferentially in the direction of the applied field, which effectively lowers the energy barriers in the forward direction. The number of conducting ions therefore increases with both temperature and electric field. Early work indicated that oxide glasses containing large amounts of alkali ions were essentially electrolytic in nature. Under a dc field, electrolysis was found to occur and Faraday's laws were obeyed. Electronic conduction was found to take place only in transition metal oxide–containing glasses. Otherwise, the transference number for the alkali ions is close to unity, the mobility of divalent and higher valence ions being considerably lower. However, ionic conduction in oxide glasses is strongly influenced by such factors as composition, temperature, and thermal history.

Glasses typically exhibit an amorphous or random network structure with characteristic short-range order of only a few atomic spaces. The creation of structural disorder occurs during fusion of the glass and, as in lithium borate glass, is presumed to result from the formation of a large excess of nearly equivalent sites, giving enhanced ionic mobility. When a second alkali is added to an alkali-silicate glass, the conductivity sharply decreases. The simple concept of a random silicon network cannot explain this anomalous behavior, and many theories have been proposed to explain it [93,101]. One such theory assumes that the glassy structure is composed of two regions: polar regions, where the alkali ions preferentially aggregate, and nonpolar regions, consisting of the glass-forming oxide. When two ions are present there will be two different types of polar regions, with each ion associated only with one type of polar region. The decrease in conductivity can be attributed to the inability of ions to exchange between the polar regions and to the obstacles caused by one polar region interacting with another ion. For a fixed concentration of charge carriers, addition of MgO, PbO, BaO, and CaO tend to decrease the conductivity. This behavior has been attributed to the "blocking" of the conducting ions by the heavier divalent atoms.

G. Electrical Porcelains

1. Introduction

Porcelains can be described as polycrystalline ceramic bodies containing, typically, more than about 10 vol % of a vitreous second phase [105–113]. This vitreous phase not only controls densification, porosity, and microsructure development, but also the dielectric, mechanical, thermal, and durability properties of the porcelain. Porcelains are usually silicate-based bodies, which can be classified as triaxial, if they are formulated primarily from silica, clay, and feldspar as batch constituents, or as non-feldspathic (e.g., steatite, zircon, forsterite), where the feldspar component is typically absent, if the clay and feldspar content is low. The compositions, properties, and uses of the different categories of porcelains are described below.

Electrical porcelains are chemically inert and durable, finding extensive use in household, laboratory, and industrial applications. In particular, they are widely used for electrical transmission line insulation due to high stability of their electrical, mechanical, and thermal properties in the presence of the harsh environments. An important characteristic of any insulation, particularly for high-voltage switching and circuit-breaking devices, is the ability to withstand arcing, which can degrade the insulation resistance leading to dielectric breakdown. To protect against electrical arcing, high softening temperature, high resistance to thermal shock, and high dielectric strength are also required. Heat dissipation is also important because the heat produced during arcing can cause the insulator to fracture. Electrical porcelains are strong in compression, with compressive strengths ranging up to 40,000 psi. The high compressive strength and physical stiffness with practically zero cold flow assures dimensional stability, which is further accentuated by the Porcelain's rigidity. These factors combine to facilitate insulator design and to reduce deformation under actual service conditions, even at elevated temperatures.

The electrical resistivity of all dense porcelain bodies is typically greater than 10^{11} Ω-cm at 25°C, but this value depends on the type and quantity of glassy phase and other impurities present in the ceramic [20,89,105]. Since in the glassy state the alkali ions are more weakly bonded than in the crystalline state, large amounts of an alkali-containing glassy phase in the porcelain ceramic will lower resistivities at high temperatures and also lead to higher dielectric losses [68]. For high-fired porcelains, such as cordierite, forsterite, low-loss steatite, zircon, and high-alumina porcelains, the glassy phase consists mainly of BaO and other alkaline earth oxides. Hence, the electrical losses are much lower than in the alkali-containing porcelains. Crystalline, glass-free ceramics (Al_2O_3, BeO, MgO, SiO_2) exhibit higher resistivities and even lower losses than the glassy-phase ceramics, particularly at elevated temperatures. Figure 18 compares volume resistivity data versus reciprocal absolute temperature for some common porcelains and oxide insulators.

2. Triaxial Porcelains

Triaxial porcelains comprise a large fraction of the commonly used porcelain insulators. They exhibit fairly low mechanical strength, fair thermal shock resis-

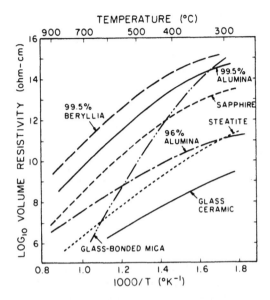

Figure 18   Plot showing volume resistivities as a function of reciprocal absolute temperature for common ceramic insulators. (From Ref. 18.)

tance, and relatively poor high-frequency characteristics (due to alkali ion migration) but they are generally satisfactory for low-frequency ($< 10^4$ Hz) use at ordinary temperatures. The triaxial porcelains are formulated from mixtures of feldspars [$(K, Na)_2O-3Al_2O_3-6SiO_2$], kaolinite clays ($Al_2O_3-2SiO_2-2H_2O$), and flint or quartz ($SiO_2$), which places these ceramic bodies in the phase system [$(K, Na)_2O-Al_2O_3-SiO_2$)] in terms of oxide constituents [108,90]. Table 21 gives the raw material composition ranges for the triaxial porcelains.

The raw material compositions for the triaxial porcelains are typically in the range of: 40–60 wt% binder (clays); 20–35 wt% flux (feldspars), and 20–30 wt% flint or quartz (SiO_2). Some filler materials such as alumina (Al_2O_3) or Zircon ($ZrSiO_4$) may sometimes be added [109]. The raw materials each contribute different characteristics to the fired porcelain. The kaolinite clay provides the main body for the ceramic and controls the firing range and distortion during sintering. It also provides the first liquid phase in the system ($\sim 900°C$), from which the main crystalline phase, mullite, is subsequently developed. In most formulations, however, up to one-half the kaolinite may be replaced by ball clay. The effect is to increase the plasticity and green strength in the formed ware, but the ball clay may also introduce alkali and other impurities. The feldspar is a low melting flux phase that reacts with other constituents to lower the liquid

Table 21 Raw Material Ranges for Triaxial Porcelain Bodies (wt %)

| Raw material | Generic | Low tension | | High tension |
		a	b	c
Kaolinite	20–35	21	25	35
Ball clay	15–30	25	20	10
Feldspar	20–35	34	33	25
Flint	20–30	20	23	25
(BaCa)CO$_3$	0–5	—	—	5
sintering temperature	1250–1330°C			1300–1450°C

formation temperature and increase the amount of liquid formed in the system [106,107,108]. The liquid permeates the porous microstructure leading to rapid densification, primarily by a viscous flow mechanism. The fluxes used commercially are mainly alkali feldspars ($NaAl_3SiO_8$, $KAlSi_3O_8$) and nepheline syenite ($Na_3KAl_4Si_4O_{16}$), from which some of the mullite phase also develops. However, the increased glassy phase and alkali content tend to degrade both the dielectric properties and mechanical strength of the porcelain.

Partial dissolution of the quartz phase above 1200°C increases the liquid phase viscosity and helps to maintain shape during firing. An increase in the quartz content usually gives improved mechanical strength, provided that the grain size is below ~ 10 μm, in order to minimize thermal expansion stresses. The silica phase also helps to reduce distortion and shrinkage [108]. The alumina phase, mainly added mainly as a filler component, serves to improve both the thermal diffusivity and firing range of the porcelain. Minor ingredients (2–5 wt%) such as talc, zircon, and the alkaline earth carbonates aid in the formation and control of the low-temperature eutectic phases, resulting in generally improved dielectric and mechanical properties of the porcelain [84]. The phase composition and microstructure can also be altered by incorporating small amounts of oxide additives, such as Fe, Ti, Cr, V and Nb, which function as mineralizers. The porcelain body may also be heat-treated under controlled conditions to re-crystallize the glassy phase, with improvement in both the elrctrical and mechanical properties.

The mullite phase is the most important crystalline constituent of the porcelain, as it imparts high temperature stability, mechanical strength, low creep rate, low thermal expansion coefficient and low thermal conductivity to the porcelain. Because of the very low interdiffusion rates for Si^{4+} and Al^{3+} within the mullite lattice, the kinetics of mullite formation depend strongly on precursor mixing.

The SiO_4 tetrahedral and $Al(OOH)_4$ octahedral layer are sequential in the kaolinite structure, allowing good mixing of the Al and Si ions at the atomic level. Hence it is the chief source of mullite formation in the triaxial porcelains. The general reaction steps can be outlined as follows. At ~550°C, dehydroxylation of the kaolinite ($Al_2O_3.2SiO_2.2H_2O$) to the metakaolin ($Al_2O_3.2SiO_2$) state occurs, followed by transformation of the metakaolin to a metastable spinel-type phase, with the release of a large amounts of silica, at ~ 950–1000°C. The spinel phase then transforms to mullite above 1075°C. On further heating to 1200°C, transformation of quartz to the cristoballite phase begins, but this phase disappears above 1400°C [20,105,108]. The forms of mullite occuring in triaxial porcelains, therefore, are the result of: a).formation of primary mullite from decomposition of the kaolinite b).formation of secondary mullite from reaction of the feldspar, clay and quartz c). the formation of tertiary mullite by precipitation from the alumina-rich liquid obtained by dissolution of alumina filler. The size and shape of the mullite crystals is determine by the fluidity of the local liquid phase from which they precipitate and grow, which is also a function of temperature and composition, determined largely by the extent of mixing of the porcelain raw materials and the type and amount of fluxes used.

3. Hard Porcelains

Typical compositions and dielectric properties for representative porcelain bodies are given in Table 21. As indicated in Table 21, a distinction must be made between the categories of triaxial porcelains described as low tension (low mullite content) and high tension or hard porcelains (high mullite content) [11,20]. The low-tension porcelains generally have looser raw material tolerances and are used unglazed, for less exacting low-voltage and low-frequency applications, such as light sockets, fuse blocks, bushings, hot plate and toaster insulation. They are usually sintered in the range 1260–1320°C and are characterized by by moderately low dielectric strengths (~40–100 V/mm) and loss factors (0.05–0.10 at 1 KHz). In contrast, the high-tension porcelains are sintered in the range 1350 to 1450°C to a high gloss finish.

Lower impurity content is achieved by reduction or elimination of the ball clay content, which is replaced by bentonite clays (~ 3.0 wt %) or alkaline earth additives. These porcelains can be considered as low-loss, high-dielectric-strength (~250–400 V/mm) materials, because of lower mobile ion content. They are, therefore, suitable for high-voltage applications such as transmission line insulators, high-voltage circuit breakers, and cutouts. Table 22 compares the important mechanical, thermal and dielectric properties for the low and high tension porcelains.

4. Porcelain Glazes

For environmentally sensitive applications, the porcelain must be dense, moisture impervious, and have minimal mechanical flaws [105,106]. This is aided by the

Table 22 Properties of Triaxial Porcelain Bodies

Property	Units	Material Low-tension porcelain	High-tension porcelain
Density	g/cm^3	2.2–2.4	2.3–2.5
Water absorption	%	0.5–2.0	0.0
Coeff. of linear thermal expansion	10^{-6}/°C	5.0–6.5	5.0–6.8
Safe operating temperature	°C	900	1000
Thermal conductivity	w/cm K	0.016–0.021	0.0084–0.021
Tensile strength	lb/in.^{2}	1500–2500	3000–8000
Compressive strength	lb/in.^{2}	25,000–50,0000	25,000–50,0000
Impact strength $^1/_2$ -in. rod	ft/lb	0.2–0.3	0.2–0.3
Modulus of elasticity	10^{-6} lb/in.^{2}	7–10	7–14
Thermal shock resistance		Moderate	Moderately good
Dielectric constant		6.0–7.0	6.0–7.0
Power factor at 1 MHz		0.010–0.020	0.006–0.010
Resistivity, room temperature	Ω-cm	10^{12}–10^{14}	10^{12}–10^{14}
Dielectric constant		5.5–7.0	6–7
tan δ		0.005	0.003
Loss factor		0.035	0.020

Source: Data from Refs. 10, 20, and 89.

use of a glaze coating, which may be a self glaze, feldspathic galze, or semiconducting brown glaze.

Self-glazing arises from the higher sintering temperature for the hard porcelains, resulting in a significantly higher vitreous phase and pore-free microstructure. Typical oxide constituents for a suitable high-temperature feldspathic glaze maturing in the range 1200–1250°C can be given (in wt %) as: SiO_2, > 70.3; Al_2O_3, 11.9; CaO, 7.9; K_2O, 4.4; BaO, 3.6; ZnO (or MgO), 1.9. A brown glaze can be developed from this feldspathic glaze by incorporating 20–35wt% calcined and ground coloring oxides (Fe_2O_3, TiO_2, MnO_2) into the glaze mixture and sintering in a reducing atmosphere to obtain some ions in a lower oxidation state. The juxtaposition of such ions as (Ti^{3+}, Ti^{4+}), (Fe^{2+}, Fe^{3+}), and (Mn^{2+}, Mn^{3+}, Mn^{4+}) in the glaze results in some n-type conduction and a surface resistivity (ρ_s) of ~ 10^7 Ω/square. This serves to reduce the corona discharge caused by ionization of moist ambient in the high potential gradient near the conductor/insulator interface.

5. Hydrophobic Coatings

The malfunction of electrical equipment due to corona discharge has been attributed to electrolytically deposited films on power line insulators, which can occur under the conditions of high air humidity, and in industrial areas where pollutants such as dust or toxic gases exist. Under such conditions, electrically conducting films may deposit on the surface of insulators, which even at the nominal operating voltage may lead to flashover. Some improvement in contamination resistance can be achieved through insulator design, which has the effect of lengthening both the conducting path and the dry zone, thereby raising the contamination level at which significant distortion and shrinkage occur [110–113].

The tendency of current leakage to increase with increase in the level of contamination is typical of all hydrophilic porcelains. To convert the hydrophilic silicate surface to a hydrophobic surface, an organic coating can be applied to the inorganic porcelain. The surface hydroxyl groups formed by hydrolysis of Si–O–Si bonds show a strong affinity for attaching additional water molecules via hydrogen bonding. The result is the formation of multiple layers of water molecules, or a water film, on the silicate surface. This plausibly explains why clean, glazed surfaces can be readily wetted by hydrophilic liquids. These hydrophilic surfaces exhibit a high surface energy, owing to the high concentration of polar molecules and molecular chains.

Several techniques have been applied to increasing the hydrophobicity of high-tension insulator surfaces, including the use of commercial resin–type varnishes, plasma-assisted polymer coating dip coating or spray coating using fluoroalkylsilanes, as well as other industrial coating processes. On application of the coating, the protective layer modifies the surface of the insulator in such a way that the polar groups in the surface are replaced by nonpolar groups. That is, the insulator surface, with a coating containing mainly nonpolar groups now behaves as a repellant toward water due to the lowered surface energy. Therefore, the water will tend to form droplets instead of spreading, thereby disrupting the conduction and current leakage [110] (Figure 19).

The type of resin generally used is a transparent varnish with a sol-gel-like structure that is deposited on the insulator by spraying. Subsequent thermal treatment which involves curing at ~160°C for ~2 h, produces an organic-inorganic network in the sprayed film. In the plasma coating method, a hydrophobic polymer film of ~1.3 μm is deposited on the surface of the insulator using a plasma reactor. Coatings based on fluorinated alkylsilanes can be prepared using dipping or spray coating and provide chemical "cross-linking" and bonding. The fluorinated alkyl silanes and plasma assisted coatings show better resistance to contamination and are more durable than varnish coatings.

Silicone coatings have found application for this purpose. In the case of glazed porcelains, even under moderate levels of contamination, a strong leakage

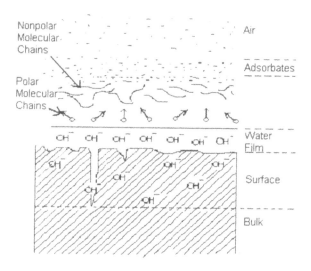

Figure 19 Schematic of hydrophobic layer formation on porcelain surface. *Source:* Ref. 110.

current is generated with an early I_{max} followed by flashover. The silicone insulators exhibit a different leakage – current behavior, where discharges practically never occur even at quite high levels of contamination. This superior contamination characteristic arises from hydrophobicity of the silicone material. The water-repellant property of this material prevents formation of water films and continuous conducting zones. The hydrophobicity thus provides a barrier to current leakage and prevents flashover across areas protected by the skirts. The maintenance-free period of such units with silicone-coated insulators can be significantly extended [110,111,113].

6. Oxide Porcelains Insulators

Most ceramic insulators are to be found in the system $MgO \cdot Al_2O_3 \cdot SiO_2$. These include the oxide end members MgO, Al_2O_3, and SiO_2; the binary compounds $MgO \cdot SiO_2$ (steatite), $2MgO \cdot SiO_2$ (forsterite), $MgO \cdot Al_2O_3$ (spinel), and $3Al_2O_3 \cdot 2SiO_2$ (mullite); and the ternary compound $MgO \cdot Al_2O_3 \cdot 2.5SiO_2$ (cordierite) [114–117]. Typical properties of these insulators are given in Tables 1 and 2. Table 23 compares raw material compositions for the porcelain insulators in this system as well as for zircon ($ZrO_2 \cdot SiO_2$). The raw materials used are principally talc ($3MgO \cdot 4SiO_2 \cdot 2H_2O$), kaolinite clays, and alkaline earth fluxes such as $BaCO_3$ and $CaCO_3$. These porcelains are of higher purity than the triaxial porcelains

Table 23 Raw Material Compositions for Non-Feldspathic Porcelain (wt %)

Raw material	Steatite	Forsterite	Cordierite	Zircon
Talc	80–90	60–70	35–45	10–20
Kaolin/clay	5–7	5–10	40–50	10–20
Mg(OH)$_2$	—	20–45	—	—
BaCO$_3$	6–7	5–7	0–2	6–8
CaCO$_3$	0–1	1–3	—	0–1
Zircon	—	—	—	55–70
Al$_2$O$_3$	—	—	12–15	—
Sintering temp. (°C)	1260–1400	1275–1375	1250–1350	1300–1400

with superior dielectric properties but are more difficult to produce due to the shorter sintering range.

(a) *Steatite* ($MgO \cdot SiO_2$). The steatite ceramics are most commonly used for high-frequency insulation due to their low dielectric losses resulting from the low alkali content. Steatites are used in variable capacitors, coil forms, electron tube sockets and general structural insulation (e.g., bushings, spacers, support bars) [18,20,89]. They can be fabricated to close dimensional tolerances using automatic dry pressing and extrusion methods, due to the large talc content. Unlike triaxial porcelains, steatite bodies require close control of the firing temperature, since the firing range for liquid formation is short in the $MgO \cdot SiO_2$ system.

The addition of kaolinite clay (<10 wt%) improves these processing characteristics, while the BaCO$_3$ provides fluxing and stabilization of the phase composition, consisting primarily of enstatite (MgSiO$_3$). Steatites exhibit excellent insulating properties, very good mechanical strength, but only fair thermal shock resistance [24]. Steatites are also characterized by a low dielectric constant and low dissipation factor, which increases modestly with temperature at low frequencies but is relatively independent of temperature. It is, therefore, a very useful insulator for both low- and high-voltage and frequency applications.

(b) *Forsterite* ($2MgO \cdot SiO_2$). The absence of alkali ions in the vitreous phase give forsterite insulators a higher resistivity and lower dielectric loss with increasing temperature than steatite bodies. This makes them suitable for high-frequency and microwave purposes but the thermal shock resistance is poor [114,115]. However, the forsterite bodies tend to be very dense and impervious, and this property, coupled with their high thermal expansion coefficients, has made these materials widely popular for ceramic vacuum tube envelopes and ceramic-metal bonding applications.

(c) *Mullite* ($3Al_2O_3 \cdot 2 \cdot SiO_2$). Mullite is a valuable ceramic material because of its low thermal expansion and resistance to spalling and deformation under load. Mullite was originally used in spark plug insulation but is now widely used for thermocouple tubing, in kiln furniture, and as a liner material in tube-type furnaces [20,89]. It is formed from a mixture of alumina and silica, but the specific composition of a given specimen depends on the method of synthesis. The theoretical composition is 71.8% alumina and 28.2% silica, but alumina content can be as high as 77% or as low as 62.6%.

Like Al_2O_3 and similar oxides, O^{2-} ion vacancies developed in the mullite structure at high temperature ($\geq 1300°C$) will lead to some ionic conduction via oxygen vacancy migration. Below this temperature, n-type electronic conduction might be expected to predominate, depending on impurity levels. Substitution for Al^{3+} ions in the (AlO_6) octahedral and (AlO_4) tetrahedral sites of mullite lattice by the transition metal ions, M^{n+}, can also make the material n-type (if $n > 3$) or p-type (if $n < 3$) conductor. Apart from this, incorporation of M^{n+} ion in interstitial sites can impart electrical conductivity to mullite lattice [118]. With such impurity dopants the resistivity of mullite may decreases by several orders of magnitude at elevated temperature.

(d) *Cordierite* ($MgO \cdot Al_2O_3 \cdot 2.5SiO_2$). Cordierite ceramics exhibit very low thermal expansions, leading to excellent thermal shock resistance, particularly in the porous state [20,117]. The materials are, in consequence, frequently used for such applications as wire and carbon resistor cores, electric heater plates, thermocouple insulators, and burner nozzles. The porosity arises from the difficulty in controlling densification, glassy phase, and composition in the cordierite phase field due to the steep liquidus curves that exist in the three-component system.

(e) *Zircon* ($ZrSiO_4$). Zircon is an important ceramic material because of its low coefficient of thermal expansion ($\sim 4.1 \times 10^{-6}$ $°C^{-1}$ between 25°C and 1400°C) as well as its low coefficient of thermal conductivity (0.051 W cm^{-1} $°C^{-1}$ at room temperature and 0.035 W cm^{-1} $°C^{-1}$ at 1000°C). It does not undergo any structural transformation until dissociation at about 1700°C according to the ZrO_2-SiO_2 system phase diagram [10,89]. In addition, it has the advantage of excellent dielectric properties, good thermal shock resistance, high mechanical strength, and ease of fabrication. These materials represent an attractive alternative to the other porcelains but insulating uses are not widespread, reflecting the higher raw material and fabrication costs for the porcelain. Cordierite-zircon bodies have also been prepared with low thermal expansion and excellent electric insulation characteristics [20].

(f) *Spodumene.* Insulators in the $Li_2O \cdot Al_2O_3 \cdot SiO_2$ system, including spodumene ($Li_2O \cdot Al_2O_3 \cdot 4SiO_2$), typically possess very low net shrinkage on sintering

Table 24 Properties of Non-Feldspathic Porcelains [10,18,20,89]

Material	High-tension porcelain	Alumina porcelain	Steatite	Fosterite	Zircon porcelain	Codierite refractories
Specific gravity, g/cm^3	2.3–2.5	3.1–3.9	2.5–2.7	2.7–2.9	3.5–3.8	1.6–2.1
Water absorption, %	0.0	0.0	0.0	0.0	0.0	5.0–15.0
Linear expansion, at 20–700°C, 10^{-6} in./(in.)(°C)	5.0–6.8	5.5–8.1	8.6–10.5	11	3.5–5.5	2.5–3.0
Safe operating temp., °C	1000	1350–1500	1000–1100	1000–1100	1000–1200	1,250
Thermal conduct. 1/cm^2)(cm)(s)(°C)	0.002–0.005	0.007–0.05	0.005–0.006	0.005–0.010	0.010–0.015	0.003–0.004
Tensile strength, psi	3,000–8,000	8000–30,000	8000–10,000	8000–10,000	10,000–15,000	1000–3500
Compressive strength ,psi	25,000–50,000	80,000–250,000	65,000–130,000	60,000–100,000	80,000–150,000	20,000–45,000
Flexural strength,psi	9,000–15,000	20,000–45,000	16,000–24,000	18,000–150,000	10,000–15,000	1500–7000
Impact strength, (1/2-in.rod) ft/lb	0.2–0.3	0.5–0.7	0.3–0.4	0.3–0.4	0.4–0.5	0.2–0.25
Modulus of elasticity, psi x 10^{-6}	7–14	15–32	13–15	13–15	20–30	2–5
Thermal shock resistance	Moderate	Excellent	Moderate	Poor	Good	Excellent
Dielectric strength min, kV/mm^{-1}	20	17	15	13	15	17
Resistivity at room temp., Ω-cm	10^{12}–10^{14}	10^{14}–10^{15}	10^{13}–10^{15}	10^{13}–10^{15}	10^{13}–10^{15}	10^{12}–10^{14}
T_e value, °C	200–500	500–800	450–1,000	Above 1,000	700–900	400–700
Power factor at 1 MHz	0.006–0.010	0.001–0.002	0.0008–0.0035	0.0003	0.0006–0.0020	0.004–0.010
Dielectric constant	6–7	8–9	5.5–7.5	6.2	8–9	4.5–5.5

Source: Data from Refs. 10, 18, 20, and 80.

(due to significant expansion of the β-spodumene phase) and a near-zero or negative thermal expansion [20]. For these materials, dielectric properties approach those of the feldspathic porcelains, although strength values tend to be lower, but thermal shock properties are excellent. The combination of these properties is advantageous for insulator applications requiring dimensional and thermal stability. Pyroceram contains both crystalline and glassy phases and the material is, therefore, a glass ceramic. Dielectric properties of these materials tend to be dictated by the glassy phase, but the crystalline phase imparts substantial mechanical strength [18,20]. These materials are used in substrate applications, and offer advantages in controlled dimensions and ease of fabrication. Compositions fall mainly in the alkali/alkaline earth-alumina-silicate systems.

Table 24 provides comparative data on the physical chemical, thermal, mechanical, and dielectric properties of the nonfeldspathic or nontriaxial porcelains bodies.

REFERENCES

1. Waser R. 'Modeling of electroceramics—applications and prospects'. J Eur Ceram Soc 1999; 19:655–664.
2. Newnham RE. 'Electroceramics'. Rep Progr Phys 1989; 52:123–156.
3. Duh J-G, Chiou S-B, Hsu WY, Yao J. Electric Ceramics. Steele BCH, ed. Vol. 36: British Ceramic Proceeding. Stoke-on-Trent, UK: Institute of Ceramics, 1985: 19–30.
4. Daniel VV. Dielectric Relaxation. New York: Academic Press, 1967 Chaps. 1, 2, 5, 6, 12, 13, and 16.
5. Zaky A, Hawley R. Dielectric Solids. New York: Dover Press, 1970, Chaps. 1, 2, and 6.
6. Blythe R. 'Electrical Properties of Polymers'. Cambridge: Cambridge University Press, 1979. Chaps. 2–7.
7. Anderson JC. 'Dielectrics'. New York: Rheinhold, 1964, Chaps. 1–8.
8. Harrop PJ. Dielectrics. New York: John Wiley and Sons, 1972, Chaps. 1–6.
9. Bogoroditskii NP, Pasynkov VV. Properties of Electronic Materials, Part I: Dielectrics. Cambridge: Boston Technical Publishers, 1967.
10. Buchanan RC. 'Properties of ceramic insulators' In: Buchanan RC, ed. Ceramic Materials for Electronics. 2nd ed. New York: Marcel Dekker, 1991.
11. Kingery WD, Bowen HK, Uhlmann DR. Introduction to Ceramics. 2nd ed. New York: John Wiley and Sons, 1976.
12. Hench LL, West JK. 'Principles of Electronic Ceramics': John Wiley and Sons, 1990. Chaps. 2–7.
13. Moulson JK, Herbert JM. Electroceramics. New York: Chapman and Hall, 1990.
14. Richerson DW. Modern Ceramic Engineering. 2nd ed. New York: Marcel Dekker, 1992., Chaps. 4, 5, and 6.

15. Tummala RR, Rymaszewski EJ, eds. Microelectronics Packaging Handbook. New York: Van Nostrand Reinhold, 1989, Chap. 4.
16. Scneider SJ, ed. Engineered Materials Handbook. 'Ceramics and Glasses'. Vol. 4: ASM International, 1991. Secs. 1, 11, 14, and 15.
17. Minges ML. Electronic Materials Handbook. 'Packaging'. Vol. 1: ASM International, 1989. Secs. 1, 3, and 4.
18. Clark FM. Insulating Materials for Design and Engineering practice. New York: John Wiley and Sons, 1976. Chap. 16.
19. Wachtman JB. 'Mechanical Properties of Ceramics'. New York: John Wiley and Sons, 1996.
20. Harper CA. Handbook of Ceramic, Glasses and Diamonds. New York: McGraw-Hill, 2001. Chaps. 1–9.
21. Chiang Y, Birnie D, Kingery WD. Physical Ceramics. New York: John Wiley and Sons, 1997.
22. Kröger A. The Chemistry of Imperfect Crystals. New York: Amsterdam/Wiley, 1964.
23. Kroger FA. 'Point defects in solids: physics, chemistry and thermodynamics'. In: Schock RN, ed. Point Defects in Minerals. Washington, D. C.: American Geophysical Union, 1985:1–17.
24. Kofsted P. Nonstoichiometry, Diffusion, and Electrical Conductivity in Binary Metal Oxides. New York: Wiley-Interscience, 1972.
25. Greenwood NN. Ionic Crystals, Lattice Defects and Nonstoichiometry. New York: Chemical Publishing Company, 1970. Chaps. 2–6.
26. Adams DM. Inorganic Solids—An Introduction to concepts in Solid State Structural Chemistry. New York: John Wiley and Sons, 1974, Chap. 5.
27. Bunget I, Popescu M. 'Physics of Solid Dielectrics'. Translation, V. Vasilescu. Vol. 19. Amsterdam: Elsevier, 1984. Chaps. 1–8.
28. Buchanan RC, Park T. Materials Crystal Chemistry. New York: Marcel Dekker, 1997. Chap. 4.
29. Wernicke R. Ceramic Monograph. Handbook of Ceramics, Supplement to Interceram. Vol. 34, 1985:1–8.
30. Blaise G. Charge localization and transport in disordered dielectric materials. J Electrostatics. Vol. 50, 2001:69–89.
31. Paranjape VV, Panat PV. 'Discrete polarization model of the polaron'. J Phys Condens Matter 1991; 3:2319–2329.
32. Kim J, Roseman RD, Buchanan RC. 'Microstructural effects on conductivity in donor doped $BaTiO_3$'. Ferroelectrics 1996; 177:255–271.
33. Roseman RD, Kim J, Buchanan RC. 'Structural phase transitions in donor doped BaTiO3 and effects on PTCR behavior'. Ferroelectrics 1996; 177:273–282.
34. Buchanan RC, Roseman RD. 'Kirk–Othmer Encyclopedia of Chemical Technology'. New York: John Wiley and Sons, 2002, Chap. 19.
35. Surplice NA. The electrical conductivity of calcium and strontium oxides. Br J Appl Phys 1996; 17:175–180.
36. Jain H, Nowick S. Electrical conductivity of synthetic and natural quartz crystals. J Appl Phys 1982; 53:477–484.

37. Evans AG. 'Perspectives on the development of high–toughness ceramics'. J Am Ceram Soc 1990; 73:187–206.
38. Dorre E, Hubner H. 'Alumina: Processing, Properties and Applications'. New York: Springer-Verlag, 1984.
39. Bradt RC, Scott WD. Mechanical properties of alumina. In: Hart LD, ed. Am. Ceram. Soc.. OH: Columbus, 1990:23–39.
40. Bae SI, Baik S. 'Critical concentration of MgO for the prevention of grain growth in alumina'. J Am Ceram Soc 1994; 77(4):135–138.
41. Harmer M. 'Use of solid solution additives in ceramics processing'. Adv Ceram 1984; 10:679–696.
42. Pappis J, Kingery WD. Electrical properties of single crystal and polycrystalline alumina at high temperatures. J Am Ceram Soc 1961; 44(9):459–464.
43. Sharapov VM, Alimov VKh, Gavrilov LE. Deuterium accumulation in beryllium oxide layer exposed to deuterium atoms. J Nucl Mater 1998; 2:803.
44. Heikinheimo L, Heikkinen J, Linden J, Kaye A, Orivuori S, Saarelma S, Tahtinen S, Walton R, Wasastjerna F. Dielectric window for reactor-like ICRF vacuum transmission line. Fusion Eng Design 2001; 55:419–436.
45. Sempolinski DR, Kingery WD, Tuller HL. Electronic conductivity of single crystalline MgO. J Am Ceram Soc 1980; 63:669–675.
46. Sempolinski DR, Kingery WD. 'Ionic conductivity and magnesium vacancy mobility in magnesium oxide'. J Am Ceram Soc 1980; 63:664–69.
47. Vest RW, Tallan NM, Tripp WC. Electrical properties and defect structure of zirconia I, monoclinic phase. J Am Ceram Soc 1964; 47(12):635–640.
48. Kumar A, Rajdeep D, Douglass DL. Effect of oxide defect structure on the electrical properties of ZrO_2. J Am Ceram Soc 1972; 55(9):439–445.
49. Gupta TK, Grekila RB. 'Electrical conductivity of tetragonal zirconia below the transformation temperature'. J Electrochem Soc Solid State Sci 1981; 128(4): 929–931.
50. Kingery WD, Pappis J, Doty ME, Hill DC. J. Am. Ceram. Soc. 1959; 42(8): 393–398.
51. Bonanos N, Slotwinski RK, Steele BCH, Butler EP. J. Mater. Sci. 1984; 19: 785–793.
52. Ramana Rao AV, Tare VB. Scripta Metallurgica. 1972; 6(2):141–148.
53. Moure C, Jurado JR, Duran P. Electric Ceramics, British Ceramic Proceeding Steele BCH, ed. Vol. 36, 1985:31–44.
54. Duh G, Chiou S-B, Hsu WY, Yao J. Electric Ceramics, British Ceramic Proceeding Steele BCH, ed. Vol. 36, 1985:19–30.
55. Chiodelli G, Flor G, Scagliotti M. 'Electrical properties of the ZrO_2-CeO_2 system'. Solid State Ionics 1996; 91:109–121.
56. Elbadraoui E, Baudour JL, Bouree F, Gillot B, Fritsch S, Rousset A. 'Cation distribution and mechanism of electrical conduction in nickel–copper manganite spinels'. Solid State Ionics 1997; 93:219–225.
57. Bradley FN. 'Chemistry, microstructure, and processing of ferrites.' In Javitz AE, ed. Materials for Magnetic Functions. New York: Hayden, 1971.
58. Standley KJ. 'Electrical properties of ferrites and garnets' In Oxide Magnetic Materials. 2nd ed. Oxford: Clarendon Press, 1972.

59. Nell KJ, Wood BJ. 'High temperature electrical measurements and thermodynamic properties of Fe_3O_4–$FeCr_2O_4$–$MgCr_2O_4$–$FeAl_2O_4$ spinels'. Am Mineralogist 1991; 76:405–426.

60. Kharton VV, Viskup AP, Kovalevsky AV, Jurado JR, Naumovich EN, Vecher AA, Frade JR. 'Oxygen ionic conductivity of Ti-containing strontium ferrite'. Solid State Ionics 2000; 133:57–65.

61. Basak D, Ghose J. 'Studies on the conduction process of cadmium-substituted copper chromite spinels'. J Solid State Chem 1994; 112:222–227.

62. Gartstein E, Mason TO. Reanalysis of wustite electrical properties. J Am Ceram Soc 1982; 65:C24–C26.

63. Harvey JW, Johnson DW. Binder systems in ferrites. J Am Ceram Soc Bull 1990; 59(8):2127–33.

64. Ullmann H, Trofimenko N, Naoumidis A, Stover D. 'Ionic/electronic mixed conduction relations in perovskite-type oxides by defect structure'. J Eur Ceram Soc 1999; 19:791–796.

65. Sprague J, Tuller HL. 'Mixed ionic and electronic conduction in Mn/Mo doped gadolinium titanate'. J Eur Ceram Soc 1999; 19:803–806.

66. Takemoto M, Miyajima T, Takayanagi K, Ogawa T, Ikawa H, Omata T. 'Properties of transition metal oxides with layered perovskite structure'. Solid State Ionics 1998; 108:255–260.

67. Waser R, Hagenbeck R. 'Grain boundaries in dielectric and mixed-conducting ceramics'. Acta Mater 2000; 48:797–825.

68. Kulwicki B. in Levinson L, ed. Adv Ceram. Vol. 1, 1981:138–154.

69. Heywang W. J. Mater. Sci. 1971; 6:1214–1226.

70. Jonker GH. Solid State Electron. 1964; 1:895–903.

71. Takada K, Chang E, Smyth DM. In Blum JB, Cannon, eds. WR, eds. Multilayer Ceramic Devices. Advances in Ceramics. Vol. 19, 1987:147–51.

72. Hill DC, Tuller HL. 'Ceramic sensors: theory and practice.' In Buchanan RC, ed. Ceramic Materials for Electronics. 2nd ed. New York: Marcel Dekker, 1991.

73. Roseman RD. 'High temperature poling effects on conducting barium titanate ceramics'. Ferroelectrics 1998; 215:31–45.

74. Roseman RD, Liu G. 'Temperature and voltage effects on microstructure and electrical behavior of donor modified $BaTiO_3$'. Ferroelectrics 1999; 221:181–185.

75. Williams W. 'Electrical properties of hard materials'. Int J Refract Metals Hard Mater 1999; 17:21–26.

76. Pearson HO. 'Handbook of Refractory Carbides, Properties, Characteristics, Processing and Applications'. Westwood, NJ: Noyes Publications, 1996. Chaps. 1–16.

77. Yanagida H, Koumoto K, Miyayama M. 'The Chemistry of Ceramics'. New York: John Wiley and Sons, 1996, Chaps. 2–4.

78. McColm IJ. Ceramic Hardness. New York: Plenum Press, 1990. Chap. 6.

79. Neudeck PG. SiC technology, Encyclopedia of Materials Science and Technology. Bushchow KHJ, Cahn RW, Flemmings MC, Iischner B, Krammer EJ, Mahajan S, eds. Vol. 9. Oxford: Elsevier Science, 2001:8508–8519.

80. Neudeck PG, Okojie RS, Chen LY. High temperature Electronics—A Role of wide Bandgap semiconductors? Proceedings of the IEEE 2002; 90(6):1065–1076.

81. Thorp JS, Sharif RI. D.C. Electrical properties of hot pressed Nitrogen ceramics. J Mater Sci 1978; 13:441–449.
82. Stevels JM. Handbuch der Physik. Vol. 20. Berlin: Springer-Verlag, 1957:353–391.
83. Schwartz M, Mackenzie JD. 'Ionic conductivity in calcium silicate glasses'. J Am Ceram Soc 1966; 49(11):582–585.
84. Isard JO. J Non-Cryst Solids 1969; 1:235–261.
85. Hsieh CH, Jain H. 'Influence of network-forming cations on ionic conduction in sodium silicate glasses'. J Non-Crystalline Solids 1995; 183:1–11.
86. Paul A. Chemistry of Glasses. London: Chapman and Hall, 1982, Chaps. 2 and 4.
87. Bansal MP, Doremus RH. Handbook of Glass Properties. New York: Academic Press, 1986, Chaps. 13–15 and 18.
88. Stevels JM. Progress in the Theory of the Physical Properties of Glass. London: Elsevier, 1964.
89. Espe W. Materials of High Vacuum Technology. Silicates. Vol. 2. Oxford: Pergamon Press, 1968, Chaps. 10–12.4.
90. Brook RJ. in Tallan NM, ed. Electrical Conductivity in Ceramics and Glass. New York: Marcel Dekker, 1974, Chap. 3.
91. Cordes NM, Baranovskii SD. 'On the conduction mechanisms in ionic glasses'. Phys Stat Sol (b) 2000; 218:133.
92. Kahnt H. 'Ionic transport in glasses'. J Non-Cryst Solids 1996; 203:225–231.
93. Kaps H, Schirrmeister F, Stefanski P. J Non-Cryst Solids 1996; 203:225–231.
94. Ravaine D. J Non-Cryst Solids 1986; 87:159–170.
95. Paul I. 'Chemistry of Glasses'. 2nd ed. New York: Chapman and Hall, 1990. Chaps. 1–4.
96. Doremus RH. Glass Science. 2nd ed. New York: John Wiley and Sons, 1994. Chaps. 15 and 16.
97. Scholze H. 'Glass Nature Structure and Properties'. New York: Springer-Verlag, 1990. Chaps. 3.1–3.6.
98. McCauley RA. Corrosion in Ceramics. New York: Marcel Dekker, 1995. Chap. 6.
99. Mayer JW, Lau SS. Electronic Materials Science: For Integrated Circuits in Si and GaAs. New York: Macmillan, 1990, Chaps. 9 and 10.
100. Meiners G. Electric properties of Al_2O_3 and AlP_xO_y dielectric layers on InP. Thin Solid Films 1984; 113:85–92.
101. Locsei B. Molten Silicates and Their Properties. New York: Chemical Publishing Company, 1970, Chaps. 3–6.
102. Tuller H. 'Ionic conduction in nanocrystalline materials'. Solid State Ionics 2000; 131:143–157.
103. Doi A. Free volumes in several ion conducting glasses. J of Non-Crystalline Solids 1999; 247:155–158.
104. Schneider SJ. Engineered Materials Handbook. 'Ceramics and Glasses'. Vol. 4: ASM International, 1991. Secs. 1, 14, 15.
105. Worrall WE. Ceramic Raw Materials. 2nd ed.. New York: Pergamon Press, 1982.
106. Dinsdale A. 'Pottery Science': Materials, Processes and Products. New York: John Wiley and Sons, 1986. Chaps. 1, 15–19.
107. Shi Y, Huang X, Yan D. Ceramics Int 1998; 24:393–400.

108. Bishai AM, Al-Khayat BHF, Awni FA. Ceram Bull 1985; 64(4):598–601.
109. Chaudhari SP, Sarkar P, Chakraborty AK. Cer Int 1999; 25:91–99.
110. Liebermann J. 'New effective ways toward solving the problem of contamination of porcelain insulators'. Refractories Ind Ceram 2002; 43(1–2):53–64.
111. Moreno VM, Gorur RS. 'Effect of long-term corona on non-ceramic outdoor insulator, housing materials' IEEE Trans. Dielectrics Electrical Insul 2001; 8(1):117–128.
112. Janick MA. 'Insulators: composites vs porcelains'. Electrical World T&D, Transmission May/June, 2001.
113. Moreno V, Gorur RS. 'AC and DC performance of polymeric high voltage insulating materials' IEEE Trans. Dielectrics Electrical Insul 1999; 6(3):342–350.
114. Cygan RT, Lasaga AC. 'Dielectric and polarization behavior of forsterite at elevated temperatures'. Am Mineralogist 1986; 71:758–766.
115. Smyth DM, Stocker RL. Point Defects and non-stoichiometry in forsterite. Phy earth Planet Inter 1975; 10:183–192.
116. Schock RN, Duba AG, Shankland TJ. Electrical conduction in olivine. J Geophys Res 1989; 94:5829–5839.
117. Schmidbauer E, Mirwald PW. Electrical conductivity of codierite. Mineral Petrol 1993; 48:201–214.

2

Highly Conductive Ceramics

Harry L. Tuller
Massachusetts Institute of Technology
Cambridge, Massachusetts, U.S.A.

I. INTRODUCTION

For many centuries ceramics were admired for their stability at elevated temperatures, chemical and abrasion resistance, and dimensional stability. So it is no surprise that initial applications for ceramics in the electrical sphere, now designated *electroceramics*, were viewed primarily from the perspective of their insulating or dielectric properties. Applications such as high-voltage standoffs, capacitors, spark plugs, and the like followed naturally. In recent decades, however, this picture of ceramics has changed rapidly. It has now become evident that ceramics exhibit the full spectrum of electrical properties spanning the gap from superconducting to insulating electronic conductors and from fast ionic to insulating ionic conductors. In many cases, it is the unique combination of desired electrical properties together with the characteristic stability of ceramics that renders them the obvious choice for many technological applications.

$La_{1-x}Sr_xMnO_3$ (LSM), used as the cathode in solid oxide fuel cells (SOFCs), is a prime example of an electroceramic that provides a unique combination of required properties. LSM satisfies the need for a metallic conductor capable of operating under a highly oxidizing environment at temperatures of nearly 1000°C [1]. Likewise, SiC is used as the heating element in furnaces capable of operating in air to 1500°C. One of the most important electroceramic applications in today's technology is the use of indium tin oxide (ITO) as a transparent electrode in displays, semiconductor light sources, and solar cells [2]. Others include the wide band gap nitrides (e.g., $Al_{1-x}Ga_xN$) from which semiconducting junction devices are fabricated. The wide band gap enables operation at high temperatures and/or high power levels as electronic devices and as light sources and detectors operating in the green, blue, and violet portions of the optical spectrum [3]. Superconductivity exhibited by some oxides above liquid nitrogen temperatures has stimulated a great deal of development in the past decade despite the more complicated processing required as compared to superconducting metallic alloys [4].

The ionic bonding of many ceramics allows for ionic diffusion and correspondingly, under the influence of an electric field, ionic conduction. For many years this contribution was ignored as being inconsequential. However, over the past approximately three decades, an increasing number of solids have been identified that support anomalously high levels of ionic conductivity. Indeed, some solids exhibit levels of ionic conductivity comparable to that of liquids. Such materials are designated as *fast ion conductors*. If the conductivity is predominantly ionic, then they are known as solid electrolytes, whereas if they support both ionic and electronic conduction, they are known as *mixed ionic electronic conductors* (MIECs). Such solids represent the underpinnings of the fields of *solid-state ionics* and *solid-state electrochemistry* that have grown in importance as our society has become more acutely concerned with efficient and environmentally clean methods for energy conversion, conservation, and storage [5]. Examples of relevant technologies include lithium batteries, solid oxide fuel cells, electrochromic windows, and automotive exhaust sensors.

Electrical conductors can be subdivided into various categories: by the nature and sign of the charge carriers, by the magnitude of electrical conductivity, by the carrier mobility mechanism, by their temperature dependence, and so on. The categories one chooses will depend on those aspects of most immediate concern for a given application. For example, for an electronics engineer, the magnitude of resistance in a circuit element or the p- versus n-type nature of carriers in junction devices will be of most immediate concern, while for someone designing a fuel cell, the mix of ionic and electronic conductivity will be important.

The full range of electrical conductivities of general interest easily covers more than 25 orders of magnitude. Figure 1 includes a listing of typical materials that fall within the three most common categories of electronic and ionic conductors. The bounds for the three categories of electronic conductors—metals, semiconductors, and insulators—are somewhat vague, as are the distinctions between ionic conductors—fast ion conductors, conventional ionic conductors, and insulators. Later in this chapter more precise definitions will be presented. Nevertheless, from the standpoint of applications, these bounds are generally satisfactory.

A fourth category of electronic ceramic conductors, nearly unknown until mid-1986, has now received much attention by the scientific community. A grow-

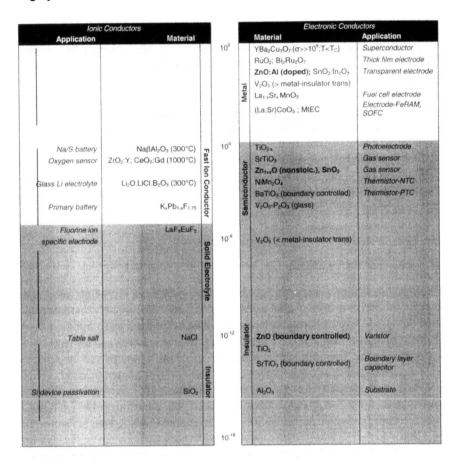

Figure 1   Logarithmic scale of the electrical conductivities of materials categorized both by magnitude and carrier type (i.e., ionic vs. electronic). Examples of the various categories are given together with some typical applications.

ing number of oxides, primarily cuprates, exhibit superconductivity, which reflects the loss of measurable resistivity below a characteristic critical temperature, T_c. The anomalously high T_c values (e.g., as high as 122 K for $TlBa_2Ca_3Cu_4O_{11}$ [6]) exhibited by these oxides have stimulated the search for improved theories of superconductivity. Likewise, high T_c values have also rendered many additional uses of these materials more feasible.

Many of the applications listed in Figure 1 rely heavily on the enhanced chemical and thermal stability exhibited by ceramics over that of metals or polymers. Other applications, such as sensors, rely, conversely, on the high sensitivity exhibited by some ceramics to their environment (e.g., atmosphere, temperature, and pressure). These are described in some detail in the chapter on ceramic sensors.

The aim of this chapter is to (1) describe the key features necessary for achieving reasonably high ionic and electronic conduction in ceramics; (2) describe how the crystal, electronic, and defect structures affect transport properties; (3) review experimental results and interpretations for a number of representative systems; and (4) discuss briefly some of the considerations important in choosing and optimizing ceramics for specific applications (transparent electrodes, energy conversion, sensor technology, etc.).

II. ELECTRONIC CONDUCTION IN OXIDES

A. Itinerant Band Model

The band theory of Block and Wilson [7] provides us with simple rules for predicting the conductive nature of solids. Partially filled bands correspond to metallic conduction, whereas completely filled or empty bands signify either semiconducting or insulating character. The distinction between an intrinsic semiconductor and an insulator is only qualitative (i.e., the magnitude of the energy gap separating the occupied valence from the empty conduction band). This scheme, which enables us to predict the density of mobile carriers contributing to conduction in a given material, has been used with great success in analyzing the electronic properties of the majority of conventional semiconductors and metals. However, as we will discover later, serious discrepancies often exist between predicted and observed properties when this theory is applied to conduction in narrow-band metal oxides.

The electrical conductivity, σ, which is the proportionality constant between current density j and electric field E, is given by

$$j/E = \sigma = ne\mu \tag{1}$$

in which n is the carrier density (number/cm^3), μ the mobility (cm^2/V-s), and q the electronic charge (1.6×10^{-19} C). The many orders of magnitude difference in σ which distinguish metals, semiconductors, and insulators generally resides in differences in n rather then μ. Thus, in the half-filled band of a metal such as sodium, a high density of electrons, on the order of 10^{22} cm^{-3*}, coexist with an equal number of unoccupied states, while in a semiconductor or insulator the

---

* This is not strictly true since only those electrons in the vicinity of the Fermi energy ($\sim kT$) have easy access to unoccupied energy levels.

ratio of occupied to unoccupied levels in the conduction band is far below unity, [e.g., n(300 K) in Si $\sim 10^{10}$ cm^{-3}].

To predict the electron occupancy of a given energy interval, one must convolute the density of states function with the Fermi function. The density-of-states function $D(E)$, describing the number of states per unit energy interval, derived for the case of a spherical Fermi surface, is given by [8]:

$$D_{c,v}(E) = m^{*3/2}_{e,h}[2(E - E_c)]^{1/2} / \pi^2 (h/2\pi)^3 \qquad (2)$$

in which m^* is the effective mass and h Planck's constant; the subscripts C and V refer to the conduction and valence band whereas e and h refer to electrons and holes, respectively. The probability of occupancy of a given energy state by an electron is given by the Fermi function [8],

$$f(E) = \frac{1}{1 + \exp[E - E_F)/kT]} \qquad (3)$$

in which E is the energy of interest, E_F the Fermi energy, and T the temperature in degrees Kelvin. Convoluting the electron and hole density-of-states functions with the Fermi function leads to the following important relations for the density of electrons in the conduction band and holes in the valence band [9]:

$$n = N_c \exp[-(E_c - E_F / kT)] \qquad (4a)$$

$$p = N_v \exp[-(E_F - E_v / kT)] \qquad (4b)$$

in which E_c and E_v are the energies at the bottom of the conduction and top of the valence band, respectively. The effective densities of states in the conduction band, N_c, and the valence band, N_v, are defined by

$$N_{c,v} = 2(2\pi m^*_{e,h}kT / h^2)^{3/2} \qquad (5)$$

in which k and h are the Boltzmann and Planck constants, respectively.

Taking the product of Eqs. (4a) and (4b), we obtain

$$np = N_c N_v \exp(-E_g / kT) \qquad (6)$$

in which E_g is the band gap energy. Note that the np product, which is independent of the Fermi energy, may be viewed as the equilibrium constant corresponding to the reaction

$$O \Leftrightarrow e' + h^{\cdot} \qquad (7)$$

which describes the thermal generation of an electron–hole pair by excitation of an electron across the band gap.

For an intrinsic semiconductor

$$n = p = n_i = (N_c N_v)^{1/2} \exp(-E_g / 2kT) \qquad (8)$$

Given the exponential dependence of free carrier density on E_g, it becomes clear how a factor of 2 in band gap could easily distinguish a semiconductor from an insulator at a given temperature T.

A certain minimum concentration of atomic defects is predicted on the basis of statistical thermodynamics in all solids. Generally, large concentrations of additional defects are added either intentionally or unintentionally during the processing of materials. Due to the distortions in potential they create in their vicinity, discrete energy levels form within the energy gap, often in the vicinity of the band edges. Impurities with an excess or deficiency of electrons relative to the host lattice may thus donate electrons to the conduction band or accept electrons from the valence band, thereby contributing to an electron or hole density far in excess of that expected by the intrinsic process described in Eq. (8). Impurity ionization energies typical in elemental semiconductors are of the order of several tens of millielectron volts, consistent with a modified hydrogenic model in which the ionization energy (13.6 eV) of the hydrogen atom is greatly reduced as a consequence of the dielectric constant, which enters the equation as ε^{-2}.

Other defect levels often exist far from the band edges. These deep levels are formed when the wave functions of the defects do not mix effectively with those of the host. Such levels contribute little directly to the excess carrier density but can serve to compensate the influence of shallow levels by trapping excess electrons or holes.

With a single type of donor D or acceptor A, two extrinsic conduction regimes are obtained as a function of temperature. At low temperatures (i.e., $kT \ll E_{D,A}$; $E_{D,A}$ = ionization energy of the donor or acceptor, respectively), the temperature dependence of carriers is given by [10]

$$n, p \approx (N_{D,A} N_{c,v})^{1/2} \exp(-E_{D,A} / 2kT) \tag{9}$$

in which $N_{D,A}$ is the donor or acceptor impurity density. For $kT \approx E_{D,A}$, all impurity levels are effectively ionized and the carrier density is temperature independent and equal to the impurity density. Finally, at sufficiently high temperatures, the intrinsic carrier density surpasses that of the impurity-generated carriers and a temperature dependence of the form given in Eq. (8) is obtained.

Predictions based on the above models often hold true for metal oxides. Thus TiO_2, with an empty Ti-3d conduction band approximately 3 eV above the largely O-2p valence band, is highly insulating in its undoped form ($\sim 10^{-12}$ S/cm), while when doped with pentavalent niobium, with an ionization energy of only ~ 0.05 eV, or rendered oxygen deficient, it becomes an n-type semiconductor (≤ 1 S/cm at room temperature) (Figure 1). Oxides with fully occupied valence bands (d^{10}) and wide s-like conduction bands, such as ZnO, CdO, Ga_2O_3, In_2O_3 SnO_2, Sb_2O_5, and their mixed oxides, exhibit exceptionally high conductivities when donor doped to degeneracy. $In_{2-x}Sn_xO_3$ (indium tin oxide, ITO) thin films have conductivities as high as $\sim 10^4$ S/cm while retaining optical transparency

[2]. On the other hand, very wide band gap oxides such as MgO ($E_g \sim 6.7$ eV), which accommodate only deep traps, remain highly insulating to temperatures above 1200°C. On the other hand, oxides such as ReO_3, which intrinsically possess a partially filled $5d$ conduction band, exhibit metallic conduction as expected from the Block–Wilson band picture.

B. Narrow-Band Conduction

In contrast to the examples in which the electronic properties of oxides follow the Block–Wilson predictions, there are many more that fail. MnO, for example, has five d electrons, resulting in a partially filled band. Contrary to expectations, rather than being metallic, this material is highly insulating. Similar arguments hold for CoO and NiO, as well as for example, among the perovskites (e.g., $LaMnO_3$, $LaFeO_3$) and corundum structure oxides (e.g., Cr_2O_3, Fe_2O_3) [11].

To understand the sources of these discrepancies, a number of features characteristic of metal oxides must be taken into account. First, as pointed out by Mott [12], when overlap between nearest-neighbor electronic wave functions is small, electron–electron correlations become sufficiently important to result in the localization of the "conduction" electrons around their ion cores, resulting in the formation of a *Mott insulator*. Since weakly extended d and f orbitals in transition and rare earth metal oxides often contribute to the formation of conduction or valence bands, it is not surprising that such materials may not satisfy predictions of classical band theory in which electron–electron correlations may be neglected.

The self-trapping of carriers in partially filled but narrow bands is visualized in terms of a lattice distortion induced by the carrier polarizing the charged ions in its vicinity. When the distortion is highly localized, the carrier and the accompanying distortion is known as a *small polaron*. Motion of this quasi-particle proceeds by an activated hopping process and takes the form [13]

$$\mu = [(1 - c)ea^2v_0/kT] \exp(-E_H/kT) \tag{10}$$

in which E_H is the hopping energy, $(1 - c)$ the fraction of sites unoccupied, a the jump distance, and v_0 the attempt frequency. As expected, small-polaron mobilities are considerably reduced in magnitude relative to scatter-limited mobilities and are generally reported to be on the order of 10^{-4} and 10^{-2} cm^2/V-s at elevated temperatures, hundreds to thousands of times smaller than for broadband conduction.

Where the distortion is inadequate to self-trap the carrier, but still serves to slow the carrier, the designation *large polaron* is applied. The large-polaron mobility at temperatures above the Debye temperature is given by [14]

$$\mu = \mu'_0 T^{-1/2} \tag{11}$$

and its magnitude is expected to be $\sim 1-100$ cm^2/V-s at elevated temperatures. Under these circumstances, the bandwidths are sufficiently wide to support itinerant electronic transport and thus agree with expectations based on the band model. Examples of both cases (i.e., small versus large polaron) will be presented shortly.

Second, because the structures of oxides vary considerably, symmetry considerations must be examined to establish the nature and degree of energy band splitting, related to the splitting of certain hybridized atomic orbitals into subbands. A tight-binding approach, as discussed by Honig and Vest [15,16], is very effective in highlighting the implications of these features for transport in oxides.

Last, the influence of nonstoichiometry on the electronic carrier density needs to be examined. Changes in stoichiometry (i.e., the metal-to-oxygen ratio), induced by changes in the oxygen or metal vapor partial pressure, can result in the creation of large numbers of donor- or acceptor-like states in the band gap and thus render a wide band gap oxide semiconducting or, in the case of certain cuprates, even superconducting (below T_c). Furthermore, the influence of donor or acceptor impurities may be markedly impacted by relatively small changes in the stoichiometric ratio. Examples of the influence of stoichiometric imbalance on electronic properties will also be provided.

C. Band Structure Determinations for Narrow-Band Oxides

Aside from the Block–Wilson approach, which examines the implications of applying periodic constraints on a quasi-free electron, a tight-binding model that begins with electrons localized near the ion cores has also been successfully applied in predicting the major features of the quantum mechanical band picture in solids. Because of the localized nature of the d and f levels in many of the oxides of interest, it is natural to examine the implications of the tight-binding model on band formation and transport in such oxides [15–17].

In the tight-binding approach one assumes that the crystal wave function may be written as a linear combination of the unperturbed atomic functions and that the presence of the neighboring atoms is reflected only as a weak perturbation in the potential seen by the atom of interest. In the solution of Schrödinger's equation, one thus obtains for the energy term the initial energy plus a small perturbation given by [18]

$$E = E_0 - \alpha - \beta \sum_m \exp(ik \cdot r_m)$$

(12)

in which E_0 is the unperturbed energy; α the "self" term, which reflects the shift in energy of the level due to the change in potential seen by the central atom; and β the overlap integral, which is a measure of how strongly the wave functions of the neighboring atoms overlap with the central atom. Given the weak interactions assumed in the tight-binding model, one generally assumes little error in truncating the summation beyond the nearest-neighbor atoms.

For a simple cubic lattice (in which $r_m = a$) one may readily show that the last term in Eq. (12) can take on values between $\pm\ 6\beta$, thus broadening the initially discrete level into a band with width 12β, as illustrated in Figure 2. Clearly, as the overlap of wave functions increases, so does the bandwidth. The effective mass may be calculated by examining the curvature of the E versus k relationship (in the vicinity of $k = 0$), i.e.:

$$m^* = h^2\ (d^2E/dk^2)^{-1} = h^2(2\beta a^2)^{-1} \tag{13}$$

One readily observes the inverse relationship between m^* and the bandwidth.

Certain rules must be obeyed in deriving the crystalline energy bands from the interaction of cationic and anionic atomic wave functions [15]. First, energy bands split under the influence of crystal fields (e.g., d levels split into a lower threefold degenerate t_{2g} level and a higher twofold degenerate e_g level under octahedral symmetry). Second, meaningful overlap resulting in bond formation requires both compatible symmetry and proximity in energy. Strong interactions lead to splitting of resulting hybrid energies into bonding and antibonding levels, with the degree of splitting proportional to the degree of interaction. Last, as indicated in Eq. (12), the widths of the bands also increase with increasing overlap. Thus, bandwidths increase as one proceeds from δ-type to π-type to σ-type orbital overlaps and for a given type of bond, as one proceeds from lower to higher lying energies. With these guidelines in hand, one can proceed to examine the energy diagrams for a number of oxide systems.

1. Band Structures and Electrical Conduction: Perovskites

For simplicity we begin with ReO_3, which crystallizes in the perovskite structure ABO_3 illustrated in Figure 3 but with the A site empty. Due to the octahedral surroundings of the Re ion, its $5d$ levels split into t_{2g} and e_g sublevels. The O_{2p} orbitals also split ($2p_\pi$, $2p_\sigma$), with the $2p_\sigma$ orbital lying lower in energy as its

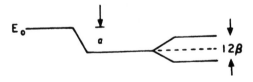

Figure 2 The degenerate electronic states of isolated atoms are observed to shift lower and broaden into a band of energies with bandwidth 12β as atoms are brought together in compound formation.

lobes point directly to the electropositive neighbors. The orientations of the various orbitals along the edges of the unit cell, which correspond to the directions of greatest overlap, are illustrated in Figure 4.

Due to their σ-type overlap (e.g., Figure 4a), a series of bonding and antibonding σ-like bands are formed from admixtures of $6p$, $6s$, $5e_g$, $2p_\sigma$, and $2s$ states. Several of the σ bands with antibonding character are illustrated in Figure 5, with $\sigma^*(e_g)$ and σ^* corresponding to the bands with predominant $5e_g$ and $6s$ character, respectively. Figure 4b illustrates the weaker π-type bonding that occurs between $5t_{2g}$ and $2p_\pi$ orbitals, resulting in the formation of the π-like bands also illustrated in Figure 5. In contrast to the schematic band diagrams of Figure 5, the higher lying σ-like bands can be expected to be wider than the π-like bands below them, in agreement with Eq. (12). This has important implications for the nature of transport in the respective bands, as we shall soon discover.

Since the number of states available for occupancy is conserved, one need only fill them in sequence to establish the position of the Fermi energy. The total number of valence electrons in the formula unit ReO_3 equals 25, of which 24 fill the lower bands of anionic character. This leaves one electron in the $5t_{2g}$ band, resulting in a partially filled band. ReO_3 is indeed metallic with conductivity within a factor of 2 of Cu at room temperature [16]. On the other hand, WO_3, with a structure closely related to that of ReO_3 but with a d^0 electronic configuration, is

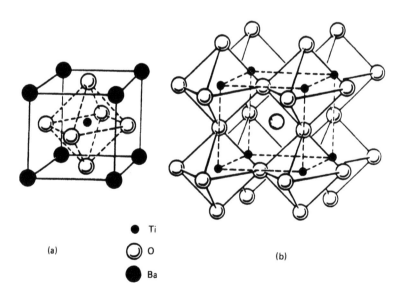

(a)

● Ti

◐ O

● Ba

(b)

Figure 3 Ion positions in ideal perovskite structure with $BaTiO_3$ used as an example. (From Ref. 17.)

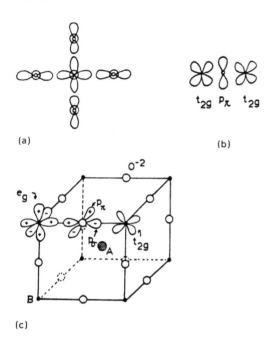

(a)

t_{2g} p_π t_{2g}

(b)

(c)

Figure 4 Orbital overlap: (a) σ bonding; (b) π bonding. (c) Relative orientations of the cation 3d orbitals and oxygen 2p orbitals, leading to both σ and π types of bonding in the perovskite structure.

found to be a semiconductor. However, as discussed below, the conductivity of WO_3 is highly sensitive to the degree of oxygen nonstoichiometry and proton and alkali ion insertion [19].

Extension of the above model to the case of the perovskites in which the body center position is filled with the A atom (ABO_3) is straightforward since the energy band diagram described above for ReO_3 is little perturbed, given the negligible overlap along the A-B and A-O bond directions (Figure 4). The bands associated with the A atoms (e.g., Sr) lie well above the other bands so far considered. Consequently, the valence electrons associated with the A atoms will fall to the lower empty d-like energy states and must be included when establishing the position of the Fermi energy.

Based on the energy scheme presented in Figure 5 for perovskites (similar considerations result in band distributions for the other crystal structures), all compounds with electronic configurations corresponding to d^1-d^5 and d^7-d^9, which correspond to either partially filled $\pi^*(t_{2g})$ or $\sigma^*(e_g)$ bands, should be metallic, whereas d^6 or d^{10} configurations should be semiconducting or insulating.

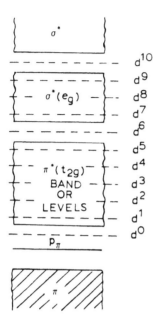

Figure 5 Predicted Fermi levels for the B-cation electronic band configurations in the perovskite structure. (After Ref. 16.)

Of course, this ignores the influence of the bandwidth on transport. Below a critical bandwidth b_c, carriers must localize and render the expected metal semi-conducting or insulating.

Goodenough [17] compiled a table classifying the energy bands found in perovskite oxides in terms of their localized or collective natures. An update of this table is presented in Figure 6, in which the compounds are grouped in columns according to their electronic configurations and in rows according to the valence states of the two sets of cations. In general, oxides with small net spin and high B-cation valence state enjoy the greatest degree of overlap.

For a band to support metallic conduction, its bandwidth b must be greater than a critical bandwidth b_c. For some compounds $b^\sigma > b^\pi > b_c$, so a partially filled t_{2g} or e_g band will lead to metallic conduction; this holds true [e.g., $SrVO_3$ ($\pi^{*1} \sigma^{*0}$) and $LaNiO_3$ (t_{2g}^6, σ^{*1})]. The latter is an example where only the upper σ^* band of the solid is wide enough to remain delocalized. For high-spin conditions and low B ion valence, both b^σ and b^π drop below b_c, causing both bands to localize and rendering oxides such as $LaFeO_3$ (t_{2g}^3, e_g^2) and $LaCrO_3$ (t_{2g}^3, e_g^0) semiconducting.

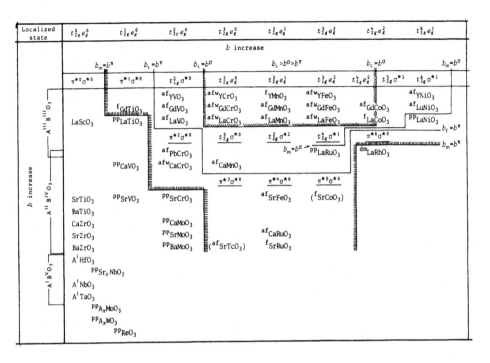

Figure 6 Energy band classification of perovskite oxides and analogs, indicating degree of localization. af, Antiferromagnetic; afw, antiferromagnetic with parasitic ferromagnetism; f, ferromagnetic; dm, diamagnetic; pp, Pauli paramagnetism. (From Ref. 22.)

The perovskite $LaCoO_3$ is interesting because it exhibits a transition from localized to collective electronic transport with either temperature ($T > 937°C$) [20] or composition [21]. Introduction of Sr, for example, to give $La_{1-x}Sr_xCoO_3$, results in the formation of Co^{4+} to maintain charge neutrality. Formation of Co^{4+} with an intermediate spin configuration of $t_{2g}^4 e_g^1$ leads to band broadening and with it metallic conduction for $0.1 < x < 0.5$ [20,22] (see also discussion of mixed-valence effects below). $La_{0.5}Sr_{0.5}CoO_3$, for example, exhibits a conductivity $\sigma > 10^4$ S/cm [23,24]. However, Van Buren and de Wit [25] interpret their thermopower measurements to indicate some carrier dependence on temperature up to $T = 500°C$ and suggest, therefore, that these materials fall somewhere between semiconductors and metals.

A familiar example of metallic conductivity induced by d-band filling is the group of alkali tungsten bronzes (e.g., Na_xWO_3) in which the alkali ions donate their outer shell electron to the initially empty d band of WO_3 [19]. Of particular note is the ability to reversibly insert and remove such ions from the material by electrochemical means at relatively low temperatures. This has led to the application of similar materials as "solid solution" electrodes, which serve simultaneously as a source and sink of reactants and as electronically conductive electrodes in solid-state batteries [26,27]. Similar considerations apply to electrochromic displays based on the insertion of, for example, protons into initially insulating and transparent WO_3 to form conducting and highly colored H_xWO_3 [28]. Note that the above examples require both high electronic and ionic conduction for their operation.

The first observation of superconductivity in oxides with moderately high T_c (13 K) was reported by Sleight et al. [29] in the system $BaPb_{1-x}Bi_xO_3$. (The more recently discovered high-T_c superconductors are discussed separately below.) The end member $BaPbO_3$, reported to be a metal [30], first appears to be another example of the failure of simple energy band models to predict electrical behavior given the filled d^{10} configuration of the Pb^{4+} ion. Sleight et al. [29] explain $BaPbO_3$ metallic conduction on the basis of Pb 6s-O 2p overlap due to the high degree of covalency, whereas Vest and Honig [16] suggest instead $\sigma^*(e_g)$ and σ^* overlap (see Figure 5).

A spectrum of electrical conductivities ranging from insulating to metallic to superconducting is thus observed for the perovskite oxides (see figure 16 in Ref. 16). The first two categories are now for the most part understood on the basis of Goodenough's criteria, as represented in Figure 6. However, these criteria relate only to intrinsic electronic properties. Many intrinsically insulating oxides (e.g., $SrTiO_3$) exhibit interesting semiconducting properties when doped or driven from stoichiometry by atmosphere treatment. These characteristics are treated in some detail in a later section.

2. Other Oxides

The electrical properties of a large number of binary monoxides, sesquioxides, and dioxides have also been reviewed by Honig [15], Vest and Honig [16], and Rao and Subba Rao [11]. These oxides are subdivided according to their d-level filling and type of conductivity. The majority of the monoxides crystallize in the rock salt structure (PdO, tetragonal), the sesquioxides in the corundum structure (Mn_2O_3, In_2O_3, Tl_2O_3, C-type cubic), and the dioxides in the rutile structure or distortions thereof. The lower symmetry of the rutile and corundum structures results in further splitting of the d levels, contributing to more difficulty in the interpretation of band structure than for the perovskites. Nevertheless, on the whole, these results also appear to satisfy Goodenough's criteria. Generally, the oxides with elements Cr, Mn, Fe, Co, and Ni tend to remain semiconducting, even when characterized by partially filled bands, due to inadequate overlap and band broadening.

D. Semiconductor–Metal Transitions

A number of the oxide systems, as illustrated in Figure 7, have been found to exhibit abrupt semiconductor–metal transitions (e.g., $V_2O_3 \sim 8$ orders of magnitude jump in conductivity at 150 K) in the vicinity of some characteristic temperature T. A large number of theories to explain this phenomenon have been proposed and are summarized by Vest and Honig [16] and Rao and Subba Rao [11]. Some, like the Mott transition model [31], may be understood on the basis of our previous discussions. Simply, one assumes that the spacing between atoms decreases as the temperature increases, thereby resulting in increased electron–electron overlap. At some critical interatomic distance a switch-over from localized to collective behavior is expected. However, most authorities, discount this model on the basis that the likelihood that many materials would lie so close to the critical atomic spacing is very small indeed.

Other models include transitions induced by crystallographic distortions or magnetic ordering. In either case, this would correspond to bands originally split due to lower crystal symmetries or magnetic ordering reuniting as the distortions in symmetry disappear or magnetic order is eliminated above the Néel temperature. A number of additional models have been proposed but are beyond the scope of this chapter. The reader is directed to Refs. 11, 15, 16, 31, 32, and 33 for further information.

E. Metal–Superconductor Transitions

1. Models

The abrupt transition of metallic to superconducting behavior was first observed in mercury at ~ 4 K in 1911 by the Dutch scientist K. Onnes. Over the next 75 years, scores of other elements and alloys were discovered to exhibit metal-superconducting transitions, with Nb_3Ge having one of the highest transition temperatures at 23 K. While oxides such as $BaPb_{1-x}Bi_xO_3$ also exhibited moderately high transition temperatures [29], as mentioned above, they did not attract a great deal of attention, particularly from technologists, given their characteristic brittle nature. However, the succession of discoveries beginning with the discovery of 40 K superconductivity in $La_{2-x}Ba_xCu_2O_4$ by Bednorz and Mueller [34] in 1986, followed by 90 K superconductivity in $YBa_2Cu_3O_7$ by Chu's group [35] in 1987, and finally the multilayered cuprates of the form $Tl_mCa_{n-1}Ba_2Cu_nO_{2n+m+2}$ and $Bi_mCa_{n-1}Sr_2Cu_nO_{2n+m+2}$ with 90–125 K superconductivity [36–38], has in recent years focused the lion's share of attention on oxide superconductors.

In contrast to normal metals, whose resistivities drop nearly linearly with decreasing temperature until they reach the finite and constant residual resistivity, materials such as mercury, NbGe, and many of the cuprates listed above exhibit

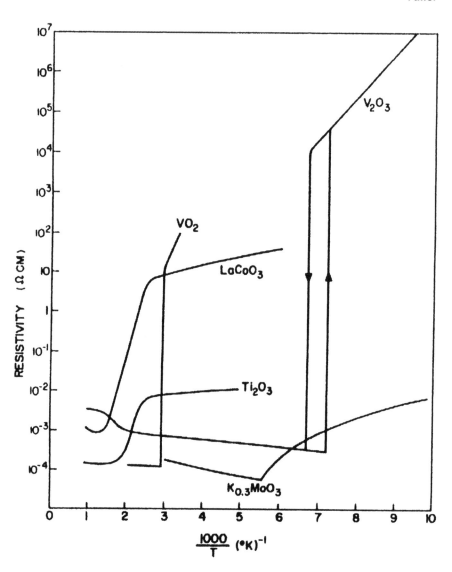

Figure 7   Resistivity vs temperature curves for a number of materials exhibiting metal insulator transitions. (From Ref. 16.)

a nearly abrupt drop to immeasurably small values of resistivity. Attempts to detect current decay in a closed loop of superconducting metal have shown [39] the resistivity of the metal to be less than 10^{-24} S/cm, i.e., approximately 10^{-18} the resistivity of copper at room temperature! This adds an additional 18 orders of magnitude to the 24 orders of magnitude in conductivity already discussed in Figure 1.

An acceptable quantum mechanical model describing the behavior of electronic charge carriers in the superconducting phase did not become available until 1957. Bardeen, Cooper, and Schrieffer proposed a model, now known as the BCS theory [40], which explains the unusual behavior as being due to electrons combining into "Cooper pairs." Electrons, which normally tend to repel each other due to coulombic forces, are proposed to form lower energy pairs in the vicinity of the superconducting critical temperature, T_c.

Froehlich had demonstrated earlier that an attractive interaction can exist between a pair of electrons of which one emits a phonon that is absorbed by the other [41]. It is interesting to note that strong electron–phonon interactions were already discussed earlier in relation to polaron formation in narrow-band oxides. In the present context, electron–phonon interactions result in markedly enhanced mobilities rather than reduced mobilities as in the small-polaron case.

Cooper found that two electrons, in the vicinity of the Fermi energy with equal and opposite momentum, will form a coupled pair if the lowering of the potential energy due to the electron–phonon interaction exceeds the amount by which the kinetic energy is in excess of twice the Fermi energy. The BCS theory extends this concept to many interacting electrons and shows that the key interaction remains localized between pairs of electrons [40]. The loss of resistance in the superconducting state is tied to the fact that any scattering event experienced by one electron in the pair occurs in an equal and opposite manner for the other electron, thereby conserving the total momentum of the pair. This is in contrast to the normal state, in which the change in momentum is due to a scattering event that randomizes the motion of the electron relative to the direction of the electric field.

An energy gap separates the bound state from the normal dissociated state and, according to BCS theory, is given by

$$E_{gap} (T = 0 \text{ K}) = 3.5T_c \qquad (14)$$

The lower the value of T_c, the lower the gap energy necessary to dissociate the Cooper pair. Electron–phonon interactions are sufficiently strong to account for T_c values as high as perhaps 30 K. The new oxide superconductors, with T_c values approaching five times that value, require stronger interactions to account for their high T_c values.

Increasing current density in the superconducting regime reflects increasing momentum. At some level of current density, it becomes energetically possible for the Cooper pairs to dissociate. The critical current density, J_c, is defined as the current density below T_c for which the solid is driven back into its normal state.

Superconductors exhibit the Meissner effect in that the magnetic flux density in the material is always zero. This is accomplished by induction of currents on the surface of the superconductor upon the application of external magnetic fields so as to cancel the flux inside. The applied magnetic field strength necessary to induce the critical current J_c and drive the material normal is called the critical field H_c. Both J_c and H_c drop toward zero as T approaches T_c, e.g.,

$$H_c \cong H_0 [1 - (T/T_c)^2] \tag{15}$$

where H_0 is the critical field at $T = 0$ K.

From the applications standpoint, all three parameters, J_c, H_c, and T_c, should be as high as possible to enable high current-carrying capacities for superconducting cables and high magnetic fields in superconducting magnets at readily accessible temperatures. Grain boundaries play a critical role in determining J_c in some superconducting oxides such as $YBa_2Cu_3O_7$. Two factors have been correlated with the observation of so-called *weak links* between grain boundaries. The critical current density is observed to decrease 3–4 orders of magnitude as the misorientation angle between adjacent grains increases from 0° to 45° [42]. Furthermore, oxygen disorder or deficiency at grain boundaries has also been correlated with *weak link* behavior [43].

2. Oxide Superconductors

Outside of the $BaBiO_3$-based superconductors, which are cubic perovskite, nearly all the high-T_c oxide superconductors are characterized by Cu-O planes or layers. Schuller and Jorgensen [44] divide these into "single-layered" $La_{2-x}R_xCuO_4$ (R = Ba, Sr, Ca), "double-layered" $RBa_2Cu_3O_{7-x}$ (R = Y, Nd, Sm, Eu, Gd, Dy, Ho, Er, Tm, Yb, Lu), and "multilayered" Tl-Cu-Ba-Cu-O and Bi-Ca-Sr-Cu-O compounds. T_c appears to increase as the number of Cu-O planes increases and their separation distance decreases. Conduction tends to be anisotropic with the higher component of conductivity oriented along the Cu-O planes.

The structure of the single-layered $La_{2-x}R_xCuO_4$ compound is shown in Figure 8. The square-planar CuO_2 planes are separated by two LaO layers. Substitution of divalent R cations for La is believed to contribute holes into the CuO_2 valence band.

The double-layered $YBa_2Cu_3O_{7-x}$ structure of Figure 9 shows two dimpled CuO_2 planes separated by an yttrium layer. Also visible are the BaO and CuO layers, which contain Cu-O chains. As T increases, oxygen is lost from the Cu'1'-O layers, ultimately leading to an orthorhombic-to-tetragonal transition at about $x = 0.5$ as the remaining oxygens disorder and move to sites between the chains.

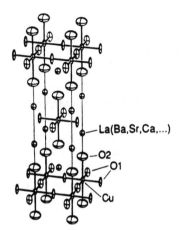

Figure 8   Structure of the "single-layered" $La_{2-x}R_xCuO_4$ compound. (From Ref. 44.)

T_c drops toward zero as x approaches 0.6. Excess oxygen above $x = 0.5$, which enters sites adjacent to Cu'1', is believed to contribute excess holes to the Cu'2'-O planes leading to superconductivity.

The majority of the superconducting cuprates conduct by holes ($Nd_{2-x}Ce_x$-CuO_4 is an exception) and a major question remains regarding whether they

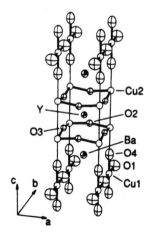

Figure 9   Structure of the "double-layered" $RBa_2Cu_3O_{7-x}$ "1–2–3" compound. (From Ref. 44.)

conduct along a Cu band (i.e., Cu^{3+}) or an oxygen band (i.e., O^-). The most commonly held belief [45] is that the highly mobile holes reside in the oxygen band. La_2CuO_4 already contains built-in holes in the form of Cu^{2+} (Cu^+ has a closed-shell configuration) but these remain relatively localized, as is consistent with the fact that La_2CuO_4 is antiferromagnetic. When Sr is substituted for La, excess holes are formed in the CuO_2 plane. Cu^{3+} is an improbable configuration in view of the strong coulombic repulsion associated with double occupancy of the narrow Cu 3d states [45]. The mobile holes therefore most likely reside in O_{2p} states. It is interesting to note that some attribute the attractive interaction of mobile carriers to antiferromagnetic order in the undoped state [46].

Many models have been proposed to explain the persistence of copper pairs to temperatures far above that explainable by electron–phonon interactions, including excitonic, spin coupling, bipolaron formation, and others [47]. Theoretical work in this area is still in a state of flux and confirmation of these models must await further experimentation. In a later section, we examine the key role of defects and nonstoichiometry in influencing the transport properties of some of the above-named cuprates.

III. DEFECT-CONTROLLED TRANSPORT

We have already encountered semiconduction due to the localization of carriers in partially filled bands as a result of small-polaron formation. In these materials, because large concentrations of carriers, albeit trapped, already find themselves in the "conduction" band, variations in impurity concentrations or stoichiometry are expected to have only secondary effects on conduction. In this section we concern ourselves instead with solids that satisfy conditions characteristic of intrinsic insulators (i.e., large gap separating filled and empty bands) but may be induced to become highly conductive by control of doping and stoichiometry. Interest in this group of semiconducting ceramics is growing due to their importance in increasing numbers of electronically active devices (e.g., varistors, thermistors, gas sensors, transparent electrodes, etc.).

A. Controlled-Valence Semiconductors

We begin by considering the case of CeO_2, which reportedly has a moderately large band gap of ~3 eV separating the O-2p valence band from the empty Ce-4f conduction band [48]. Applying Eq. (8), we calculate an intrinsic carrier density of $\sim10^{14}$ cm^{-3} at 1273 K. This may be compared with the $2 \times 10^{20} cm^{-3}$ excess electrons obtained by doping with approximately 1% niobium or uranium [49,50], a value roughly 6 orders of magnitude greater than the intrinsic level. Conductivities as high as ~1 S/cm are obtained in this way. Given the strong temperature dependence of Eq. (8), the discrepancy between intrinsic and extrinsic grows rapidly larger as the temperature is decreased.

Assuming full ionization of the Nb or U donors, the weakly activated temperature dependence of the conductivity is ascribed to an activated, small-polaron mobility with hopping energy of approximately 0.3 eV, in agreement with results obtained for nonstoichiometric CeO_2 [13,50]. Because the electrons in the Ce-4f band are localized, it becomes convenient to think of them as forming Ce^{3+} ions in a sea of Ce^{4+} ions with transport occurring by an activated exchange of electrons from "occupied" Ce^{3+} sites to "empty" Ce^{4+} sites. Furthermore, following this line of reasoning, a maximum in the conductivity should occur when the concentrations of Ce^{3+} and Ce^{4+} ions are equal, followed by a drop back toward zero as all of the Ce ions become trivalent. Under these circumstances,

$$\sigma \propto c(1 - c) \tag{16}$$

in which c is the fraction occupied (e.g., Ce^{3+}) and $(1 - c)$ the fraction unoccupied (e.g., Ce^{4+}). Clearly, conditions of $c = 0$ (empty band) or $c = 1$ (full band) should result in no conduction.

The above illustrates the importance of inducing mixed valence in the cations to ensure conduction in narrow-band insulating oxides. In CeO_2, mixed valence can result from doping with a higher valent dopant such as Nb^{5+}. To maintain charge neutrality, the ionized donor is compensated by a negative charge, the self-trapped electron. In terms of presently accepted defect notation [51], this condition is given by

$$[Nb^{\cdot}_{Ce}] = [Ce'_{Ce}] = [e'] \tag{17}$$

in which subscripts refer to lattice sites and dots (positive) and primes (negative) refer to net charge relative to the lattice. The square brackets denote defect concentration.

Next consider FeO,* which possesses a filled (t_{2g}) band by virtue of its $3d^6$ electronic configuration. Since all the Fe atoms are Fe^{2+}, motion of an electron from one Fe ion to its neighbor would require the formation of an Fe^+, Fe^{3+} pair, a process requiring the elevation of an electron across the $\pi(t_{2g})$-$\sigma(e_g)$ band gap. If, on the other hand, a monovalent acceptor such as Li were added to the crystal, it would induce the formation of Fe^{3+} ions or effectively holes in the π (t_{2g}) band. This neutrality relation is given by

$$[Li'_{Fe}] = [Fe^{\cdot}_{Fe}] = [h^{\cdot}] \tag{18}$$

and corresponds to the case in which term c in Eq. (16) is close to 1. This method of creating mixed valence for conduction by altervalent doping is known as the

---

* For the purposes of this argument, we ignore the fact that strictly stoichiometric FeO does not exist.

controlled valence technique [52] and has been applied extensively to many ceramic systems including NiO:Li [52], CoO:Li [52], TiO_2:Nb [53], $SrTiO_3$:Nb_{Ti} [54], $SrTiO_3$:La_{Sr} [55], $SrTiO_3$:Y_{Sr} [56], UO_2:Y_u [57], $Y_3Fe_5O_{12}$:Ca_{Fe} [58], and $LaCrO_3$:Sr_{La} [59], as well as many others. In the early work of Verwey and coworkers on the NiO-Li_2O system [52], they demonstrated the ability to control the p-type conductivity by nearly two orders of magnitude.

It is important to emphasize that the controlled-valence approach is important not only for solids with filled bands but at times for those with partially filled bands as well. Examples include NiO and CoO, with $3d^8$ and $3d^7$ configurations, respectively, which exhibit little conduction as long as the Ni and Co are divalent. Apparently, by creating a fraction of trivalent ions by doping with Li, we enable electron transfer between ions without engaging strong electron–electron correlation effects.

A similar situation was pointed out earlier for the La_2CuO_4, system in which the Cu^{2+} ($3d^9$) state was inadequate to support metallic conduction. However, substitution of Sr on La sites, which is compensated by holes, leads to metallic conduction and ultimately to superconductivity at $T \sim 40$ K. Figure 10 illustrates the fact that while hole compensation is predominant in $La_{2-x}Sr_xCuO_4$ for x values up to 0.2–0.3 (depending on processing conditions) [60], another compensation mechanism, probably oxygen vacancy formation, takes over for larger values of x [61,62].

While many conductive oxides are characterized by narrow conduction or valence bands, and so the concept of controlled valence is appropriate, those with wide bands can be treated in a manner similar to that of conventional semiconductors in which dopants are used simply to increase the electron or hole population in the conduction or valence band, respectively, as discussed in the following section.

B. Doped Semiconductors

As mentioned above, a number of oxides with closed shell d^{10} cations, such as ZnO, CdO, In_2O_3, and SnO_2, exhibit exceptionally high electronic conductivities when appropriately doped. This is a consequence of the wide conduction band that supports high electron mobilities (~ 100 cm^2/V-s) and the ability to dope these materials into degeneracy. The most familiar of these oxides is Sn-doped In_2O_3 (indium tin oxide, ITO), with a resistivity of 1–2×10^{-4} Ω-cm, commonly used as a transparent electrode. Due to the limited natural abundance of In, much attention has been directed to the development of transparent electrodes based on donor-doped ZnO given that Zn is inexpensive, abundant, and harmless. Donor dopants examined to date include group III elements Al, Ga, In, and B, and group IV elements Si, Ge, Ti, Zr, and Hf substituting on the Zn site as well as F

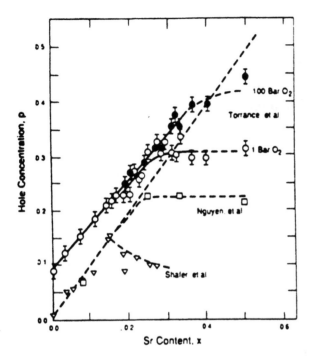

Figure 10 Hole concentration versus Sr concentration in $La_{2-x}Sr_xCuO_4$. Hole concentration is determined by a titration technique. (From Ref. 60.)

substitution on the O site. These dopants result in resistivities comparable to those attained in ITO [63].

In recent years the focus of attention has been expanded to include ternary and quaternary systems. These generally include the above-mentioned cations plus Ga. Figure 11 illustrates the band gaps of many of the key transparent conductive oxides (TCOs) as a function of the average valence of the cation. A large band gap is essential for transparency in the visible. Freeman et al. [64] have examined a series of quaternary systems examining in some detail the role of crystal and defect structure in controlling phase stability and generating free carriers. Table 1 lists some of these systems, including their crystalline structure and composition limits.

Interest in the potential for transparent p-n junction devices has stimulated interest in transparent p-type conductors. Some success has been achieved with Cu- and Ag-based oxides such as $AgInO_2$, $SrCu_2O_2$, and $CuAlO_2$ [65].

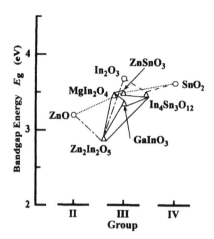

Figure 11   Band gap energies of transparent conductive oxides: binary (○) and ternary compounds (Δ). (From Ref. 63.)

C. Nonstoichiometry

1. Semiconductors

Three mechanisms for defect formation in oxides must be considered simultaneously. These are (1) thermally induced intrinsic electronic and intrinsic ionic disorder (e.g., Schottky and Frenkel defects), (2) redox, and (3) impurity-induced defects. The first two categories of defects are predicted from statistical thermody-

Table 1   Chemical Formulae, Crystal Structures, and Solution Ranges of TCO Phases

System	Phase	Formula
Zn-In-Ga	Layered intergrowths	$In_{t-x}Ga_{1+x}O_3(ZnO)_k$:
		$k = 1$ ($-0.34 < x < 0.06$)
		$k = 2$ ($-0.54 < x < 0.3$)
		$k = 3$ ($-1 < x < 0.42$)
Ga-In-Sn	T-phase	$Ga_{3-x}In_{5+x}Sn_2O_{16}$ ($0.3 < x < 1.6$)
Zn-In-Sn	Bixbyite	$In_{2-2x}Sn_xZn_xO_3$ ($0 < x < 0.4$)
Cd-In-Sn	Spinel	$Cd_{1+x}In_{2-2x}Sn_xO_4$ ($0 < x < 0.75$)
Cd-In-Sn	Bixbyite	$In_{2-2x}Sn_xCd_xO_3$ ($0 < x < 0.34$)

namics [66], and the latter form to satisfy electroneutrality. Examples of typical defect reactions in the three categories are given in Table 2.

Intrinsic electronic disorder [Eq. (19)] has already been discussed. Two types of intrinsic ionic disorder are represented in Eqs. (20) and (21), with the first corresponding to a pair of cations and anions, leaving the interior of the crystal—Schottky disorder—and the second to an anion leaving a normal site—Frenkel disorder. Generally, only one of these two mechanisms need be considered for a given material, the choice being determined by the crystal structure (i.e., open structures encourage Frenkel disorder, whereas close-packed structures encourage Schottky disorder). Both of these mechanisms leave the stoichiometric balance intact.

Oxidation–reduction behavior, as represented by Eqs. (22) and (23), results in an imbalance in the ideal cation-to-anion ratio and thus leads to *nonstoichiometry*. In both cases, equilibration with the gas phase by exchange of oxygen between the crystal lattice and the gas phase results in the generation of electronic carriers.

Consider the two-dimensional representation of an oxide with cubic rock salt structure illustrated in Figure 12a. On the left, oxygen from the gas phase enters the lattice interstitially, becomes negatively charged O''_i, and thereby creates two holes to maintain charge neutrality. On the right, on the other hand, oxygen leaves the lattice, thereby freeing its two electrons while simultaneously creating a vacant oxygen site, $V_O^{\bullet\bullet}$. The redox reactions (22) and (23) are written with the assumption that the defect states are fully ionized. As with conventional acceptors and donors, the electronic carriers tend to recombine with the atomic defects as the temperature is lowered. For the doubly charged species considered here, two energy levels per defect are formed in the forbidden gap between conduction and valence band. These are sketched in Figure 12b.

Deviations from stoichiometry in the direction of oxygen excess (MO_{1+x}) or deficiency (MO_{1-x}) thus form defect states that act identically in every way to impurity-related acceptor or donor states, respectively. Some of the special

Table 2 Typical Defect Reactions

Defect reactions	Mass action relations	
$0 \Leftrightarrow e' + h^{\bullet}$	$n \cdot p = K_e(T)$	(19)
$MO \Leftrightarrow V_M'' + V_o^{\bullet\bullet}$	$[V_M''] \cdot [V_o^{\bullet\bullet}] = K_S(T)$	(20)
$O_O \Leftrightarrow V_o^{\bullet\bullet} + O_i''$	$[V_o^{\bullet\bullet}] \cdot [O_i''] = K_F(T)$	(21)
$1/2O_2 \Leftrightarrow O_i'' + 2h^{\bullet}$	$[O_i''] \cdot p^2 = K_{OX}(T)P_{O_2}^{1/2}$	(22)
$O_O \Leftrightarrow V_o^{\bullet\bullet} + 2e' + 1/2O_2$	$[V_o^{\bullet\bullet}] \cdot n^2 = K_R(T)P_{O_2}^{-1/2}$	(23)
$N_2O_3 (MO_2) \Leftrightarrow 2N_M' + 3O_O + V_o^{\bullet\bullet}$	$[N_M']^2 \cdot [V_o^{\bullet\bullet}]/a_{N_2O_3} = K_N(T)$	(24)

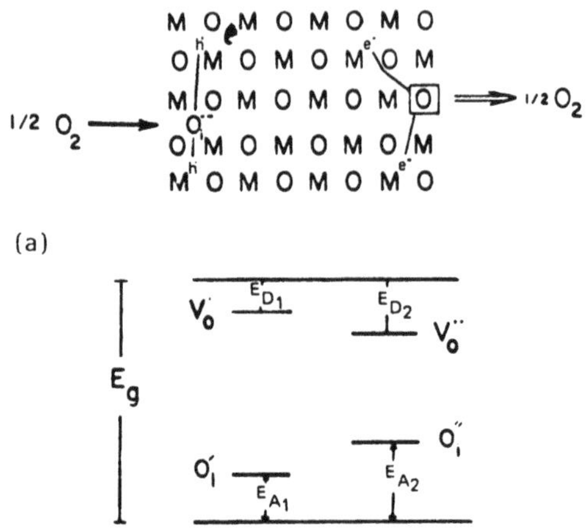

(a)

(b)

Figure 12 (a) Schematic of a two-dimensional square lattice. Entry of excess oxygen into lattice on left results in formation of an oxygen interstitial and two holes, whereas exit of an oxygen atom into the gas phase at the right results in formation of an oxygen vacancy and two electons. (b) Representation of the energy gap E_g separating valence and conduction bands. The two upper discrete energy levels in the gap correspond to oxygen vacancy donor states, whereas the two lower levels correspond to oxygen interstitial acceptor states.

features associated with such reactions are as follows: (1) the nature of the semiconductor can be changed from p type to n type under isothermal conditions by control of the oxygen partial pressure (or, equivalently, the metal vapor pressure); (2) the atomic defects that make up the acceptor and donor states can be quite mobile under some operating conditions, in contrast to the situation in conventional semiconductors. The latter aspect has special relevance for our upcoming treatment of ionic and mixed ionic–electronic conduction in oxides.

Given that most oxides of interest here have relatively large band gaps (e.g., TiO_2, $E_g \sim 3$ eV) and correspondingly large Schottky and Frenkel energies, the likelihood of intrinsic disorder controlling electrical behavior over an extensive range of conditions is practically nil. Therefore, either redox (atmosphere) or impurities will control the defect equilibria, depending on the stoichiometry of the oxide. Furthermore, we will discover that the controlled-valency concept discussed earlier can also be markedly influenced by variations in stoichiometry, i.e., electronic compensation induced by altervalent impurities may be converted to ionic compensation [see Eq. (24)], thereby rendering a material initially semiconducting, insulating. This principle is instrumental in the formation of grain boundary–controlled active devices [67].

In general, the electrical behavior of solids depends on defects formed in response to both impurities and deviations from stoichiometry. At or near stoichiometry, impurities predominate, whereas under strongly reducing or oxidizing conditions, defects associated with deviations from stoichiometry control. To characterize the electrical response of a metal oxide to temperature and atmosphere excursions, a series of simultaneous reactions of the form represented by Eqs. (19)–(24) must therefore be considered. Furthermore, assuming the addition of an acceptor impurity N'_M, a representative electroneutrality equation for the case considered in Figure 12 would be

$$2[O''_i] + [N'_M] + n = 2[V_O^{\cdot\cdot}] + p \tag{25}$$

Note that intrinsic Frenkel disorder is assumed to predominate, so that Eq. (20) may be ignored in subsequent discussions. Furthermore, it may be recognized that the remaining equations are not independent of each other (i.e., $K_{ox}K_R = K_F K_e^2$). One may therefore remove one of the equations [e.g., Eq. (22)] in solving for the unknown defect densities.

A piecewise solution to such problems is commonly attempted by sequentially choosing conditions for which only one term on either side of Eq. (25) need be considered. For example, under heavily reducing conditions, Eq. (25) may be simplified to read

$$n = 2\,[V_O^{\cdot\cdot}] \tag{26}$$

Combining this with Eq. (23), one obtains

$$n = 2[V_O^{\cdot\cdot}] = [2K_R\,(T)]^{1/3}\,P_{O_2}^{\,-1/6} \tag{27}$$

and from Eqs. (19) and (21),

$$p = K_e(T)[2K_R\,(T)]^{-1/3}\,P_{O_2}^{\,1/6} \tag{28}$$

$$[O''_i] = K_F\,[K_R\,(T)/4]^{-1/3}\,P_{O_2}^{\,1/6} \tag{29}$$

The three other defect regimes most likely to occur with successively increasing P_{O_2} include $2[V_O^{\cdot\cdot}] = [N'_M]$, $p = [N'_M]$, and $p = 2[O''_i]$. A diagram depicting the atmosphere dependencies of the defects over the four defect regimes is presented in Figure 13. Here the carrier densities of the individual defects have been multiplied by their mobilities to give the partial conductivities as a function of P_{O_2}.

We emphasize the predominant partial conductivity in each defect regime in Figure 13 with a solid curve. The two solid curves represent two different ways in which the total conductivity may vary with atmosphere. The upper curve

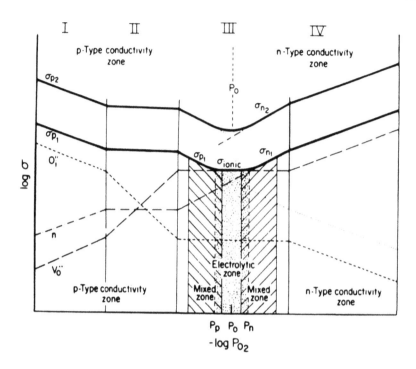

Figure 13 Conductivity isotherms. Solid curves correspond to the following cases: (1) μ_n, $\mu_p > \mu(V_O^{\cdot\cdot})$; (2) μ_n, $\mu_p \gg \mu$ $(V_O^{\cdot\cdot})$. In case 1, five conductivity zones are obtained with the central zone (electrolytic) corresponding to predominantly ionic conduction. P_p and P_n correspond to the P_{O2} values at which $\sigma_p = \sigma_{ion}$ and $\sigma_n = \sigma_{ion}$, respectively, while P_0 corresponds to the P_{O2} at which $\sigma_n = \sigma_p$. Overlap of electrolytic and mixed zones is used to illustrate imprecise nature of these zones. (From Ref. 140.)

corresponds to the case where μ_n, μ_p are sufficiently greater than $\mu(V_O^{\cdot\cdot})$ that even in the defect regime where $V_O^{\cdot\cdot}$ is the predominant defect (i.e., $2[V_O^{\cdot\cdot}] = [N'_M]$), the total conductivity remains electronic. In the lower curve, the carrier mobility inequality is not nearly so pronounced, so that at the P_{O_2} $(= P_0)$ at which electronic defects are at a minimum (i.e., $n = p$) conduction is predominantly ionic. Under these circumstances the oxide acts as a solid electrolyte, the topic of a later section. Aside from the electrolytic zone, the neighboring zones on either side are designated as mixed zones within which both ionic and electronic conductivities are of comparable magnitude.

Focusing on the upper curve of Figure 13 for the moment, we make the following observations:

1. Where they are P_{O_2} dependent, electrons and holes always exhibit opposite P_{O_2} dependencies, as might be expected from Eq. (19). Electrons follow a $P_{O_2}^{-1/x}$, holes a $P_{O_2}^{1/x}$ dependence.
2. In each defect regime, the majority carrier exhibits a unique P_{O_2} dependence.
3. The controlled-valence feature holds in only one of the defect regimes, namely, where $p = [N'_M]$.

The first observation enables us to identify the type of electronic carrier (n or p), whereas the second gives us the defect regime. By establishing the bounds of the controlled-valence regime, it allows us to predict the conditions of temperature and P_{O_2} for which the material will remain highly conducting. In general, it should be noted that only one or two of these defect regimes are experimentally accessible at a given temperature. By following the above guidelines, it becomes possible to identify the defect regimes one is operating within and with the aid of the defect relations, to derive the controlling equilibrium constants, $K(T)$.

To illustrate the above predictions, conductivity data for single-crystal $Ba_{0.03}Sr_{0.97}TiO_3$ are shown between 800°C and 1100°C at intervals of 50°C in Figure 14 [68] (note that the log P_{O_2} axis is the reverse of that in Figure 13). An obvious feature of the data is the minimum in σ at elevated P_{O_2}, indicating a transition from p-type to n-type conductivity. This places the data in the defect regime described by the simplified neutrality relation $[N'_M] = 2[V_O^{\cdot\cdot}]$ with $P_o \sim 10^{-1}$ Pa at 800°C. The solutions for n and p in this region are given by

$$n = (2K_R (T)/[N'_M])^{1/2} P_{O_2}^{-1/4} \tag{28a}$$

$$p = K_e (T)([2K_R(T)/ [N'_M])^{-1/2} P_{O_2}^{1/4} \tag{29a}$$

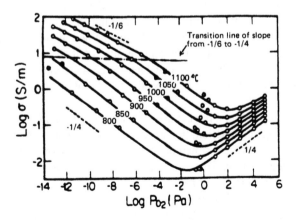

Figure 14 Measured and calculated conductivities of $Ba_{0.03}Sr_{0.97}TiO_3$. Dots are experimental data; lines are calculated values. (From Ref. 68.)

The observed $+1/4$ slope at high P_{O_2} and $-1/4$ slope at low P_{O_2} are in agreement with these predictions. Thermoelectric power measurements, performed concurrently with the electrical conductivity measurements [68] show a sharp transition from n to p type at the same P_{O_2} values at which the conductivity goes through its minima, again strengthening the applicability of the defect model.

In the upper left-hand corner of Figure 14, corresponding to reducing conditions at the highest temperatures, the slope is observed to decrease from $-1/4$ toward $-1/6$. This we interpret as signifying the transition from one defect regime to the bordering defect regime characterized by the neutrality relation, $n = 2[V_O^{\cdot\cdot}]$ [see Eq. (27)]. A similar transition in slope is observed in the thermoelectric power.

Given a well-defined defect model, it is possible to derive a number of the equilibrium constants and carrier mobilities from these data. From a plot of the conductivity minimum versus $1/T$ [see Eq. (8)] a value for the thermal band gap of 3.35 ± 0.4 eV (at 0 K) is obtained. From a combined analysis of the electrical conductivity and thermoelectric power following the approach of Jonker [69], mobility values for the electrons and holes ($\mu_e = 0.23$ cm^2/V-s and $\mu_h = 0.10$ cm^2/V-s at 1000°C) were obtained [70]. Finally, from a plot of log σ_n versus $1/T$ at constant P_{O_2} in the $P_{O_2}^{-1/4}$ regime, an expression for K_R [see Eq. (28)] given by

$$K_R(T) = 1.03 \times 10^{69} \exp(-5.18 \pm 0.03 \text{ e V}/kT) \text{ cm}^{-9} \text{ atm} \qquad (30)$$

was obtained [68].

The influence of atmosphere on σ in this material, and in other titanates is notable. For example, σ (1100°C) in Figure 14 increases nearly three orders of magnitude, to ~ 3 S/cm in going from the conductivity minimum to $P_{O_2} \cong 10^{-17}$ atm. Even higher conductivities may be obtained by annealing in H$_2$ [56]. In practice, these high conductivities in the titanates may be maintained by quenching to room temperature [71,72]. For this reason and because of their relative stability in basic solutions, they have been of great interest as photoanodes in photoelectrochemical cells [73–75] and more recently in photovoltaic cells [76,77].

In many oxides, quenching does not retain the high-temperature conductivities. To satisfy conditions for high conduction, first the excess donors or acceptors induced in the material as a result of nonstoichiometry must not be allowed to diffuse out of the solid during cooling. Thus, the more rapid the diffusivity of $V_O^{\cdot\cdot}$ in SrTiO$_3$, for example, the more rapid need be the quench. Also, upon cooling, the donor or acceptor states will deionize, resulting in a greatly depressed conductivity. The high dielectric constants, characteristic of the titanates, ensure low ionization energies at least for the donor centers. The same does not hold true for many other oxides.

An interesting consequence of the first requirement being only partially fulfilled is the existence of a number of technically important grain boundary–controlled electronic devices. Consider that diffusion generally occurs much more rapidly along grain boundaries than through the bulk. Thus, during cooldown, while the donor or acceptor defects may be maintained in the core of the grain, partial equilibration may occur in the vicinity of the grain boundaries, resulting in a depletion of carriers. Such depletion layers result in the formation of barriers that block the flow of carriers between conductive grains. Examples of devices that exhibit such features include varistors (ZnO) [67,78], PTC thermistors (BaTiO$_3$) [79], and boundary layer capacitors (SrTiO$_3$). Detailed discussions concerning a number of these devices may be found in other chapters of this book.

2. Superconductors

Unlike conventional superconducting alloys (e.g., Nb$_3$Sn), the electronic properties of the cuprates are highly dependent on processing and dopants. The cuprates readily exchange oxygen with the gas phase at processing and annealing temperatures, typically in the range of 300–1000°C. As in other highly nonstoichiometric oxides, such redox reactions can lead to large excursions in the free carrier density. YBa$_2$Cu$_3$O$_{7-x}$ (YBCO) becomes metallic, and at $T = 90$ K superconducting, only for values of x close to zero. At larger values of x the hole density is depressed, leading to semiconducting or insulating properties.

Redox phenomena in YBCO remain active to temperatures as low as 300–400°C. This is illustrated in Figure 15 [80], in which the log of the conductivity of YBCO is plotted against reciprocal temperature at five oxygen pressures ranging from 10^{-4} to 1 atmosphere. Upon heating, YBCO loses oxygen and thereby depresses the hole density, the predominant electronic charge carrier, reaching a minimum in conductivity 2–3 orders of magnitude lower than the low-temperature value. Obviously, the lower the P_{O_2}, the greater the degree of reduction and the more depressed the conductivity at a given temperature. The p-type conductivity shows a rather steep $P_{O_2}^{1/2}$ dependence, with a transition in σ to n-type at high T and low P_{O_2} as indicated above. Defect models have been proposed [81,82] to explain these phenomena.

C. Extrinsic Broad-Band Semiconductors

Although large numbers of well-known oxides possess broad s- and p-derived conduction bands with large associated carrier mobilities (e.g., MgO [83] and Al$_2$O$_3$ [84]), only a handful can be made highly conducting. This is related to the difficulty in forming shallow donor or acceptor levels in many oxides [83].

Large-gap oxides that can be prepared as highly conducting semiconductors include In$_2$O$_3$ [85], SnO$_2$ [86], ZnO [87], Tl$_2$O$_3$ [88], and CdO [89]. Conductivi-

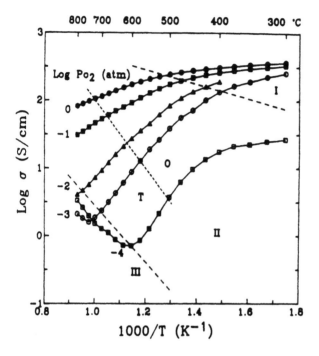

Figure 15 Log σ versus 1000/T for YBCO measured at a series of controlled P_{O_2} values as indicated. The different conduction regimes designated by roman numerals are described in the text. The letters O and T refer to the orthorhombic and tetragonal phases, respectively. (From Ref. 80.)

ties as high as 10^4 S/cm have been reported for a number of these systems, while mobilities of 1200 cm^2/V-s were reportedly measured on CVD-prepared antimony-doped SnO$_2$ single crystals [90].

Carrier densities up to levels of 10^{20} cm^{-3} may be induced in either one of the two ways described above. Well-known examples of controlled valence include Sn-doped In$_2$O$_3$ [91] and Sb-doped SnO$_2$ [92,93]. High densities of carriers may also be induced by deviations from stoichiometry. In the case of ZnO this is believed to be by the creation of excess Zn in the form of zinc interstitials [94], while in SnO$_2$ oxygen vacancies are believed to be the source of shallow donors [90].

The combination of large band gaps and high carrier mobilities provides for relative transparency in the visible even for conductive materials. The high carrier mobility and therefore reduced carrier density requirement provides for a lower plasma frequency. This ensures low absorption in the visible even at electrical conductivity levels of $\sim 10^2$–10^3 S/cm. ITO films are especially notable examples of transparent electrodes.

IV. IONIC CONDUCTION IN OXIDES

Optimized ionic conduction is a well-known characteristic of molten salts and aqueous electrolytes wherein all ions move with little hindrance within their surroundings. This leads to ionic conductivities as high as 10^{-1}–10^1 S/cm in molten salts at temperatures of 400–900°C [95]. In contrast, typical ionic solids possess limited numbers of mobile ions that are hindered in their motion by virtue of being trapped in relatively stable potential wells. Ionic conduction in such solids easily falls below 10^{-10} S/cm for temperatures between room temperature and 200°C. In the following sections we examine the circumstances under which the magnitude of ionic conduction in solids approaches that found in liquid electrolytes. Solids that fall in this category are most popularly called *fast* ion conductors (FICs), although other terms, such as *superionic* and *optimized* ionic conductors, are also used sometimes. Obviously, the expression *solid electrolyte* also describes this class of materials, but it is often also applied to ionic conductors with low or intermediate levels of ionic conductivity.

Motion of ions is described by an activated jump process for which the diffusion coefficient is given by [96]

$$D = D_0 \exp(-\Delta G/kT) = \gamma(1 - c)Za^2 \nu_0 \exp(\Delta S/k) \exp(-E_m/kT) \quad (31)$$

where a is the jump distance, ν_0 the attempt frequency, and E_m the migration energy. The factor $(1 - c)Z$ defines the number of neighboring unoccupied sites, while γ includes geometrical and correlation factors. Note the similarity to Eq. (10). Since the ion mobility is defined by $\mu = qD/kT$ and the density of carriers of charge q is N_c, where N is the density of ion sites in the sublattice of interest, the ionic conductivity becomes

$$\sigma_{ion} = \gamma(Nq/kT)^2 c(1 - c)Za^2 \nu_0 \exp(\Delta S/k) \exp(-E_m/kT)$$
$$= (\sigma_0/T) \exp(-E/kT) \quad (32)$$

This expression shows that σ_{ion} is nonzero only when the product $c(1 - c)$ is nonzero. Since all normal sites are fully occupied (i.e., $c = 1$) and all interstitial sites are empty (i.e., $c = 0$) in a perfect classical crystal, this is expected to lead to highly insulating characteristics.

The classical theory of ionic conduction in solids is thus described in terms of the creation and motion of atomic defects, notably vacancies and interstitials. The energy band diagram used earlier to describe intrinsic electronic disorder may be equally well applied to describe intrinsic ionic disorder with the notable difference being that the lower band now corresponds to, for example, normally occupied sites and the upper band normally unoccupied interstitial sites rather than energy states. Similarly, the designated energies now correspond to the formation of, e.g., a Frenkel pair rather than an electron–hole pair. The concentration of ionic carriers is a sensitive function of the Frenkel energy [see Eq. (21)] and might therefore distinguish between solids with measurable ionic conductivity (i.e., a solid electrolyte from those that are highly insulating even at relatively elevated temperatures). Similar comments are applicable to solids with predominantly Schottky disorder [i.e., Eq. (20)].

As with electronic conductors, addition of altervalent impurities may result in a marked enhancement of carriers above that of intrinsic levels. The considerably reduced ionization energies now correspond to the energies required to dissociate impurity–defect pairs as compared to intrinsic defect generation. For example, E_A might correspond to the energy required to dissociate an acceptor–anion vacancy pair or E_D the energy to dissociate a donor–anion interstitial pair. Such dissociative effects have been extensively reported in both the halide and oxide literature [97].

In general, therefore, two energies contribute to ionic conduction: a defect energy, E_D, which may be related to the Frenkel or Schottky formation energy or to a dissociation energy, and a migration energy, E_m. The value of E in Eq. (32) therefore takes on different values in three characteristic temperature regimes. These include:

1. $E = E_m + E_A/2$: extrinsic associated regime at low T
2. $E = E_m$: extrinsic fully dissociated regime at intermediate T
3. $E = E_m + E_F/2$: intrinsic defect regime at elevated T (e.g., for Frenkel equilibrium)

For optimized ionic conduction to exist, two criteria must be satisfied simultaneously. First the term c in Eq. (32) must approach 1/2. This corresponds to nearly all of the ions on a given sublattice being mobile. Second, the crystal structure must be so arranged as to enable easy motion of ions from one equivalent site to the next. This is reflected in exceptionally low values for the migration energy E_m. In the next section we discuss the conditions under which these criteria are satisfied.

V.   FAST ION CONDUCTION

A number of routes leading to exceptionally high ion carrier densities in solids have been identified over the last several decades. These are subdivided into two major categories below (i.e., structurally disordered solids and highly defective solids). An important new development in recent years is the focus on the role of interfaces in creating ionic disorder localized in the vicinity of the boundaries.

For nano-sized structures, these disordered regimes may represent a large fraction of the overall volume of the material.

A. Structurally Disordered Crystalline Solids

In contrast to the idealized picture of crystal structures presented above, many solids exist in which a sublattice of sites is only partially occupied. Strock [98] came to such a conclusion in the 1930s in relation to the Ag sublattice in the high-temperature form (α phase) of AgI. More recent neutron diffraction studies [99] differ with regard to the number of Ag sites that are in fact equivalent. The special feature of partial occupancy of sites is nevertheless retained.

Other notable systems characterized by sublattice disorder include Nasicon ($Na_3Zr_2PSi_2O_{12}$), sodium β-alumina (1.2 Na_2O–0.11 Al_2O_3), and $LiAlSiO_4$, which exhibit fast ion transport in three, two, and one dimensions, respectively. Hundreds of other structurally disordered conductors may be found listed in review articles on the subject [100,101]. Figure 16 illustrates the log $\sigma - 1/T$ relations for representative FICs.

By virtue of the coexistence of occupied and unoccupied sites on a given sublattice, FIC conductors resemble metals in that the order of 10^{21}–10^{22} cm^{-3} carriers already exists at low temperatures. Similarities between FICs and liquid electrolytes are also often noted, the most important of which is that the disordered sublattice in the solid resembles the disordered nature of ions in a liquid. For this reason one often hears the term *lattice melting* used to describe phase transitions in solids that relate to the conversion of a conventional ionic conductor to a FIC (e.g., β to α transition in AgI at approximately 150°C). Nevertheless, most investigators now believe that transport in FICs occurs via correlated jumps between well-defined sites rather than the liquid-like motion characteristic of aqueous or molten salt electrolytes.

The major structural features of FICs may be subdivided into the following two elements:

Framework: a highly ordered, immobile sublattice that provides continuous open channels for ion transport

Mobile carrier sublattice: a random distribution of carriers over an excess number of equipotential sites

Although perhaps not explicitly stated, the framework must provide tunnels of dimensions comparable to those of the mobile ions so that major network ion displacements and thus significant strain contributions to the migration energy do not occur. Aside from strain effects, Flygare and Huggins [102] have demonstrated theoretically the importance of coulombic and polarization interactions between the mobile and network ions in establishing the magnitude of the migration energy and in turn the ion mobility. They showed that ions of intermediate

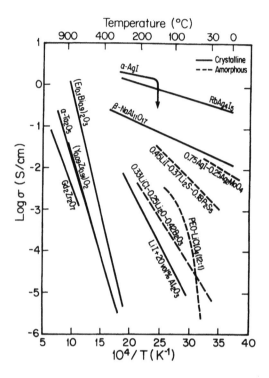

Figure 16 Temperature dependence of ionic conductivity of a number of representative FICs, illustrating the fact that some of these materials achieve ionic conductivities approaching those of liquids only at elevated temperatures.

radius would be expected to be most mobile given that small ions tend to "stick" to walls of the tunnels due to high polarization energies, whereas large ions are limited in their motion by repulsive or equivalently large strain contributions. Thus, for each structure an optimal mobile ion size exists, as has since been experimentally verified in, for example, sodium β-alumina [103].

The latter feature of optimized ion size matched to an immobile ion network with well-defined geometry and polarizability is perhaps not as well emphasized in some of the FIC literature as it should be. It is, for example, possible to find structurally disordered solids with their associated high intrinsic carrier densities that nevertheless exhibit unexceptional magnitudes of ionic conductivities [e.g., σ (25°C) ~ 10^{-2} S/cm for Na$^+$ in β-alumina, whereas σ (25°C) ~ 10^{-11} S/cm for H$_3$O$^+$ in β-alumina [104]. It is for this reason that certain authors prefer the term *optimized ion conductors* to FICs to emphasize the importance of optimizing both carrier density and ion mobility for obtaining exceptional levels of ionic conductivity at temperatures well below the melting point of the solid. This brings us to a description of a number of transport parameters that are characteristics of FICs.

Although no precise criterion now exists for categorizing FICs, the following conduction parameters are useful in emphasizing the major features:

1. Unusually large magnitude of ionic conductivity ($\sigma > 10^{-2}$ S/cm) well below the melting point
2. Low activation energies ($E \sim 0.05$–0.5 eV)
3. Low pre-exponentials [$\sigma_0 \sim 10^2$–10^4 (S/cm)K]

The first two features have already been discussed above, i.e., the combination of low migration energies and high $c(1 - c)$ terms in Eq. (32) leads to anomalously high ionic conductivities. Based on the large $c(1 - c)$ term, it is therefore all the more surprising to find the pre-exponential σ_0 to be orders of magnitude lower in FICs than in conventional conductors. Since the terms γ, N, Z, and a in Eq. (32) can be expected to differ little in various solids, this leaves only ν_0 and ΔS as the possible sources of this discrepancy. Given the disordered nature of these solids and the shallow potential wells for the mobile ions, the low values for these parameters are not at all surprising.

An important feature, so far ignored, has been correlation effects between carriers that must arise as the concentration of mobile ions increases to the levels observed in FICs. For example, calculations by Wang et al. [105] have demonstrated that cooperative motions of ions can lead to significantly lower calculated migration energies than those based on consideration of isolated jumps alone.

Next we examine the pyrochlore system, which can be induced to exhibit intrinsic oxygen ion disorder. Pyrochlores are compounds having the general formula A(III)$_2$B(IV)$_2$O$_6$X, where X is an anion that also can be oxygen. The pyrochlore crystal structure is a superstructure of the defect fluorite lattice with exactly twice the lattice parameter. The pyrochlores represent a particularly interesting system because the degree of disorder can be varied almost continuously from low to high values using solid-solution systems. By replacing, for example, half the Zr^{4+} ions in ZrO_2 with Gd^{3+}, one out of eight oxygens must be removed to maintain charge neutrality, resulting in the chemical formula $Gd_2Zr_2O_7$. Although oxygen vacancies occur at random throughout the anion sublattice in an ideal defect fluorite (e.g., YSZ), they are ordered onto particular sites in the pyrochlore structure. Thus, one properly views these as empty interstitial oxygen sites rather than oxygen vacancies. As a consequence, nearly ideal pyrochlore oxides, such as $Gd_2Ti_2O_7$, are ionic insulators [106].

Two types of disorder are important in pyrochlores: an antisite disorder on the cation sublattices in which A and B cations switch positions, and a Frenkel-like disorder on the anion sublattice. A study of Gd_2O_3-ZrO_2 solid solutions by Burggraaf and coworkers [107] showed a local maximum in ionic conductivity corresponding to the $Gd_2Zr_2O_7$ compound. Moon and Tuller [108] demonstrated that disorder and, thereby, the ionic conductivity could be varied in an almost continuous manner by varying the r_A/r_B ratio. Intuitively, one might expect that as the radius ratio approaches unity, antisite disorder increases, and with the cation environments of the oxygen ions becoming more homogeneous, exchange between regular and interstitial sites also becomes more favorable. This was confirmed by studying compositions in the system $Gd_2(Zr_xTi_{1-x})O_7$, in which the r_A/r_B ratio was varied from 1.74 to 1.47 as x was increased from 0 to 1.

Figure 17 illustrates the large increase in the ionic conductivity, more than 4 orders of magnitude at 600°C, as zirconium is systematically substituted for titanium [108]. Similar results were obtained in the $Y_2(Zr_{1-x}Ti_x)O_7$ system. Neutron diffraction studies on this system with $x = 0.3, 0.45, 0.6,$ and 0.9 by Wuensch and coworkers [109] confirm the systematic disordering of the oxygen sublattice with the oxygen vacancy fraction on the regular site increasing from 0.006 ($x = 0.3$) to 0.043 ($x = 0.45$) and finally to 0.078 ($x = 0.6$). At $x = 0.9$, all three oxygen sites are one-eighth empty, which is consistent with the structure having reverted back to defect fluorite. This case study illustrates that no sharp distinctions necessarily differentiate fast ion conductors from conventional ionic conductors.

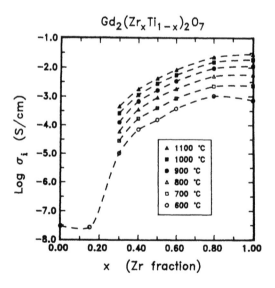

Figure 17 Log σ versus mole fraction of zirconium in $Gd_2(Zr_xTi_{1-x})_2O_7$ at 600, 800, and 1000°C. The sharp rise in ionic conductivity for $x > 0.2$ should be noted (From Ref. 108.)

Other important intrinsically disordered oxygen ion conductors are based on Bi_2O_3. At 730°C [110], the low-temperature semiconducting modification transforms to the δ phase which is accompanied by an oxide ion conductivity jump of almost 3 orders of magnitude. This is tied to the highly disordered fluorite-type structure, where a quarter of the oxygen sites are intrinsically empty, and to the high polarizability of the bismuth cation. Takahashi and Iwahara [111] stabilized the high-temperature δ phase to well below the transition temperature by doping with various oxides, including rare earth oxides such as Y_2O_3. In all cases, the addition of stabilizing ions results in a decrease rather than an increase in conductivity.

High oxide ion conductivity was also discovered above 570°C in the Aurivillius-type γ phase of $Bi_4V_2O_{11}$ [112,113], where one -quarter of the oxygen sites coordinating V^{5+} were empty. Partial substitution of vanadium by lower valence cations, such as copper, nickel, or cobalt, led to a new family of so-called BIMEVOX compounds [114], with a remarkably high oxygen ion conductivity at moderate temperatures. The copper-substituted compound has an oxide ion conductivity above 200°C, which is about 2 orders of magnitude higher than other oxide ion conductors. The variable valence of the transition metal ions does, however, introduce significant electronic conductivity.

In the next section, studies on structurally disordered FICs are discussed, which in some ways resemble even more closely analogous liquid electrolytes.

B. Glasses

One of the often-mentioned criteria for FIC in solids (see above) is the existence of a highly ordered framework that provides channels for the ready motion of ions in the complementary, disordered sublattice. Reports of FIC in glasses [115] raised serious doubts concerning the relevance of this feature. The amorphous state is viewed as being liquid-like. Glasses are known to lack long-range order, with short-range order typically extending to at most a few atom spacings. Because the covalently bonded framework in oxide glasses is characterized by variable bond angles and lengths, it is often referred to as a random network. Although amorphous structures are generally open and rigid, they do not provide a coherent network with well-defined and interconnected polyhedra. Although highly oriented channels may be helpful in FIC, they are not essential as demonstrated by the existence of FIC in glasses.

Fast ionic conductivity is observed in many glasses containing small cations with mole fraction greater than about 0.20, such as silver, copper, lithium, and sodium [116,117]. These glasses typically contain network formers (e.g., SiO_2, B_2O_3, P_2O_5, or GeS_2), network modifiers (e.g., Ag_2O, Li_2O, Cu_2O, or Ag_2S), and a dopant compound, largely halides (e.g., AgI, CuI, and LiCl). The network structure and, therefore, its physical and chemical properties can be substantially modified by addition of the modifier. Dopant salts, on the other hand, do not strongly interact with the network, but *dissolve* into the interstices of the glass structure. A number of phenomenological trends have been shown to be fairly general. Ion conduction increases (1) in the order K, Na, Li, Ag, (2) with modifier concentration, (3) with halide addition in the order Cl, Br, I, (4) in sulfide versus oxide glasses, and (5) in correlation with decreasing density and glass transition temperature.

Increased ion conductivity can result from two sources: increased carrier density and/or carrier mobility. Some authors speculate [118] that the large salt anions enhance ionic transport in glasses by lowering electrostatic and/or polarization barriers. In the so-called *weak electrolyte theory*, the observed activation energy for conduction is thus believed to be made up, in large part, of an association energy that is reduced on addition of the halide anions [119]. An alternate model attributes the increased conductivity to major changes induced in the glass structure by the additives as reflected in changes in glass transition temperature, T_g, and the density ρ. In this latter model, a large fraction of the carriers is already assumed to be unassociated and free to move, but with increased ionic mobility driven by structural changes. Here [120–123], the predominant influence of the halide addition is believed to impact the strain component of the migration energy. Figure 18, for example, illustrates the influence of chlorine additions on a series of glasses based on the diborate composition, in which the total alkali content is maintained constant. The nearly linear increase of log σ with chlorine additions for the lithium and sodium glasses is simply a reflection of the corresponding nearly linear decrease in activation energy. Complementary measurements of glass transition temperature and density point to a weakening and dilation of the network upon chlorine additions.

Another transport–structure correlation important in conducting glasses is known as the the mixed-alkali effect, the source of which has been of great controversy for many years [124–126]. The most recent models focus on local relaxations of the structure about the mobile ions, thereby creating separate pathways appropriate for the various-sized ions. The mismatch in sites for the two ions results ultimately in an overall decrease in conductivity of the mixed system [126].

High ionic conductivities are accompanied by low activation energies: about 0.2–0.3 eV for Ag ion conductors and 0.4–0.7 eV for alkali conductors. Typical values for σ_0 are about 5×10^4 S/cm for Ag and 1–10×10^5 S/cm for alkali conductors. These values for σ_0 cover the range between the lowest values obtained for the best crystalline FICs (10^2–10^3 S/cm) and the much higher values obtained for classical ionic conductors (e.g., NaCl; 10^6 S/cm K). This suggests that the distinction between FIC and classical conductors occurs only in degree rather than form.

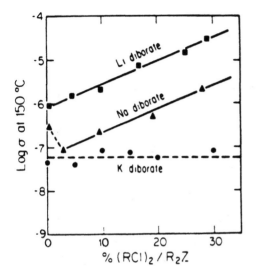

Figure 18 Log σ vs. chloride substitution for $Li_2B_2O_7$, $Na_2B_2O_7$, $K_2B_2O_7$ glasses [122].

Furthermore, in contrast to predictions based on Eq. (32), σ_0 in many FIC glasses is virtually insensitive to composition variation in which the number of mobile interstitial cations is presumably increased and for which the ionic conductivity increases by orders of magnitude. A dramatic case in point was reported by Kawamoto et al. [127], in which σ_0 was found to be independent of Ag content in As_2S_3-, $GeS-GeS_2-$, and P_2S_5-based glasses under conditions for which $\sigma(25°C)$ increased by approximately 6 orders of magnitude. Needless to say, the classical picture that predicts σ_0 to be proportional to carrier concentration appears to be far of the mark in this case. Figure 19 compares a sampling of the large variety of FIC oxide glasses against a number of well-known crystalline FICs. This figure reemphasizes the wide spectrum of ionic conductivities that lie between so-called ionic insulators and the most idealized FICs.

C. Highly Defective Solids

Anomalously high concentrations of ionic carriers may be induced in intrinsically insulating solids. In the following we briefly discuss two approaches for generating such *highly defective solids.*

We already know that ionic defect densities may be greatly enhanced above intrinsic levels by doping with altervalent impurities. However, the solubility limit of such impurities is often limited to only tens or hundreds of ppm. This

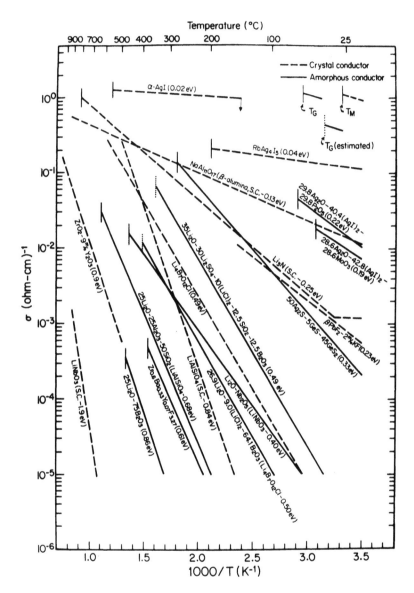

Figure 19 The ionic conductivities as a function of reciprocal temperature of a number of fast Ag, Li, and F conducting glasses are compared with some of the better known crystalline FICs. Glass transition and melting temperatures are indicated where appropriate. (From Ref. 115.)

corresponds to roughly 10^{17}–10^{18} defects/cm^3, a value 10^3–10^4 times smaller than in typical FICs. However, many compounds exist in which the solubility limit is extensive, reaching the 10–20% level even at reduced temperatures. Perhaps the most familiar example of such a system is stabilized zirconia, which due to its wide solid solubility with lower valent cations, such as Ca^{2+} and Y^{3+}, exhibits exceptionally high oxygen ion conductivity ($\sigma \sim 10^{-1}$ S/cm) at temperatures approaching 1000°C.

As discussed above, high carrier densities must be coupled with high ion mobilities in order to attain high magnitudes of ionic conduction. For example, the cubic fluorite structure exhibited by stabilized zirconia (ZrO_2), ceria (CeO_2), and thoria (ThO_2) supports high oxygen ion mobility due to the low fourfold coordination of cations around the oxygens, coupled with the interconnected nature of the face-shared polyhedra that surround the oxygen sites. Migration energies as low as ~ 0.6 eV are reported for oxygen vacancy motion in ceria-based solid solutions [128]. High fluorine ion mobility is also observed in fluorite CaF_2 and related crystal systems.

More recently, Ishihara demonstrated that very high oxygen conductivity can be achieved in the perovskite $LaGaO_3$ by acceptor doping on both the La and Ga sites [129]. The solid solution $(La_{1-x}Sr_x)(Ga_{1-y}Mg_y)O_3$ exhibits ionic conductivity levels above that of ZrO_2 and CeO_2, e.g., 3×10^{-1} S/cm at 850°C.

Perovskites also support some of the highest proton conductivities at elevated temperatures. The most popular of these are ABO_3-type compounds with A = Ba or Sr and B = Ce or Zr. Upon acceptor doping, e.g., $SrCe_{0.95}Yb_{0.05}O_3$, oxygen vacancies are generated as in the gallate above. However, in the presence of moisture, water is adsorbed and protons are generated [130]:

$$H_2O + V_o^{\cdot\cdot} + O_o \Leftrightarrow 2OH^{\cdot} \tag{33}$$

Given the high proton mobility, this is sufficient to induce large proton conductivity. Perovskite-related structures with the general formula $A_3B'B''O_9$ also exhibit high protonic conductivity [131]. Atomistic calculations simulating proton diffusion in numerous perovskite-type oxides are reported by the group of Catlow [132].

Ionic conductivities do not generally increase linearly with foreign atom additions. At the levels of defects being discussed here, defect–defect interactions become important, generally leading to defect ordering. This results in a maximum in ionic conductivity at some level of doping that depends on the particular system being investigated. Nowick and coworkers [133,134] have demonstrated, in a series of studies, that the deviations from ideality are caused initially by composition-dependent activation energies rather than pre-exponentials, a feature already observed earlier for a number of FIC glasses.

The formation of ionic defects that accompany excursions in composition away from stoichiometry due to redox reactions may also be large. For example,

CeO_2 may be readily reduced to $CeO_{1.8}$ at 1000°C [135], resulting in oxygen vacancy concentrations of 5×10^{21} cm^{-3} (i.e., $c = 0.9$). Although such excursions from stoichiometry are generally accompanied by electronic defects of the same order of magnitude [see Eq. (26)], this need not be the case, as was demonstrated for the ThO_2-CeO_2 system [136].

D. Interfacial Ionic Conduction and Nanostructural effects

Interfaces can significantly modify the ionic conductivity of polycrystalline or composite materials and thin films. One origin of enhanced interfacial diffusivity is the disordered nature of the interface core where defect formation and migration energies are generally notably decreased. The interface core can therefore act as a short-circuiting pathway for diffusion at decreased temperature. Alternatively, modified levels of ionic conductivity near interfaces may result from space–charge regions formed near interfaces to compensate charged defects and impurities segregated to surfaces, grain, and phase boundaries. Grain boundaries, for example, serve as source and sink for impurities and point defects and thus often take on a net negative or positive charge relative to the grains. To maintain overall charge neutrality, a space charge of opposite charge forms in the grains adjacent to the grain boundaries with width related to the Debye length L_D given by

$$L_D = (\varepsilon_r \varepsilon_0 kT/Tq^2 n_b)^{1/2} \tag{34}$$

in which n_b is the majority charge carrier concentration within the grain, $\varepsilon_r \varepsilon_0$ the dielectric constant, k the Boltzmann constant, T the temperature, and q the electron charge. Depending on the sign of the charge at the interface, a depletion or accumulation of mobile ions in the vicinity of the boundary will form. Liang provided one of the first demonstrations of enhancement in the $LiI:Al_2O_3$ system [137]. In this work, the Li ion conductivity in LiI was enhanced by nearly 2 orders of magnitude by the addition of Al_2O_3 as a second phase.

The defect concentration profile in the space–charge region can be expressed as [138].

$$c_i/c_i^\infty = \exp[-q_i(\phi - \phi^\infty)/kT] \tag{35}$$

The bulk concentration (c_i^∞) is a function of temperature, chemical potential, and doping. The local concentration in the space–charge region (c_i) depends on the difference between the bulk and the local electrical potential (ϕ^∞ and ϕ). For positive values of ϕ, the concentrations of all negative defects are increased by the exponential factor, whereas those of the positive defects are decreased by the same factor and vice versa for negative values.

Interfaces in systems with very small lateral dimensions, e.g., thin films as well as polycrystalline materials with very small grains, can be expected to have an enhanced effect on ionic conduction. Apart from the ionic conductivity en-hancement because of the increased interface area, an additional conductivity increase can be observed if the grain size approaches the Debye length. In this case, the space–charge regions overlap, and the defect densities no longer reach bulk values, even at the center of the particles [138]. In the limit of very small grains, local charge neutrality is nowhere satisfied, and a full depletion (or accu-mulation) of charge carriers can occur with major consequences for ionic and electronic conductivity. Maier [138] has calculated the supplementary enhance-ment, corresponding to the nanosize factor g, where c^* is the defect concentration in the center of the grain:

$$g = (4\lambda/L)[(c_0 - c^*)/c_0]^{1/2} \tag{36}$$

For a large effect ($c_0 >> c^*$), $g = 4\lambda/L$, and with a grain size $L = 0.4\lambda$, the conductivity is enhanced by an additional order of magnitude. Such a mesoscopic effect was recently demonstrated for artificially modulated heterolayers of the solid ionic conductors CaF_2/BaF_2 [139].

VI. MIXED IONIC ELECTRONIC CONDUCTORS

In this chapter, we first focused on the conditions necessary to achieve high levels of electronic conductivity. This revolved around questions related to the partial occupation of energy bands and the ability of the electronic carriers to delocalize. We then examined conditions necessary to support high levels of ionic conductiv-ity in ionically bonded oxides. Here the emphasis was on the availability of partially occupied sublattices with the ions occupying relatively shallow potential wells. In some oxides, conditions are such that both electronic and ionic conduc-tion are supported simultaneously. Such compounds are known as *mixed ionic electronic conductors (MIECs)* [140,141].

The three major components of the elemental solid oxide fuel cell (SOFC) include the cathode, the electrolyte, and the anode. While the solid electrolyte is selected so that it conducts only ions to ensure as close to ideal Nernst open-circuit potential, the electrodes must support the reduction–oxidation reactions that occur at the electrolyte–electrode–gas interfaces. For example, when current is being drawn, the following reaction occurs at the cathode:

$$1/2O_2 + 2e' + V_O^{\bullet\bullet} \Leftrightarrow O_O \tag{37}$$

This reaction is accelerated if the cathode can provide both electrons, as in a typical current collector, as well as ionic defects. An example of such a MIEC cathode is $La_{1-x}Sr_xCoO_3$, which has an electronic conductivity of >100 S/cm and an oxygen ion conductivity of >1 S/cm at temperatures above 800°C [141].

Unfortunately, this material is unstable in contact with yttria-stabilized zirconia, the electrolyte of choise.

Recent success in achieving both high oxygen ion conductivity and electronic conductivity in perovskite oxides illustrates some key concepts discussed earlier in this chapter. The system $(La_{1-x}Sr_x)(Ga_{1-y}Mg_y)O_3$ (LSGM) was mentioned above as exhibiting one of the highest oxygen ion conductivities ($\sim 3 \times 10^{-1}$ S/cm at 850°C) as a consequence of high levels of acceptor doping (Sr and Mg) on both the La and Ga sites. As a consequence, it is now being considered as one of several candidates as the electrolyte in solid oxide fuel cells. Experiments have shown that MIEC in a fuel cell electrode contribute to reduced overpotentials [142]. It has also been recognized that a single-phase *monolithic* fuel cell structure would benefit from the minimization of chemical and thermal–mechanical degradation [142,143]. Consequently, an electrode based on LSGM would satisfy all requirements. Long et al. proposed to add a transition metal in solution that would introduce an additional 3d *conducting* band within the wide band gap of the initially electronically insulating gallate [144]. As expected, as the Ni content in the system $La_{0.9}Sr_{0.1}Ga_{1-x}Ni_xO_3$ (LSGN) increased, the electronic conductivity increased finally reaching about 50 S/cm—this without decreasing the already high ionic conductivity. Figure 20 illustrates schematically the broadening of the initially discrete impurity level with increasing x due to increasing levels of wave function overlap as expected from Eq. (12). Improved electrode performance was indeed observed with the LSGM–LSGN interface [145].

VII. APPLICATIONS

Highly conducting ceramics are rarely of interest in applications for their highly conducting nature alone. Conventional metals, easily fabricated into complex geometries, readily satisfy this criterion. Conductive ceramics are essential in applications where the material must stand up against a highly corrosive environment, often at elevated temperatures. Examples include fuel cell electrodes operating at 1000°C in oxidizing environments [5,142] and electrodes operating in molten Na-S batteries [146].

Metallic oxides are often used in the fabrication of conductive thick films. Oxides are required since conventional metals oxidize under the high-firing conditions required for substrate adhesion and microstructure development. Metallic oxides commonly used in these applications include RuO_2 and $Bi_2Ru_2O_7$ [147]. Conductive oxide thin-film electrodes have recently begun to penetrate into integrated circuits. In particular, the development of nonvolatile ferroelectric memories (FeRAMs) requires that high-quality ferroelectric films be deposited onto electrodes that are resistant to oxidation, provide appropriate crystallographic orientation, and contribute to improved resistance to switching fatigue. Multicom-

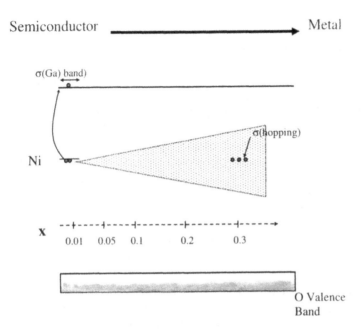

Figure 20 Schematic band energy band diagram for the system $La_{0.9}Sr_{0.1}Ga_{1-x}Ni_xO_3$ in which the Ni 3d derived discrete impurity level is seen to broaden into an impurity *conducting* band as x increases. For $x = 0.5$, the conductivity is found to be metallic [144].

ponent conductive metal oxides, such as $SrRuO_3$ and $(La,Sr)CoO_3$, have exhibited good performance [148].

The combination of metallic conduction and optical transparency, as discussed above, can be achieved with wide band gap oxides that have been doped to degeneracy while maintaining the plasma edge in the IR part of the spectrum. The major applications include flat panel displays, films on office building, and electrochromic windows for thermal management. Efforts are now being expended in identifying alternate transparent electrodes that use lower cost or less toxic components, are easier to process, have higher conductivities and/or lower optical losses. An intriguing direction is the development of p-type transparent electrodes, which in combination with the common n-type electrodes could be used to prepare transparent p-n junction–based devices [2,63–65].

The strong temperature and atmosphere dependence of the electrical conductivity of some oxides has already been noted. This makes semiconducting oxides such as SnO_2, $NiMn_2O_4$, and others attractive as sensing elements. Chapter 6 on *ceramic sensors* discusses these applications in some detail. With the advent of micromachined or microelectromechanical systems (MEMS), the interest in integrating functional ceramic films onto such devices has increased dramatically [149,150].

Fast ionic conducting ceramics and MIECs are finding extensive application in various solid state electrochemical devices. Some of these include fuel cells and electroyzers [1,101,142], high energy density Li batteries [27,101], electrochromic windows [19], and auto exhaust sensors [151]. Some of the most important applications of solid-state electronics and solid-state ionics and their categorization by type and magnitude of conductivity (e.g., dielectric, semiconducting, metallic, and superconducting) are illustrated in Figure 21. The figure also emphasizes that solids need not be strictly ionic or electronic, but can and often do exhibit mixed ionic–electronic conductivity.

Semiconducting titanate electrodes are being investigated as the active element in semiconductor–liquid junction photovoltaic cells. A major feature of these materials is their stability under conditions where conventional semiconductors corrode [76,77].

The unique electrical and magnetic properties of superconductors lead either to enhanced or unique device characteristics. Low electrical loss even up to microwave frequencies holds promise for improved microwave devices [152] as well as at low frequencies for power transmission. Josephson tunnel junctions, used as memory devices, exhibit extremely high switching speeds (10^{-12} s) and markedly

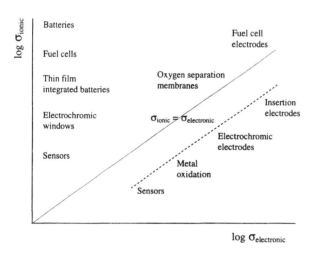

Figure 21 Illustration of typical applications for ionic and electronic conductors as a function of the magnitude of conductivity. Examples of applications requiring mixed conduction fall within the quadrant bound by the two axes. (From Ref. 155.)

lower power dissipation than semiconductor devices. The combination of low resistance and low operating temperatures results in extremely low electrical noise levels in superconducting materials. This contributes in part to the high sensitivity of SQUIDs, or *s*uperconducting *qu*antum *i*nterference *d*evices, to extremely low magnetic fields. The high current densities possible in electromagnets with super-conducting coils, without significant power dissipation, allow for the generation of exceptionally high magnetic fields (5–25 T). Such high-field magnets are central to the operation of magnetic resonance imaging devices in medical applications and large-scale particle accelerators [153,154]. Many additional applications are being seriously considered today, including magnetically levitated trains, improved motors, generators, and transformers, and others too numerous to list in this short summary [4,152–154]. While the high operating temperatures of the ceramic materials offer great potential savings on cryogenics, much progress must still be made in processing these highly anisotropic and brittle materials.

VII. CONCLUSIONS

Although ceramics are often more commonly known for their insulating properties, in reality they exhibit an extensive range of electrical properties, competing with metals and liquid electrolytes for the highest electronic and ionic conductivities, respectively. Although progress has been made in the last decades in clarifying the source of the often complex electrical behavior observed in many ceramics, much remains to be done. Indeed, one can point to few oxides for which the electrical behavior comes even remotely close to the understanding one now has for the conventional semiconductors such as Si or GaAs. A great deal more activity is now being generated in these materials both by those interested in gaining an improved fundamental understanding and by the many groups interested in their application. Based on this attention, one can expect significant progress in this field in coming years.

ACKNOWLEDGMENT

Support from the National Science Foundation (Grant No. DMR-0228787) and ARO–MURI under grant DAAD-0101–0566 for topics related to this work are appreciated.

REFERENCES

1. Minh NQ, Takahashi T. Science and Technology of Ceramic Fuel Cells. Amsterdam: Elsevier Science, 1995:118.

2. Ginley DS, Bright C. MRS Bull 2000; 25:15.
3. Gallium Nitride (GaN) II Pankove JI, Moustakas TD, eds. In Semiconductors and Semimetals Vol 57. San Diego: Academic Press, 1999.
4. Wesche R. High Temperature Superconductors: Kluwer Academic Boston, 1998.
5. Knauth P, Tuller HL. J Am Ceram Soc 2002; 85:1654.
6. Sheng ZZ, Hermann AM. Nature 1988; 332:138.
7. Wilson AH. Theory of Metals. 2nd ed. Cambridge: Cambridge University Press, 1954.
8. Elliot S. The Physics and Chemistry of Solids. Chichester: John Wiley and Sons, 1998:295.
9. Kittel CK. Introduction to Solid State Physics. 5th ed. New York: John Wiley and Sons, 1976.
10. Wang S. Fundamentals of Semiconductor Theory and Device Physics. Englewood Cliffs, NJ: Prentice Hall, 1989:208.
11. Rao CNR, Subba Rao GV. Phy Status Solidi, (a) 1970; 1:597.
12. Mott NF. Proc Phys Soc (Lond) 1949; A62:416.
13. Tuller HL, Nowick AS. J Phys Chem Sol 1977; 38:859–867.
14. Appel J. Solid State Phys 1968; 21:193.
15. Honig JM. In Electrodes of Conductive Metallic Oxides, Part A. Trasatti S, ed. Amsterdam: Elsevier Science, 1980:1–96.
16. Vest RW, Honig JM. Highly Conducting Ceramics and the Conductor–Insulator Transition. In Electrical Conductivity in Ceramics and Glass, Part B, Marcel Dekker. 1974:343–452.
17. Goodenough JB. Prog Solid State Chem 1971; 5:145.
18. McKelvey JP. Solid State and Semiconductor Physics, Harper and Row. 1966.
19. Granqvist CG. Electrochromism and Electrochomic Devices. In The CRC Handbook of Solid State Electrochemistry. Gellings PJ, Bouwmeester HJM, eds. Boca Raton: CRC Press, 1997:587–615.
20. Raccah PM, Goodenough JB. Phys Rev 1967; 155:932–943.
21. Raccah PM, Goodenough JB. J Appl Phys 1968; 39:1209–1210.
22. Tamura H, Yoneyama H, Matsumoto Y. Phisicochemical and Electrochemical Properties of Perovskite Oxides. In Electrodes of Conductive Metallic Oxides, Part A. Trasatti S, ed: Elsevier Sciences Amsterdam, 1980:261–299.
23. Jonker GH, Van Santen JH. Physica 1953; 29:120.
24. Takeda Y, Kanno R, Noda M, Yamamoto O. J Electrochem Soc 1987; 11:2656.
25. van Buren FR, de Wit JHW. J Electrochem Soc 1979; 126:1817–1820.
26. Reimers JN, Dahn JR. J Electrochem Soc 1992; 139:1862–1870.
27. Julien C, Nazri GA. Solid State Batteries: Materials Design and Optimization, Kluwer Academic Boston. 1994.
28. Green M. Ionics 2000; 5:161–70.
29. Sleight AW, Gillson JL, Bierstedt PE. Solid State Commun 1975; 17:27–28.
30. Suzuki M, Enomoto Y, Murakami T, Inamura T. Jpn J Appl Phys 1981; 20(4(Suppl.)):13–16.
31. Mott NF. Adv Phys 1967; 16:49.
32. Adler D, Brooks H. Phys Rev 1967; 155:826.

33. Mott NF. Rev Mod Phys 1967; 12:328.
34. Bednorz JG, Mueller KA. Z Phys B 1986; 64:189.
35. Wu MK, Ashburn JR, Torng CJ, Hor PH, Meng RL, Gao L, Huang ZJ, Wang YQ, Chu CW. Phys Rev Lett 1987; 58:908.
36. von Schnering HG, Walz L, Schwerz M, Becker W, Hartweg M, Popp T, Hettich B, Muller P, Kampf G. Angew Chem Int Ed Engl 1988; 27:574.
37. Maeda H, Tanaka Y, Fukutoni M, Asano T. Jpn J Appl Phys 1988; 27:L209.
38. Sheng ZZ, Hermann AM. Nature (London) 1988; 332:138.
39. Rose-Innes AC, Rhoderick EH. Introduction to Superconductivity. 2nd ed. Oxford: Pergamon Press, 1978:8.
40. Bardeen J, Cooper LN, Schrieffer KR. Phys Rev 1957; 108:1175.
41. See Ref. 39, p. 118.
42. Dimos D, Chaudhari P, Mannhart J, LeGoues FK. Phys Rev Lett 1988; 61:219.
43. Char K. MRS Bull 1994; 19:51.
44. Schuller IK, Jorgensen JD. MRS Bull 1989; 14:27.
45. Emery VJ. MRS Bull 1989; 14:67.
46. Aharmy, et al. A. Phys Rev Lett 1988; 60:1330.
47. Halley JW, ed. Theories of High Temperature Superconductivity. Reading, MA: Addison-Wesley, 1988.
48. Lübke S, Wiemhöfer H-D. Solid State Ionics 1999; 117:229.
49. Naik IK, Tien TY. J Electrochem Soc 1979; 126:562.
50. Stratton TG, Tuller HL. J Chem Soc, Faraday Trans 2 1987; 83:1143.
51. Kröger FA, Vink HJ. In Solid State Physics Seitz F, Turnbull D, eds. Vol. 3. New York: Academic Press, 1956:307.
52. Verwey EJW, Haaijman PW, Romeijn FC, van Oosterhaut GW. Philips Res Rep 1950; 5:173.
53. Gantron J, Marucco JF, Lamasson P. Mater Res Bull 1981; 16:579.
54. Frederkse JPR, Thurber WR, Hosler WR. Phys Rev 1964; 134:A442.
55. Balachandran U, Eror NG. J Electrochem Soc 1982; 129:1021.
56. Hui SQ, Petric A. J Electrochem Soc 2002; 149:J1.
57. Dudney NJ, Coble RL, Tuller HL. J Am Ceram Soc 1981; 64:627.
58. Larson PK, Metselaar R. Phys Rev B 1976; 14:2520–27.
59. Meadowcraft DB. Br J Appl Phys (J Phys D) Ser 2 1969; 2:1225–1233.
60. Cheong SW, Thomson JD, Fisk Z. Physica C 1989; 158:109.
61. Tsai M-J, Opila EJ, Tuller HL. In High Temperature Superconductors: Fundamental Properties and Novel Materials Processing Christen D, Narayan J, Schneemeyer L, eds. Materials Research Society. Vol. 169. Pittsburgh, 1990:65–68.
62. Opila EJ, Tuller HL. J Am Ceram Soc 1994; 77:2727–37.
63. Minami T. MRS Bull 2000; 25:38.
64. Freeman AJ, Poeppelmeier KR, Mason TO, Chang RPH, Marks TJ. MRS Bull 2000; 25:45.
65. Kawazoe H, Yanagi H, Ueda K, Hosono H. MRS Bull 2000; 25:28.
66. Kröger FA. The Chemistry of Imperfect Crystals. 2nd ed. North-Holland, Amsterdam, 1974.
67. Tuller HL. J Electroceramics 1999; 4(S1):33–40.

68. Choi GM, Tuller HL. J Am Ceram Soc 1988; 71:201.
69. Jonker GH. Philips Res Rep 1968; 23:131.
70. Choi GM, Tuller HL, Goldschmidt D. Phys Rev B 1986; 34:6972–6979.
71. Tuller HL, Kowalski JM, Gealy FD. Study of Redox Kinetics of Semiconducting Photoelectric Interfaces. In Grain Boundaries and Interfaces in Ceramics. Advances in Ceramics Yan MF, Heuer AH, eds. Vol. 7. Columbus, OH: American Ceramic Society, 1984:281–293.
72. Odekirk B, Balachandran U, Eror NG, Blaken JS. Mater Res Bull 1982; 17:199.
73. Kowalski JM, Tuller HL. Ceram Int 1981; 7:55.
74. Campet G, Dare-Edwards MP, Hamnet A, Goodenough JB. Nouv J Chim 1980; 4:501.
75. Maruska HP, Ghosh AK. Solar Energy Mater 1979; 1:237.
76. Levy B. J Electroceram 1997; 1:239.
77. Nazeeruddin MK, Kay A, Rodicio I, Humphrey-Baker R, Muller E, Liska P, Vlachopoulos N, Gräzel M. J Am Chem Soc 1993; 115:6382.
78. Sukkar MH, Tuller HL. In Grain Boundaries and Interfaces in Ceramics. Advances in Ceramics Yan M, Heuer AH, eds. Vol. 5. Columbus, OH: American Ceramic Society, 1984:71.
79. Wernicke R. Phys Status Solidi (a) 1978; 47:139–144.
80. Choi GM, Tuller HL, Tsai MJ. In Proc. Symp. High Temperature Superconductors, Materials Research Society Symposium Brodsky MW, Dynes RC, Kitizawa K, Tuller HL, eds. Vol. 99. Pittsburg: Materials Research Society, 1988:141.
81. Riess I, Porat O, Tuller HL. J Superconductivity 1993; 6:313–316.
82. Maier J, Tuller HL. Phys Rev B 1993; 47:8105–8110.
83. Sempolenski DR, Kingery WD, Tuller HL. J Am Ceram Soc 1980; 63:669.
84. Dutt BV, Hurrell JP, Kröger FA. J Am Ceram Soc 1975; 59:920.
85. Groth R. Phys Status Solidi 1966; 11:69.
86. Van Daal HJ. Solid State Commun 1968; 6:5.
87. Li PW, Hagemark KI. J Solid State Chem 1975; 12:371.
88. Shukla VN, Wirtz GP. J Am Ceram Soc 1977; 60:253.
89. Koffyberg FP. J Solid State Chem 1970; 2:176.
90. Fonstad CG, Linz A, Rediker PH. J Electrochem Soc 1969; 116:1269.
91. Smith FTJ, Lyn SL. J Electrochem Soc 1981; 128:2388.
92. Marley JA, Dockerty RC. Phys Rev 1965; 140:A304.
93. Paria MK, Maiti HS. J Mater Sci 1982; 17:3275.
94. Sukkar MH, Tuller HL. Defect Equilibria in ZnO Varistor Materials. In Grain Boundaries and Interfaces in Ceramics. Advances in Ceramics Yan M, Heuer AH, eds. Vol. 5. Columbus, OH: American Ceramic Society, 1984:71.
95. Paria MK, Maiti HS. J Mater Sci 1982; 17:3275.
96. Goodenough JB. Skeleton Structures. In Solid Electrolytes. Hagenmuller P, Van Gool W, eds. New York: Academic Press, 1978:393.
97. Hladik J, ed. Physics of Electrolytes. Vol. 1. New York: Academic Press, 1972.
98. Strock LW. Z Phys Chem, B25 1932; 32:132.
99. Cava R, Wuensch BJ. Solid State Commun 1977; 24:411–416.
100. Kudo T. In The CRC Handbook of Solid State Electrochemistry. Gellings PJ, Bouwmeester HJM, eds. Boca Raton: CRC Press, 1997:195.

101. Knauth P, Tuller HL. J Am Ceram Soc 2002; 85:1654.
102. Flygare WH, Huggins RA. J Phys Chem Sol 1973; 34:1199.
103. Wittingham MS. Electrochim Acta 1975; 20:575.
104. Farrington GC. Sensors and Actuators 1981; 1:329.
105. Wang J, Kaffari M, Choi D. J Chem Phys 1975; 63:772.
106. Kramer SA, Tuller HL. Solid State Ionics 1995; 82:15–23.
107. van Dijk T, de Vries KJ, Burffraaf AJ. Phys Status Solidi A 1980; 58:115–125.
108. Moon PK, Tuller HL. Solid State Ionics 1988; 28–30:470–474.
109. Wuensch BJ, Eberman KW, Heremans C, Ku EM, Onnerud P, Yeo EME, Haile SM, Stalick JK, Jorgensen JD. Solid State Ionics 2000; 129:111–133.
110. Harwig HA, Gerards AG. J Solid State Chem 1978; 26:265–274.
111. Takahashi T, Iwahara H. Mater Res Bull 1978; 13:1447–1453.
112. Kendall KR, Navas C, Thomas JK, zur Loye H-C. Chem Mater 1996; 8:642–649.
113. Boivin JC, Mairesse G. Chem Mater 1998; 10:2870–2888.
114. Abraham 141 F, Boivin JC, Mairesse G, Nowogrocki G. Solid State Ionics 1990; 40–41:934–937.
115. Tuller HL, Button DP, Uhlmann DR. J Non-Cryst Solids 1980; 42:297.
116. Tuller HL, Barsoum MW. J Non-Cryst Solids 1985; 73:331–350.
117. Fusco FA, Tuller HL. Fast Ion Transport in Glasses. In Superionic Solids and Solid Electrolytes: Recent Trends. New York: Academic Press, 1989:43.
118. Souquet JL. Solid State Ionics 1981; 5:77.
119. Ravaine D, Souquet JL. Phys Chem Glasses 1977; 18:27–31.
120. Button DP, Tandon RP, Tuller HL, Uhlmann DR. J Non-Cryst Solids 1980; 42: 297.
121. Button DP, Tandon RP, Tuller HL, Uhlmann DR. Solid State Ionics 1981; 5:655.
122. Fusco FA, Tuller HL, Button DP. Lithium, Sodium, and Potassium Transport in Fast Ion Conducting Glasses: Trends and Models. In Proc. Symp. Electro-Ceramics and Solid State Ionics. Tuller HL, Smyth DM, eds. Pennington, N.J.: Electrochemical Society, 1988:167.
123. Button DP, Moon PK, Tuller HL, Uhlmann DR. Glastech Ber 1983; 56K:856.
124. Isard JO. J Non-Cryst Solids 1968–1969; 1:235–262.
125. Day DE. J Non-Cryst Solids 1976; 21:343–372.
126. Ingram MD. Glastech Ber Glass Sci Technol 1994; 67:151–155.
127. Kawamoto Y, Nagura N, Tsuchihasi S. J Am Ceram Soc 1974; 57:489.
128. Wang DY, Park DS, Griffiths J, Nowick AS. Solid State Ionics 1981; 2:95.
129. Ishihara T, Matsuda H, Takita Y. J Am Chem Soc 1994; 116:3801–3803.
130. Iwahara H, Esaka T, Uchida H, Maeda H. Solid State Ionics 1981; 3/4:359–363.
131. Nowick AS, Du Y. Solid State Ionics 1995; 77:137–146.
132. Cherry M, Islam MS, Gale JD, Catlow CRA. J Phys Chem 1995; 99:14614–14618.
133. Wang DY, Park DS, Griffiths J, Nowick AS. Solid State Ionics 1981; 2:95.
134. Gerhart-Anderson R, Nowick AS. Solid State Ionics 1981; 5:547.
135. Tuller HL, Nowick AS. J Electrochem Soc 1979; 126:209.
136. Fujimoto HH, Tuller HL. Mixed Ionic and Electronic Transport in Thoria Electrolytes. In Fast Ion Transport in Solids. Vashista P, Mundy JN, Shenoy GK, eds. Amsterdam: Elsevier/North-Holland, 1979:649.

137. Liang CC. J Electrochem Soc 1973; 120:1289.
138. Maier J. Prog Solid State Chem 1995; 23:171–263.
139. Sata H, Eberman K, Eberl K, Maier J. Nature (London) 2000; 408:946–948.
140. Tuller HL. Mixed Conduction in Nonstoichiometric Oxides. In Non-Stoichiometric Oxides. Sorensen OT, ed. New Yorks: Academic Press, 1981:271.
141. Riess I. Electrochemistry of Mixed Ionic-Electronic Conductors. In The CRC Handbook of Solid State Electrochemistry. Gellings PJ, Bouwmeester HJM, eds. Boca Raton: CRC Press, 1997:223.
142. Tuller HL. Materials Design and Optimization. In Oxygen Ion and Mixed Conductors and their Technological Applications. Tuller HL, Schoonman J, Riess I, eds. The Netherlands: Kluwer Academic, 2000:245.
143. Kramer SA, Spears MA, Tuller HL. U.S. Patent No. 5,5403,461, 1995.
144. Long NJ, Lecarpentier F, Tuller HL. J Electroceram 1999; 3:4:399–407.
145. Lecarpentier F, Tuller HL, Long N. J Electroceram 2000; 5:225–230.
146. Weiner SA, Cairns EJ. Materials Problems in Rechargeable Batteries. In Solid State Chemistry of Energy Conversion and Storage. Goodenough JB, Wittingham MS, eds. Washington, D.C.: American Chemical Society, 1977:635–710.
147. Pike GE, Seager CH. J Appl Phys 1977; 48:5152.
148. Jones RE, Desu SB. MRS Bull 1996; 21:55.
149. Tuller HL, Mlcak R. J Electroceram 2000; 4:415–425.
150. Tuller HL, Mlcak R. Curr Opin Solid State Mater Sci 1998; 3:501–504.
151. Moseley PT, Tofield BC, eds. Solid State Gas Sensors. Adam Hilger, 1987.
152. Defense Science Board, *Report of the Defense Science Board Task Force on Military System Applications of Superconductors*, U.S. Department of Defense, October 1988.
153. Foner S, Schwartz BB, eds. Superconductor Materials Science: Metallurgy, Fabrication, and Applications. New York: Plenum Press, 1981.
154. VanderSande JB. Superconductivity: A Guide for Industrial Applications. Vol. 2. Cambridge, MA: Innovation 128, 1990.
155. Tuller HL. J Phys Chem Solids 1994; 55:1393–1404.

3

Ceramic Capacitor Materials

Thomas G. Reynolds III
Niceville, Florida, U.S.A.

Relva C. Buchanan
University of Cincinnati
Cincinnati, Ohio, U.S.A.

I. INTRODUCTION

The phenomenal growth of integrated circuit technology has been accompanied by increased consumption of discrete circuit components such as resistors and capacitors. Ceramic capacitors account for a major segment of this volume, usage of which is currently more than 300 billion pieces annually. This vigorous component technology is traceable to the Leyden jar of 1745, the first capacitor design to have enjoyed widespread use. In its most advanced, dry design, with metal foil electrodes in contact with glass dielectric, this early device had volumetric efficiency and loss characteristics that are poor by contemporary standards. Nevertheless, it figured in pioneering experiments on the nature of static electricity, including Franklin's lightning investigations, and some others of questionable value.[*]

The capacitor's basic function, energy storage, has been broadened since the early experiments, to include the blocking of direct current or the coupling

---

[*] In a demonstration before members of the court of Louis XV of France, 700 monks with joined hands allegedly were caused to jump simultaneously by a Leyden jar discharge [1].

of ac circuits. In bypass applications, the capacitor separates the ac and dc portions of a mixed signal. Alternating currents are also separated by capacitors according to frequency, and the charge–discharge characteristics of resistance–capacitance combinations are applied in timing circuits. Physically large-scale tasks, such as high-energy storage and power factor correction, fall to larger and different types of capacitors than those discussed in this chapter.

The properties of capacitor dielectric materials that determine in detail the manner in which the energy storage function is performed are as follows:

1. Dielectric constant relative to that of vacuum; commonly known as the relative dielectric constant or permittivity, which measures the response of the dielectric's polarization mechanisms to an electric field.
2. Dissipation factor, power factor, and loss factors. These measure the net inefficiency of the polarization process(es).
3. Insulation resistance: A measure of the efficiency of dc blocking.
4. Temperature, frequency, and field strength dependencies of the above parameters.

This chapter is confined to capacitors whose dielectrics are ceramic materials. Although there are specific applications that use steatite, glass, and porcelain compounds, the majority of ceramic capacitors are based, in large part, on the properties of barium titanate and solid solutions with other oxides (i.e., $SrTiO_3$, CaO, ZrO_2), and trace amounts of specific compounds (e.g., MnO, Nb_2O_5, rare earth oxides) that are used to optimize the properties. A major development has been the introduction of base metal electrode systems, which have significantly reduced cost as a result of replacing the silver/palladium alloys with nickel metal electrodes. Concurrent with materials development there have been several parallel developments in the fabrication of the ceramic capacitors: miniaturization of sizes and "footprint" on the circuit board; higher capacitance per unit volume as a consequence of thinner dielectric layers; high-voltage capacitors, thin-film capacitors, and others.

II. HISTORICAL BACKGROUND

In a sense, the very glass material that served as the Leyden jar's dielectric persisted in use as a dielectric into the twentieth century. The properties of electrical porcelain are largely determined by the composition and content of the glass in its microstructure. Porcelain insulators figured importantly in early electrical technology as isolating and structural members. Steatite ceramics were then developed for the same applications. Containing feldspar as a fluxing constituent, they were still subject, although to a lesser degree, to high-loss behavior similar to that of porcelain. For capacitors used in the first radio circuitry in the early twentieth century, the premium dielectric materials were paper and mica. Besides having the best available loss characteristics, these materials could be fabricated into thin sheets to provide high capacitance, in spite of their modest dielectric constants.

Thurnauer [2,3] described steatite ceramics used in Germany around 1920 that were fabricated with alkaline earth oxide fluxes, instead of alkaline oxides (feldspar), giving them low-loss behavior. It was a natural step to develop such materials as substitutes for mica and paper. A challenge to ceramic technology, therefore, was to develop ways of fabricating these materials into very thin cross-sections; the extruded tubular steatite ceramic capacitor made its appearance as a mica capacitor replacement. However, such has been the commercial hardiness of the mica capacitor that it is still used for specialized applications. Many of the mica applications were preempted by ceramics during World War II, when the mica sources became uncertain, and the product never regained its prewar share of the market.

Another mineral displaced by a ceramic because of wartime shortages was soapstone or block talc. Easily machineable and fired into vacuum-tube insulators, the mineral became a strategic material. It was, however, largely replaced by ceramic-processed steatite, which is basically of the same composition, and which has found application in fixed tubular, trimmer, and other capacitor types.

Thus, displacement of natural dielectric materials by ceramics, under the technical demands of a growing electronics industry, has been a key pattern in ceramic capacitor development. Another pattern is that the development of ceramic dielectrics technology has been influenced by, and in turn has influenced, the miniaturization of the devices, equipment, and systems in which capacitors are used. The increase in available dielectric constant from less than 10 for porcelains to, effectively, 100,000 in grain boundary barrier layer capacitors has been instrumental in this size reduction.

Finally, the technology of fabricating ceramics has kept pace with the wide range of material properties, providing geometries that not only capitalize on these properties but are appropriate to the equipment assembly methods used. Hybrid circuits, for example, employ chip capacitors with closely defined electrical characteristics and physical geometries, which are installed in place by automated assembly methods. This trend continues with the evolution and major use of ceramic chip capacitors in surface mount technology (SMT).

The layer capacitors has been instrumental in this size reduction. Additionally, suitable capacitors for thin- and thick-film circuits have been developed. Where minimization of the dielectric loss is of foremost importance, as in high-frequency circuits, ceramic capacitors have been developed with exceptionally low dissipation factors and large time constants.

III. FERROELECTRICITY IN CAPACITOR TECHNOLOGY

The discovery of ferroelectricity in barium titanate in the 1940s made available for ceramic capacitor design dielectric constants up to 2 orders of magnitude greater than previously known. This class of ceramic materials has so dominated capacitor technology that the standards of performance incorporate many ferroelectricity-related idiosyncrasies.

In ferroelectric materials, preexisting electric dipoles, the presence of which is predictable from the crystalline symmetry of the material, interact at a distance, generally not predictably, to spontaneously polarize subvolumes of the crystal or ceramic mass. The spontaneously polarized regions, with a single direction of polarization, are called *domains*, after ferromagnetics terminology. The relative orientation relationships of the domains to one another is governed by the crystal symmetry. In the barium titanate family of materials, on which most ceramic capacitors are based,* the crystal structure is the cubic or pseudocubic perovskite structure. The spontaneous polarization can orient parallel to any of the pseudocube unit cell edges, so that adjacent domains have 180° or near-90° relationships to one another. However, the domain structure of a titanate ceramic as fabricated can be quite complex. This results from the random, sintered microstructure and strain patterns that develop on cooling through the Curie temperature, T_c (Figure 1b), due to anisotropic dimensional changes.

The domains are also orientable by an externally applied electric field, the effect of which is to increase the component of polarization in the field direction. At the Curie temperature the spontaneous polarization disappears, but significant effects of the applied field on permittivity persist for at least 50°C higher. If the applied field is removed, some of the regions that were oriented retain the new orientation while others revert. Much of the behavior of a ceramic capacitor can be described in relation to the spontaneously polar character or the associated ferroelectric hysteresis loop behavior (Figure 1a).

With increasingly high ac field the dielectric constant ($\delta P/\delta E$) first increases, then decreases as the dipolar contribution increases, then saturates. When increasingly dc biased, the dielectric shows a decreasing low-signal ac dielectric constant, since the bias serves increasingly to repress domain reversibility. Increasing frequency also leads to a decrease in dielectric constant in the form of a relaxation, centered in the gigahertz range (Figure 1b), when the spontaneous polarization lags behind the applied frequency. Coincident with this relaxation change, the dissipation factor passes through a maximal value.

---

* Relaxator-type dielectrics, to be discussed, form also in the tungsten bronze structure.

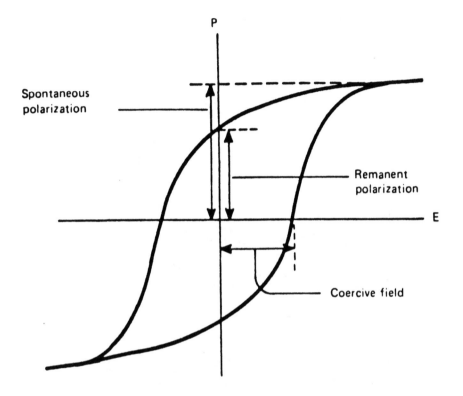

P

Spontaneous
polarization

Remanent
polarization

E

Coercive field

Figure 1a A ferroelectric hysteresis loop.

IV. DIELECTRIC PROPERTIES OF MULTIPHASE SYSTEMS

Except in research studies, orientation of microstructure to enhance electrical characteristics has not been fully utilized in ceramic dielectrics. Two major exceptions are the orientation of ferrite permanent-magnet materials and the poling of ceramics to produce piezoelectric transducers. In these cases, there are accompanying effects on the dielectric properties, but not of a magnitude useful for applications in capacitors. However, efforts to rationalize the characteristics of dielectric systems in terms of constituent phase properties and their microstructural arrangement are numerous. The case of microstructures in which there are large differences of conductivity between the phases is most important to capacitors and is taken up in the section on grain boundary barrier layer capacitors.

The other major microstructural case is that in which the permittivities of the constituent phases differ significantly but their conductivities are the same.

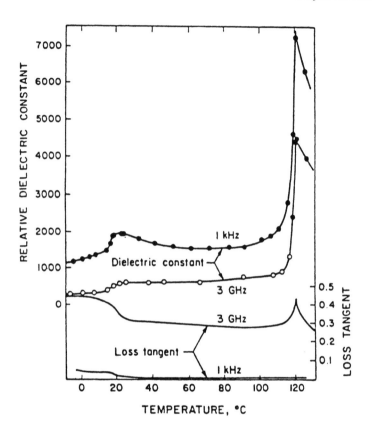

Figure 1b   Relative dielectric constant and loss tangent for polycrystalline BaTiO$_3$.

Actually, the prevalence of minor low-frequency relaxation phenomena is evidence that this pure condition does not usually apply.

A simplification is to discuss this subject in relationship to a system of only two phases. Maxwell's treatise [4] addresses the problem of the dielectric constant of a mixture of two isotropic phases, given their dielectric constants and the geometry of the mixture. The analysis deals first with planar parallel arrangements of two phases oriented either normal to or parallel to the specimen electrodes. More realistic, however, is the case where phase 1 is continuous and has dispersed in it phase 2 in the form of spheres. The spheres are assumed to be separated, enough that they do not disturb the lines of flux around one another. Although this is strictly true only at infinite dilution, Maxwell's relations have some applicability at up to 10–50% volume concentration of phase 2. For $(\varepsilon_r)_2 > (\varepsilon_r)_1$,

$$(\varepsilon_r)_T = (\varepsilon_r)_2 \frac{(\varepsilon_r)_1 + 2(\varepsilon_r)_2 - 2(1-\upsilon_2)\left[(\varepsilon_r)_2 - (\varepsilon_r)_1\right]}{(\varepsilon_r)_1 + 2(\varepsilon_r)_2 + (1-\upsilon_2)\left[(\varepsilon_r)_2 - (\varepsilon_r)_1\right]} \text{ for phase 2 continuous} \tag{1}$$

and

$$(\varepsilon_r)_T = (\varepsilon_r)_1 \frac{(\varepsilon_r)_2 + 2(\varepsilon_r)_1 - 2(1-\upsilon_2)\left[(\varepsilon_r)_1 - (\varepsilon_r)_2\right]}{(\varepsilon_r)_2 + 2(\varepsilon_r)_1 + (1-\upsilon_2)\left[(\varepsilon_r)_1 - (\varepsilon_r)_2\right]} \text{ for phase 1 continuous} \tag{2}$$

where υ is the volume fraction of one phase. The shape and dimensions of the disperse phase do not enter into this calculation. The relationships have had application in such systems as resins loaded with a high-ε_r titanate powder or a partially devitrified glass. Mitoff [5] has provided a useful discussion of the extent to which these Maxwell relationships can be modified for applicability where there are disparate values of permittivity of the two phases, deviations from sphericity of the disperse phase, and where both phases are continuous instead of only one. In general, the most accurate predictions can be made for the most dilute dispersions.

Figure 2 is a graphical representation of the behavior obtained when the two phases have widely differing dielectric constants ($\varepsilon_{r1}/\varepsilon_{r2} \sim 1000$). The extreme curves represent laminated structures arrayed parallel and normal to the electrodes. Within these extremes fall the cases of dilute dispersions of spheres of one phase in a matrix of the other. Curves for the latter terminate in the dashed sections, where the condition of dilute dispersion no longer applies.

The diagonal straight line of Figure 2 represents a widely used empirical relationship due to Lichtenecker [6]:

$$\log \varepsilon_{rT} = v_1 \log \varepsilon_{r1} + v_2 \log \varepsilon_{r2} + \cdots \tag{3}$$

In use of this relationship there is no concern at all for the physical geometry of the system.

A summary of the extensive literature on modeling of heterogeneous dielectric systems is given by Van Beek [7]. The many physical models that have been proposed and analyzed have had mainly academic interest and have rarely played a role in the design or realization of a ceramic capacitor. In fact, Lichtenecker's empirical mixing rule seems to have been the most frequently used.

V. BASIC CERAMIC DIELECTRIC MATERIALS

A. Porcelain and Steatite

Classical porcelain and steatite have little remaining place as the dielectrics of modern electronic circuit capacitors, but they leave some heritage, namely:

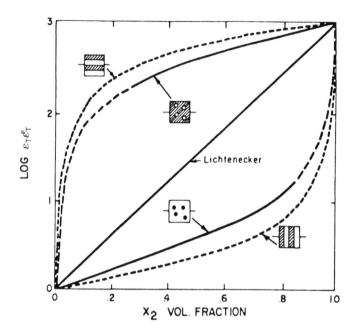

Figure 2   Plot of Maxwell equation for dielectric mixtures with $\varepsilon_{r1}/\varepsilon_{r2} = 1000$. (After Ref. 6.)

1. The dogma that alkaline oxide–free compositions are essential for high resistivity and low dissipation factor came from the development of low-loss porcelains and steatites. It is adhered to in specifying the raw materials of dielectrics of other types, as well as of metallization, soldering fluxes, coatings, and encapsulants.
2. The use of steatite as a diluent in dielectric formulations with graded dielectric constants enables a series of capacitors with fixed dimensions but graduated capacitance values to be fabricated.

B. Rutile

Electrical porcelains and steatite ceramics have dielectric constants in the range 5–7 and positive temperature coefficients of capacitance. The move to blends of steatite with rutile was thus a natural one; it opened the way to 10-fold higher dielectric constant and to the possibility of lowered positive temperature coefficients or negative ones. Single-crystal rutile's properties had been known since 1902: dielectric constant of about 170 along the tetragonal c axis and 80 in the a directions. The ceramic mixtures, typified by steatite-TiO_2, have near-zero temperature coefficients of capacitance for mixtures with dielectric constants of 15–20. The dissipation factor of rutile is typically two to five times that of low-loss steatites, and their blends have intermediate loss values.

C. Barium Titanate

The discovery of ferroelectric barium titanate opened the present era of ceramic dielectric materials. The high dielectric constant was first described in the United States in 1942 [2]. Recognition of $BaTiO_3$ as a new ferroelectric compound followed, possibly independently, in several countries [8–10]. A question of historical interest is posed by the lapse of a decade or more between active development of titania-based dielectrics and the discovery of $BaTiO_3$. A case can be made that there may have been earlier syntheses of the material without recognition of its dielectric character.

Barium titanate has become the basic ceramic capacitor dielectric material in current use. As earlier with rutile, the main mode of its use has been in combination with other materials. Primary application requisites are a high capacitance and, to the degree possible, stable capacitance over the temperature range of use of the component. Practical temperature ranges of use have been specified as from $-55°C$ to $125°C$, or segments within that range. Since pure, crystalline $BaTiO_3$ has a peak permittivity at about $130°C$ (see Figure 1b) and another, secondary peak at $10°C$ [11], corrective modification is required for practical application. The formulation of a titanate-based dielectric is to a degree empirical, but compositions in use are based on several principles.

1. Solid Solutions

Solid solution with an isostructural compound broadens the Curie peak [12]. This was first perceived in $BaTiO_3$-$SrTiO_3$ solid solutions (Figure 3), and many such combinations are now known. The amount of peak broadening accompanying solid solution does not of itself enable a useful capacitance versus temperature characteristic. However, limited solubility or nonequilibrium processing can promote the formation of two or more solid solutions, each with its Curie peak. When a phase with a high permittivity peak coexists with a lower permittivity material, their summed permittivity is less sharply peaked. Further contributing to flattening of permittivity versus temperature for $BaTiO_3$ is that the smaller peak associated with the $10°C$ tetragonal-to-orthorhombic phase transition tends to be shifted upward in temperature as the main, cubic-to-tetragonal peak shifts downward (see Figure 1b).

2. Fine-Grained Materials

Reducing the grain size of the barium titanate ceramic below about 1 μm in diameter has a flattening effect on capacitance versus temperature. The origin of

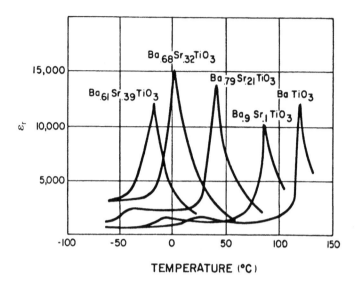

Figure 3   Effect of solid solution on ε_r versus temperature characteristic. From Ref. 12.)

this effect is in the nature of the symmetry transition associated with the 130°C permittivity peak of $BaTiO_3$. The microstructure of a large-grained ceramic contains "90°" twins formed to relieve the strains generated when the c axis elongates on passing from the cubic to the tetragonal symmetry; in the strained state the highest permittivity is displayed.

The domain boundary planes have surface energy proportional to the square of the grain diameter; the strain energy responsible for domain formation, a volume effect, is related to the cube of the grain diameter. As indicated schematically in Figure 4, with decreasing grain size a critical size, D_c, is reached where it is energetically less costly to support an elastic strain than to relieve it by twin formation. The ceramic in which this situation exists will have average grain size of the order of 1 μm or less and be untwinned; that is, each grain will comprise a single domain with one direction of spontaneous polarization, be stressed, and tend to cubic symmetry [13]. The permittivity of a ceramic in this condition will be greater than when unstrained, having a room temperature value of about 2000, compared with 1400 for large-grained, freely twinned barium titanate. To take advantage of this effect, $BaTiO_3$ starting material of very fine grain size and processing that minimizes grain growth are required.

3.   Additives

A number of additives have been associated with the ability to flatten the permittivity versus temperature characteristic of barium titanate ceramic. Nickel oxide

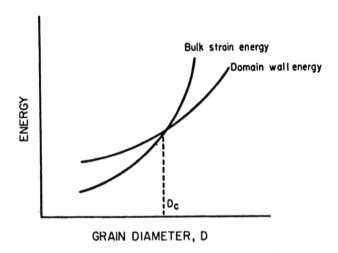

Figure 4   Limiting grain size for ferroelectric domain formation.

and bismuth stannate have been described to function in this way [14,15]. The additives are aliostructural and are effective in small quantities. Their function is attributed to grain growth inhibition accomplished by lodging in the grain boundaries. The terms "peak shifting" and "peak depressing" have been applied to the functions just described and is due to Coffen [14].

Plessner and West [15] early enumerated materials combinations that provide usefully flattened, high-permittivity versus temperature characteristics, and others have extended the list and demonstrated also the utility of nonstoichiometric substitutions:

$BaTiO_3$ and $MgSnO_3$
$BaTiO_3$, $SrTiO_3$, and $CaTiO_3$
$BaTiO_3$, $CaSnO_3$, and CaO
$BaTiO_3$ and $CaZrO_3$
$BaTiO_3$ and $CaSnO_3$
$BaTiO_3$ and $Bi_2(SnO_3)_3$

The latter combination, although still widely used, has been a victim of precious metal economics. The cost-reducing change to palladium-based alloys in multilayer capacitor electrodes from platinum-gold and the fact of intermetallic compound formation between bismuth and palladium have largely reduced the use of bismuth-containing Curie peak depressors [see Section VII.C.3(a)].

In brief, high-permittivity dielectrics for modern ceramic capacitors are based on (1) ferroelectric barium titanate and its isomorphs and their solid solutions, and (2) ceramic processes that lead to nonequilibrium mixtures of such phases. The resultant locally inhomogeneous microstructure possesses a broadened, flattened permittivity–temperature characteristic, which becomes further flattened and elevated if grain growth during sintering can be sufficiently inhibited.

D. Relaxor Dielectrics

Relaxor dielectrics comprise a class of ceramic compositions originally described by Smolenskii and Isupov [18]. They are characterized by high dielectric constants and a broad peak in the permittivity versus temperature behavior, even in the absence of additives and even in single crystals. This behavior is attributed to a microstructure of small (~100nm), ordered regions, too small to yield the sharp phase transition of normal ferroelectrics [19]. As a result, some spontaneous polarization and associated ferroelectric behavior are retained over a very broad temperature range. Compositions in the tungsten bronze structure system $Pb(Mg_{1/3}Nb_{2/3})O_3$-$PbTiO_3$-$Ba(Zn_{1/3}Nb_{2/3})O_3$ have diffuse peaks with heights corresponding to permittivities of 25,000 [20]. A large number of specific dielectric compositions of this type are described in the patent literature. An attractive secondary characteristic is that dense ceramics are achievable at sintering temperatures low enough for applicability with silver-palladium alloy electrodes in multilayer capacitors.

VI. PERFORMANCE CATEGORIES OF CERAMIC CAPACITORS

In the United States, the Electronic Industries Association has defined three categories of ceramic capacitors [21] having the following characteristics:

Class 1: Relatively low capacitance with low dielectric loss, low temperature coefficient of capacitance, and low rate of aging of the capacitance value. Their specified temperature range of use is −55°C to 85°C and temperature coefficients of capacitance values range from "zero" to + 75,000 or − 75,000 ppm/°C, with tolerances on the temperature coefficient ranging from ± 30 ppm/°C to ± 2500 ppm/°C. This class of capacitors uses relatively low permittivity dielectrics ($\varepsilon_r \leq 100$). The most commonly specified class 1 temperature coefficient is 0 ± 30 ppm/°C.

Class 2: Based on high-permittivity dielectrics ($\varepsilon_r \leq 15,000$), primarily ferroelectric barium titanate. Because of the prevalence of ferroelectric character, important specified parameters are the voltage coefficient of capacitance, voltage coefficient of series resistance, voltage coefficient of dissipation factor, and rate of capacitance decrease upon aging.

Subclasses of class 2 are based on the temperature coefficient of capacitance and the temperature range of use. Temperature ranges of use are specified according to limits: as low as $+ 10$, $- 30$, or $- 55°C$ and as high as $+45$, $+65$, $+85$, $+105$, or $+125°C$. Within these spans, maximal specified deviations from the room temperature capacitance value may range from $+1$ to $\pm 22\%$ and from $+22$, -33% to $+22$, -82%. Some commonly specified temperature coefficients are:

> Z5U: less than $+ 22\%$ and -56% deviation from the room temperature capacitance over the range $+10°C$ to $+85°C$.
>
> Y5U: ditto, except over the temperature range $-30°C$ to $+85°C$.
>
> X7R: less than $\pm 15\%$ change between $-55°C$ and $+ 125°C$.

Class 3: These capacitors are characterized by very high capacitance values derived from electrode barrier layers or grain boundary barrier layers, and low-voltage-withstanding ability.

VII. VARIETIES OF CERAMIC CAPACITORS

A. Film Capacitors

Wafer or disk discrete ceramic capacitors have dielectrics at least 100 μm thick bearing electrodes with thickness of about 25 μm. In thick-film capacitors, both the electrode material and the dielectric have thickness similar to that of the electrodes in discrete capacitors. Their formulations include low-melting glass binders that enable adherence to a substrate to be developed at well under 1000°C. Usually the electrode and dielectric thick-film materials are formulated as paints and deposited in the desired pattern by the silk screen process.

In thin-film capacitors all layers, dielectric and metal, are deposited on a substrate by vacuum evaporation or cathode sputtering in thicknesses about 2 orders of magnitude less than those of thick films. The dielectric layer may be initially deposited as a metal and require an oxidation or anodization treatment to produce an oxide film.

MOS (metal-oxide semiconductor) or MNOS (negative-channel metal-oxide semiconductor) capacitors are discrete-component versions of capacitors incorporated into integrated circuits. A single-crystal silicon wafer is gold diffused on one face for conductivity and provided with an oxide film (or oxide-nitride film) on the other. The oxide is then coated with a 1-μm-thick evaporated film of aluminum. However, these are not, strictly ceramic materials and are not discussed further.

Ferroelectric thin-film memories have recently been developed as components for random access memory applications (RAMs). Films fabricated from $KNbO_3$ or $PbZr_{0.6}Ti_{0.4}O_3$ (PZT) have been the major ferroelectric materials investigated [22,23]. The films are prepared mainly by sputtering, but in the latter case, sol-gel techniques have been applied as well. The sol-gel-derived PZT films (from solution precursors) are typically postdeposited onto standard silicon IC (integrated circuit) substrates by spinning, followed by heat treatment as low as 450°C, to effect densification. PZT films have been the most widely developed and film thicknesses typically are in the range of 100–300 nm. Mechanisms associated with switching times and switching kinetics in these films are reasonably well understood, as are the failure modes that cause aging (fatigue and reduced memory retention) in these materials. The failure modes are associated with domain pinning, caused by the trapping of various point defects on grain boundaries and domain walls [25]. *Depinning* of the domains in the *nominally* 5-V-switched materials can be effected with a higher voltage pulse ($\sim$10 V, but below the breakdown voltage), which is applied periodically to the device after a given number of duty cycles ($\sim 10^6$). In consequence, performance characteristics for the ferroelectric random memory devices (FRAMs) can be maintained at an acceptable level of reliability.

    A significant advantage of the ferroelectric memories is their radiation hardness. This is combined with high speed (30 ns read/erase/write operation), 1-3V standard silicon logic (a function of film thickness), very high bit density (2 $\times$ 2 μm cell size), and complete nonvolatility (no standby power needed). Although still in the niche market phase, FRAMs are expected to significantly impact applications involving electrical read-only memories (EEPROMs).

1. Thin-Film Capacitors

Thin-film capacitors figure only modestly in the discrete-component market. Most commonly they are seen in hybrid circuits that employ resistors and conductors also made by thin- and/or thick-film methods and active device chips. The nature of thin-film processes dictates batch processing, so the product tends to be more costly than ceramic wafer or thick-film components. Consequently, thin-film capacitors are most likely to be found in applications where compactness and reliability are important, operating voltage is low, and the limited range of capacitance values and higher cost is acceptable.

    Silicon monoxide, a common vacuum-deposited thin-film dielectric, has room temperature permittivity around 6 and is typically deposited in 5- to 30-μm thicknesses. The deposited material can contain silicon and silica as secondary phases. Deposition conditions govern the proportions of the three phases and the net capacitor characteristics. Evaporated aluminum on a glass or glazed alumina ceramic substrate makes the most common base electrode and is also used as a counterelectrode in approximately 5-μm thicknesses. Typical characteristics of such a structure would be 0.01 μF/cm^2, 25-V working voltage, and 25Ω-F RC product. The temperature coefficient of capacitance (TCC) is 100 ppm/°C or less over the capacitor operating temperature range. Details of the procedure for depositing smooth SiO for dielectric use tend to be customized, but evaporation source temperatures can be as high as 1600°C. As deposited, the layered device may be significantly stressed, and annealing below 700°C is done to reduce this and stabilize the electrical characteristics. Pinholes are a cause of failure by shorting; therefore, areas larger than 1 cm^2, which might be too defect susceptible, are not commonly made. SiO films as produced are bipolar (i.e., function as capacitors in either polarity).

On the other hand, sputtered or evaporated and anodized tantalum oxide provides a unipolar capacitor, necessitating for some applications two capacitors to do the work of a single bipolar. Glass or glazed alumina substrates provide the necessary smoothness and high-temperature resistance.

In one published procedure [26] a 4-μm-thick tantalum metal film is first deposited by cathode sputtering on a lapped or glazed 99% alumina substrate. The geometry of the sputtered device is defined by use of a metal mask or by removing undesired material in a photoresist mask-and-etch operation. The tantalum film is then made the anode in an electrolyte solution containing a platinum cathode. Anodization is conducted for approximately 30 min including a period of reversed current to improve the quality of the Ta_2O_5 film. A counterelectrode of gold or palladium-gold is applied in the same manner as the base electrode. Such a capacitor, formed at 200 V, has a capacitance density of 0.06 μF/cm^2 and TCC of 150–250 ppm/°C. The sixfold increased capacitance density over that mentioned earlier for silicon monoxide is due to the higher permittivity (ε_r = 21.2) of anodic tantalum oxide. Numerous possibilities exist for modifying the capacitor construction and the dielectric material composition. For example, either of two crystalline forms of tantalum can be produced, which react differently to anodization and to carbon and oxygen additives.

A major rationale for the use of these thin-film capacitors is that their positive TCC is opposite in sign to, and of about the same magnitude as, the temperature coefficient of resistance of thin-film resistors. In combination they can then be used to construct precise, stable RC circuits.

Titanium oxide, silicon dioxide, aluminum oxide, and aluminum silicate make alternative films suitable for capacitors. Sometimes the high dielectric constant of tantalum oxide is tempered by applying a second film, such as SiO, to give a duplex dielectric. This is done to provide precise lower capacitance values without having to deviate from an adopted geometry. Also, a duplex film is less vulnerable to pinhole defects than a single layer. This type of capacitor would have an evaporated gold counterelectrode and capacitance density of about 0.01 μF/cm^2.

Thin films of titanate ceramics have been prepared by evaporation or reactive sputtering, using both metal and oxide precursor materials. Potential applications include the developing ferroelectric memories and sensors.

2. Thick-Film Capacitors

Designers of thick-film hybrid circuits have the option of using prefabricated discrete capacitors or of constructing them in place on the circuit substrate using the same screen-printing process and equipment that are used for the resistors and conductors of thick-film hybrids. The convenience and low cost of thick-film capacitors are offset, however, by their limited range of capacitance values and working characteristics. This is due to constraints on process conditions imposed by the nature of thick-film processing. A thick-film capacitor is composed of substrate, screened base electrode, dielectric layer, and counterelectrode. Ideally, each layer is processed at a lower temperature than the one beneath, limiting interaction and preserving the dimensions and compositional identity of each. With some materials systems one attempts to avoid interaction of the layers despite the use of a single firing for the entire capacitor, or for the dielectric and counterelectrode. Process temperatures are in the range 600–900°C. Dielectric pastes maturable within this range incorporate, as do the conductive and resistive pastes, ingredients of low-melting glasses. When the glass is nonreactive and has a typically low dielectric constant, its volume fraction must be minimized so that the high-permittivity phase is continuous and a maximal capacitance level is maintained. The reason for this constraint is discussed in Section IV. Yet, for cohesion of the dielectric and a strong bond to the substrate, at least 10 vol % of the bonding material is required. Another approach is to employ the glassy material in higher proportion but of a composition that will devitrify on firing. Then, if the crystalline product is itself a high-permittivity material, a fairly high net permittivity may be realizable without sacrifice of the mechanical bond.

   An example of the interaction problem is thick-film capacitors fabricated from a $BaTiO_3$-based ($\varepsilon_r = 9000$ ceramic Y5U) dielectric with 10% of a glass frit of molar composition $60CdO-20B_2 O_3-20SiO_2$ [26]. Fired at 900°C into discrete disks, the mixture exhibited permittivity of about 40. The same mixture was used as a thick-film capacitor dielectric in the following manner: A silver paint bottom electrode was screened and fired on 96% alumina substrates. The dielectric paste and a counterelectrode were then applied in successive screenings and fired in one operation under the same conditions as previously. The resultant capacitance corresponded to about 40% lower effective permittivity (i.e., $\varepsilon_r \sim 25$) as a thick-film capacitor than as a sintered, subsequently electroded disk. This is attributable to reaction during cofiring of the dielectric and the counterelectrode. The temperature coefficient of capacitance was less than $\pm 20\%$ over -55°C to $+125$°C, and room temperature insulation resistance was about 10^{12} Ω.

   The patent literature is replete with examples of thick-film capacitor materials compositions and processing [27–30]. Current art is typified by U.S. patent 3,666,505 [29]: A calcined mixture of $BaTiO_3$ and 1–5 mol % of such oxides as ZrO_2, Al_2O_3, SiO_2, or Nb_2O_5 is mixed with 1–5% of Fe_2O_3 and 1–5% of lead bismuth borosilicate glass. A double application of the paste is applied to a pre-fired platinum-gold electrode on an alumina substrate. The dielectric and a Pt-Au counterelectrode are fired together at 1050°C. A dielectric constant of 1400 and dissipation factor of 0.025 are typical. The ε_r value varies less than 20% over −25°C to +150°C.

B. Single-Layer Discrete Capacitors

Figure 5 illustrates the common ceramic disk capacitor. Disks, rectangular plates, and tubes with Electronic Industries Association class 1 and class 2 performance are manufactured by substantially classical ceramic methods, highly automated. The powdered raw materials are batched and mixed, then pressed, cast, or extruded into the desired shape. Binders and plasticizers of types and in concentrations appropriate to the forming method are added to provide formability and green strength. Sintering to a dense ceramic is a final dielectric fabrication step, followed by application of electrodes and leads and packaging or protective coating. Class 3 capacitors receive somewhat modified handling and will be taken up separately in Section VII.D.

The dielectric powder materials can be purchased ready for batching and forming or prepared from the constituent oxides or their precursor compounds.

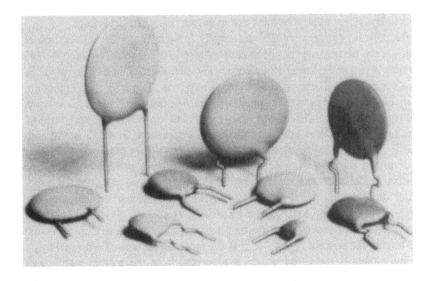

Figure 5 Ceramic disk capacitors. (Courtesy of Centralab Inc.)

In the latter case it is customary to provide a lower temperature calcination heat treatment followed by crushing and/or ball milling before the final forming step. Where the final dielectric properties depend on a particular, nonequilibrium mixture of phases in the final product, the selection of starting raw materials (i.e., whether titanate compounds or the elemental oxides or carbonates, etc.) and the specific conditions of this part of the process are quite critical.

An important trend in fine ceramic technology is to employ special powder preparation methods based on chemical synthesis which produce fine, nonagglomerated material with narrow particle size distribution [31–33]. Such powders lend themselves to sintering at lower temperatures than are normally required for densification and produce a ceramic of very uniform grain structure. The implications of this capability with respect to dielectric materials can be profound; the ability to produce a controlled and reproducible powder alone should be of great importance.

When disks and plates are to be pressed, an organic binder/lubricant is included in the pressing powder. For very large quantities of material, spray drying produces powder suitable for pressing without an excess of fine fractions. The compaction into disks or wafers is done in steel or carbide dies in high–speed automatic presses. In a less common form of powder compaction, the powder is sprinkled continuously into the ''pinch'' of a pair of steel rolls and a continuous sheet of compacted material emerges. This process is best suited to thicker slabs and disks. Extrusion of an appropriately plasticized powder mass into strip, tubes, or rods is another continuous process, which is then followed by baking to a hard condition and cutting into the desired lengths or slices. However, such extrusion process is limited to relatively thick components, e.g., disk capacitors.

Because multilayer ceramics are mostly formed by continuous casting of ceramic sheets, that process has been adopted also to the production of capacitors comprising a single thickness of dielectric. The cast sheet is then cut or punched into wafers or disks. These processes are all based on a ceramic slurry that contains, in addition to the powdered dielectric material, organic binder(s), a solvent, and plasticizers. These ingredients are mixed and dispersed by ball milling. Acrylic resin systems are commonly used. The suspension, de-aired, is cast continuously onto a moving substrate and drawn under a doctor blade, or the doctor blade is drawn over the slurry. In either case, a precise thickness of the cast suspension is defined. The most common substrate is a silicone-treated paper strip and various polymer sheets such as polyethylene. An earlier design of one such piece of equipment is described by Howatt et al. [34]. Modern equipment is of precision construction, casting tape automatically at rates of 1–10 ft/min and removing the solvent rapidly so that the tape can be wound onto reels.

Subsequently, the ceramic tape is separated from the substrate and disks or wafers are punched or cut from the sheet, which is described as being in a ''green'' or ''leather-hard'' condition. An interesting but little used variation is to wind a strip of the green dielectric onto a mandrel in a tube that ultimately serves as a cylindrical capacitor [36].

Ceramic firing of dielectrics of the type described is done in air atmosphere at temperatures of 900–1300°C; an arrest in the heating schedule at 300–400°C permits elimination of the organic contents.

Electrodes are applied to single-thickness capacitors in a subsequent heat treatment at a lower temperature so that there is no interaction with the dielectric. Thus, the sintering requirements of the electrode material and dielectric material can be addressed separately. Electrode materials will be taken up later in this chapter.

C. Multilayer Capacitors

The largest class of ceramic capacitors produced, in numbers and in value, is the multilayer type. Relative to a capacitor with a single thickness of dielectric, multilayer construction affords greater capacitance density at the cost of lowered operating voltage. Small multilayer units or chips are physically compatible with solid-state semiconductor components in hybrid circuit construction, where the low-voltage rating is not of consequence. Figure 6 is a cut-away view of a multilayer capacitor. Although there are still a number of leaded ceramic capacitors, they are mostly used in specialty applications (e.g., high voltage or very large value capacitance).

1. Basic Multilayer Fabrication Methods

The multilayer ceramic capacitor was initially the result of efforts to replace mica with ceramic dielectrics. In mica capacitors, thin mica splittings are electroded,

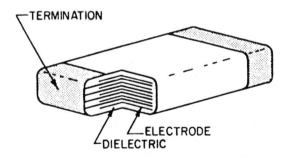

Figure 6   Cutaway view of multilayer ceramic chip capacitor. (Courtesy of Vitramon, Inc.)

stacked, and electrically connected in parallel. The first mica substitute materials to be made into multilayers were low-melting and devitrifying glasses developed by DuPont during World War II [36,37].

Two general fabrication procedures were employed in this early work: (1) Alternate layers of slurries of ceramic dielectric powder and of silver powder are deposited on a substrate by successively spraying or casting the dielectric slurry and screen printing the electrode. Following drying and sintering, a monolithic laminated structure is obtained whose laminae can be connected in parallel to form a capacitor. (2) In a related process, the dielectric is cast into individual sheets as described in Section VII.B, dried, and imprinted on one side with silver electrode paint. Following drying, the electroded green strips are stacked, compacted into a slab, fired, and parallel connected.

The low-melting dielectric glasses for these components derived compositionally from vitrifiable enamels used in decorating bottles and other glassware. These materials consist of inorganic colored pigments dispersed in and ultimately bonded by low-melting glasses based on lead borosilicate. Melting in the low range 500–800°C, the vitrifiable colors can be matured without deformation of the glass articles to which they are applied. More important from the monolithic capacitor standpoint is that silver powder makes a compatible electrode for use with these glasses, remaining solid while the glass dielectric fuses, but being close enough to its 960°C melting point to sinter readily into conductive layers.

A 1946 patent of Deyrup [38] provides the following example of the composition (in mol %) of one of these vitreous materials: 23.2 PbO, 44.0 SiO_2, 2.7 K_2O, 2.5 Na_2O, 2.3 Li_2O, 10.4 NaF, 8.7 MgO, and 6.2 SrO. The composition is formulated from oxides, carbonates, and fluorides, is melted at 1000–1200°C and quenched in water, is pulverized into fine powder, and is sintered into its finished form at 740°C. The material is described as having a 1-MHz loss tangent of 2 $\times$ 10^{-4} and a permittivity of about 10. The glass crystallizes to a large degree during the 740°C firing.

An interesting aspect of this dielectric material is emphasized in the patent—the low dielectric loss despite the substantial content of alkali-metal ions. The latter traditionally are regarded as mobile and agents of dielectric loss. In the light of current understanding the contradiction is attributable to the lead ion or to pairing of alkali modifier ions so that their usual mobility in the glass network is greatly reduced.

Modern versions of this 35-year-old dielectric composition are still in use. The capacitors are, in fact, mica replacement units characterized by permittivity around 10, low dielectric loss, high insulation resistance, small temperature coefficient of capacitance (90 ppm/°C), and maintenance of these characteristics to 100 MHz or higher. The partially devitrified glass is white and is sometimes referred to as a porcelain dielectric. Actually, in composition, microstructure, and electrical properties, such materials are far removed from the classical clay-flint-feldspar porcelain.

The driving force for ceramic multiplayer capacitors has been the introduction of SMT and the automatic assembly of printed circuit boards using this method. In addition, advances in semiconductor technology have resulted in lower and lower operating voltages. This trend is coupled with increasing emphasis on portable, battery-powered devices where small size and low-voltage operation are desired. The ceramic capacitor, which started as the rather large 2220,* has been continually reduced in size.

Starting from this development, the evolution of the multilayer capacitor has been in response to two needs: (1) dielectric materials, which enable greater capacitance density and reduce component size; and (2) lower cost electrodes. Thus, starting with the glass-based dielectrics, which were low melting and could be cosintered with silver electrodes, the trend was to higher permittivity titanates—which, however, necessitated higher sintering electrode material combinations of platinum, palladium, and gold. As the costs of these metals mounted during the 1970s and 1980s, ceramic fabrication stratagems were devised to reduce the electrode materials costs. Such stratagems included (1) simple minimization of the quantity of metal employed in an electrode by optimizing its composition, paint characteristics, and application technique; (2) use of glass-derived dielectrics which fire at sufficiently low temperatures to be compatible with a silver-palladium alloy of as much as 90 wt % silver content; (3) use of base metal electrodes, enabled either by employing firing atmospheres reducing to the electrode metal or by infiltrating the capacitor's electrode regions, after ceramic sintering, with the molten base metal; (4) use of relaxator-type dielectrics, many of whose compositions have firing temperatures low enough for use with silver-palladium; and (5) further development of the grain boundary barrier layer capacitor to provide multilayer-equivalent capacitance density, or greater, in single-layer ceramics with silver or base metal electrodes. Furthermore, there is effort to combine multilayer and barrier layer technologies. These variations on the multilayer capacitor concept are now taken up individually.

(a) *Lamination Stacking.* Thurnauer [2] describes the observation by U.S. technical teams in Germany after the close of World War II of a pilot plant layout for producing thin ceramic sheet for capacitors by continuous casting and doctor blading. A similar method was under development in the United States during the war and was patented in 1949 [35]. The method was used to produce thin sheets of ceramic barium titanate and of a number of related materials: TiO_2, $MgTiO_3$, $CaTiO_3$, $SrTiO_3$, and $BaTiO_3 + MgZrO_3$. The process described in the

---

* 2220 is a standard EIA classification where $xxyy = 0.xx$ in. $\times$ $0.yy$ in. As a result, 2220 = 0.22 $\times$ 0.20 in.

patent and in the 1947 Howatt et al. paper [34] contains the basic elements of modern titanate multilayer technology.

The solvent-resin system employed in this process, which has enjoyed remarkable longevity and had many offspring in the capacitor technology, was composed as follows: (1) A polymer mixture consisting of 78.9 wt % Albalyn polyester (methyl abietate), 15.8% Ethocel 50 (ethylcellulose), and 5.3% DuPont M95 resin, a natural varnish. (The reasons behind this complex resin mixture may have been to maximize mutual solubility, to provide the necessary strength and flexibility to the film, and to spread pyrolysis during firing over a wide range of temperature.) (2) Diethyl oxalate plasticizer—13.2% of the total resin content. (3) Staybelite (triethylene glycol dihydroabietate), a cross-linking agent for the Albalyn used in equivalent amount as the polyester. (4) Toluene as the system solvent.

These organics were mutually dissolved, with heating, and ball milled with the dielectric power at an appropriate milling fluidity. For titania, the ratio of organic medium to dielectric powder was 34 wt %. Following milling and de-airing, the suspension was permitted to flow through a slit onto a moving stainless steel belt that carried it under infrared drying lamps. Firing of lengths of the strip (100–800 μm thick) was at 1350°C for 4 h. In general, it was the practice to apply electrodes by firing on a commercial silver paint at 600°C. However, a momentous step into multilayer technology is described as follows [34]:

> It was also found possible to combine the firing of the ceramic pieces and the electrodes. ··· The dry sheets (4 × 4 cm squares) were coated on both sides with a gold-platinum paste ··· and stacked several ··· squares high. ··· These were fired successfully into one dense block in a single operation. ··· The objection to the one-fire process was that the high firing temperature of the ceramic body necessitated using the expensive noble metal paste.

This experiment in effect teaches a current manufacturing process. In it layers of green dielectric are individually electroded, stacked, and pressed, with or without heat, into a slab or block. The backed or calcined blocks are cut into individual multilayer capacitors ready for final firing. Accurate alignment of the electrodes of the stacked laminae is necessary for maximal capacitance and can be problematical when the thin, deformable green ceramic sheets are handled. This basically is the multilayer capacitor process in major current use. Similar processing is also used for the construction of electronic modules in which as many as 50 layers of alumina ceramic substrate bearing thick-film resistors and conductors are laminated into a block with conductive "vias" linking the layers (see Chapter 10).

As improvements in milling, dispersion of the milled powder, and casting were made, it has been possible to improve the film quality and reduce the grain size of the fired ceramic, resulting in a reduction of the dielectric thickness re-

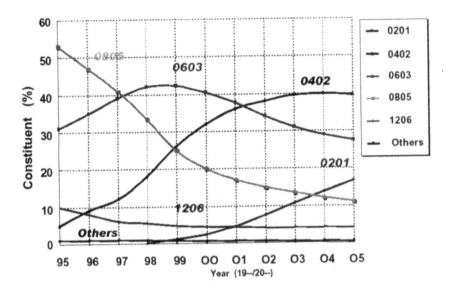

Figure 7 Actual and projected market share for various types of capacitors, showing the marked trend toward smaller, denser capacitor units.

quired for a certain applied voltage. The size evolution of ceramic capacitors can be seen in Figure 7. Currently 0402 and 0201 (0.020 × 0.010 in.) capacitors are being handled, placed, and soldered on in production quantities. Typically the height dimension of ceramic capacitors is no more than the width and in many cases considerably smaller than the lateral dimension. Otherwise there are problems with picking and placing the capacitor in the proper orientation. The phenomenon of tombstoning, where the capacitor stands on end during soldering, is caused by an imbalance of the surface tension forces at each end of the capacitor as the solder is melted. Although it would be expected that smaller capacitors are less subject to this problem because of the very small distance between electrodes, the much smaller mass of the smaller sizes renders them very sensitive to very small differential forces.

Figure 7 gives the family of curves that show the evolution of sizes versus years. Current state of the art is for fired dielectric layers to be 2–5 μm thick and the number of layers can be in excess of 300, resulting in very high volumetric efficiencies. This method of fabricating ceramic green (unfired) sheet, screen printing electrodes, stacking/laminating, and firing lends itself to highly automated manufacturing efficiencies and tight process control.

Typically 'green sheet' is cast using either a doctor blade or a vertical casting method. The doctor blade approach uses a horizontally moving substrate, and the thickness of the cast film is controlled by a knife edge that is separated by a controlled distance from the substrate. Control of film thickness in vertical casting is accomplished by careful control of the rheology of the ceramic slurry, speed of the substrate as it passes vertically through the slurry, and by temperature and humidity of the casting environment. In general, it appears that vertical casting results in higher casting speeds.

Fabrication of multiplayer ceramic capacitors by this methods is usually done in large sheets, and the capacitors are singulated prior to firing by sawing or knife-edge separation using methods and techniques that are proprietary to the manufacturers. An edge margin is necessary on the individual capacitor to prevent the exposure of internal electrodes and to give sufficient tolerance for efficient manufacturing. While such edge margins can be easily achieved on large parts with minimal effect on the electrode area, as volume manufacturing has moved to the 0603 (0.060 × 0.030) and smaller sizes, stacking and laminating of the individual layers requires the use of intelligent (e.g., vision assisted, or precise mechanical registration) stacking equipment. Although the methods mentioned below may still be used in some applications, green sheet casting appears to be the primary method for fabrication of ceramic multiplayer capacitors, with some manufacturers making more than 15–20 billion pieces per month by this method.

(b) *Spray Deposition.*   The limiting thinness of dielectric elements in a capacitor made as just described is imposed by the tendency of the resin-bonded dielectric sheets to deform and misalign the stacks. Alternative processes avoid this problem by casting, spraying, or dip-coating layers of dielectric slurry onto a platen or substrate.

The spray process was first described in connection with the devitrifying glass-type dielectric [41]. Subsequently, at least one U.S. manufacturer has applied it with titanate ceramics and noble metal electrodes. In the process, platens pass successively under a spray head that deposits dielectric layers, through a dryer to a silk-screening station where a multiple-capacitor electrode pattern is applied, through a dryer, and back for the next spray coat.

(c) *Buildup Process.*   In place of spraying, a process successively casting thin layers of ceramic slurry onto a platen with intermediate silk screening of electrodes is also practiced [42]. The dielectric slurry has flow characteristics that permit casting a 250-μm-thick layer repeatedly. Following more or less superficial drying, the cast surface is smooth and receptive of the silk-screened electrode paint pattern.

In modern monolithic capacitor production, methods of the types described produce hard, dry, green dielectric and metal laminate stacks measuring 50,000–100,000 mm^2 in area. From these are cut or sawn individual capacitors ranging in area from approximately 150 to 2 mm^2. The potential for producing large numbers of parts is obvious. Depending on the electrode configuration printed into a stack, individual capacitors can be cut out with exposed electrodes at the ends, to be subsequently metallized and soldered, or with a margin of dielectric material all around so that the electrodes are entirely internal. In the latter case it is then necessary to drill lead holes or otherwise gain access to the opposing sets of electrodes.

2. Infiltrated Electrodes

Rutt [43–45] developed an alternative method that makes it possible to have in one component multilayer construction, a high-sintering "conventional" barium titanate-based dielectric, and low-melting base metal electrodes. This is done by first accomplishing ceramic maturation and subsequently putting the electrodes in place. In the laminate assembly step, laminae of the dielectric material are alternated with layers of the dielectric material precursor compounds rather than electrode material. When these are cofired, the dielectric sinters to a high density while the precursor layer materials combine into chemically similar compounds but in a very porous state. In addition, the organic binder content of the porous layer is made greater to enhance its ultimate porosity, and additional pyrolyzables, such as carbon, can be included for the same purpose. Following firing, the multilayer structures are evacuated, then immersed in a molten low-melting solder-like alloy under 10–15 atm pressure. This causes the alloy to intrude into the porous layer regions to provide electrodes.

The following example composition is from the second-cited Rutt patent [43].

Composition of the dielectric strata: 93 w/o $BaTiO_3$ and 7 w/o $Bi_2O_3 \cdot ZrO_2$, mixed and uncalcined, both materials about 1.5 μm average particle size. 100 g of this powder is mixed with 4 g ethylcellulose binder, 3 g butyl benzyl phthalate, 2 ml acetic acid defoamer, and 30 ml dichlorethane solvent.

Composition of the porous strata: 66.9 w/o $BaCO_3$, 27.1 TiO_2, 3.32 Bi_2O_3, 2.64 ZrO_2. The TiO_2 has an average particle size of 5–10 μm, the other solids 1–2 μm. One hundred grams of this dielectric power is mixed with a medium containing 14.5 g acrylic resin, 1.3 g ethylcellulose, 1.6 g lecithin dispersant, and 83 ml pine oil. The unreacted precursors, the high binder content, and the relatively large particle size all lead to the high porosity of these layers following firing. The two different solvent and binder systems help maintain the individualities of the two layers. The dielectric layer slurry is cast into 50-μm-thick sheets, and an electrode pattern is silk screened onto them using the porous layer paint. The layers are assembled into a multilayer stack and pressed in a die at 85°C and 28 kg/cm^2 to consolidate the stack. The resultant slab is then cut into individual capacitors with the porous layer edges exposed on at least two sides. The high content of binder material is eliminated in an 86-h heating to 420°C; the temperature is then raised more rapidly for final firing at 1260°C.

In a variation of the process, the multilayers are first impregnated with silver nitrate solution. This is thermally decomposed in place to provide a silver electrode prestructure, which is then more readily wetted by the solder-like alloy.

Example electrode alloy compositions range from pure lead to alloys of lead, aluminum, copper, zinc, bismuth, tin, cadmium, and others. The capacitors and electrode metal are maintained apart from one another in a closed vessel heated above the melting temperature of the metal. A rough vacuum is applied, the capacitors are immersed, and the pressure is raised to about 14 kg/m^2 to force the metal into the porous ceramic [45].

Currently this method offers no advantages in the fabrication of ceramic capacitors. The reduction in thickness of dielectric and electrode layers as well as the overly complicated fabrication process make infiltrated electrodes an interesting but commercially unuseable development. In addition, the electrode material of choice was lead or a high-lead-containing alloy, which is unacceptable because of environmental concerns with lead.

3. Low Sintering Temperature Dielectrics

Another approach to less costly construction materials in multilayer capacitors is to use silver-based electrodes in conjunction with dielectric ceramics having intermediate sintering temperatures. Such compositions require higher firing than the fusible glasses already described, but lower than conventional titanate materials. That is, they are sinterable in the range 900–1150°C.

(a) *Electrode Alloys.*   As introduction to dielectrics with lower sintering temperatures, it is appropriate to digress to the subject of electrode materials for ceramic dielectrics in general. The original application of the classes of materials that now serve as electrodes for ceramic capacitors was for decorating dinnerware and for bonding metals to ceramics or glasses. Gold, platinum, silver, and palladium, in the form of finely divided powders, in admixture with low-melting glasses have been used individually and in various combinations for these purposes. For multilayer capacitor electrodes the glass content is generally omitted and the metal powders must be heated to the range where atomic mobility is sufficient to produce sintering and densification in a reasonable time. Sintering temperatures of platinum, palladium, gold, and silver are about 1250, 1100, 750, and 650°C, respectively. The densification of platinum or combinations of platinum-palladium or platinum-gold therefore takes place compatibly with the firing of titanate ceramics, and these were, in fact, the original electrode metal combinations of choice. Silver or its alloys with gold, palladium, or platinum would be appropriate with lower-firing ceramics or, with very little or no alloying, with low-melting crystallizing glasses.

Because sintering shrinkages of the metal and ceramic powders differ, delamination of the multilayer structure would be commonplace in the absence of any but mechanical bonding. Fortunately, the noble metal powders, especially Pt and Pd, undergo surface oxidation and reaction with the ceramic to provide a chemical bonding mechanism. This is particularly effective in titanate capacitors with nickel electrodes due to the latter's partial oxidation and solution in the titanate (see Section VII.C.4).

The reactivity of dielectric with electrode necessitates empiricism in the formulation of dielectric compositions. The dielectric properties are altered by the interaction. The magnitude of the effect depends on the materials, the ratio of electrode thickness to dielectric thickness, and the conditions under which they are sintered. Whereas the base metal case has been quite thoroughly documented, the effect in noble–electroded capacitors, where not cataclysmic, has tended to be discounted. However, Figure 8 illustrates, the effect of reaction during air firing of various noble metal electrodes, screen printed, with a 50-μm-thick BX dielectric. The intrinsic dielectric behavior is represented by the curve for the dielectric with sputtered silver electrodes.

In practice, electrode metals are used in the form of fine, generally submicronmeter-diameter powders produced by precipitation with reductants from aqueous solution. All binaries in the system Pt-Au-Pd-Ag form continuous solid solutions, so that binary combinations are available with sintering characteristics

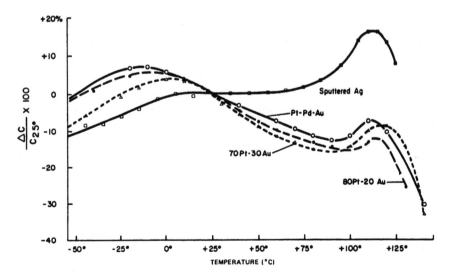

Figure 8 Effect of electrode-dielectric interaction on capacitor characteristics. (From Ref. 47.)

to suit a range of ceramic processing. The binary metal combination can be simply a powder mixtures, but if coprecipitated under suitable conditions can be indistinguishable from, and is described as, a solid solution [48]. The powders are suspended in a solution of one or more organic resins with a net rheology suitable for silk screening.

In the Ag-Pd binary system, compositions with palladium as the minority alloying element provide a desirable combination of low cost and high liquids temperatures [49]. Such electrode compositions are always formulated with concern for "silver migration," an electrolytic phenomenon associated with the conjunction of silver, electric field, and moisture [69]. In this process, silver goes into water solution at the anode, migrates to the cathode, and plates out in dendritic form. Similar, related mechanisms operate whether the silver is initially present as exposed metal, is encapsulated in a glassy matrix, or is dissolved in a borosilicate glass. According to Miller [49], the rate of silver migration is one-tenth as fast in an 80:20 silver-palladium alloy as in silver.

An effect of the trend to use of palladium in electrode alloys has been to displace bismuth from many dielectric formulations. Compounds such as bismuth stannate or bismuth zirconate have had wide use as Curie peak suppressors [15]. However, reaction of the bismuth with palladium to form Bi-Pd intermetallic compounds occurs below the titanate capacitor sintering range [49]. The reaction is exothermic and disruptive of the not-yet-coherent multilayer structure. When the upper temperature stability limit of PdO is depressed by a low ambient oxygen partial pressure, the Bi-Pd interaction effects in a capacitor are ameliorated. This is believed to be due to preferential solution of the palladium in silver [50] and suggests a way of circumventing this electrode–dielectric materials incompatibility.

An obvious step further in the direction of low-cost electrode materials is to make use of base metals. Developments in this direction are taken up below. Attempts to employ combinations of base and noble metals have not been followed by significant use [51].

(b) *End Termination Materials.* In addition to the material that comprises the electrodes of a multilayer capacitor, another metallization is used to connect its two sets of electrodes in parallel. This material is applied to the exposed electrodes at each end of the capacitor. It may consist of the electrode material itself, cofired with the capacitor. More commonly, however, a low-cost silver-frit combination is applied to already fired capacitors and subsequently fired at a lower temperature, (600–900°C) than the capacitor internal electrode. Such material is vulnerable, in subsequent mounting and processing, to dissolution by molten solder combination. Protective platings, a barrier layer of copper or nickel, or coatings of silver-bearing solders are sometimes employed to prevent this [52].

In the case of nickel barrier layer, it is necessary to plate an additional layer of pure tin or a lead-tin alloy to enhance the wettability of the end termination in subsequent soldering procresses.

(c) *Glass-Reacted Capacitors.* To permit substitution of low-cost silver-based electrode alloys for platinum or palladium-based alloys, a dielectric densi-fying at about the sintering temperature of the alloy is needed. For example, an alloy of composition 70% Ag-30% Pd melting at 1225°C would sinter in the range 700–900°C. Maher [53–55] and others have developed titanate compositions that meet this requirement through the use of glass formers that foster liquid-phase sintering. The novelty of these compositions is that ceramic sintering at the desired low temperature is accomplished with a small enough content of the glass-forming constituents that the high permittivity of the titanate constituent(s) is not substantially compromised. An example of the compositions used for a high-permittivity dielectric is as follows:

1. A low-melting glass of composition (in wt %) 36 CdO, 23 Bi_2O_3, 25 PbO, 5 ZnO, 1 Al_2O_3, 5 B_2O_3, 5 SiO_2 is prepared by mixing the constituents, melting in a platinum crucible at 980°C, crushing, and ball milling in water to less than 1 μm average particle size.
2. A "barium titanate" (presumably a zirconate- and strontia-modified $BaTiO_3$, to shift and broaden the permittivity versus temperature characteristic) is mixed and reacted at 1350–1400°C, then crushed and milled.
3. The products of compositions 1 and 2 are combined in proportions up to 10 wt % glass, fabricated into thin sheets, and fired at 1100–1150°C for 1 h.

The resultant ceramic has an ε_r of 1550 and a dissipation factor of 0.015–0.017 at 1 kHz using a 40 V/mm measuring field. The temperature coefficient of capacitance is ±10% over −55°C to 125°C.

Maher [55] describes compositions in which the same glass phase is used with dielectric combinations from the system BaO-TiO_2-Nd_2O_3 to provide a series of dielectrics with graduated positive and negative temperature coefficients of capacitance. An NPO composition in this family has an ε_r value of 68. Again, the process sequence is such that the titanate portion is well reacted before the glass phase is introduced. Capacitor firing temperatures are no higher than 1150°C, permitting the use of Ag-Pd electrodes in the composition range of 15–40 at % Pd.

Guiding principles with the flux-sintered compositions are to minimize the content of flux to begin with and to select for fluxing those constituents that can be molecularly assimilated by the high ε_r phase(s), thereby lessening the dilution effect due to the glass. Burn [57] has examined a variety of types of glass compositions for this purpose in the system $CdO\text{-}Bi_2O_3\text{-}PbO\text{-}B_2O_3$, measuring their efficacy by the ability to cause grain growth during sintering and in the process to be homogeneously assimilated so that they do not remain to function as diluents. SiO_2 and GeO_2 were found to be useful alternative glass formers for this purpose; ZnO, Cu_2O, and Li_2O are alternative glass modifiers also readily taken up by the perovskite lattice, whereas Nb_2O_5, Y_2O_3, Sb_2O_3, and WO_3 are glass modifiers that are less perovskite compatible.

Many of the resultant reactive glass additives, in amounts well under 5 wt %, enable the titanate ceramics to be readily densified. An example (in wt %):

> *Glass:* 1.5 ZnO, 0.56 $BO_{1.5}$, 1.5 $BiO_{1.5}$
> *Ceramic:* 96.44 $BaTi_{0.903}Zr_{0.097}O_3$

This combination is sinterable at 1100°C in multilayer structures with 30 wt % Pd–70 wt % Ag electrodes. The reacted glass flux dielectric in combination with silver-palladium electrodes in a multilayer capacitor is commercially important and in use by several manufacturers.

A somewhat differently oriented use of low-melting additives is described by Payne and Park [57]. For example, the simple combination of $BaTiO_3$ and $Pb_5Ge_3O_{11}$ is said to provide in a 950°C firing a dielectric with a permittivity of 1210 and a dissipation factor 0.012. The lead germanate flux is described as being, in amounts up to 10 vol %, segregated entirely in the grain boundaries. This is unexpected; in view of the level of permittivity, significant assimilation of the flux phase would have been more usual. A broadened composition of fluxing material is given as $(PbO)_x(GeO2)_{y-z}(SiO_2)_z$ [58], where $1 < x \le 6$, $1 < y < 3$, $3 \le (y + z)$. This type of composition is more chemically resistant than $Pb_5Ge_3O_{11}$.

(d) Glass Ribbon Capacitors. Several methods have been advanced for producing multilayer ceramic capacitors by stacking laminae of glass ribbon and metal foil and subjecting the stack to hot pressing. In the latter step, the layers are plastically compacted into a dense, coherent body and the glass is substantially crystallized into a high-permittivity crystalline dielectric.

The attraction of such a process is that technology exists for fast, continuous production of thin, uniform glass ribbon. In addition, a sufficiently low hot-pressing temperature would permit the use of low-cost aluminum foil electrodes.

The success of such a process hinges on the ability to formulate a glass with a minimum of network–forming constituents so that conversion to the crystalline dielectric approaches 100%. Compositions have been designed for this purpose [60,61], based on barium titanate, such as 30–40 (in mol %) BaO, 15–40 TiO_2, 9.5–26 SiO_2, 7–25 $AlO_{1.5}$. The high content of residual glass-forming constituents presumably accounts for the limited success of this concept. However, there is current commercial production of multilayer capacitors by a closely related method. The dielectric is instead a noncrystallizing glass ribbon with a permittivity of about 20. The electrodes are of aluminum foil and the entire unit is encapsulated in glass with the leads emerging. This product has performance characteristics in a class with the mica replacement, glass-derived capacitors with silver electrodes discussed in Section VII.C.1.

(e) *Relaxator Dielectrics.*   An additional class of dielectric materials are the relaxator ferroelectrics. As indicated, these were first studied by Smolenski and coworkers [61,62] and represent a class of materials that exhibit properties significantly different from those of normal ferroelectrics. These differences may be summarized as follows [64]:

1. Although the weak field dielectric permittivity reaches a high peak value at the Curie temperature (20,000), it is frequency dependent and shifts to higher temperatures with increasing frequency.
2. At low temperature the material is ferroelectric, but as the temperature approaches the Curie point, the hysteresis loop closes and becomes simply nonlinear. There is no sudden loss of the spontaneous polarization at the Curie point as would be expected at a first-order phase transition.
3. There is no evidence of optical anisotropy or X-ray line splitting, which would indicate a first-order transition.
4. The materials exhibit a ferroelectric-to-paraelectric transition with a diffuse character attributed to microinhomogeneities in composition. These stem from disorder of the ions occupying octahedral positions in the unit cells. The broad peak of dielectric constant versus temperature that results from the characteristics described above is of interest as the basis for a capacitor dielectric and is illustrated in Figure 9.

Relaxator behavior was early recognized as a widespread phenomenon among ternary oxide perovskites. Mixtures of two or more relaxator materials offer the opportunity to place their dielectric constant peaks, which individually are broad, so that they are influential over the entire temperature range of use of the intended capacitor. Typical of the relaxator ferroelectric materials is lead magnesium niobate (PMN, $Pb(Mg_{1/3}Nb_{2/3})O3$). The charge neutrality on the B site is maintained by having the proper ratio of magnesium and niobium. The low sintering temperature of relaxators, 900–1200°C, makes them particularly attractive because of the possibility of using high silver content electrodes. The combination of high dielectric constant and low sintering temperature has created interest in relaxators as candidates for multilayer capacitors. Studies have been carried out on a large number of compounds in the lead-based perovskites. These can be described as [65]:

$$A(B_1B_2)O_3$$

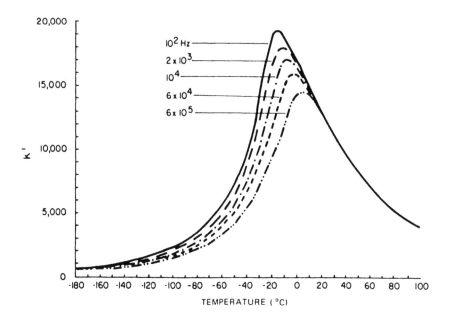

Figure 9 Relaxation behavior in single-crystal lead magnesium niobate. (From Ref. 85)

where $A = Pb$; $B_1 = Mg^{2+}, Zn^{2+}, Ni^{2+}, Fe^{3+}, Sc^{3+}$; $B_2 = Ti^{4+}, Nb^{5+}, Ta^{5+}$, W^{6+}. Iizawa et al. [86] describe compositions based on $Pb(Fe_{2/3}W_{1/3})O_3$-$PbZrO_3$ plus MnO_2 or CeO_2, which sinter at 950°C and exhibit room temperature ε_r of 5000–8000, tan δ of 0.013–0.019, insulation resistance of 10^{10} Ω-cm, and TCC over 20–85°C of less than 30°C.

Tanei et al. [86] list 99.5–99.95 wt % Pb $(Fe_{2/3}W_{1/3})_{1-x}Ti_xO_3$ plus 0.05%–0.5% MnO_2. In these compositions the purpose of $PbTiO_3$ is to provide an elevated Curie point, and MnO_2 is described as an acceptor dopant to decrease tan δ and increase resistivity. However, peak suppression as an additional function of the added MnO_2 is also possible. This ceramic, following calcination at 650–850°C, is sintered at 1000°C or lower. Room temperature dielectric properties are ε_r ~2000, tan δ $\leq$0.05, and volume resistivity of ~10^9 Ω-cm. Jang et al. [82] have combined relaxator compounds to provide in one dielectric two broad ε_r versus temperature peaks as follows: $Pb(Mg_{1/3}Nb_{2/3})O_3$-$PbTiO_3$-$Ba(Zn_{1/3}Nb_{2/3})O_3$. The heights of the maxima of ε_r versus temperature are controllable by the composition. However, the firing temperatures indicated for these materials are 1280°C or higher, necessitating the use of higher cost electrode compositions for capacitor applications.

Yonezawa et al. [65,82] reported the fabrication of $Pb(Fe_{1/2}Nb_{1/2})O_3$ ceramics having a room temperature dielectric constant of 20,000 and a sintering temperature below 1000°C. Subsequently, there has been a high level of scientific and patent activity that has encompassed almost all of the possible B-site cation substitutions. Kato and coworkers [66] prepared ceramic multilayer capacitors having a dielectric thickness of 30 μm from a composition in the pseudoternary system $Pb(Mg_{1/3}Nb_{2/3})O_3$-$PbTiO_3$-$Pb(Ni_{1/2}W_{1/2})O_3$. By introducing 0.1 wt % MnO_2 to a 70:20:10 composition that sintered at 1050°C, they reported the fabrication of capacitors having a calculated dielectric constant of more than 20,000. However, the material was too temperature dependent for normal capacitor use.

In order to better control the temperature characteristic of the relaxor ferroelectrics, Takahara [67] proposed the use of a system of macroscopically mixed crystals of PFW and PFN. They reported measurements on pressed pellets fired in air at 950–1000°C of about 12,000 and a reduced temperature dependence. The processing of relaxator materials is not as advanced as for the $BaTiO_3$-based materials. In addition to problems caused by the volatility of lead in the compound, processing using normal mixed-oxide solid-state reaction leads to the formation of a stable pyrochlore phase. Swartz and Shrout [68] have studied the reaction mechanism in detail, which appears to proceed with the formation of an intermediate lead niobate pyrochlore phase at temperatures of 700–800°C, followed by reaction of this phase with the excess lead oxide and magnesium oxide to form the perovskite (schematically shown below):

$$PbO + Nb_2O_5 + (MgO) \rightarrow \text{cubic pyrochlore}$$

$$\text{Cubic pyrochlore} + PbO + MgO \rightarrow \text{perovskite} + \text{pyrochlore}$$

Even extended calcination times resulted in the formation of only 83% of the perovskite phase. This dilution causes the dielectric properties of the relaxator phase to be considerably reduced. The authors in turn proposed a "novel" fabrication route by presintering the MgO and Nb_2O_5 components to form $MgNb_2O_6$. This intermediate compound is then reacted with PbO to form $Pb(Mg_{1/3}Nb_{2/3})O_3$. Experiments using a 2% excess of MgO resulted in the formation of 100% perovskite phase at 800°C for 4 h. The formation of the relaxator compound remains an area of active investigation both for standard ceramic fabrication [69–72] and for nonconventional preparative methods such as thermal spray decomposition [73], sol-gel processing [74], and hydrolysis and coprecipitation [75–77].

As indicated, substantial patent activity in the composition family of lead-based relaxators has been reported. These formulations, related to specific capacitor designations, are summarized in Table 1. Although there is relatively limited commercial use of the relaxator materials, their properties, in multilayer and thin-film form, have been studied and reported on extensively in the technical literature. Without defining the relaxator system, except to say that it was composed of lead-containing complex perovskites similar to those developed by Yonezawa

Table 1 Patent Relaxator–Based Dielectric Systems

Composition	Class	Inventor	Ref.
PLZT-Ag	X7R	Sprague	U.S. Patent 4,027,209 (1973)
PMW-PT-ST	X7R	DuPont	U.S. Patent 4,048,546 (1973)
PFN-PFW	Y5V	NEC	U.S. Patent 4,078,938 (1978)
PMN-PFN	Y5V	TDK	U.S. Patent 4,216,102 (1980)
PMN-PMT	—	TDK	U.S. Patent 4,216,103 (1980)
PFW-PZ	Z5U	TDK	U.S. Patent 4,235,635 (1980)
PFN-PFW-PZN	Y5V	NEC	U.S. Patent 4,236,928 (1980)
PMN-PT	Y5V	TDK	US. Patent 4,265,668 (1981).
PMN-PFN-PMW	Y5V	TDK	U.S. Patent 4,287,075 (1981)
PFW-PT-MN	Z5U	Hitachi	U.S. Patent 4,308,571 (1981)
PMN-PZN-PT	Z5U	Murata	U.S. Patent 4,339,544 (1982)
PFN-PFM-PNN	Z5U, Y5V	Ferro	U.S. Patent 4,379,319 (1983)
PMW-PT-PNN	Z5U	NEC	U.S. Patent 4,450,240 (1984)
PFN-BaCa(CuW)-PFW	Y5V	Yoshiba	U.S. Patent 4,544,644 (1985)
PMN-PZN	Z5U	STL	U.K. Patent 2,127,187a (1984)
PMN-PFN-PT	Z5U	STL	U.K. Patent 2,216,575 (1984)
PMN-PZN-PFN	Z5U	Matsushita	Japan Patent 59–107959 (1984)
PMN-PFW-Pt	—	Matsushita	Japan Patent 59–203759 (1984)
PNN-PFN-PFW	Y5V	Matsushita	Japan Patent 59–111201 (1984)
PMN-PFN-PbGe	Z5U	Union Carbide	U.S. Patent 4,550,088 (1985)
PZN-PMN-PT-BT-ST	Z5U	Toshiba	Japan Patent 61–155245 (1986)
PZN-PT-BT-ST	X7R	Toshiba	Japan Patent 61–250904 (1986)
PMN-PLZT	ZSU	MMC	U.S. Patent 4,716,134 (1987)
BT-PMN-PZN	X7R,X7S	Toshiba	U.S. Patent 4,767,732 (1988)

[65], Hishiyama and coworkers [78] reported on a series of multilayer capacitors having the Y5V temperature characteristic and dielectric constants of approximately 20,000. These capacitors, with operating voltages of 100, were fabricated in the 0850 size with a maximum capacitance value of 1 F. There are, however, a few negative factors associated with the use of relaxator materials for capacitor applications. Large electrostrictive effects present in some compositions may be reflected in instability of the capacitance value. Tungsten and niobium oxides are also more costly constituents than zirconates and titanates. Also, by definition, the relaxator character is associated with strong frequency dependencies of ε_r and tan δ at the diffuse phase transitions. With a conventional ferroelectric-based material, relaxations tend to be smaller and to occur at higher frequencies.

The low sintering temperature of the lead relaxators has prompted the use of these materials in the design of an integrated ceramic substrate with built-in resistive and capacitive elements [81,82]. Multilayer ceramic cofire technology was employed with an alumina/lead/borosilicate glass as the low-ε_r substrate that sintered at 900°C. Because of the low firing temperature, Ag-Pd and Au conductors could be screen printed in conjunction with RuO_2 resistor pastes. The high-ε_r dielectric system used was in the binary system PFN-PWN [$Pb(Fe_{2/3}Nb_{1/3})O_3$-$Pb(W_{1/2}Nb_{1/2})O_3$], which also cofired at 900°C.

A further attempt to take advantage of the high capacitance values of the relaxator materials has been to use them as thin-film capacitors or, as above, to incorporate them into multilayer ceramic substrates. Quek and Yan [79] prepared $Pb(Fe_{1/2}Nb_{1/2})O_3$ by a sol-gel technique that was spin coated onto Si wafers. Subsequent firing at 300°C for 2 h in O_2 resulted in dense films of 0.8 μm having a dielectric constant of 81. Although this value is roughly three times higher than for Ta_2O_5 films currently used in IC applications, it is lower than the results obtained in pressed pellets of the same materials. The difference can be attributed to the possible formation of an SiO_2 layer parallel to the PFN thin film and to the smaller grain sizes developed. The dielectric constant of 81 for the PFN films compares with values of $\approx$ 2000 for bulk samples at 25°C.

4. Base-Metal-Electrode Multilayer Capacitors

The concept of ceramic dielectric compositions and processing compatible with base metal electrodes (BMEs) for cofiring is fairly old [87,88]. The original intent was to eliminate the need for noble metals in multilayer capacitors. Most capacitors to that time used electrodes Ag-Pd alloys, Ag-rich alloys being used for low temperature fired dielectrics and Pd-rich alloys being used for high firing temperatures.

Although the Ag and Pd costs were not prohibitively high during the early development of BME capacitors, these precious metal electrodes contributed significantly to the overall cost of multilayer ceramic capacitors. Furthermore, they imposed an upper limit on the expansion of the capacitance range because of the high layer counts and associate high electrode costs. The palladium price was stable in the range of $80–90/oz until mid-1997 when it increased to the level $200/oz and began a steep climb to over $1000/oz in Jan 2001. Although the price for Pd subsequently declined, the instability in pricing triggered increased interest and activity in base metal electrode use.

BME ceramic capacitors using nickel as the electrode material were first investigated by Herbert in 1963 (88). Subsequently, a patent was issued in that year for his composition, which used manganese oxide to inhibit nickel reduction (87). In order to prevent oxidation of the nickel it was necessary to fire in a reducing atmosphere. Oxidation–reduction equilibrium curves, well known from geological studies (89), indicate that the added MnO_2 would be more stable than Ni, and consequently would inhibit the oxidation of Ni. Since it is a necessary condition that the electrode material have very little or no interaction with the dielectric, these effects needed to be investigated for the Ni system. Burn and Maher [90] presented analytical data, e.g., indicating essentially no solubility of NiO in $(BaTiO_3)_{0.865}(CaZrO_3)_{0.135}$ fired at 1350°C with an oxygen pressure of 10^{-11} atm. With increasing oxygen pressure to 10^{-7} atm, NiO solubility up to 0.30 mol % was found. Such conditions render Ni suitable as an electrode.

Of greater importance is the need to prevent or inhibit the reduction of the Ti^{4+} cations in the barium titanate ceramic. In order to be able to use a metallic nickel electrode, it is necessary to fire in a reducing atmosphere under temperature and atmosphere conditions that are below the Ni/NiO equilibrium line (Figure 10). [The equilibrium oxygen partial pressure reactions for Ni/NiO and Ti_2O_3/TiO_2 as a function of temperature are described in "Dielectric Materials for Base Metal Multilayer Ceramic Capacitors", Y Sakabe, T. Takagi, K. Wakino and D.M. Smyth, *Advances in Ceramics*, Vol 19, 1987, p. 105.]

Because of the close proximity of the Ni/NiO and the TiO2/Ti_2O_3 equilibrium oxygen partial-pressure boundaries, it would be practically impossible to maintain the firing conditions needed to maintain metallic nickel without reducing the TiO_2 unless a reduction inhibitor can be added. Reduction of the Ti^{4+} would

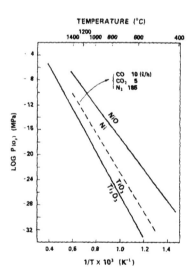

Figure 10   Equilibrium oxygen partial pressure for Ni/NiO and Ti_2O_3/TiO_2 reactions as a function of temperature. The dashed line is that of the CO-CO_2 gas mixture with N_2 carrier gas.

result in increased dielectric losses, lowered insulation resistance, and, in the extreme case, semiconducting behavior making the material unusable as a capacitor.

The stoichiometry or A/B ratio of large A-site (e.g., Ba^{2+}, Ca^{2+}, Pb^{2+}) to small B-site (e.g., Ti^{4+}, Zr^{4+}) cations has been identified by Eror et al. [93] as basic to the successful fabrication of BME multilayer capacitors. The critical lower limit of this large-to-small cation ratio has been defined as 0.95. The Error, Burn, and Maher patent [93] established 10^{-7} atm as the maximal oxygen pressure permissible with firing temperatures up to 1400°C.

The effect of dopants, impurities, and defect chemistry has been reviewed in detail by Sakabe et al. [94]. Dielectric compositions of $BaTiO_3$ doped with various acceptor-type cations (e.g., Co, Cr, Fe, Mn, Ga, and Ni) were studied in conjunction with Ni electrodes. In previous work, Sakabe and coworkers [95] had demonstrated a composition of the type $\{(Ba_{1-x}Ca_x)O\}_m \{Ti_{1-y}Zr_y)O_2$, where $1.005 < m < 1.03$, $0.02 < x < 0.22$, $0 < y < 0.20$ which resisted reduction when fired in a H_2/N_2 mixture (1.5: 100) A patent was issued on this material system in 1978 [96].

Careful control of the composition so that the divalent cations (Ba, Ca) are in slight excess is necessary to maintain the high resistivity under reducing conditions. The effect of calcium on retarding the reduction of Ti^{4+} is clearly seen in Figure 11 [95].

The period from the mid-1970s to the 1990s was a very active time in the development of BME compositions, and many patents were issued for compositional variations compatible with base metal electrodes [97–116].

Production of ceramic capacitors having Z5U temperature characteristic composition began in 1983 in Murata. These first components were leaded, epoxy-coated capacitors having a 25-V rating and 66 layers of 33-μm dielectric. Following continuous development of compositions for Ni-electrode, the Y5V temperature characteristic was introduced in 1988. Current technology in the Z5U temperature characteristic is of the order of 350 layers of approximately 3 μm thickness. The more temperature-stable X7R temperature characteristic was introduced into mass production by Murata in 1993.

Because of the continuing concern about degradation of the insulation resistance at elevated temperatures (hot IR) for X7R fired in a reducing atmosphere, a series of studies were undertaken to evaluate the effects of rare earth oxide doping on X7R dielectrics [117]. In these studies it was reported that $BaTiO_3$ doped with Dy_2O_3, CO_2O_3, MgO, MnO_2, $BaCO_3$, and glass frit while maintaining a dielectric constant of ±3500 had increased breakdown voltage and a mean time to failure, based on accelerated life test, almost 2 orders of magnitude greater than a control material not doped with rare earth oxides. The results of these and other studies of the complex defect chemistry and charge compensation within the $BaTiO_3$ system has been a stable and reliable BME ceramic dielectric.

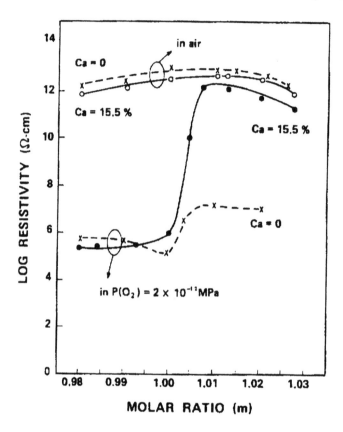

Figure 11 Resistivity of $[(Ba_{1-x}Ca_x)O]_m$ TiO_2 ceramics sintered in air and in an atmosphere of 2×10^{-12} MPa O_2 at 1400°C for 2 h as a function of molar ratio m.

Currently ceramic capacitors using BMEs are offered by a wide range of manufacturers (Murata, AVX, TDK, etc.), and they constitute more than 90% of ceramic capacitor manufacturing. The use of these systems has significantly reduced the cost and allowed the expansion of the capacitance range to $10+$ μF with 100 μF in the near future. This development allows ceramic capacitors to compete with tantalum capacitors in many applications.

D. Barrier Layer Capacitors

Ceramic capacitors of the types discussed up to this point store energy by means of electronic, ionic, and orientational polarization, with a possible inadvertent and minor contribution from interfacial polarization. The latter is evident as an increase in energy dissipation on descending through power frequencies toward dc. Increasingly, however, components are becoming available whose function is based *primarily* on polarization at interfaces between grains and/or between electrode and dielectric. By their nature these capacitors are limited to low operating voltages. The growth of interest in them derives largely from the growth of solid-state circuitry with compatible low operating voltages.

The theory of the dielectric properties of multiphase systems as developed by Maxwell [4] has already been discussed. This was expanded by Wagner [118] to cover the ac case and has been simplified for applicability to ceramic systems by Volger [118]. When the phases of a ceramic have differing conductivities, imposition of an electric field causes more charge to be removed from the more conductive phase than can be taken up by the less conductive phase. The phase interfaces thereby become charged or polarized. Outwardly this is evidenced by the flow of a decreasing current when a dc voltage is applied or by an opposite current when the voltage is lifted.

If one makes assumptions, as follows, regarding a ceramic dielectric material:

One species of grain
Uniform grain size
Continuous grain boundary
Grain boundary thickness $<<$ grain diameter
Grain boundary permittivity $=$ grain permittivity
Grain boundary resistance $>>$ grain resistance

and $d_g/d_{gb} << \rho_{gb}/\rho_g$, where d_g and d_{gb} are the grain and grain boundary dimensions, respectively, and ρ_{gb} and ρ_g their resistivities, then a simple two-layer model is applicable (Figure 12). This predicts the following dispersion behavior of the grain boundary barrier layer system below its relaxation frequency:

1. $\varepsilon_{reffective} = \varepsilon_{rintrinsic} (d_g/d_{gb})$
2. $\rho_{effective} = \rho_{gb} (d_{gb}/d_g)$
3. $RC_{effective} = RC_{intrinsic} (d_g/d_{gb})^{1/2}$

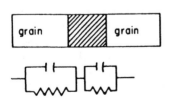

Figure 12 Equivalent circuit of grain boundary barrier-layer capacitor (From Ref. 118.)

Above the relaxation frequency, all characteristics assume the intrinsic values.

The first commercial barrier layer capacitors were described in a 1950 patent [121] as a layer of insulating titanate ceramic bearing on one side an electrode layer of reduced, conductive titanate and on the other a metal electrode. The structure was produced from a barium strontium titanate material containing 0.5 wt % of didymium oxide as a "reduction catalyst." A ceramic disk of this composition is first sintered in air, then refired at a lower temperature in air with one face in contact with a carbon slab. Effective dielectric constants of 20,000 were claimed, with resistance over 20 MΩ.

The above-described component has developed into a widely manufactured type of disk or wafer capacitor. Present practices are generally as follows. The green ceramic is fired in air to full density with insulating character. It is then heated in hydrogen or forming gas and reduced homogeneously to a semiconductive body. A thin layer at the surfaces is then reoxidized to insulating condition by heating in air. Finally, silver paste (metal plus glass) electrodes are applied to the two faces. It is possible to combine the sintering and reducing steps into one; similarly, the reoxidation and electrode firing steps are combinable into one.

Commercial capacitor products of this technology are characterized by low RC, about $0.1-1$ $M\Omega$-μF depending on size, rather high dielectric loss (tan δ of $0.05-0.10$ at room temperature, 1 kHz), and low to moderately high voltage ratings. In spite of their history of use, the physical characteristics of these components have not been as yet definitively described. Schottky barrier-like behavior has been described when one of the two junctions is removed by grinding [122]. However, dielectric layers produced by more extensive reoxidation appear to have normal dielectric characteristics. Furthermore, a reaction of the glass frit and/or of the silver of the counterelectrodes with the dielectric layer seems to have beneficial effects on the latter's voltage-withstanding ability and on other dielectric characteristics.

A subsequent development was the combination of atmosphere reoxidation of the reduced titanate with counterdoping by electron acceptor ions [121,122]. These treatments were intended to provide electrode barrier layer p-n junction capacitors, but to a greater degree, it turned out, they provide active junctions at the contacts between grains. Specific experiments with n-type $BaTiO_3$ counterdoped with undiffused copper oxide show no contribution to capacitance by an electrode barrier effect [123]. A detailed description for the first commercial production of grain boundary barrier layer (GBBL) capacitors is given by Waku [123]. To a commercial-grade barium titanate is added up to 0.5 mol % Dy_2O_3 to help reduce it to a semiconducting state and small quantities of Al_2O_3 and SiO_2 intended to promote grain growth. (For high capacitance from series-connected GBBLs, the grain size needs to be large; for high-voltage-withstanding ability the applied field needs to be distributed over the many boundaries of a fine-grained ceramic.) The green ceramic disks are fired in nitrogen. A slurry of copper oxide is then painted onto each disk face and fired in air at 1000–1100°C. Copper ions diffuse throughout the specimen, mainly via a liquid phase. Then, with silver paste applied, capacitance corresponding to a room temperature permittivity of about 20,000 is measured in a $BaTiO_3$ material whose intrinsic ε_r value is 1000. For grains of 20–40 μm, the boundary layer thickness is thereby indicated to be (20–40 μm) $\times$ (1000/20,000) or about 1–2 μm.

A succeeding development in GBBL process technology has been to incorporate both n and p dopants in the titanate lattice initially so that they can accomplish their functions in a single firing [124,125]. A typical composition is the following [128]: $Ba(Ti_{0.9}Sn_{0.1})O_3$ + 0.1 mol % Dy_2O_3 + 0.4 wt % SiO_2 + 0.5 mol % $BaCO_3$ + 0.6 mol % CuO.

Specimens of this mixture are fired in hydrogen and cooled in air. The electroding procedure is not specified. The "n" and "p" dopants, Dy^{3+} and Cu^{2+}, are both batch constituents. The dysprosium presumably occupies the Ti^{4+} sites, whereas copper confines itself to the grain boundary regions, counterdoping the n-type grains to produce insulating $Ba(Ti, Sn)O_3$. The importance of large grains is stressed, and the silica additive and barium nonstoichiometry are for the purpose of promoting grain growth.

Heywang [127] has accounted for the characteristics of GBBL capacitors in terms of a Schottky barrier between n-type conductive grains. The p-dopant content must be high enough to give a thick grain boundary region. The p-type semiconductive grain boundary has low conductivity because of the depth of the acceptor energy states available for conduction. A Philips group have refined the Heywang model to one where n-i-n, rather than n-p-n, junctions determine the capacitor characteristics [128–133].

Most recently, strontium titanate has become a commercially significant GBBL material. As a linear dielectric, $SrTiO_3$ is free of the shortcomings of ferroelectric $BaTiO_3$-based materials: the capacitance value does not age, is not voltage dependent, is not highly temperature dependent, and the dissipation factor is low. Also, $SrTiO_3$ is intrinsically capable of higher conductivity than $BaTiO_3$ [134,135]. Thus, its Maxwell–Wagner dispersion comes at about 1 GHz, a higher frequency than for $BaTiO_3$. A composition and procedure from the patent literature are as follows [136]:

Base material: $SrTiO_3$ + 0.8 wt % Nb_2O_5 (or Ta_2O_5) + 0.2 wt % SiO_2 + 3.0 wt % ZrO_2. Sinter in 99% N_2 + 1% H_2 for 2 h at 1450°C.

Fire silver paint containing a frit of 50 wt % Bi_2O_3, 10 wt % B_2O_3, 40 wt % PbO for 2 h at 1180°C. Nb_2O_5 is the n dopant. The prolonged and exceptionally high-temperature electrode firing indicates that an electrode paint constituent, presumably bismuth, serves as counterdopant. The capacitor has an effective ε_r value of 23,000, tan δ value of 0.008, insulation resistance of 5×10^{10} Ω, and 2.4% capacitance decrease between room temperature and 80°C.

A model of this $SrTiO_3$ ceramic with semiconductive grains, resistive or insulating grain surfaces, and an insulating phase in the grain boundaries has been developed with results that in good agreement with the device properties [137,138]. Burn and Neirman [137] attribute the grain semiconductivity to donor electrons produced when oxygen is lost during low pO_2 firing as well as to those due to donor impurities. A room temperature grain resistivity of less than 6×10^4 Ω-cm was estimated. Impregnation of the grain boundaries by 5 wt % of a mixture of 90 wt % Bi_2O_3 and 10 wt % Cu_2O added to the silver metallizing paint imparted overall apparent resistivity around 10^{11} Ω-cm. Figure 13 demonstrates the coincidence of maxima of grain size and apparent dielectric constant, a GBBL characteristic.

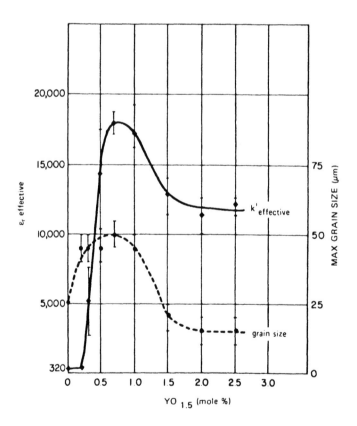

Figure 13   Dependence of $\varepsilon_{reffective}$ on grain size in a GBBL capacitor. (From Ref. 116.)

E. Multilayer GBBL Capacitors

The concept of combining multilayer geometry with GBBL microstructure to obtain exceptionally high-capacitance density is initially attractive but contains several inherent problems: (1) GBBL capacitor microstructures are large grained to maximize capacitance. But a multilayer device with excessively thick laminations dictated by the grain size is limited in bulk capacitance. (2) The chemistry of a multilayer GBBL capacitor becomes even more complex due to interaction of the cofired dielectric and electrodes. Schmelz [141] has reported that Pd^{2+} or Pt^{2+} counterdopes n-type $BaTiO_3$, an effect that would reduce an intended GBBL device to an ordinary multilayer one. Infiltrated electrodes or the low-temperature electroding methods discussed earlier might solve this problem. There has been work along these lines [140] in which the GBBL chemistry is first built into the dielectric powder and a low-temperature multilayer process is then practiced. However, these concepts have not been demonstrated commercially.

F. Thin-Film Ceramic Capacitors

The miniaturization trend of discrete multiplayer ceramic capacitors has reached the 0201 (0.6 × 0.3 mm) level, and it would appear that any further reduction of capacitor "footprint" on the PCB results in negligible improvement.

   Thus, it becomes necessary to investigate the possibility of fabricating the capacitor via thin-film methods for those capacitance values that are low and require low operating voltage. For the sake of this discussion we will ignore the large body of work associated with materials for ferroelectric memories and charge sources for dynamic random access memory (DRAM).*

   Traditional thin film dielectrics have in the past focused on SiO_2 and SiN_x ($\varepsilon = 5-7$), but in planar capacitor arrangement these can only achieve 50–500 pF/mm^2 [143]. The driving factors for the use of thin-film capacitors in the range of 10–50 pF are lower equivalent series inductance, lower capacitance values for circuit elements in decoupling, signal filtering, and impedance matching as operating frequencies increase. The lower inductance is a consequence of the device either being single layer or, if multilayer, having a much more inductance-friendly electrode structure and charge flow. A further potential benefit is the possibility of incorporating the capacitor structure in or on the active circuit and have it operate at low voltage. The easiest and most common materials are SiO_2 and Si_3N_4, which have dielectric permittivities of 5–7, followed by Ta_2O_5 ($\varepsilon_r =$

---

* For detailed review of the operating and materials requirements for future DRAM, see D.E. Kotecki, et al. (Ba, Sr)TiO_3 dielectrics for future stacked-capacitor DRAM. *IBM J Res Dev* 43(3), May 3, 1998, pp. 367–381.

Table 2 Relative Permittivity and Temperature Coefficient of Permittivity for Various
Materials (After Ref. 142)

Material	Relative permittivity (er)	Temp coeff, of permittivity (ppm/K)
SiO_2, Si_3N_4	5–7	20–40
Ta_2O_5	21.5	100
TiO_2	85	–720
$Ba_{(1-x)}$ SF_xTiO_3 (BST) (td)	250–500	
Pb (Zr,Ti) O_3 (PZT) (td)	400–1500	–55
Pb (Mg, Nb) O_3 (PMN) (td)	>1500	

21) and TiO_2 ($\varepsilon_r = 85$). However, all of these have performance drawbacks
(Table 2) [142]. The trade-offs between size and capacitance can be easily understood from an analysis by Ulrich and Leftwich [143].

A variety of methods have been investigated for the formation of ceramic
thin-film capacitors.

1. Metal-organic deposition (spin coating)
2. Metal-organic chemical vapor deposition
3. Sputtering
4. Pulsed laser ablation

Metal-organic deposition (MOD) has been an attractive approach because it offers
the possibility of spin coating and the apparent ease with which stoichiometry
can be controlled in the precursor stage. In order for such an approach to be
useful for integration of the capacitor on the surface of an IC, it is necessary to
spin it on an already fabricated and passivated IC and achieve a conversion
temperature (organometallic precursor to crystalline oxide layer) to prevent reactions on the silicon IC and its interconnections. In a series of experiments, Baumert
and colleagues [144] investigated the preparation and low-temperature calcina-
tions of $Ba_{0.7}Sr0_.3TiO_3$ prepared from carboxylate precursors [145,146] and spin
coated to yield fully reacted films of the order of 0.13–0.16 μm with Pt electrodes
on Si. While the samples postprocessed at 600–650°C in oxygen for 1–2 h exhibited polycrystalline structure and a specific capacitance of 20 nF/mm^2, samples
reacted at 450–500 C were not fully reacted and resulted in specific capacitances
of 1–7 nF/mm^2. Kinetic studies of time to constant weight loss indicate that
reaction at temperatures of 450°C is not achievable within a reasonable time.

Single-layer thin films of lead lanthanum zirconium titanate (PLZT) have
been investigated by the MOD method. It has been reported by Dimos [142] that

1-μm films have been obtained that had a dielectric constant in excess of 900 and high breakdown field strength ($E_b > 900$ kV/cm) resulting in a capacitance per unit are of 8–9 nF/mm^2.

MOCVD (metallo-organic chemical vapor deposition) techniques have been investigated by a number of people [147–149]. Sakabe and coworkers investigated this method [150–153] for the fabrication of multilayer thin film barium strontium titanate ceramic capacitors using 3 metallo-organic sources for the cations. These sources were mixed independently and introduced into the reaction chamber in argon and oxygen gases. The substrate was held at 650°C, and alternate layers of Pt electrode material were deposited. This process was repeated until 13 dielectric layers were deposited. A postdeposition anneal at 650°C resulted in improved performance, and multilayer ceramic capacitors with 12 BST layers of 0.26 μm were formed. A maximal dielectric constant of $\varepsilon_r = 1000$ was achieved for these films.

Sputtering and pulsed laser ablation (PLD) have been investigated for film formation. However, sputtering offers too low a deposition rate and PLD, while giving apparently good films, resulted in highly variable dielectric properties [154].

Because of the difficulties and potential yield losses of incorporating complex oxide structures on already finished integrated circuits, it is the more usual approach to prepare thin-film capacitors on a separate substrate and then mount them directly on the silicon chip using direct-attachment methods.

Viii. CAPACITOR PERFORMANCE PARAMETERS

The parameters whose values together describe the performance of a capacitor are as follows:

1. Capacitance and its dependence on temperature, voltage, and frequency
2. Dissipation factor and its dependence on temperature, voltage, and frequency
3. Insulation resistance and its short-term dependence on temperature and voltage
4. Response of insulation resistance to long-term testing under dc field and elevated temperature
5. Maximal permissible operating voltage
6. Stability of dielectric characteristics with aging in the absence of an electric field

In the following section, these parameters are discussed in greater detail.

A. Capacitance

Dielectric materials differ from one another in the numbers and types of microscopic charged entities available to respond to an applied electric field. Total response to the field is the polarization P, proportional to the number of charged entities N, the sum of their individual polarizabilities α_T, and the electric field(s) they experience:

$$P = N\alpha_T E_1 \tag{4}$$

P in ceramic dielectric materials can be made up of four components: electronic, atomic, orientational, and interfacial polarizations.

Electronic polarization P_e occurs within atoms when the electric field slightly displaces the electron clouds of the atoms relative to the nuclei. The electronic polarizability α_e of an atom is given by

$$\alpha_e = \frac{\varepsilon_0\, 4\pi R^3}{3} \tag{5}$$

where R is the atomic radius. Materials with large atoms therefore have higher electronic polarization. There is no temperature dependence except that due to thermal expansion effect on density (ρ). This would make for a slight negative temperature coefficient of dielectric constant:

$$\frac{d\varepsilon_r}{dT} = -\frac{1}{3}\frac{(\varepsilon_r-1)(\varepsilon_r+2)}{\rho}\frac{dp}{dT} \tag{6}$$

where $1/3\, dp/\rho dT$ is the linear thermal expansion coefficient. Electronic polarization can respond to frequencies into the optical and ultraviolet range (i.e., beyond the frequency range of interest for capacitors). The electronic contribution to ε_r is given by Maxwell's rule from electromagnetic theory as the square of the refractive index. Dielectrics having electronic polarization only are linear dielectrics. Only organic compounds such as polyethylene are in this category, with permittivity values less than 5.

Atomic or ionic polarization is molecular in scale but can also be considered instantaneous in comparison with the periods of power and radio frequencies. In ionic polarization, crystal lattice ions of opposite sign separate under the influence of the external field. The oscillation periods are greater than those of electrons because of atomic inertia. The dielectric constant of a dielectric material with atomic polarizability begins to decrease with increasing frequency in the infrared region. The temperature dependence of atomic polarization is positive; for alkali halides its magnitude is about the same as for electronic polarization.

Dipolar or orientational polarization P_0 occurs when molecules possess a permanent electric moment that is alignable by the applied electric field. This mechanism is most pronounced in liquids and gases, but is also important in polymers with long polar molecules and in the special cases of electrets and ferroelectric materials. The limiting frequency for orientational polarization in liquids and gases is in the microwave region, about 10^{10} Hz. In ceramic dielectrics such moments would not contribute to ε_r beyond radio frequencies. Under dc potential long-period polarizations can be built up. Temperature is a strong influence, having an inverse relationship to relaxation time.

Ferroelectricity [155,156] can be considered a special case of orientational polarization, characterized by the interaction and spontaneous alignment of existing dipole moments. Their polarization is reversible by an applied field. Rather than rotating, the dipoles align and reverse align by cooperative ion displacements. This polarization mechanism relaxes in the infrared.

In ferroelectrics, a strong temperature dependence of dielectric constant is associated with crystallographic phase transitions, which must be suppressed or circumvented in order to make practical capacitors. The dielectric constant of materials of this class is also highly voltage sensitive, decreasing with increased dc field but increasing with an increasingly strong ac signal.

Interfacial polarization occurs in multiphased materials when the phases have different conductivities. For example, GBBL capacitors, contain conductive grains separated by insulating grain boundary regions. Because of this heterogeneity an internal distortion of the imposed electric field is produced which from outside is evidenced as a large dielectric constant. The dimensional scale in interracial polarization is much greater than atomic or molecular; the field-distorting charges can span a grain diameter or the dielectric thickness.

The response times of interfacial polarizations are determined by differences of dielectric constant and conductivity of the phases. When all phases are highly resistive, the polarization may occur only at very low frequencies; when one component has a very low resistance, as in the GBBL capacitor, the dielectric relaxation can be in the high radio frequency range. The effect of voltage can be to increase, decrease, or not affect ε_r, depending on the matrix material characteristics. Similarly, temperature increase can either raise or lower the permittivity. Figure 14 provides a representation of ε_r versus frequency for a hypothetical dielectric material that possesses all four types of polarization.

B. Energy Dissipation

The lower section of Figure 14 depicts the absorption of electrical energy for each type of dielectric polarization as a function of frequency. Each process has an associated energy loss. The energy losses have peak values in the relaxation frequency ranges of their polarization mechanisms. But even at frequencies well below or well above the relaxation of any polarization mechanism, a finite energy loss or inefficiency is associated with any real energy storage mechanism.

In an ideal capacitor, application of a dc field is followed by the flow of a dielectric absorption current that falls to zero as the energy storage function

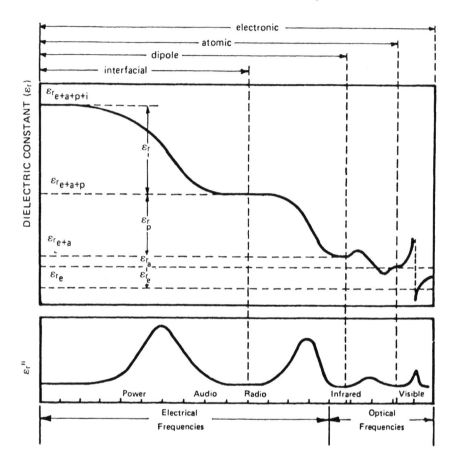

Figure 14 Dielectric polarization mechanisms and their frequency dependence. (From Ref. 122.)

saturates and there is no energy loss. In an ac field with ideal loss-less capacitor, the absorption current leads the voltage by 90° and there is no dielectric loss.

In a real capacitor an increment of the dc current flow during absorption is not recoverable on discharge. Under ac, the phase angle between current and voltage is something less than 90°, the angular deviation from 90° being called the loss angle, δ. Tangent δ is a measure of efficiency of the capacitor and is termed the dissipation factor. Commonly, tan δ × 100, called the "percent dissipation," is used.

In addition to losses due to inefficiency of the polarization mechanisms, the finite conductivity of any real dielectric material adds an i^2R component to its loss mechanisms. Insulation resistance will be taken up later. The shapes of the ideal dissipation factor and permittivity spectra were developed by Debye [157] based on a spherical dipolar molecule orienting and reorienting in a viscous medium. The loss factor goes through a maximum for every polarization mechanism due to the dissipation of field energy as frictional heat when the bound charges or dipoles or space charges move toward the polarized condition. The absorption peak is greatest for materials showing the greatest change in dielectric constant in passing through a dispersion region. Generally, the greater the number of polarization mechanisms in a given frequency region, the greater the dissipation factor of a dielectric.

C. Insulation Resistance

The energy gap for electronic conduction is so large in the compounds constituting the classical ceramic dielectrics—alumina, steatite, mica, and porcelain—that the most important conduction mechanism is ion migration. But generally even at the highest temperatures of usual capacitor use, ion conduction is not significant. In the case of glasses, however, or of a crystalline material with an impurity glassy phase, electrolytic conductivity by ions can be substantial. For example, a commercial soda-lime-silica glass has conductivity of about 10^{-8} $(\Omega\text{-cm})^{-1}$ at 150°C. This type of glass is avoided in capacitors. Rather, devitrifying glasses are employed, or alkali-containing glasses whose alkali ions have been immobilized by associations with other network or modifier ions.

Conductivity behavior is of particular importance in the case of barium titanate, the prototypical material of most ceramic dielectrics. $BaTiO_3$ has a smaller band gap (3.2 eV) than do the classical materials mentioned above, but is intrinsically a good insulator. However, its crystal structure lends itself to several varieties of defects, any of which can lead to high electronic conductivity. Usually the carriers are n type, their presence associated with oxygen vacancies [158,159]. The oxygen vacancies, in turn, are due to dissociation at elevated processing temperatures, or to aliovalent acceptor dopants present in the ceramic as raw materials impurities, or acquired during processing. For example, Al_2O_3, providing trivalent aluminum in substitution for Ti^{4+}, is a common $BaTiO_3$ raw material impurity and can also be acquired from attrition of mill jars or grinding media. Early in the technology of barium titanate–based capacitors, without the benefit of knowledge of its solid-state chemistry, it was recognized that addition of niobia is beneficial in increasing the resistance and prolonging the lifetime under dc load.

D. Field-Induced Degradation

More than by the short-term value of its resistance on application of voltage, a capacitor is judged by the stability of the resistance value under sustained dc field and at elevated temperature. Figure 15 is a useful qualitative depiction of the varieties of response to electric field and temperature that are possible.

At low temperature and high field, the short horizontal line at the lower right of the diagram represents the weakly temperature-dependent intrinsic breakdown

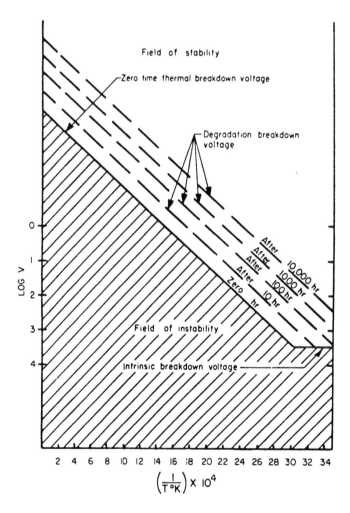

Figure 15   Ceramic dielectric degradation and breakdown under voltage. (From Ref. 126.)

strength. At higher temperatures, however, the breakdown voltage decreases rapidly with increasing temperature. An additional important aspect is that the breakdown voltage at a given temperature is time dependent (i.e., a mode of failure by degradation of the dielectric is recognized). In this mode the resistance of the capacitor decreases continuously with time until runaway i^2R heating causes thermal breakdown. The activation energy for conduction remains constant whatever the duration of the degradation process. This form of degradation has been a significant problem in the commercialization of high dielectric constant titanate ceramic dielectrics. As an example of the early magnitude of the problem, Gruver et al. [160] described a 1951 titanate-based capacitor dielectric formulation with an initial ε_r value of 100. Under 1.1 kV/mm at 150°C, the resistivity of this material decreased by 1 order of magnitude in 500 h; in contrast, a current commercial material [161] with a ε_r value of 1000–1500 is described as showing no measurable resistance change in 500 h at 125°C and 10 kV/mm.

Aspects of the detailed mechanism of the degradation process are not clear. In general terms, it can be described as follows: Oxygen vacancies are produced both by the quenching-in of dissociation occurring at high temperature during firing of the ceramic and by an electrolytic process in which vacancies are injected at the anode and diffuse under field toward the cathode. Ionic conduction has been reported to occur at as low as 100°C in $BaTiO_3$ [162]. The oxygen vacancies' ($+2$) charge is compensated by $3d$ electrons of Ti^{4+}, producing some content of Ti^{3+} ions. This introduces an additional, electron-exchange conduction mechanism.

E. Dielectric Strength

The rated operating voltage of a capacitor is an engineering parameter that takes into account the intrinsic dielectric strength of the dielectric material as well as the contribution of ambient temperature to thermal breakdown. A design factor to ensure function over the specified lifetime is included.

The intrinsic dielectric strength is measured at low temperatures where current heating is not a factor. The mechanism of breakdown is the emission of electrons under the applied field, which in turn free more electrons, ultimately producing an electron avalanche that punctures the test specimen. The field at which breakdown occurs is nearly temperature independent. Experimental determination of the breakdown field strength is complicated by edge effects or other inhomogeneity of the field experienced by the test specimen, by the geometry and materials of the electrodes used, the specimen thickness, the presence of pores or phase boundaries in the ceramic, and so on. The most reliable tests are performed with the specimen immersed in a dielectric fluid and using a specimen cross-section especially contoured to give uniform field. More relevant to the conditions under which capacitors are used, dielectric breakdown tests commonly are carried out at room or elevated temperature. In these cases, the field strength at which breakdown occurs is governed by leakage resistance and the rate of heat generation in the test specimen relative to its rate of heat dissipation. Breakdown values therefore can vary widely depending on the test conditions. Even with nominally identical conditions, microstructural variations in test specimens necessitate care that breakdown test results be statistically meaningful.

In summary, the maximal voltage withstanding ability of a ceramic capacitor is governed not only by the intrinsic breakdown strength of its dielectric material but also by the capacitor's design and construction.

F. Aging and Piezoelectricity

Two related capacitor performance characteristics, piezoelectricity and aging, stem from the ferroelectric nature of $BaTiO_3$. Piezoelectricity by definition, accompanies ferroelectricity; that is, in the crystal symmetry classes in which ferroelectricity can occur, so does piezoelectricity. In a ceramic with randomly oriented grains and domains of $BaTiO_3$, there is no net piezoelectric effect. However, should the capacitor be subjected to a dc field of the order of 1 kV/mm, such as is experienced in a voltage-withstanding test, domain alignments can impart a net crystallographic orientation to the ceramic. Then the piezoelectric effect, direct and inverse, is experienced. The direct piezoelectric effect can be objectionable in low-voltage circuits, generating spurious voltages when the capacitor experiences a shock or acceleration. Preventives of the condition are to avoid field strengths strong enough for poling, to anneal out any residual polarization, or to employ ceramic compositions that display minimal piezoelectric characteristics. Examples of such compositions are fine-grained, nearly cubic $BaTiO_3$ and a ceramic whose ferroelectric phases have Curie points below the operating temperature range.

The piezoelectric effect in dielectric materials is subject to aging, an effect discussed below in relation to permittivity.

G. Aging of Dielectric Characteristics

Capacitors of Electric Industries Association's class 2 are subject to a decrease of permittivity and of dissipation factor with time. This "aging" is a ferroelectric phenomenon and begins following the last heating above the Curie temperature. This may occur when end terminations are affixed, when polymer encapsulants cured, or when capacitor chips are soldered into a circuit.

As a manifestation of ferroelectricity, aging is most pronounced in the higher permittivity dielectrics. A typical aging characteristic for an X7R-type capacitor is shown in Figure 16. A capacitor with the Z5U characteristic, having three to four times higher permittivity, would age three to four times faster than

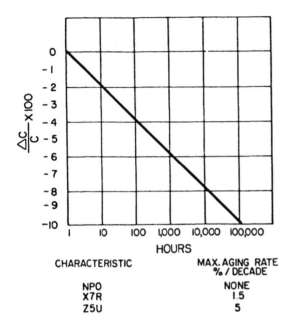

CHARACTERISTIC	MAX. AGING RATE % / DECADE
NPO	NONE
X7R	1.5
Z5U	5

Figure 16   Aging of capacitance in ceramic capacitors. (From Ref. 129.)

the X7R. The tan δ value also ages, decreasing several times faster than the capacitance. Solid-solution dielectrics with axial ratios close to unity age less. Fine-grained dielectrics age less than coarse-grained material of the same composition.

The aging phenomenon has been the subject of study since the application of barium titanate in capacitors. This is understandable because the phenomenon effectively determines the tolerance within which a capacitance value can be guaranteed. The present perception of aging is primarily that of a relaxation of mechanical stresses. The stresses are generated by the strain of the cubic-to-tetragonal symmetry phase change; their relaxation takes place through the nucleation and growth of 90° domains [13]. The means of suppressing aging, therefore, are to inhibit domain nucleation by limiting the grains to submicrometer size and/or to reduce the c/a unit cell ratio and the associated intergrain stresses through the use of solid-solution compositions.

The aging of BZT (solid solution of $BaTiO_3$ and $BaZrO_3$) have been investigated extensively by Buchanan and coworkers [164,165], who found that fine-grained samples (approx 0.5 μm) doped with zirconium at the 2 wt % level exhibited decreased aging.

The strong interest in BME systems has prompted Hennings and colleagues to reexamine the BZT system with regard to materials doped with MnO (a standard dopant in BME systems) [165]. Thin solid samples were prepared using the reaction of solid oxides, fired under conditions simulating the firing of BME, i.e., 136°C, 6 h in moist argon/hydrogen (98%/2%). Following thinning the samples, which had a range of Zr and MnO doping of 1 wt %, were subjected to a variety of partial reoxidation regimes in order to simulate behavior in a BME system. Samples that were annealed in air had the lowest aging rate and those annealed at 1×10^{-14} showed the highest aging rate. This was attributed to the variable valence of the Mn ions. In general, it appeared that aging was lowest for the lowest zirconium substitutions and for those samples that were annealed at the highest pO_2. Their conclusion was that for BME, applications parts should be annealed at the highest possible pO_2 consistent with maintaining the metallic Ni electrode.

H. Low-Inductance Capacitors

The development of μ-processors has resulted in the development of a specialized class of low equivalent series inductance capacitors (low ESL) to decouple the power supply and provide a charge source close to the processor. As the switching speeds increase, lower inductance and more total capacitance are required to switch the millions of transistor gates.

The standard design of the ceramic multilayer capacitor with the terminations on the ends has a ESL of approximated 1000 pH, and this value is too large for effective decoupling. The inductance in the circuit limits the delivery of charge to the circuit. In 1990 IBM developed the D-Cap [167,168]. This design, based on a vertical alignment of the electrodes and self-canceling current paths, resulted in very low inductance. In addition, it was designed to be compatible with the IBM C4 attachment process. However, this design relied on 100% platinum electrodes and as a consequence has been very expensive to produce. Other designs have evolved that also provide lower inductance, and because they are implemented in the BME material system, they offer significant cost advantages. These designs incorporate the more conventional and easy-to-manufacture horizontal electrode structure. A variety of electrode structures are currently being produced.

By placing the electrodes along the long edge of the capacitor, the ESL is roughly cut in half. This is the simplest termination structure modification (LLL, Murata Manufacturing Co.; other manufacturers offer similar designs). If the terminations along each of the long sides are alternately connected to the positive and negative layers in the capacitor the ESL is further reduced. Typical structures are made with four electrodes (two positive and two negative on each long side). These structures are illustrated in Figure 17. It should be noted that increases in microprocessor speed result in lower switching voltages. As a consequence, the

A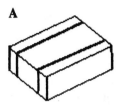

B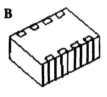

Figure 17 Different designs for equivalent series inductance capacitors (low ESL). (a) LLL design. (b) LLA design.

dielectric thickness can be reduced, more capacitance can be achieved in the same package, and/or the physical size of the decoupling capacitor can be reduced. These changes also contribute to a lowering of the ESL.

Other geometrical configurations of electrodes result in a further lowering of the inductance to the range of 50–80 pH. The ultimate goal is to reduce the ESL to the range of 10–20 pH at which point the "loop inductance" (inductance of the circuit within the PCB) will also become a major factor.

To get the capacitor closer to the μ-processor, designers are actively looking at embedding the decoupling capacitor(s) in the PCB. However, these approaches place very strong requirements on the thickness, shape, and overall size of the capacitors to be embedded. In addition, the higher operating temperature of high-frequency μ-processors puts additional stress on the capacitor/PCB material system. Such capacitor designs are still highly experimental and proprietary to the manufacturers; however, they represent a new and continually emerging area for the application of ceramic capacitors.

IX. PACKAGING OF CERAMIC CAPACITORS

Some degree of protection against the use environment is required for all ceramic capacitors. The protection can be applied to individual capacitors, to capacitor-containing subcircuits, or to entire circuits. In detail, the purposes of packaging are:

1. To exclude contaminants, particularly mobile ions, that can contribute to the surface conductivity or, if absorbed by a less than totally dense ceramic, to the bulk conductivity. In addition, where a coincidence of silver metallization, moisture, and electric field might lead to the generation of silver ions and their injection into and migration in and on the ceramic, coating or packaging excludes moisture and prevents this degradation. Proton injection has been implicated in the long-term degradation of dielectrics [126] and waterproof packaging forestalls that process as well.

2. To protect against mechanical mishandling, thermal shock, detachment of leads, and so on. Where the dielectric possesses some residual piezoelectricity, a shock-absorbent package insulates it from voltage-generating stresses.

3. To standardize. The package commonly takes forms that have become standard for other electronic components. This enables the use of common equipment and methods for placement on and attachment to circuit boards. Packages may take one or several of the following forms:

a. "Self-packaging," employed with multilayer capacitors in which the electrodes are totally internal except where their alternate sets of edges are exposed. These surfaces are metallized with a glass frit/powdered metal combination that serves as packaging as well as in a current-collecting function. The two metallized faces usually are then solder coated, as the final fabrication step for a leadless capacitor, or during attachment of soldered leads, or when a leadless capacitor is soldered to its circuit board. Ceramic capacitors of the self-packaged type may also be impregnated or coated with a silicone resin or other moisture repellent.

b. Axially leaded multilayer capacitors can be packaged in cylindrical sealed glass tubes of the type used for diodes and discrete transistors. Insertion of the capacitors into the tubes and bonding of the glass to the wire lead at each end of the device are done automatically.

c. Dip or conformal insulating coatings are used very commonly on single-layer (i.e., disk or wafer) leaded capacitors. Phenolic or epoxy thermoset coatings are most used in this low-cost packaging method. They are applied by immersing the part in a slurry of the monomer, draining, and heating to cross-link the coating material. An alternative approach is the fluidized-bed technique. The dry powdered coating material is fluidized in an air stream and allowed to impinge on the preheated component. The powder particles fuse together to coat the capacitor. Sometimes additional heating is needed to smooth the coat or cure the polymer.

d. When the capacitor is to be a leaded device with precisely defined exterior shape and dimensions, the packaging method may be one of the following:

1. *Potting in a premolded case.* Injection-molded cases of epoxy or silicone resin are common. After insertion of the body of the leaded capacitor, the case is filled with a liquid potting resin, such as an epoxy. Commonly, an intermediate elastomer coating is applied to the part before potting. This takes up strains due to differences of thermal contraction of the ceramic and the potting material that might otherwise be great enough to detach end termination metallization or soldered-on leads. The shock-absorbing encapsulant also helps prevent generation of spurious voltages, as described previously.

2. *Molding in dies using high- or low-pressure techniques.* In liquid resin molding, the resin formulation is injected at 5–50 psi into a hot ($\leq 100°C$) mold containing the capacitor. In compression or transfer molding, the temperature is at least 50°C higher and the parts being jacketed can experience 10,000 psi.

Phenol formaldehyde, epoxy, and silicone resins are the most common. Capacitors can be packaged individually or in multiple units in industry-standard integrated circuit housings (single in-line, dual in-line, DIL).

Although these molding methods are useful, the great majority of ceramic components are fabricated to be attached by surface mount technology (SMT; see below). Molded components are simply too expensive and offer no advantages.

A. Capacitors for Automated Hybrid Circuit Assembly

Printed circuit board technology, which is the standard for microelectronic circuits, mounts discrete resistors, capacitors, and semiconductor devices, in the form of "chips," on both sides of substrates containing printed wiring patterns. The substrates themselves are usually multilayer organic or ceramic boards. Placement of the chips is automated and an adhesive holds them in place until they are soldered *en masse*. SMT has emerged as the dominant packaging mode, especially for low- to moderate-cost (and moderate-speed) hybrid circuit applications [168]. The high-speed placement of chip capacitors and resistors places a premium on tight dimensional tolerances. An inherent thermal expansion mismatch exists with the substrate and other components in this packaging mode, but thermal stresses are minimized by keeping the lateral dimensions of the componennts small.

Although 60Sn40Pb eutectic solder has been used for many years, high-temperature solders are currently being employed to solder integrated circuits (e.g., microprocessors) and board components in one "pass" through the soldering operation. To prevent leaching of the termination material into the solder a barrier layer of Ni is normally plated between the end termination and the solderable surface. This barrier layer is typically 2 μm. The outer surface of the electrode is usually barrel plated (electroplated) as a solder layer (Pb/Sn) or a pure tin (Sn) later. Typically pure tin is more than adequate in providing a wettable surface for the solder and avoids the use of lead in plating baths and lead material in the capacitor termination. A variety of soldering techniques are employed. These include wave soldering of the chips to the substrate, the chips being held in position with a nonconductive epoxy. A more common technique, solder paste reflow, uses prefluxed solder paste to hold the chips in position prior to reflow. Heat sources employed include infraned heat transfer, vapor phase reflow, and direct heating of the substrate. Advantages of the SMT technique are that it permits rapid pick and placement of components (including IC devices), develops strong solder bonds, and allows ready access for testing as well as for replacement of defective components.

Other techniques used for chip capacitor attachment include the use of nonconductive thermal epoxies to fix the capacitor body to the substrate, prior to electrode attachment, typically by soldering. Direct-wiring bonding of the chip using an external heat source combined with thermal compression or ultrasonic bonding may also be employed. These bonding schemes, however, do not provide for the high throughput and reliability of the SMT technique.

In line with these assembly concepts, capacitor manufacturers offer components in forms ready for loading onto automated assembly equipment. In older forms of packaging for automatic insertion, leaded capacitors held between strips of adhesive tape are prepared as "sticks" or wound into rolls. In the insertion machines the leads are cut free of tape, bent to shape, and the parts are "stuffed" into their boards and subsequently soldered in place. The most prevelant form for SMT operation currently is a perforated paper tape. The tape is three layered. Holes punched in a thick middle layer accommodate the chips being dispensed, with adhesive tape placed below and on top to contain the chips. The strip with its components is wound onto standard film reels, a 7-in.-diameter reel containing about 4000 capacitor chips. During insertion the upper (cover) tape is removed and the individual parts are picked up with a vacuum pipette and placed on the printed circuit board. Most frequently the electrical pads on the PCB are pretinned and the capacitor is held in place by a drop of glue/epoxy until soldering.

In the last several years a bulk cassette method of packaging has been proposed by Murata Manufacturing Co., Ltd (Fig. 18). Because of the increasing

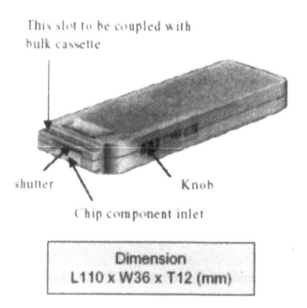

Figure 18 Illustration of bulk cassette method of capacitor packaging. (Courtesy of Murata, Inc.)

miniaturization of ceramic capacitors, current offerings of surface mount technology include 0603, 0402, and 0201 sizes (see footnote for explanation of EIA size codes). The 7-in. reel has a fixed pitch and the number of components per reel remains constant even though the components themselves have become minute (0201 = 0.51 mm / 0.25 mm). The bulk cassette technology offers several distinct advantages for board assembly: smaller storage space of inventory (1 cassette holds: 15,000 0603s), longer machine time between loading of cassettes, no downtime change over (because of buffer built into the placement machine), less scrap (reels and paper). However, modification of the mounting machines is necessary to effectively use the bulk cassette. Levinson et al. [169,170] have presented lucidly the mix of science and empiricism that goes into the formulation of ceramic dielectrics to meet EIA or military capacitor specifications. In terms of volume of use, the most important categories of dielectrics are the Z5U, X7R/ BX, and the NPO. Payne [140] has estimated the relative use of these in the United States as follows: NPO, 10%, BX/X7R, 40%, and Z5U, 50%. Most of these materials are based on BaTiO$_3$. The NPO type is the class least likely to display any vestigial ferroelectric characteristics. Figure 19 illustrates the dependence of permittivity on temperature of representatives of the three classes of materials.

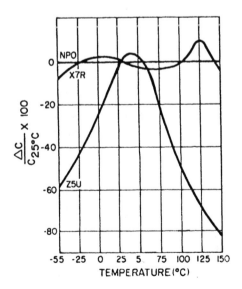

Figure 19 Temperature coefficient of capacitance versus temperature for NPO, X7R, and Z5U capacitors. (From Ref. 129.)

B. Z5U Dielectrics

For maximum capacitance the $BaTiO_3$ Curie peak has been shifted close to room temperature and broadened. Shifting is accomplished with solid-solution additives (e.g., $SrTiO_3$) and alkaline earth zirconates and stannates. Broadening is accomplished by superimposition of a number of noncoincident peaks and by depressor additives, for example, aliovalent materials, which tend not to form solid solutions but to concentrate in grain boundaries: bismuth stannate, titanate, zirconate, or oxide; titanates of iron, nickel, copper, or manganese; and magnesium zirconate and sodium niobate. Typical Z5U formulations are presented in Table 2 and X7R/BX formulations are presented in Table 3.

C. NPO Dielectrics

Originally based on TiO_2, these compositions now tend to include $BaTiO_3$ to counter the negative temperature coefficient of rutile and to increase the permittivity. To a degree, rare earth oxides are used for the same purpose. NPO dielectrics appear microstructurally to be, not solid solutions of $BaTiO_3$, but other titanate compounds [171]. As such, they do not display ferroelectric characteristics.

The compositions of Tables 2 to 4 represent basic capacitor dielectric compositions, in most of which the characteristics of ferroelectric $BaTiO_3$ have been

Table 3 Z5U Dielectric Compositions

	Composition (wt %)			
	1	2	3	Comment
$BaTiO_3$	84–90	65–80	72–76	Base material
$CaZrO_3$	8–13	—	—	Shifter
$MgZrO_3$	0–3	—	—	Depressor
$SrTiO_3$	—	7–11	5–8	Shifter
$CaTiO_3$	—	7–11	4–6	Depressor
$BaZrO_3$	—	7–11	7–10	Shifter
$CaSnO_3$	—	—	2–4	Shifter
Other[a]	1–3	8–13	0–3	
E_r (25 °C, 1 kHz)	5700–7000	5500–6500	11,500–13,000	
tan δ	≤ 0.03	≤ 0.03	≤ 0.03	

[a] Among the miscellaneous additives are acceptor materials such as Nb_2O_5, Sb_2O_5, Ta_2O_5, or rare earth oxides, to compensate n-type depressors and impurities.

Source: Ref. 102.

Table 4 X7R/BX^a Dielectric Compositions

	Composition (wt %)			
	1	2	3	Comment
$BaTiO_3$	90–97	85–92	86–94	Base material
$CaZrO_3$	2–5	4–8	—	Shifter
$BaCO_3$	0–5	—	—	Stoichiometry adjustment
$SrTiO_3$	—	3–6	—	Shifter
Bi_2O_3	—	—	5–10	Depressor, flux
Other	2–5	1–4	2–6	See note a in Table 3
E_r (25°C, 1 kHz)	1600–2000	1800	1400–1500	
tan δ	<0.025	<0.025	<0.015	

^{a} X7R is the EIA designation; BX is the U.S. military designation for capacitors with substantially the same characteristics. The principal difference is more rigorous BX requirements for life-test performance and voltage coefficient of capacitance.
Source: Ref. 130.

shaped to the required properties. They are the most widely used types of compositions. However, more recent approaches based on relaxator materials, crystallizing glasses, base metal-compatible dielectrics, and grain boundary polarization effects, which have been described in the text, have gained considerably in commercial acceptance.

Table 5 NPO Dielectric Compositions

	Composition (wt %)		
	1	2	3
$BaTiO_3$	41–49	39–47	15–21
TiO_2	48–54	41–47	26–34
ZrO_2	—	8–13	—
$Nd(CO_3)_4$	—	—	59–45
Other	3–7	2–5	Small
E_r (25°C, 1 kHz)	35	30	62
tan δ	<0.002	<0.002	<0.002

Source: Ref. 130.

REFERENCES

1. Encyclopedia Britannica. 1980; 10:136.
2. Thurnauer H. Am Ceram Soc Bull 1977; 56:219.
3. Thurnauer H. Am Ceram Soc Bull 1977; 56:861.
4. Maxwell JC. Electricity and Magnetism. Vol. 1. Oxford: Clarendon Press, 1892.
5. Mitoff SP. Adv Mater Res 1968; 3:305.
6. Lichtenecker K. Phys Z 1909; 10:1005.
7. Ven Beek LKH. Prog Dielectr 1967; 7:71.
8. Wul BM, Vereschagen FL. Dokl Akad Nauk SSSR 1946; 48:634.
9. Okazaki K. Am Ceram Soc Bull 1982; 61:932.
10. Von Hippel AR, Merz WJ. Phys Rev. Dielectrics and Waves. Vol. 76. New York: John Wiley and Sons, 1949:1221.
11. Bunting EN, Shelton GR, Creamer AS. J Res Natl Bur Std 1947; 38:337.
12. Buessem WR, Cross LE, Goswami AK. J Am Ceram Soc 1966; 49:33.
13. Wainer E, Wentworth C. J Am Ceram Soc 1952; 35:207.
14. Coffeen WW. J Am Ceram Soc 1954; 37:480.
15. Plessner KW, West R. Prog Dielectr 1960; 2:167.
16. Subbarao EC. Non-stoichiometric Compounds. New York: Academic Press, 1954: 296 Cited on p. 90, end of line 8.
17. Rutt TC. Am Ceram Soc Bull 1968; 47:873.
18. Smolenskii GA, Isupov VA. Dokl Akad Nauk SSSR 1954; 97:653.
19. Swartz S, Shrout T, Otteson C. Am Ceram Soc Bull 1982; 61:359.
20. Jang SJ, Cross LE, Uchino K. J Am Ceram Soc 1981; 64:209.
21. Electronic Industries Association RS-198, American Standard Requirements for Ceramic Dielectric Capacitors, Classes 1 and 2. New York: American Standards Association, 1958.
22. Scott JF, Raz de Araujo CA. Science 1989; 246:1400–1405.
23. Tuttle BA. Mater Res Soc Bull 1987; 12(7):40–45.
24. Schulze WA, Gururaja TR. Mater Res Soc Bull 1987; 12(7):47–52.
25. Berry RW, Hall PM, Harris MT. Thin Film Technology. New York: Van Nostrand Reinhold, 1968:371.
26. Biggers JV, Marshall GL, Stickler DW. Solid State Technol 1970; 13:63.
27. Hoffman LC. U.S. Patent 3,656,984, 1972.
28. Hoffman LC, Nakayama T. U.S. Patent 3,666,505, 1972.
29. Bacher RJ. U.S. Patent 4,071,881, 1978.
30. Bacher RJ. U.S. Patent 4,089,038, 1978.
31. Barringer E, Bowen HK. Am Ceram Soc Bull 1982; 61:336.
32. Ring TA. Mater Res Soc Bull 1987; 12(7):34–38.
33. Phillips DS, Vogt GJ. Mater Res Soc Bull 1987; 12(7):54–58.
34. Howatt GN, Breckenridge RG, Brownlow JM. J Am Ceram Soc 1947; 30:237.
35. Rodriguez A. U.S. Patent 3,004,197, 1961.
36. Deyrup AJ. U.S. Patent 2,389,420, 1945.
37. Ballard KH. U.S. Patent 2,395,442, 1946.
38. Deyrup AJ. U.S. Patent 2,413,549, 1946.
39. Howatt GN. U.S. Patent 2,486,410, 1949.

40. Bradford CI, Weller BL, McNeight SA. Electronics Dec. 1947; 20:106.
41. Lee PW, Weller BL. U.S. Patent 2,779,975, 1957.
42. Rutt TC. U.S. Patent 3,679,950, 1972.
43. Rutt TC. U.S. Patent 4,071,880, 1978.
44. Werring W, Schnoller M, Wahl H. Ferroelectrics 1986; 68(1–4):145–156.
45. Rutt TC, Stynes JA. IEEE Trans Parts, Hybrids, Packag PHP-9 1973:144.
46. Goodman TG. Adv Geram 1981; 1:230.
47. Hoffman LC. U.S. Patent 3,390,981, 1968.
48. Miller LF. Proc Electron Components Conf 1968:52.
49. Newnham RE. U.S. Dept. of Commerce, 1982.
50. Padwal S, deBruin HJ. J Electrochem Soc 1982; 129:1921.
51. Sheard JE. U.S. Patent 3,872,360, 1975.
52. Piper J. U.S. Patent 3,612,963, 1971.
53. Maher G. U.S. Patent 3,619,220, 1971.
54. Maher G. U.S. Patent 3,682,766, 1972.
55. Maher G. U.S. Patent 3,811,937, 1974.
56. Burn I. J Mater Sci 1982; 17:1398.
57. Payne DA, Park SM. U.S. Patent 4,158,219, 1979.
58. Payne D, Park SM. U.S. Patent 4,218,723, 1980.
59. Herczog A, Stookey SD. U.S. Patent 3,195,030, 1965.
60. Herczog A, Layton M. U.S. Patent 3,490,887, 1970.
61. Smolenski GA, Isupov VA. Dokl Akad Nauk SSSR 1954; 97:653.
62. Smolenski GA, Agranovuskaya AI. Sov Phys Solid State 1959; 1:1429–1437.
63. Eric Cross L. Ferroelectrics 1987; 78:241–267.
64. Shrout TR, Halliyal A. Am Ceram Soc Bull 1984; 63(6):808–810.
65. Yonezawa M, et al, Proc. 1st Meeting on Ferroelectric Materials and Their Applications, Kyoto, Japan, 1977.
66. Koto J, et al, Proc. 1st Meeting on Ferroelectric Materials and Their Applications, Kyoto, Japan, 1977.
67. Tarahara H, Kiuchi K. Adv Ceram Mater 1986; 1(4):346–349.
68. Swartz SL, Shrout TR. Mater Res Bull 1982; 17:1245–1250.
69. LeJeune M, Boilot JP. Ceram Int 1982; 8(3):99–103.
70. Shrout TR, Halliyal A. Am Ceram Soc Bull 1987; 66(4):704–711.
71. Fu S, Chen G. Ferroelectrics 1988; 82:119–126.
72. Chen G, Fu S. J Mater Sci 1988; 23:3258–3262.
73. Kakegawa K, et al. Am Ceram Soc Commun 1988; 71(1):C49–C52.
74. Dey K, et al. IEEE Trans Ultrasonics 1988; 35(1):80–81.
75. Feng K, et al, Schulze W. Adv Ceram Mater 1988; 3(5):468–472.
76. Dgohara T, et al. J Mater Sci Lett 1988; 7:867–869.
77. Kakegawa K, et al. J Mater Sci Lett 1988; 7:230–232.
78. Hishiyama T, et al. NEC Res Dev July 1988:no. 90, 29–35.
79. Quek HM, Yan MF. Ferroelectrics 1987; 74:95–108.
80. Fuji S, et al. U.S. Patent 4,574,255, 1986.
81. Utsumi K, et al. Ferroelectrics 1986; 68(1–4):157–179.
82. Yonezawa M. Ferroelectrics 1986; 68(1–4):181–189.
83. Smolenskii GA, Agranovskaya AI. Sov Phys Solid State 1960; 1:1429.

84. Bokov VA, Myl'nikova IE. Sov Phys Solid State 1961; 3:613.
85. Iizawa O, Fujiwara S, Ueoka H, et al. U.S. Patent 4,235,635, 1980.
86. Tanei H, Ikegami A, Taguchi N, et al. U.S. Patent 4,308,571, 1981.
87. Herbert JM. U.S. Patent 2,750,657, 1956.
88. Herbert JM. Trans Br Ceram Soc 1963; 62:645.
89. Wicks CE, Block FE. US Bur Mines Bull 1963; 605.
90. Burn I, Maher GH. J Mater Sci 1975; 10:633.
91. Chan NH, Smyth DM. J Electrochem Soc 1976; 123:1584.
92. Daniels J. Philips Res Rep 1976; 31:505.
93. Eror NG, Burn I, Maher GH. Method of forming a ceramic dielectric body, U.S. Patent 3,920,781, 1975.
94. Sakabe Y, Takagi T, Wakino K, Smyth DM. Dielectric materials for base–metal multilayer ceramic capacitors. Adv Ceram 1987; 19:103–115.
95. Sakabe Y, Minai K, Wakino K. High-dielectic ceramics for base-metal monolithic capacitors. Jpn J Appl Phys 1981; 20(Suppl. 20–4):147–150.
96. Sakabe Y, Seno H. Method for making a monolithic ceramic capacitor employing a non-reducing dielectric ceramic composition, U.S. Patent 4,115,493, 1978.
97. Sakabe Y, et al. Laminated ceramic capacitor, U.S. Patent 4,451,869, 1984.
98. Nisioka G, et al. High permitivity ceramic composition, U.S. Patent 4,477,581, 1984.
99. Fujino M, et al. Non-reducible dielectric ceramic composition, U.S. Patent 4.859,641, Aug 22, 1989.
100. Mori Y, et al. Non-reducing dielectric ceramic composition, U.S. Patent 4,959,333, Sept 25, 1990.
101. Nishioka G, et al. Method for producing non-reducible dielectric ceramic composition, U.S. Patent 4,988,468, Jan 29, 1991.
102. Wada N, et al. Method for production of nonreducible dielectric ceramic composition, U.S. Patent No. 5,232,880, Aug 3, 1993.
103. Sano H, et al. Non-reducible dielectric ceramic composition, U.S. Patent 5,248,640, Sep 28, 1993.
104. Ohkubo S, et al. Non-reduction type dielectric ceramic composition, U.S. Patent 5.204,301, Apr 20, 1993.
105. Sano H, et al. Non-reducible dielectric ceramic composition, U.S. Patent 5,264,402, Nov 23, 1993.
106. Sano H, et al. Non-reducible dielectric ceramic composition, U.S. Patent 5,322,828, Jun 21, 1994.
107. Nishiyama T, et al. Non-reducible dielectric ceramic composition, U.S. Patent 5,397,753, Mar 14, 1995.
108. Sano H, Sakabel Y. non-reducible dielectric ceramic composition, U.S. Patent 5,510,305, Apr 23, 1996.
109. Wada N, Kohno Yoshiaki. Production of nonreducible dielectric ceramic composition, U.S. Patent 5,310,709, May 10, 1994.
110. Nishiyama T, et al. Nonreducing dielectric ceramic composition, U.S. Patent 5,268,342, Dec 7, 1993.
111. Buehler BE. Monolithic base metal electrode capacitor, U.S. Patent 3,757,177, 1973.

112. Eror NG, Burn I, Maher GH. Method of forming a ceramic dielectric body, U.S. Patent 3,920,781, 1975.
113. Dorrian JF. Base metal electrode capacitor and method of making the same, U.S. Patent 4,097,911, 1978.
114. Dorrian JF. Base metal electrode capacitor and method of making the same, U.S. Patent 4,241,378, 1980.
115. Dirstine RT. Method of making ceramic dielectric for base-metal electrode capacitor, U.S. Patent 4,386,985, 1983.
116. Sakabe Y, Hamaji Y, Nishiyama T. New barium titanate based material for MLCs with Ni electrode. Ferroelectrics 1992; 133:133–138.
117. Hamaji Y, et al. Effects of rare earth oxides on X7R dielectrics, Proceedings of the 7th US–Japan Seminar on Dielectric and Piezoelectric Ceramics, Nov 1995, Tsukuba, Japan) Wagner KW, ed. Vol. 2, 1914:371.
118. Volger J. Prog Semiconduct 1960; 4:207.
119. Roup RR, Butler CE. U. S. Patent 2,520,376, 1950.
120. Roup RR, Kilby JS. U. S. Patent 2,841,508, 1958.
121. Saburi O. U.S. Patent 3,299,332, 1967.
122. Noorlander W. U.S. Patent 3,386,856, 1968.
123. Waku S. U.S. Patent 3,473,958, 1969.
124. Brauer H, Kuschke R. U.S. Patent 3,569,802, 1971.
125. Brauer H, Kuschke R. Ger. Patent 1,614,605, 1974.
126. Waku S, Nishimura A, Murakami T, Yamaji A, Edahiro T, Uchidate M. Rev Electr Commun Lab 1971; 19:665.
127. Heywang W. J Mater Sci 1971; 6:1214.
128. Daniels J, Härdtl KH. Philips Res Rep 1976; 31:489.
129. Daniels J. Philips Res Rep 1976; 31:505.
130. Hennings D. Philips Res Rep 1976; 31:516.
131. Daniels J, Wernicke R. Philips Res Rep 1976; 31:544.
132. Ihrig H, Puschert W. J Appl Phys 1977; 48:3081.
133. Wernicke R. Phys Status Solidi 1978; 47:139.
134. Frederikse HP, Thurber WR, Hosler WR. Phys Rev 1964; 134:A442.
135. Frederikse HP, Hosler WR. Phys Rev 1967; 161:822.
136. Ozawa K. Japan early patent disclosure 77–10596, 1977.
137. Burn I, Neirman S. J Mater Sci 1982; 17:3510.
138. Wernicke R. Adv Ceram 1981; 1:261.
139. Schmelz H. Powder Metall, Int 1975; 7:176.
140. Park SM, Payne DA. Am Ceram Soc Bull 1979; 58:732.
141. Klee M, et al. Ferroelectric thin films for integrated passive components. Philips J Res 1998; 51(3):363–387.
142. Dimos D, Mueller CH. Perovskite thin films for high-frequency capacitor applications. Annu Rev Mater Sci 1998; 28:397.
143. Ulrich R, Leftwich M. Sizing integrated thin-film capacitors. Elec Engin Times Asia 01 Nov 2001.
144. Baumert BA, et al. A study of barium strontium titanate thin films for use in bypass capacitors. J Mater Res 1998; 13(1):197–204.

145. Klee M, et al. J Appl Phys 1992; 72:1566.
146. Klee M, et al. in. Science of Technology of Electroceramic Thin Films 1995: 99–115.
147. Chem CS, et al. Appl Phys Lett 1994; 64:3181.
148. Buskirk PCV, et al. Jpn, J Appl Phys 1996; 35:2520.
149. Dimos D, et al. Thin-film decoupling capacitors for multichip modules. Microelec Relia 1996; 36(4):559–560.
150. Sone S, et al. Jpn J Appl Phys 1996; 35:5089.
151. Takeshima T, et al. Jpn J Appl Phys 1997; 36:5870.
152. Takeshima T, Sakabe Y. New Ceram 1998; 11:45.
153. Sakabe Y, et al. Multilayer ceramic capacitors with thin (Ba,Sr)TiO$_3$ layers by MOCVD. J Electroceram 1999; 3(2):115–121.
154. Neill DO, et al. Thin film ferroelectrics for capacitor applications. J Mater Sci 1998; 9:199–205.
155. Lines ME, Glass AM. Principles and Applications of Ferroelectrics and Related Materials. Oxford: Clarendon Press, 1977.
156. Jaffe B, Cook WR, Jaffee H. Piezoelectric Ceramics. New York: Academic Press, 1971.
157. Debye P. Polar Molecules. New York: Chemical Catalog Company, 1929.
158. Chan NH, Smyth DM. J Electrochem Soc 1976; 123:1584.
159. Long SA, Blumenthal RN. J Am Ceram Soc 1971; 54:577.
160. Gruver RM, Buessem WR, Dickey CW, et al. U.S. Air Force Tech. Rep. AFML-TR-66-164. 1966:112.
161. Maher GH. Proc. Electron. Components Conf:391, IEEE Press, 1977.
162. Glower D, Heckman RC. J Chem Phys 1964; 41:877.
163. Armstrong TR, Buchanan RC. Influence of core-shell grains on the internal stress state and permittivity response of zirconia-modified barium titanate. J Am, Ceram Soc 1990; 73(5):1268–1273.
164. Marice AK, Buchanan RC. Preparation and stoichiometry effects on microstructure and properties of high purity BaTiO3. IEEE Spectrum 1987; 74:61–70.
165. Weber U, Greuel G, Boettger U, Weber S, Hennings D, Waser R. Dielectric properties of Ba(Zr,Ti)O3-based ferroelectrics for capacitor applications. J Am Ceram Soc 2001; 84(4):759–766.
166. Oberschmidt JM, Humenik JN, A low inductance capacitor technology, Proc. 40th Electronic Components Conference, IEEE, 1990.
167. Humenik N, et al. Low-inductance decoupling capacitor for the thermal conduction modules of the IBM Enterprise System 9000 processors, IBM. J Res Dev Sept 1992; 36(5):935–942.
168. Tummala R, Rymaszewski EJ, eds. New York: Van Nostrand Reinhold, 1989:Chap. 15.
169. Levinson S, Anderson HU, Payne W, et al. Multilayer Ceramic Capacitors: Notes for Professional Development Program. New York: AIChE, 1980:3-66–3-76.
170. Levinson S, Anderson HU, Payne W, et al, AIChE. 1980:4–162.
171. Bolton RL. Temperature compensating ceramic capacitors in the system baria—rare earth oxide—titania, PhD thesis, University of Illinois. 1968.

4

Piezoelectric and Electro-optic Ceramics

Robert W. Schwartz
University of Missouri–Rolla
Rolla, Missouri, U.S.A.

John Ballato and Gene H. Haertling
Clemson University
Clemson, South Carolina, U.S.A.

I. INTRODUCTION

Ceramics encompass a wide range of polycrystalline inorganic nonmetallic materials including materials that are described as electronic, or electrical, ceramics. The spectrum of electronic ceramics includes insulators, ferrites, capacitors, substrates, varistors, PTCRs (positive temperature coefficient of resistance materials), ferroelectrics, pyroelectrics, piezoelectrics, and electro-optic ceramics. All of these materials display properties that are intimately related to their crystal structure, and how the ionic or atomic species that are present react to changes in environmental conditions, such as the application of a voltage or mechanical stress.

Our specific interest in this chapter is centered on those electronic ceramics that demonstrate piezoelectric and electro-optic effects. These materials are highly specialized, being prepared from specifically formulated compositions (often not found in nature), processed under controlled conditions, and fabricated into complex shapes with application-specific engineered properties. Devices and applications that depend critically on piezoelectric, electro-optic, and related effects include:

Loudspeakers	Gas igniters	Clocks/wristwatches	Ink jet printers
Ultrasonic cleaners	Ultrasonic imaging	Ultrasonic cutting	Nebulizers
Thermal imaging	Intrusion alarms	Knock sensors	Delay lines
Ignition systems	Ocean floor mapping	Smoke detectors	Resonators
Fans	Strain gauges	Passive and active sonar	
Global position systems	Motion sensors	Optical routers for the Internet	

Topically, the chapter is broken down as follows: Introduction, Phenomena (piezoelectric and electro-optic), Materials (single-crystal, polycrystalline, electro-optic), Processing, Properties, Devices, Applications, and several appendices that provide greater detail regarding some of the more fundamental aspects of these materials that are covered more generally in the main body of the chapter. This chapter is intended to serve as an introductory overview of the structural aspects, processing, properties, and applications of piezoelectric and electro-optic ceramics. While it is by no means an exhaustive treatise on these materials, it should provide an adequate perspective for readers wishing to explore the subject, and sufficient references are included for further study of the topic.

II. PHENOMENA

A. Piezoelectric Materials

1. Background

The term *piezoelectric* comes from the Greek "piezo," meaning "to press," which has lead to the general description of the piezoelectric response of materials as "pressure electricity." Cady has succinctly defined piezoelectricity as "electric polarization produced by mechanical strain in crystals belonging to certain classes, the polarization being proportional to the strain and changing sign with it" [1]. Piezoelectricity is treated as a linear and reversible property, with a magnitude that is dependent on the magnitude of the stress [2]. Piezoelectricity was originally discovered by Pierre and Jacque Curie in 1880 [3], who found that electric charges developed on the surface of quartz, and other crystals, when they were mechanically stressed. In fact, this is still the simplest description of piezoelectricity: the development of an electrical charge (or voltage) under the application of a mechanical pressure. This characteristic is referred to as the direct piezoelectric effect; however, in many applications, piezoelectric materials are also utilized for the reverse effect, i.e., the deformation (strain) of a material when exposed to an applied electric field. This response is called the converse piezoelectric effect, and it is useful for actuator applications. In contrast, the direct piezoelectric effect is used in what are typically referred to as sensor applications.

The crystal symmetry requirements for the existence of piezoelectricity in a material are discussed below. In addition to quartz, other naturally occurring piezoelectric materials include tourmaline, among other minerals [4]. Today single-crystal piezoelectrics, such as quartz, are widely used for frequency control and time keeping applications. However, polycrystalline piezoelectrics, most often lead zirconate titanate (PZT)–based compositions, are at least as important and find use in applications that range from ultrasonic [5] to sonar imaging [6]. While not confined to a single crystal class, structure, or composition, many of the important polycrystalline materials exist in the perovskite structure [7], which is discussed in detail below. It is also worth noting that materials that crystallize in the perovskite and layered perovskite structures may demonstrate other interesting properties, including electronic conductivity [8], ionic and electronic conductivity [9], and superconductivity [10].

Despite the fact that piezoelectricity was discovered more than 100 years ago and commercially important compositions based on PZT or barium titanate ($BaTiO_3$) have been known for more than 50 years [3], the field of piezoelectric ceramic research is far from stagnant, with exciting developments continuing to this day. Some examples of recent progress include single-crystal relaxor ferroelectrics that demonstrate exceptionally high strain response and coupling coefficients [11–14], templated growth techniques for the synthesis of materials with single-crystal-like properties [15], stress-biased actuators [16–18], the development of modeling approaches to account for the nonlinear response of these materials [19,20], the discovery of a new crystalline phase in the lead zirconate-lead titanate ($PbZrO_3$-$PbTiO_3$; PZT) compositional system that is reported to correlate with the outstanding dielectric and piezoelectric properties of some of the materials within this system [21,22], and last, but by no means least, microelectromechanical systems (MEMS) [23,24]. Accelerometers, microcavity pumps, microsurgical instruments, and other piezoelectric-based devices form part of the nanomaterials revolution that is ongoing and that utilize thin-film materials to form integrated functional devices. These aspects of piezoelectric devices are covered in depth in Chapter 8 of this text.

2. Crystallographic Considerations

Because polycrystalline piezoelectric ceramics display behavior that is more complex than single crystals, we begin our consideration of the effects of symmetry on material response by focusing on single-crystal materials. Single crystals are macroscopic homogeneous solids in which ordered periodic arrays of atoms (or ions) exist, and the smallest repeating unit of the crystal is called the unit cell. One aspect of materials that is frequently overlooked is the relationship between behavior at the atomic level, and material properties at the macroscopic, physical-world level. Especially for single crystals, the behavior at the atomic level dictates the macroscopic behavior. For example, while you might measure the bulk density of a single crystal in a laboratory experiment, it is also possible to calculate the density of the material from an atomistic perspective by knowing the types and number of atoms within the unit cell of the material, and the unit cell parameters. We might use this approach to calculate the density of a material at the atomic level in amu/$\mathring{A}^3$, from which we can determine the theoretical density in g/cm^3 or kg/m^3. In an analogous manner, we might use the Young modulus of silicon to estimate the bond strength of a Si-Si bond, and so forth. As a final example, along these same lines, we might view the macroscopic thermal expansion of a solid as resulting from an increase in the bond length as a function of temperature.

Our insight into the phenomenon of piezoelectricity is improved by considering not only what we observe at the macroscopic level, but also relating this behavior to what is occurring at the atomic level. For example, while we might apply a macroscopic pressure (σ, stress in psi, N/m^2, etc.) to the face of a crystal, or an electric field (E in kV/cm, V/m, etc.) to a pair of electrodes on opposing crystal faces of a piezoelectric material, the effects of these external stimuli are manifested at the atomic level. How the ions respond to this change in external conditions defines the response of the material. The symmetry of the material constrains the ions (or atoms) within the crystal to occupy positions in a specific repeating relationship to each other, thus building up the structure, or lattice, of the crystal. Furthermore, the symmetry of a crystal's internal structure is reflected in the symmetry of its external properties, and this symmetry determines whether or not it is possible for piezoelectricity to exist.

The elements of symmetry that are utilized by crystallographers to define symmetry about a point in space, typically the central point of a unit cell, are (1) a center of symmetry, (2) axes of rotation, (3) mirror planes, and (4) combinations of these [25,26]. If we apply the restriction that during a symmetry operation, such as rotating the crystal about a particular axis, or during a series of symmetry operations, one point must remain unmoved, the result is only 32 unique combinations of symmetry operations, called point groups. Because we have utilized a fixed (unmoved) point in our study of symmetry, and because we describe crystals by the directions that are drawn to intersect at a common point, it is possible to describe the symmetry of any given crystal according to one of the 32 *point groups*. Crystals are thus classified as belonging to one of 32 *crystal classes*, each of which corresponds to a particular point group [25]. These 32 crystal classes are frequently grouped within seven basic crystal systems, which in order of decreasing symmetry are cubic, hexagonal, rhombohedral, tetragonal, ortho-rhombic, monoclinic, and triclinic.

The relationship between the 32 crystal classes and piezoelectric behavior is shown in Figure 1. The notation that is used here to describe the various crystal

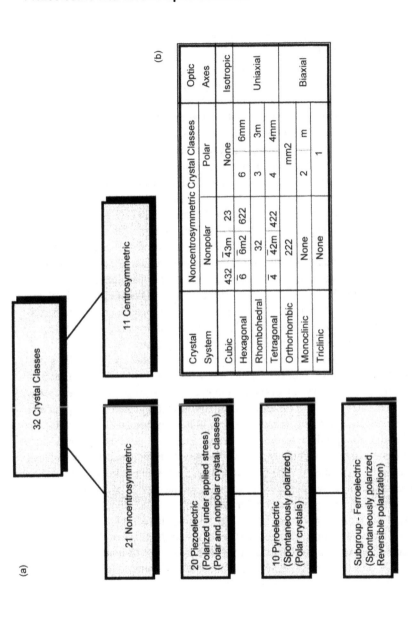

(b)

Crystal System	Noncentrosymmetric Crystal Classes		Optic Axes
	Nonpolar	Polar	
Cubic	432 $\bar{4}$3m 23	None	Isotropic
Hexagonal	$\bar{6}$ $\bar{6}$m2 622	6 6mm	Uniaxial
Rhombohedral	32	3 3m	Uniaxial
Tetragonal	$\bar{4}$ $\bar{4}$2m 422	4 4mm	Uniaxial
Orthorhombic	222	mm2	Biaxial
Monoclinic	None	2 m	Biaxial
Triclinic	None	1	Biaxial

(a)

32 Crystal Classes

11 Centrosymmetric

21 Noncentrosymmetric

20 Piezoelectric
(Polarized under applied stress)
(Polar and nonpolar crystal classes)

10 Pyroelectric
(Spontaneously polarized)
(Polar crystals)

Subgroup - Ferroelectric
(Spontaneously polarized,
Reversible polarization)

Figure 1 (a) Relationship between crystal classes and piezoelectric, pyroelectric, and ferroelectric properties. (b) Specific crystal classes for piezoelectric and pyroelectric materials together with their general optical response. Note: 432 is not piezoelectric.

classes is referred to as conventional notation, although alternative notation forms may also be seen in the literature [25]. Of the 32 crystal classes, 21 do not possess a center of symmetry (a necessary condition for piezoelectricity to exist). Of these, 20 are piezoelectric i.e., they become polarized when subjected to an applied stress. One crystal class, 432, although lacking a center of symmetry, is not piezoelectric because of other combined symmetry elements. Ten crystal classes of the 20 that are piezoelectric form a subset referred to as pyroelectric. These are materials that have a spontaneous polarization at some temperature even in the absence of an applied stress. An alternative description of these materials is that they possess a polar axis, i.e., an axis that demonstrates different properties at either end. The fact that this polarization is temperature dependent allows for the use of these materials in infrared (temperature) sensing and imaging [27,28] because the change in polarization with temperature allows for the generation of current flow in an external circuit. A further subset of pyroelectric materials are ferroelectric materials, which are defined as materials that display a spontaneous polarization whose direction may be reversed by the application of an applied electric field of magnitude lower than the breakdown strength of the material. The defining characteristic of a ferroelectric material is the polarization–electric field hysteresis loop. Nearly all of the polycrystalline piezoelectric ceramics that are commonly employed are also ferroelectric in nature.

As discussed above, the basic concept of piezoelectricity is simple: a mechanical stress, or strain, induces an electrical response such as the buildup of charge on the faces of a crystal, or the development of a voltage across the crystal. It can be seen that a lack of a center of symmetry is all-important for the presence of piezoelectricity when one considers that a homogeneous stress is centrosymmetrical and cannot produce an unsymmetrical result (e.g., a vector quantity such as electric polarization, which has positive and negative ends) unless the material lacks a center of symmetry. The typical response of a piezoelectric material to an applied stress is shown in Figure 2, where a net movement of the positive and negative ions with respect to each other has occurred. As a result of these ionic motions, at the atomic level, an electric dipole is produced, while at the macroscopic level, a polarization (charge/area) develops on the faces of the crystal. For piezoelectricity, the effect is linear and reversible, at least over small ranges for stress [2,29], and the sign of the charge produced is dependent on the direction of the stress, i.e., whether it is tensile or compressive.

Examples of piezoelectric crystals are α-quartz (point group 32), Rochelle salt (222), β-quartz (622), and gallium arsenide ($\bar{4}$3m), and the polarization–stress response is described as the direct piezoelectric effect. A review of Figure 1 indicates that all of these crystals are members of nonpolar crystal classes. That is, they are crystals that do not possess a polarization in the absence of an applied electric field, but only become polarized (develop an electric dipole) under applied stress.

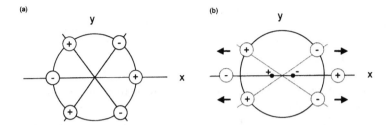

Figure 2 Schematic of (nonpolar) α-quartz (a) prior to and (b) following the application of a tensile stress along the x direction. In (a) the angles and distances of the charged species are such that there is no net dipole. In (b) displacement processes have occurred leading to a separation of the positive and negative centers of charge. (After Ref. 29b.)

We also see that there are 10 crystal classes out of a possible 20 that are designated as pyroelectric. This group of crystals possesses the unusual characteristic of being permanently polarized within a given temperature range. Unlike the more general piezoelectric classes, which only exhibit a polarization under stress, pyroelectric materials develop this polarization spontaneously and form permanent dipoles in the structure, even without an applied stress. This polarization also changes with temperature, hence the term *pyroelectricity*. Pyroelectric crystals are often called polar materials, referring to the unique polar axis of the crystal. Examples of these materials include lithium niobate (3m), barium titanate (single crystal, 4 mm), and poled polycrystalline lead zirconate titanate (equivalent to 6mm). In these and other pyroelectric materials, the length of the polar axis, and thus the magnitude of the dipole moment, varies with temperature. Present-day pyroelectric devices utilize this effect in such applications as intrusion alarms [30], thermal imaging [27,31], and geographical mapping. While the original development of pyroelectric systems was for military applications, over the past 5 years these systems have also found use in police and fire department applications [32]. Both standard bulk ceramic and thin-film devices have been developed [33].

As for the piezoelectric response in nonpolar materials such as Rochelle salt, in materials such as barium titanate ($BaTiO_3$; BT) and lead zirconate titanate ($PbZrO_3$–$PbTiO_3$; PZT), which possess a permanent electric dipole, the magnitude of the dipole is altered by the application of a mechanical stress, resulting in a change in the charge on the surface of the material. Thus, despite the presence of the spontaneous polarization, as for nonpolar materials, the piezoelectric response of pyroelectric materials may again be described as originating from the ionic displacements that result from applied stress. In addition to this *intrinsic* response, in polycrystalline materials, switching effects related to reorientation of the electric dipole also occur. Because these effects are not an inherent property of the material, they are referred to as *extrinsic* effects. This response mechanism is equally important in dictating the piezoelectric response of polycrystalline materials and is discussed further below.

Because of the technological importance of $BaTiO_3$ and PZT, and the fact that these materials are representative of other important perovskite compounds, their crystallography is further considered. Both $BaTiO_3$ and PZT display a range of crystalline forms depending on such external parameters as temperature, composition, mechanical stress, and electric field [7,34]. The behavior of these materials is also typified by a decrease in symmetry with decreasing temperature. For example, in the absence of electrical and mechanical forces, at temperatures above the Curie temperature ($T_c \sim 125°C$ for $BaTiO_3$ and $T_c \sim 490°C$ for $PbTiO_3$), the materials possess a centrosymmetric cubic structure that demonstrates paraelectric behavior (linear dielectric response). In this temperature regime, because of the centrosymmetric crystal structure, no piezoelectric or pyroelectric responses are evidenced. However, upon cooling below T_c, free-energy considerations [34,35] result in a decrease in crystal symmetry, and for titanium-rich PZT compositions and $BaTiO_3$, a tetragonal 4mm structure develops. For single-crystal materials, this is the point group that we utilize to describe electromechanical response. However, for poled polycrystalline ferroelectrics, the symmetry elements are an axis of rotation of infinite order in the direction of poling (the polar axis) and an infinite set of planes parallel to the polar axis as reflection planes [36]. In crystallographic notation, this symmetry would be described as ∞mm, which is equivalent to the dihexagonal polar crystal class 6mm (class 26). Therefore, while we describe the electromechanical response of single-crystal $BaTiO_3$ according to 4mm symmetry, the response of poled polycrystalline $BaTiO_3$ is described by 6mm symmetry. For ferroelectric electro-optic ceramics, these same symmetry conditions hold, with the poling axis being colinear with the optic axis.

The perovskite crystal structure and the ionic displacements within the unit cell that result from the cubic–tetragonal transformation are shown schematically in Figure 3 for $BaTiO_3$. In the cubic state, above the Curie transition temperature, the centers of positive and negative charge are coincident and no electrical dipole is present. In the ferroelectric state, below the Curie temperature, the centers of positive and negative charge are offset, with the Ba^{2+} and Ti^{4+} species displaced in one direction and the O^{2-} species being offset in the opposite direction. These displacements result in an electric dipole (polar axis) in the unit cell. For tetragonal materials, such as $BaTiO_3$ and titanium-rich PZT compositions, upon cooling through the Curie point, there are six equivalent <001> directions along which the polarization may develop. In contrast, in zirconium-rich PZT compositions, where a rhombohedral structure is formed, there are eight possible polarization orientations corresponding to the <111> family of directions. Finally, in $BaTiO_3$ between about $-100°C$ and $0°C$, where an orthorhombic structure is formed,

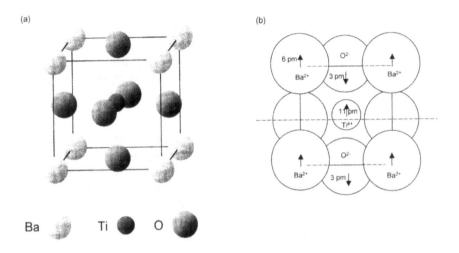

Figure 3   (a) Pseudocubic representation of the perovskite barium titanate and (b) schematic of the ionic displacements in the barium titanate unit cell at 25°C that result in its piezoelectric, pyroelectric, and ferroelectric behavior. (After Ref. 30.)

there are 12 possible polarization orientations corresponding to the <110> family of directions [37].

While the atomic displacements shown in Figure 3 may not appear to be particularly large, calculation of the dipole magnitude associated with an individual unit cell yields excellent agreement with the reported polarization of BaTiO₃ at room temperature [38]. Using the reported magnitudes of the Ba^{2+}, Ti^{4+}, and O^{2-} displacements, which are 6 pm, 11 pm, and 3 pm, respectively [30], and the unit cell parameters of 0.3994 nm (a axis) and 0.4038 nm (c axis) to determine the unit cell volume, a theoretical polarization of 15.5 μC/cm² is estimated. It is these ionic displacements that result in the polar nature (spontaneous polarization) of BaTiO₃ over particular temperature ranges; similar displacements occur in the PZT family of materials.

The presence of the spontaneous polarization in the absence of an electric field results in the description of these materials as polar, or pyroelectric (the subset of piezoelectric materials introduced above). Thus, though we utilize PZT for its piezoelectric response, the fact that it possesses a polar axis is also key in defining its response to applied stress or electric field. Finally, it will be seen that in addition to its piezoelectric and pyroelectric response, materials such as BaTiO₃ and PZT compounds are also ferroelectric i.e., the orientation of the electric dipole can be reversed by applying an electric field. This is particularly important in a process called poling, which is used to prepare polycrystalline ceramic bodies with a preferred (macroscopic) dipole orientation and, thus, piezoelectric and pyroelectric response. Ferroelectrics are a further subset of pyroelectric materials, although materials that are ferroelectric in nature are identified empirically, in part, through the measurement of a saturated hysteresis loop.

3. Thermodynamic Aspects

The above discussion suggests that the basic concept of piezoelectric (electromechanical) response is quite simple, and from many perspectives it may be viewed as such. However, a more complete understanding of piezoelectricity also requires an understanding of the interrelationships among the mechanical, electrical, and electromechanical properties of these materials. Mechanical properties describe the stress–strain response of a material, using either the compliance or stiffness as the coefficient that relates these properties, depending on which property we treat as dependent and which we treat as the external parameter that is being varied. The electrical properties of a material describe its electric field—dielectric displacement response (for our purposes, we can consider this to be polarization), and the coefficient that relates these properties is either the permittivity or the impermeability. Finally, if we are interested in describing the pyroelectric response of a material, we can talk about the polarization that develops as a function of a change in temperature by using the pyroelectric coefficient.

All of these properties bear a thermodynamic relation to one another that is well illustrated using a construct called a Heckmann diagram, which is shown in Figure 4 [7,39]. In this diagram, mechanical stress (σ), electric field (E), and temperature (T) are represented by the outer three corners of the triangle. The corresponding dependent properties—strain (S), dielectric displacement (D), and entropy (S)—are shown in the inner three circles of the triangle. These parameters are linked, respectively, by the compliance (s), permittivity (ε) and heat capacity (actually C/T). Each corner of the triangle illustrates a *principal* effect, and the three corners of the triangle thus represent the mechanical, electrical, and thermodynamic properties of the material. Later we will see how this diagram may be

───▶

Figure 4   (a) Heckmann diagram showing the interrelationships among the mechanical, electrical, and thermal properties of materials. Also shown are coupled properties, including piezoelectric and pyroelectric response, as well as thermal expansion. Large circles represent intensive variables (forces); small circles represent extensive variables (displacements). Heavy arrows represent principal properties; dotted lines represent coupled properties. (b) Heckmann diagram including principal electrical, mechanical, and thermal property relationships and electromechanical coefficients. Note that asterisks indicate indirect effects.

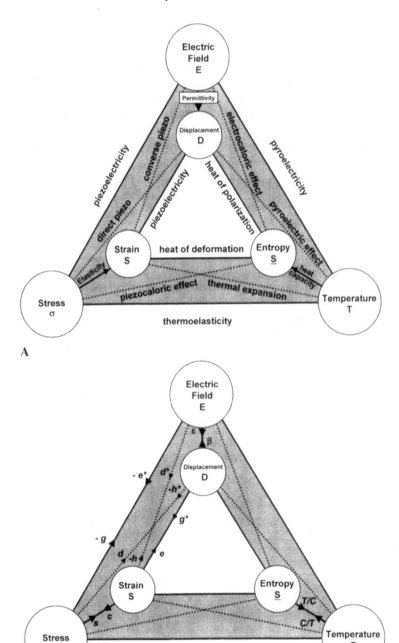

A

B

used in conjunction with standard thermodynamic notation to describe piezoelectric behavior.

Our division of material properties according to mechanical or electrical effects, etc., enables us to describe how a material has responded to a specific change in external conditions. However, the Heckmann diagram clearly shows that all of these material properties are interrelated and, therefore, must have a common origin. It is the underlying thermodynamic aspects of these materials (actually, all materials) that dictates their behavior. Hence, it is appropriate to use a thermodynamically based treatment to describe the electrical, mechanical, or electromechanical response of a material, etc. For example, we might represent the thermal behavior as the change in entropy (S) for a given change in temperature as:

$$dS = \frac{C}{T} dT \tag{1}$$

where C is the heat capacity. Using the same approach that was employed for (1), we could also describe the mechanical strain resulting from an applied stress, in terms of compliance, which is the coefficient that relates the strain to stress. Finally, in an exactly analogous fashion, we can describe the change in dielectric displacement (polarization) in terms of the applied electric field as:

$$dD = \varepsilon dE \tag{2}$$

This expression shows that ε, the permittivity, is the thermodynamic coefficient that relates the change in dielectric displacement to the change in applied electric field. We may also express ε as:

$$\varepsilon = \left(\frac{\partial D}{\partial E} \right) \tag{3}$$

where we have not yet begun to worry about what additional external parameters are held constant, or the directional nature of the response. This expression is also simplified from the perspective that we have not yet considered that we need to treat these coefficients, as well as the properties that they relate, as tensors, and not simply scalar quantities. We must utilize tensor notation because of the anisotropic nature of piezoelectric materials and the fact that we can apply mechanical stresses, or electric fields, in different crystallographic directions, and may thus generate electrical and mechanical responses in different directions. This aspect of material behavior is important to a more complete understanding of piezoelectrics and it will be further discussed shortly.

By now this method of describing the behavior of materials may begin to appear familiar, since it is the same approach that is used in classical thermodynamics. We may use one of the basic thermodynamic relationships, e.g., the Gibbs energy, as a way to introduce the concept of coupled property relationships, such as piezoelectricity, which represents the coupling of electrical and mechanical properties. Using the same approach, we could also consider pyroelectricity, which represents a coupling between thermal and electrical effects. Other coupled properties are also shown in the Heckmann diagram of Figure 4.

To develop the desired expression for piezoelectricity, consider first the expression for the Gibbs energy:

$$dG = -SdT + VdP \qquad (4)$$

where G is the Gibbs energy, V the molar volume, and P the pressure; all other terms have the same meanings as above. Considering our earlier expressions for the dielectric displacement and the permittivity, from the equation for Gibbs energy, we may write:

$$S = -\left(\frac{\partial G}{\partial T}\right)_P \qquad (5)$$

$$V = \left(\frac{\partial G}{\partial P}\right)_T \qquad (6)$$

where the subscripts denote the extensive parameter held constant [40]. These relations signify that entropy is the coefficient that relates the change in the Gibbs energy to a change in temperature and molar volume is the coefficient that relates the change in Gibbs energy to the change in pressure.

Given this example, we are now prepared to write an equation for the electromechanical response of a piezoelectric material. For the direct piezoelectric effect, we want to express the charge, or polarization, that is developed as a result of an applied stress. Since we are talking about the polarization developed, we must also consider the effects of an applied electric field. We may therefore write:

$$dD = \left(\frac{\partial D}{\partial \sigma}\right)_E d\sigma + \left(\frac{\partial D}{\partial E}\right)_\sigma dE \qquad (7)$$

Inserting our definition above for the permittivity and introducing d as the piezoelectric coefficient for the direct piezoelectric effect,

$$d = \left(\frac{\partial D}{\partial \sigma}\right)_E \qquad (8)$$

we may write:

$$dD = d\sigma + \varepsilon dE \qquad (9)$$

which shows the change in polarization as function of both an applied stress and an applied electric field. This equation highlights, as well as quantifies, the polarization response of piezoelectric materials when subjected to an applied stress or applied electric field.

4. Piezoelectric Coefficients

A variety of piezoelectric relations are employed to describe the electromechanical response of these materials under a variety of boundary conditions, but the above approach leads to two of the most commonly used expressions. These describe the dielectric displacement that is developed when the material is subjected to a change in stress, or the resulting strain that develops under an applied electric field:

$$D = d\sigma + \varepsilon^{\sigma} E \tag{10}$$

$$S = s^{E}\sigma + dE \tag{11}$$

where σ is the applied stress, ε^{σ} is the permittivity at constant stress (also called the free permittivity), E is the applied electric field, S is the strain, s^{E} is the compliance under closed-circuit conditions, and the other parameters have the same meaning as before. Depending on the boundary or operational conditions, other important equations that describe electromechanical response include:

$$\sigma = c^{E}S - eE \qquad \sigma = c^{D} S - hD \tag{12ab}$$

$$E = -hS + \beta^{s} D \qquad E = -g\sigma + \beta^{\sigma} D \tag{13ab}$$

$$D = eS + \varepsilon^{s}E \qquad S = s^{D}\sigma + gD \tag{14ab}$$

where c^{E} and c^{D} are the closed and open-circuit compliances, respectively; ε^{σ} and ε^{s} are the free (constant stress) and clamped (constant strain) permittivities, respectively; and β^{σ} and β^{s} are the free and clamped impermeabilities, respectively. The other parameters may be ascertained from consideration of Eqs. (8) and (9).

    While Eqs. (10) and (11) show the basic relationships for the direct and converse piezoelectric effects, they still do not go far enough in terms of describing the response of a piezoelectric material to a change in external mechanical or electrical boundary conditions. Because of the inherent asymmetrical nature of piezoelectric materials (the noncentrosymmetric structure and the presence of the polar axis), as well as the fact we will need to be able to predict different response directions, a tensor-based perspective is frequently employed to describe the piezoelectric response of these materials. This method allows accurate description of the response of both single-crystal and poled polycrystalline piezoelectric ceramics. An excellent source of information on the details of the approach may be found in *Physical Properties of Crystals* by J. F. Nye [39], although the basic aspects of the method are given below.

To consider the directionality of the response of piezoelectric materials, a subscript notation is included in the above equations, along with the superscript notation that is already shown that describes the applicable boundary conditions. This requires that we define the crystallographic directions of the material, which is done according to the method shown in Figure 5. When applied to mechanical behavior, the subscripts, 1, 2, and 3 define normal stress (or strain) directions in the material, whereas the subscripts 4, 5, and 6 define shear stresses (strains) about 1, 2, and 3 axes, respectively. For single crystals, the polar axis is taken as the 3-direction. For poled polycrystalline bodies, the poling direction is taken to be the z or 3 direction. Frequently in the use of these materials, electrodes are applied to the "z" crystal faces, which are perpendicular to the "3" direction and an electric field (E_3) is applied parallel to the polar axis, i.e., the "3" direction.

The subscripts that we will use to describe the directionality of both the response and the applied field, or stress, will appear on each of the terms in the above expressions. For example, when Eq. (10) is used to describe the response of a piezoelectric material to stress or electric field, D will appear as D_i, where

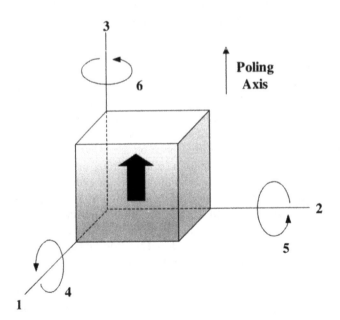

Figure 5 Notation of axes for piezoelectric materials including single crystals and poled polycrystalline bodies.

i may have values of 1, 2, or 3; d will appear as d_{ij}, where i and j may have values of 1, 2, or 3; σ will appear as σ_j, where j may have integral values between 1 and 6, indicating normal (1, 2, or 3) or shear (4, 5, or 6) stresses; ε will appear as ε_{ii}, where i may have values of 1, 2, or 3; and E will appear as E_i, where i may have values of 1, 2, or 3. Because of the symmetry of the materials in which we are interested, the ε coefficient we will encounter will typically be either ε_{r11} or ε_{r33}. (These are also frequently represented in the literature by K_{11} or K_1, and K_{33} or K_3, with the appropriate boundary condition indicated as a superscript.) A similar approach is used to include the appropriate subscript notation into Eq. (11).

We may now describe completely the response of piezoelectric materials through the following expressions that include information on boundary conditions as well as the directionality of the response:

$$D_i = d_{ij}\sigma_j + \varepsilon_{ii}^{\sigma} E_i \tag{15}$$

$$S_j = s_{ij}^{E}\sigma_j + d_{ij} E_i \tag{16}$$

The piezoelectric coefficient d_{ij} indicates the electromechanical response of the material. For Eq. (15), the piezoelectric coefficient shows the electrical output for a given mechanical input, or in Eq. (16), it shows the mechanical output for a specific electrical input. The first (i) subscript represents the electrical direction and the second (j) subscript represents the mechanical direction. For the direct piezoelectric effect, a mechanical stress σ; results in a charge on the i faces of the materials; for the converse effect, an electric field applied along the i direction induces a deformation along the j direction.

It is worth noting that this approach (the use of two subscripts to denote the piezoelectric d coefficient) is referred to as the *reduced notation* description of piezoelectricity, and that it gives an accurate description of electromechanical response yet greatly simplifies the mathematics that must be carried out compared to the *full-notation* description. In the reduced notation approach, mechanical stress and strain are described as first-rank tensors (σ_1 through σ_6 and S_1 through S_6, respectively), when in reality, these are second-rank tensor properties, and should be described in terms of two axes, one for the direction of the force and one for the face to which the force is applied. Thus, in the full-notation description, two subscripts are required for stress and strain and three are required for the d coefficient. A brief overview of the basis for the reduced notation, and the relationship between reduced and full notation piezoelectric coefficients, is given in Appendix A. Additional details are given in Refs. 2 and 39.

The expressions shown as Eq. (15) and (16) allow for calculation of the dielectric displacement or strain of piezoelectric materials when subjected to stress or an applied electric field. Useful simplifications in describing the piezoelectric response of single-crystal and poled polycrystalline materials result from the symmetry considerations shown in Appendix B, and sample calculations using this approach are shown in Appendix C. Finally, it is also important to note that the piezoelectric charge coefficient $d_{ij} = (\partial D_i/\partial \sigma_j)_E$ and piezoelectric strain coefficient $d_{ij} = (\partial S_j/\partial E_i)_\sigma$ are numerically equal. While this is not obvious from the above equations because of the different units that must be associated with these coefficients (C/N for the charge coefficient and m/V for the strain coefficient), a thermodynamic proof that

$$\left(\frac{\partial D_i}{\partial \sigma_j}\right)_E = \left(\frac{\partial S_j}{\partial E_i}\right)_\sigma \tag{17}$$

is given in Appendix D.

In addition to the charge and strain d coefficients, many other property coefficients (mechanical, electrical, and electromechanical) are used to describe the behavior of piezoelectric materials. Typical examples of frequently used properties in describing electromechanical response are given below.

d_{31} = piezoelectric strain coefficient
    └─→ stress or strain direction stress
    └─→ displacement or electric field direction (electrodes are perpendicular to the 3-axis)

d_{33} = charge or strain coefficient; longitudinal charge density/applied stress, C/N (strain developed/applied field, m/V)

d_{31} = charge or strain coefficient, lateral

S^E = compliance at constant electric field (electrodes shorted)

S^D = compliance at constant displacement (open circuit)

ε_r^σ = dielectric constant at constant stress (free)

ε_r^S = dielectric constant at constant strain (clamped)

ε_O = dielectric constant of free space, $\varepsilon_O = 8.85 \times 10^{-12}$ F/m

S_{31} = mechanical compliance, lateral, m^2/N

g_{31} = voltage coefficient, lateral, V-m/N (open-circuit electric field/applied stress)

g_{15} = voltage coefficient, shear, V-m/N (open-circuit electric field perpendicular to 1-axis/applied shear stress around 2-axis)

k_{33} = electromechanical coupling factor, longitudinal, no units

k_{31} = electromechanical coupling factor, lateral, no units

k_t = electromechanical coupling factor, thickness, no units

k_p = electromechanical coupling factor, planar (radial), no units (thin disk only)

ν = Poisson's ratio of cross contraction; typically about -0.3, no units

ρ = material density, kg/m^3

The piezoelectric charge coefficients (d_{33} and d_{31}) and the voltage coefficients (g_{33}, g_{31}, and g_{15}) describe the electromechanical characteristics of the piezoelectric material. High d coefficients are desirable for those materials utilized in motional or vibrational devices, such as sonar and sounders. Using the Heckmann diagram shown in Figure 4b, or the definitions given above and the version of the Heckmann diagram in Figure 4a, it is easily shown that the g coefficient is related to the d coefficient via the dielectric constant according to:

$$g = \frac{d}{\varepsilon_r \varepsilon_0} - \left(\frac{\partial E}{\partial \sigma} \right) = \left(\frac{2S}{\partial D} \right)$$

(18)

High g coefficients are desirable for materials intended to produce voltages in response to mechanical stress, such as gas igniters and phonograph pickups. Other relationships among the various electromechanical properties may also be derived through the use of the Heckmann diagram. Finally, typical values of piezoelectric coefficients for quartz, barium titanate, and poled polycrystalline lead zirconate titanate are shown below in the section on properties.

Another property of importance in the description of piezoelectric materials is the coupling coefficient, which describes the efficiency of the conversion of electrical to mechanical energy (or mechanical to electrical energy). The coupling coefficient may also be considered as a measure of the strength of the electromechanical response of the material.

Mathematically, the coupling factor may be expressed in general terms as:

$$k_{eff}^{2} = \frac{\text{electrical energy converted to mechanical energy}}{\text{input electrical energy}}$$

(19)

$$k_{eff}^{2} = \frac{\text{mechanical energy converted to electrical energy}}{\text{input mechanical energy}}$$

(20)

The coupling coefficient is typically determined by impedance spectroscopy and measurements of the minimal and maximal impedance frequencies associated with the resonance of the piezoelectric. While these are not exactly the resonance and antiresonance frequencies, they are more easily determined and thus are more frequently used [2]. We may therefore approximate the effective coupling coefficient as

$$k_{eff}^{2} \approx \frac{f_a^2 - f_r^2}{f_a^2}$$

(21)

where f_a is the antiresonance frequency (the frequency at which the reactance is zero) and f_r is the resonance frequency (the frequency at which the susceptance is zero). It may be seen from Eq. (21) that the greater the separation of the resonance and antiresonance frequencies, the higher the coupling coefficient. Typical coupling coefficients for perovskite materials are in the range of 0.21

(k_{31} for barium titanate) to about 0.70 (k_{33} for soft lead zirconate titanate) poly-crystalline ceramic compositions. The generally higher coupling coefficients of the PZT materials is one of the main reasons why PZT has largely replaced BT in piezoelectric applications. However, barium titanate remains the key constitu-ent for the multilayer ceramic capacitor industry (see Chapter 3 in this text). Recent investigations of single-crystal relaxor ferroelectrics, such as PMN-PT [11], have uncovered materials with coupling coefficients as high as 0.90 that are capable of generating high strains. These materials have received great atten-tion and offer the promise of sensors with significantly greater performance.

A review of the literature on piezoelectric ceramics will also reveal other coupling coefficient expressions based on material compliance under closed- or open-circuit conditions. The most common of these relationships is expressed by

$$S_{33}^E = \frac{S_{ss}^D}{1 - k_{33}^2} \qquad (22)$$

which illustrates that as the coupling coefficient increases, the closed-circuit com-pliance decreases in magnitude compared to the open-circuit compliance. Another commonly encountered expression involving the coupling coefficient is

$$d_{33} = k_{33} \, (\varepsilon_{33}^\sigma S_{33}^E)^{1/2} \qquad (23)$$

This expression is particularly interesting to consider because it captures concisely the interrelationship among the different properties of piezoelectric materials that are important in their use. That is, the equation shows mathematically that the electromechanical properties of piezoelectrics (d_{33}) are directly related to their electrical (ε_{33}^σ) and mechanical (S_{33}^E) properties. In this context, the coupling coefficient may be thought of as how strongly mechanical and electrical responses are linked. This mixing of effects (i.e., not simply mechanical or simply electrical, but rather, electromechanical) provides great insight into the fundamental nature of materials. Many material properties are interdependent, although the effect may be weak; sometimes a mixed effect is desirous (e.g., piezoelectricity), and sometimes we try to mitigate it (e.g., nonlinear frequency conversion in ultrahigh-capacity optical fibers). It is instructive to derive either of the above relationships, since additional insight may be gained into why, *physically*, piezoelectrics behave differently under different boundary conditions.

5. Single Crystal Versus Polycrystalline Behavior

(4) Overview of Microstructural Considerations. Single crystalline materials are characterized by a uniform polarization throughout the crystal. Based on this characteristic, the electromechanical response is easily described from the crystallographic properties of the material. When one now considers a ceramic

consisting of a multitude of tiny, piezoelectric crystallites, each with random orientation, quite a different situation from that of a single crystal is obtained. This random orientation causes the overall ceramic to be piezoelectrically inactive, with no electromechanical response detectable until some means is found to polarize the ceramic as a whole entity. Prior to the attainment of a polarized body, the random orientation of each of the piezoelectric crystallites results in local electromechanical responses that cancel one another. That is, while one crystallite may demonstrate an elongation for a given applied voltage, at the same time another crystallite within the body will undergo an equivalent shrinkage. In contrast, the polarized ceramic possesses a net macroscopic polarization that will respond piezoelectrically to an applied electric field or stress.

To obtain a polarized body with piezoelectric crystallites, techniques such as extrusion, hot forging, or directional recrystallization must be used during fabrication. However, a much easier method, called ''poling'' (discussed further below), is typically employed. The use of this process requires special compositions that are not only piezoelectric but also ferroelectric, i.e., materials that demonstrate reversibility of the spontaneous polarization under an applied electric field. Most of the compositions employed for transducer applications are, in fact, both piezoelectric and ferroelectric, and poling is the process that is nearly always used to obtain a ceramic body with a net macroscopic polarization.

One of the key aspects of ceramic microstructure that is important both in the poling process and in defining the electromechanical response characteristics of piezoelectric ceramics is the domain structure of the material. Domains are regions of uniform polarization within the ceramic. To put this definition into more physical terms, a domain is a region, typically within a single crystallite of the polycrystalline body; for example, in PZT in which all of the titanium atoms have been displaced from their central position in the same direction. This leads to the frequent description of ferroelectricity as a cooperative ordering phenomenon.

Because most processing routes utilize elevated temperatures during processing, the crystallographic phase present at the sintering temperature will be the cubic, paraelectric phase. Upon cooling through the Curie transition temperature, the ferroelectric phase is formed. For barium titanate, this occurs at 125°C, and at this temperature electrical dipoles are formed within the material. Also at this temperature there is an elongation along one of the crystallographic directions (the polar c axis) and a reduction in dimension along the other directions (the a axis for tetragonal materials). As with any material, piezoelectric ceramics will tend to attain the lowest Gibbs energy configuration. In the present case, this means that the material will attempt to minimize the electrostatic and mechanical strain energy associated with the phase transformation into the ferroelectric state. Therefore, the domain structure that is developed is the result of both of electrical and mechanical considerations, as well as the crystallographic constraints that are present. For tetragonal barium titanate, this means that electrical dipoles will have either a parallel or an antiparallel arrangement, or will be at right angles to each other. In tetragonal barium titanate, the domain walls are thus 90° or 180°. In contrast, rhombohedral ferroelectics, e.g., zirconium-rich PZT compositions, have 71°, 109°, or 180° boundaries due to the <111> nature of the polar axis. A typical domain structure for barium titanate is shown in Figure 6 together with a schematic of the associated domain orientations.

The fact that the domains can be realigned by an electric field leads to ferroelectricity and the ability to pole the ceramic, making it piezoelectrically active. It also leads to significantly greater electromechanical response, which is described as "extrinsic" in nature. Unfortunately, this behavior also leads to nonlinearities and hysteresis in electromechanical response as a function of electric field that for most actuator designers and users of piezoelectric ceramics present numerous challenges. Suffice it to say that domains play a key role in the observed behavior of piezoelectric materials and that effects related to domain structure have been intensively studied [41–43]. It is also worth noting briefly that the changes in domain configuration under mechanical stress are important and contribute to both the mechanical and electromechanical response of the PZT and BT materials. This behavior also leads to the description of these materials as being "ferroelastic" as well as ferroelectric.

Because of the importance of both the ferroelectric and ferroelastic nature of these materials, a few additional comments on their behavior is appropriate. The typical changes in sample dimensions and polarization with applied electric field are shown in Figure 7 below. The trace labeled as the virgin curve shows the changes that occur upon the initial application of an electric field such as that utilized during the poling process. This process promotes the realignment of domains in the direction of the applied field. Upon removal of the field, the ceramic body is left with a preferred domain alignment that is required for piezoelectric response. The polarization that is present after the removal of the field is called P_r, the remanent polarization. The process also results in a permanent deformation of the ceramic body with an elongation along the poling direction and a decrease in the lateral dimensions (those perpendicular to the field) due to the anisotropy of the unit cell and domain reorientation. Application of a strong electric field in the reverse direction promotes both 90° and 180° domain switching with the development of a net macroscopic polarization in the opposite direction. The field at which the polarization returns to zero is described as the coercive field, which is described by E_c. The changes in polarization that occur when the material is subjected to the electric field are paralleled by changes in strain. Both responses have intrinsic (unit cell) and extrinsic (domain switching) contributions, although the ferroelectric hysteresis loop and the S–E butterfly loop are directly related to the domain structure and switching characteristics of the material. The polarization–electric field (P–E) hysteresis loop is one of the defining characteristics of a ferroelectric. Further details are given below in the section on properties.

(b)

180° domain wall

90° domain wall

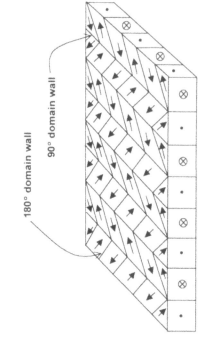

(a)

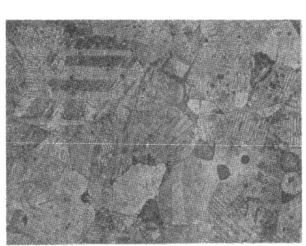

Figure 6 (a)Typical herringbone domain structure and (b) schematic of domain configuration for a barium titanate ceramic. Domain wall characteristics are indicated. [(a) after Ref. 43b.]

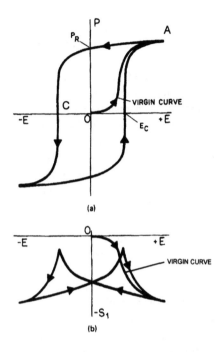

(a)

(b)

Figure 7 (a) Polarization–electric field hysteresis loop and (b) strain–electric field butterfly loop for a ferroelectric ceramic. Note that the strain (S_1) is shown for a direction perpendicular to the poling direction. O indicates the origin; A is at approximately $2 \times E_c$; and C corresponds to $E = 0$.

In using a piezoelectric ceramic, typically subcoercive field voltages are utilized to avoid concerns regarding depoling of the ceramic and the loss of piezoelectric activity. Also, the d coefficient is treated as a linear parameter by considering the strain–field response near zero electric field as shown in the figure. The reader is referred to the IEEE standards on ferroelectricity and piezoelectricity for further detail on measurements and definitions of ferroelectric and piezoelectric constants [2,36,44].

(b) Intrinsic and Extrinsic Response Mechanisms. In single-crystal materials, the observed piezoelectric response is due solely to the intrinsic response of the material at the unit cell level. The intrinsic response mechanism, also called the lattice contribution, originates from atomic displacements within the individual unit cells of the material. Upon the application of an electric field (or mechanical stress), charged species within the lattice are displaced from their zero-field positions. Because of the anionic and cationic nature of the species, they are displaced

in opposite directions, and deformation of the crystal, at both the atomic and macroscopic levels, results. The resulting strain from this effect is the intrinsic response mechanism.

Unlike single crystals, the piezoelectric response of poled polycrystalline ceramics consists of two contributions: intrinsic and extrinsic, which are shown schematically in Figure 8. The extrinsic response mechanism is the strain that results principally from non-180° domain wall motion [29,43]. The basic aspects of this effect are shown in Figure 8b, where the domain orientation preferred under the applied electric field is shown to grow at the expense of the other domain by 90° domain wall translation. In tetragonal ferroelectric materials, 90° domain wall motion of this nature (also called switching for higher electric fields) results in strain components both parallel and perpendicular to the electric field. Great attention is given to the extrinsic response mechanism because it is typically the larger electromechanical response mechanism in poled polycrystalline ceramics such as BT- and PZT-based materials [7]. The reason for this is the relatively large anisotropy present in BT and PZT materials. For example, in BT the c/a ratio is ~ 1.009 (anisotropy $\sim 0.9\%$). In PZT ceramics, the crystalline anisotropy is even larger, and for compositions near the morphotropic phase boundary is in the range of 2%. Therefore, comparatively large strains can be obtained from apparently minor changes in domain population and the sum of the intrinsic and extrinsic contributions defines the overall response of poled polycrystalline piezoelectrics. Domain wall translation also results in nonlinearities in the mechanical (compliance or stiffness) and electrical (dielectric) behavior of piezoelectric ceramics [45–47].

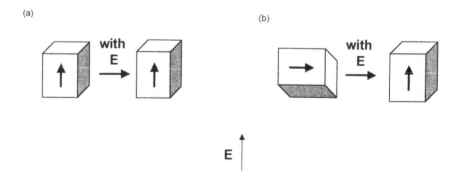

Figure 8 (a) Intrinsic (shown for single unit cell) and (b) extrinsic (shown for single domain) response mechanisms in ferroelectric ceramics. Note the field direction indicated at the bottom of the figure and that the intrinsic response magnitude, as well as the magnitude of the anisotropy of the unit cell, are exaggerated.

Because of the importance of domain effects on response, these aspects of ferroelectric ceramics have been studied in some depth. The temperature dependence of the extrinsic mechanism has been characterized by Zhang and coworkers [48], while other investigators have looked at the nonlinear nature of the response due to changes in domain wall translation contributions [29]. Detailed considerations of the effects of domain wall motion on piezoelectric response may be found in articles by Damjanovic and Hall, where a Rayleigh approach is used to account for the variations in the extrinsic contribution as a function of field (and other parameters) [29,43,49]. These articles also effectively highlight the interrelationships among dielectric, electromechanical, and mechanical properties that are expected based on the thermodynamic relationships shown in the Heckmann diagram.

A typical analysis of extrinsic contributions to dielectric permittivity is illustrated in Figure 9 for a hard PZT ceramic. The figure illustrates three regimes: the low-field regime, $E_0 < 0.4$ kV/mm, in which the relative permittivity, ε_r, is a nearly field-independent constant, an intermediate Rayleigh region where ε'_r increases linearly with E_0; and a high-field regime ($E_0 > E_c$) in which domain switching processes become evident. In the low-field region, the magnitude of the electric field is insufficient to promote significant domain wall motion and the behavior is dictated (primarily) by the intrinsic ionic response and reversible domain wall vibration [29]. In the high-field region, the dielectric constant shows a dramatic increase due to 90° and 180° domain switching processes that are associated with electric fields above the coercive field. In the intermediate regime, ε_r may be expressed using a Rayleigh relationship [29,43,49] of the form:

$$\varepsilon_r^*(E_o) = \varepsilon_r^* (0) + \alpha_d E_o \tag{24}$$

where $\varepsilon_r^* (0)$ is a field-independent term and α_d is a constant, in this instance, the dielectric Rayleigh constant that describes the contribution of larger scale irreversible domain wall translation as a function of field magnitude [29,49]. Other authors have used an analogous method to describe domain wall motion contributions to the piezoelectric coefficients of materials [49]. Finally, still others have looked at the effects of boundary conditions on the balance of intrinsic and extrinsic contributions to piezoelectric response [45,46,51–53]. Analyses in these areas are being used to develop control algorithms to account for the nonlinear nature of the piezoelectric response of polycrystalline ceramics. In some of these investigations, new finite element models of ferroelastic polycrystalline bodies and piezoelectric multilayer actuators [20] have been developed that account for nonlinearities and switching effects. The importance of extrinsic effects on the electromechanical performance of devices such as polydomain single-crystal actuators [54] and stress-based actuators, e.g., Thunder, Rainbow, and composite actuators, have also been investigated [16,17,20,55].

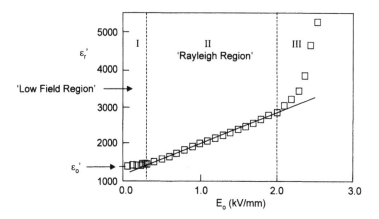

Figure 9 Illustration of the dependence of the dielectric permittivity as a function of electric field for Ferroperm PZ26, a hard PZT ceramic. (After Ref. 50.)

6. Electrostriction

In addition to piezoelectricity, electrostriction, a phenomenon that occurs in all materials, is a second mechanism that may contribute to electrically induced deformation. Electrostriction may be described simply as the mechanical deformation of a material under an applied electric field. From this brief description, electrostriction may seem analogous to piezoelectricity; however, there is no corresponding "direct" effect in electrostriction. That is, while piezoelectric materials demonstrate both mechanical-to-electrical and electrical-to-mechanical responses, electrostriction implies only the electrical-to-mechanical response. Another difference is that there are no crystallographic constraints on the electrostrictive response, while only 20 of 32 crystal classes exhibit piezoelectricity. Even glasses and liquids are subject to electrostrictive strains [39]. A final significant difference between the electrostrictive and piezoelectric responses is that electrostriction is quadratic, i.e., proportional to the square of the applied field, and as such, the response is independent of the polarity of the applied field. As such, electrostriction represents a nonlinear coupling between the elastic and electric response of materials. Typical deformations resulting from electrostriction are illustrated in Figure 10. This strain–electric field response may be compared to the piezoelectric response shown in Figure 7b.

Assuming the absence of stresses that would cause a mechanical deformation, and using the full-notation approach, the magnitude of the electrostrictive response in anisotropic materials may be expressed by [39]:

$$S_{ij} = \gamma_{ijkl}E_kE_l \tag{25}$$

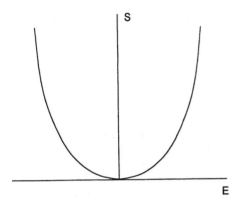

Figure 10   Strain–electric field response of electrostrictive materials.

or

$$S_{ij} = Q_{ijkl}P_kP_l \tag{26}$$

where S_{ij} is the strain, γ_{ijkl} and Q_{ijkl} are electrostrictive coefficients, E_k and E_l represent electric field, and P_k and P_l denote polarization. Electrostrictive strains may thus be expressed in terms of either the polarization or the electric field. The electrostrictive coefficients are fourth-rank tensors since they relate strain (a second-rank tensor) to electric field. If we express the total strain in terms of the piezoelectric and electrostrictive coefficients, we may write:

$$S_{ij} = d_{ijk}E_k + \gamma_{ijkl}E_kE_l \tag{27}$$

For materials that are strongly piezoelectric, e.g., PZT compositions near the morphotropic phase boundary [56], the piezoelectric effect (the d coefficient) usually contributes greater than 90% of the total strain response of the material. However, other cubic PLZT compositions, such as 9065 (also called 9/65/35; 9 mol % La on the A site and a Zr/Ti molar ratio of 65:35) demonstrate a strong electrostrictive response since they develop substantial polarizations under an applied electric field. Relaxor ferroelectric materials such as lead magnesium niobate (PMN) and its solid solutions with lead titanate (PMN-PT) also demonstrate strong electrostrictive responses. Therefore, electrostrictive contributions also become important at high electric field.

7. Equivalent Circuit Modeling

Piezoelectric materials are finding progressively greater use in modern frequency control and communication devices, particularly ones with ever diminishing dimensions such as those in microelectromechanical and micro-optoelectromechanical systems (MEMSs, MOEMSs). These devices require performance uniformity to an extent never imagined in the past. This necessitates the ability to characterize and measure precisely the properties of the materials utilized in these devices. The material properties of direct consequence to performance are the dielectric, piezoelectric, and elastic constants. Modern measurements are electrical in nature, and the characterization utilizes highly accurate equivalent electrical networks in order to describe the physical processes involved. This section summarizes traditional, though limited in precision, equivalent circuit models, as well as newer, exact versions.

(a) *The Butterworth–van Dyke Circuit.* The traditional Butterworth–van Dyke (BVD) equivalent circuit [57–59] of a resonant system that is driven by a capacitive electric field has been used for many years to model piezoelectric resonators [60,61]. The traditional BVD circuit is given in Figure 11. A distinction is made depending on the direction of the field excitation with respect to the plate thickness; the distinction is represented by the differences in Figure 11a and 11b, where TE refers to a thickness excitation that occurs when electrodes are placed on the major plate surfaces to produce a thickness-directed field and LE refers to placement of electrodes on the sides of the plate yielding a lateral excitation. For simplicity, assume the resonator structure is a piezoelectric plate

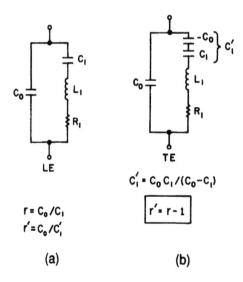

Figure 11   Traditional Butterworth–van Dyke equivalent circuit analogs for modeling piezoelectric resonators: (a) lateral excitation (LE) and (b) thickness excitation (TE).

resonator having an infinite lateral extent. This limits the case to a one-dimensional approximation and leads to simplified expressions for the circuit parameters in terms of the material constants of the structure. In this case, C_o represents the static capacitance of the structure in the absence of piezoelectricity (measured away from resonance where the material is piezoelectrically clamped); C_1 is the *dynamic* or *motional* capacitance and represents the elastic stiffness of the resonant mode; and R_1 and L_1 represent the impact of internal friction and inertial mass on the resonance behavior, respectively. For the simple plate modes described by the BVD circuit [60,62], it is found that the capacitance ratio, $r = C_0/C_1$, of the Figure 11 circuit is related to the piezoelectric coupling factor, k, by the relation: $r = \pi^2/8k^2$. In specific regard to the focus of this chapter and book, it is important to note, as shown in the figure, that the capacitance ratios r and r' for LE and TE-driven piezoelectric resonators differ by unity. Depending on the magnitude of the piezoelectric coupling factor, k, the use of the traditional BVD network of Figure 11a (LE), when the TE circuit is proper, can result in a significant error. As an example: for low-coupling substances such as AT-cut quartz ($k = 8.80\%$), r and r' differ by only 0.6%, which is typically of no consequence. However, for a poled ceramic where k can be on the order of 51%, it is found from the above expression that $r = 4.74$ and $r' = 3.74$. Here the difference is greater than 20%, which, is quite significant when considering high-coupling piezoceramic materials. More in-depth discussions on the role of excitation direction on resonator performance and equivalent circuit modeling are available in Refs. 63 and 64.

    INCLUSION OF LOSS INTO EQUIVALENT CIRCUIT ANALOGS. In comparison to single-crystal piezoelectric materials such as quartz, polycrystalline ceramics typically exhibit higher losses. The phenomenological effects of loss may be included in equivalent circuit analogs by two simple approaches [65,66]. Internal friction within the piezo material is accounted for by making the elastic stiffness complex: $c^E \rightarrow c^* = c^E + j\omega\eta$; the imaginary part is the acoustic viscosity term. This has the effect of making the velocity and wave number complex, and leads, in the lumped BVD description, to the presence of the R_1 term. The second approach is used to account for dielectric and ohmic losses. These are accommodated by making the dielectric permittivity complex $\varepsilon^s \rightarrow \varepsilon^* = \varepsilon' - j\varepsilon''$, and combining ohmic and dielectric loss terms; the total effect is additive, so that the total effective σ is the ohmic conductivity σ plus the dielectric losses, $\omega\varepsilon''$. In the traditional BVD circuit these losses appear as a shunt resistor across the static capacitor C_0. When the LE version of the BVD circuit is taken, the further addition of the shunt resistor constitutes the ceramic resonator equivalent circuit almost universally used at low frequencies. Both viscous and dielectric mechanisms are incorporated using two time constants, each of which is a real number. The "motional" time constant is $\tau_1 = R_1C_1 = \varepsilon/c$, where c is the lossless piezoelectrically stiffened elastic stiffness for the mode in question. The "static" time constant is τ_0 = $R_0 C_0$ = $\varepsilon'/(\sigma + \omega\varepsilon'')$, which reduces to $\tau_0 = \varepsilon'/\sigma$ in the dc limit.

LUMPED CIRCUITS. Unfortunately, traditional equivalent circuits, such as the BVD, suffer from a variety of shortcomings that prevent adequate resonator characterization for the purpose of extracting material coefficients, as well as for use in electronic frequency–temperature compensation in oscillator applications. However, exact electrical representations of electromechanical systems can be devised using distributed acoustic transmission line models, which then can be recast in terms of lumped circuit parameters through a partial fractions expansion of tangent functions describing the transmission line (similarly to above, loss can be included by making these tangent functions complex) [66,67]. This approach of distributed transmission lines and lumped electrical circuits also provides the ability to directly model functionally graded piezoelectric materials such as Rainbow and Thunder actuators [68]. Simplification of the exact network description using transmission lines yields the equivalent circuit shown in Figure 12, which appropriately models the high-frequency performance of ceramic resonators (TE case).

(b) *Circuit Element Values.* Consider a thickness-excited piezoelectric resonator of thickness, $2h$, coated with electrodes of area, A, mass density, ρ_e, and thickness, h_e, on the major plate surfaces. The plate is of mass density, ρ, piezoelectric coupling coefficient, k, complex dielectric permittivity and total effective conductivity in the thickness direction of ε^* and σ, respectively, effective piezoelectric constant, e, and piezoelectrically stiffened elastic constant and viscosity governing the modal velocity and attenuation, c and η, respectively. The requisite values of ε^*, σ, e, η, and c are obtained in a straightforward manner from the material tensors. Under these circumstances, the circuit element values of Figure 12 are approximately those reported in Table 1. It should be further noted that C_0 and k^2 are, most generally, complex. This renders the lumped circuit elements

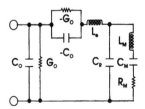

Figure 12   Equivalent circuit for a high-frequency, thickness-excited ceramic piezoelectric resonator.

Table 1 Circuit Element Values for a Thickness Excited Piezoelectric Element

$C_0 = \varepsilon' \, A/2h$	$R_0 = 1/G_0 = \tau_0/C_0;$	
$C_M = 8 \, k^2 \, C_0/M^2\pi^2;$	$C_R = C_1(\pi^2/8 - \Sigma 1/N^2);$	Σ = sum on odd $N \le M;$
$L_M = h^2/2v^2C_0k^2;$	$L_e = \mu h^2/v^2C_0k^2 = 2\mu$ $L_M;$	
$R_M = \tau_1/C_M;$	$k^2 = e^2/\varepsilon'C.$	

complex, yet the electromechanical effects can nearly all be accounted for in the same circuit form using modified values.

B. Electro-optic Phenomena

1 Background

Although electromechanical applications come first to mind when considering piezoelectric materials, there exist many commercially useful effects that also arise in these materials when light is incident upon and propagates through them. In piezo-, pyro-, and ferroelectricity, the respective phenomena arise from the response of electrically charged species (electrons, ions, dipoles, etc.) in an asymmetrical atomic potential to an applied electric field. Light, being an electromagnetic wave, can also act as an "applied" electric field, therein causing electronic and ionic polarizations (depending on the frequency) that modify the nominal dielectric properties of the material. At high frequencies, i.e., in the optical regime, the dielectric and optical properties—as quintessentially defined by the dielectric constant, ε_r, and refractive index, n, respectively—become equivalent as defined by $\varepsilon_r^2 = n$. Therefore, some of the crystallographic considerations introduced above, most notably the polar axis, are also important in defining the optical and electro-optic response of the materials.

2. Pockels and Kerr Effects

Electro-optic effects refer to the ways in which electric fields applied to a material may affect the velocities through which light waves propagate. Given the speed of light, c, and the velocity of a light wave in a material, v, the value of c/v for each wave is defined as the refractive index for that wave. The refractive indices of a material are functions of direction within an anisotropic material and are obtained by drawing an optical ellipsoid called the *indicatrix*. Good sources of more detailed information on the optical indicatrix are found in optical mineralogy texts [69,70].

The indicatrix, with ellipse axes oriented along the principal crystallographic axes, is defined by

$$\frac{x_1^2}{n_1^2} + \frac{x_2^2}{n_2^2} + \frac{x_3^2}{n_3^2} = 1 \tag{28}$$

where $n_i = \sqrt{\varepsilon_{r_i}}$ with ε_{r_i} being the principal dielectric constants. Therefore, n_i are the principal refractive indices. Along these principal directions, Eq. (28) often is recast in terms of reciprocal dielectric constants (i.e., $B_i = (\varepsilon_{r_i})^{-1} = (n_i^2)^{-1}$), called *dielectric impermeabilities*, B_i, such that

$$B_1 x_1^2 + B_2 x_2^2 + B_3 x_3^2 = 1 \tag{29}$$

In the general case of optical waves propagating at an arbitrary angle to the principal axes, the impermeability tensor is defined as

$$B_{ij} = \varepsilon_o \partial E_i / \partial D_j \tag{30}$$

Electro-optic phenomena relate to changes to this impermeability tensor resulting from the presence of the electric field and commonly are written as:

$$\Delta B_{ij} = r_{ijk} E_k + R_{ijkl} E_k E_l \tag{31}$$

where E_k and E_l are the externally applied electric fields, r_{ijk} are the *linear electro-optic* (also called Pockels) *coefficients*, and R_{ijkl} are the *quadratic electro-optic* (also called Kerr) *coefficients*. This is a more convenient form than a power series representation of the dielectric constant, or polarization, with applied field for describing the specific optical properties of anisotropic materials. Most common ferroelectric materials used in linear electro-optic applications fall into the cubic *43m* (e.g., GaAs, ZnS, InP), tetragonal *42m* (e.g., KDP, ADP), or trigonal *3m* (e.g., LiNbO$_3$, LiTaO$_3$, β-BaB$_2$O$_4$ [BBO]) crystallographic point groups.

The reduced matrices for the Pockels coefficients of these point groups are given in Table 2. The Pockels effect is represented by the first term in Eq. (31) and defines the linear change in the dielectric impermeability tensor with an externally applied electric field. Hence, it also is known as the "linear electro-optic effect" and, as mentioned, is possible only in noncentrosymmetrical crystals (with few special exceptions that are discussed below). Whereas Eq. (31) defines the modified optical indicatrix, the linear electric field dependence often is rewritten as

$$n_i \cong n_o - \frac{1}{2} r_{ij} n_o^3 E_j \tag{32}$$

where n_i is the refractive index along one of the principal axes. Equation (32) is the more useful relation for determining the modified refractive index given its

Piezoelectric and Electro-optic Ceramics

239

Table 2 Reduced Pockels (r_{ij}) and Kerr Coefficient (R_{ij}) Matrices for More Commonly Used Electro-Optic Materials

$$
\begin{bmatrix} 0 & 0 & 0 \\ 0 & 0 & 0 \\ 0 & 0 & 0 \\ r_{41} & 0 & 0 \\ 0 & r_{41} & 0 \\ 0 & 0 & r_{41} \end{bmatrix}
\qquad
\begin{bmatrix} 0 & 0 & 0 \\ 0 & 0 & 0 \\ 0 & 0 & 0 \\ r_{41} & 0 & 0 \\ 0 & r_{41} & 0 \\ 0 & 0 & r_{63} \end{bmatrix}
\qquad
\begin{bmatrix} 0 & -r_{22} & r_{13} \\ 0 & r_{22} & r_{13} \\ 0 & 0 & r_{33} \\ 0 & r_{51} & 0 \\ r_{51} & 0 & 0 \\ -r_{22} & 0 & 0 \end{bmatrix}
\qquad
\begin{bmatrix} 0 & 0 & r_{13} \\ 0 & 0 & r_{13} \\ 0 & 0 & r_{33} \\ 0 & r_{51} & 0 \\ r_{51} & 0 & 0 \\ 0 & 0 & 0 \end{bmatrix}
$$

Cubic $\bar{4}3m$ Tetragonal $\bar{4}2m$ Trigonal $3m$ Conical symmetry

$$
\begin{bmatrix}
R_{11} & R_{12} & R_{12} & 0 & 0 & 0 \\
R_{12} & R_{11} & R_{12} & 0 & 0 & 0 \\
R_{12} & R_{12} & R_{11} & 0 & 0 & 0 \\
0 & 0 & 0 & R_{11}-R_{12} & 0 & 0 \\
0 & 0 & 0 & 0 & R_{11}-R_{12} & 0 \\
0 & 0 & 0 & 0 & 0 & R_{11}-R_{12}
\end{bmatrix}
$$

Isotropic Symmetry ($\infty\infty m$)

nominal (i.e., zero-field) value, n_o,* the appropriate Pockels coefficient, and the applied electric field strength. Typical Pockels coefficients range from 10^{-12} to 10^{-10} m/V (i.e., 1–100 pm/V), so that for a field of 10^6 V/m (e.g., 1 kV across a 1-mm sample), refractive index changes on the order of 10^{-6} to 10^{-4} are produced [71].

The Kerr effect represents the quadratic dependence of the refractive index on an applied field. Although the linear electro-optic effect typically outweighs the quadratic dependence in magnitude, the Kerr effect is present in all materials and therefore always is available as a method to modify, and even control, optical properties. The Kerr dependence often is written as

---

* This zero-field value is not to be confused with the *ordinary refractive index*, also commonly denoted n_o, which corresponds to the index experienced by a wave traveling along the optic axis in uniaxial or biaxial crystals. In tetragonal and trigonal crystals (uniaxial), classes to which many of the ferroelectric crystals belong, the indices along x_1 and x_2 are this ordinary index whereas those along x_3 are the *extraordinary index, n_e*.

$$n_i \cong n_o - \frac{1}{2}R_{ij}n_o^3 E_j^2 \tag{33}$$

Typical Kerr coefficients range from 10^{-18} to 10^{-14} m^2/V^2 in crystals and 10^{-22} to 10^{-19} m^2/V^2 in liquids. Accordingly, a 10^6 V/m field will produce changes in the refractive index on the order of 10^{-6} to 10^{-2} in crystals and 10^{-10} to 10^{-7} in liquids [71]. See Table 1 for the reduced Kerr coefficient matrix for isotropic glasses and liquids ($\infty\infty m$ symmetry) that make up many Kerr devices.

3. Additional Optical Behavior of Ferroelectric Materials

The discussion so far has centered only on electro-optic contributions to a change in the impermeability tensor, ΔB_{ij}. However, since linear electro-optic materials are necessarily piezoelectric, the *piezo-optic effect* also can play a role in applied electric field modifications to B_{ij} that are manifest as changes to the material's refractive index (this is true as well for the case of electrostriction and Kerr effects). Equation (31) can be expanded to include the influence of the piezoelectric strain induced by the applied electric field. Neglecting higher-than-first-order terms, Eq. (31) is then written as

$$\Delta B_{ij} = r_{ijk} E_k + \pi_{ijkl}T_{kl} \tag{34}$$

where the second term defines the *photoelastic effect* with π_{ijkl} being the *piezo-optic coefficients* that contribute to changes in the impermeability tensor under an applied stress, T_{kl}. Linear elasticity permits the photoelastic contribution to be recast in terms of strain, rather than stress, in which case the second term in Eq. (34) becomes $p_{ijmn}S_{mn}$ where p_{ijmn} are called the *elasto-optic coefficients* and are related to the piezo-optic coefficients by $p_{ijmn} = \pi_{ijkl}c_{klmn}$.

Mechanical constraints on the material also play an important role in its electro-optic properties. When a crystal is clamped so that the strain is zero, the measured changes in the refractive index are considered then to be the primary, or true, contributions from the electro-optic effect. Under stress-free boundary conditions, the converse piezoelectric effect yields a strain that causes a contribution to ΔB_{ij} through the photoelastic effect. This added component is considered the secondary, or false, effect so that the measured electric field dependence on the refractive index is really the sum of the primary and secondary effects [39,72]. Measurements at high frequencies (away from piezoelectric resonance), where the piezo strains are exceedingly small, can quantify the primary contribution. Mechanical constraints on a system also can modify its ferroelectric behavior. Such stress-biasing effects, depending on the directionality of the stress field with respect to that of the polarization, can preferentially orient domains or lead to their depolarization over time [73,74].

Polycrystalline ferroelectric ceramics lack a linear electro-optic response if their polarization vectors are randomly oriented at the grain level, as would be the case prior to poling. The random orientation can be broken, either by the application of an electric (poling) field or by the application of a uniaxial tensile stress. The ferroelastic response of ferroelectric materials makes it possible to use a uniaxial strain to obtain a uniaxial birefringence and thus a linear electro-optic response. This occurs by means of an antiparallel alignment of microscopic dipoles that results when a biasing strain is applied, even though the average macroscopic remnant polarization vector remains zero. Thus, Pockels nonlinearities can be obtained in materials with either electric or ferroelastic poling. Most efforts in transparent electro-optic materials have focused on lanthanum-doped PZT solid solutions [4]. These materials possess acceptable optical transparency and moderately strong electro-optic coefficients in comparison to single-crystal values. A good treatment and comparison of single crystalline and polycrystalline materials is given by Agulló-López et al. [72]. Unpoled ceramics may be of interest for their quadratic dependence. In this case, the Kerr coefficients follow the isotropic tensor matrix ($\infty \infty$m). Poled polycrystalline ceramics take on a conical symmetry equivalent in tensor form to 6mm symmetry elements. Table 2 gives the reduced matrices for the isotropic Kerr coefficients and the Pockels coefficients for the point groups into which most of the common electro-optic materials fall (including poled ceramics).

4. Fundamentals of Nonlinear Optical Interactions

The same crystallographic symmetry considerations that allow ferroelectricity also provide constraints on the nonlinearities in the material's optical properties. The electric field that polarizes the material is now that of the incident electromagnetic radiation. At optical frequencies, the material's dielectric response is defined more appropriately in terms of the refractive index, n, as $n = \sqrt{\varepsilon_{ij}\mu_{ij}/\varepsilon_o\mu_o} = \sqrt{1 + \chi_{ij}}$ (for nonmagnetic materials where $\mu_{ij} \cong \mu_0$, or else the magnetic susceptibility also contributes to the refractive index). In a crystal of arbitrary anisotropy, the linear dielectric properties at optical frequencies are specified by the six independent components of the symmetric, second-rank, permittivity tensor. By a suitable rotation of axes, stemming from the solution of an eigenvalue problem [75], the permittivity tensor may be diagonalized. The diagonal terms (the eigenvalues) are the squares of the principal indices of refraction. The permittivity tensor is often given a geometrical interpretation by forming the "representation quadric" from its components and the crystal coordinate axes: $\varepsilon_{ij}X_iX_j = 1$, where X_i and X_j are points along the optical indicatrix. This quadratic surface is usually an ellipsoid for most materials. It has the very desirable property of yielding simply the refractive indices of the two possible plane waves corresponding to a given propagation direction. These are found from the intersection of the quadric surface with the plane perpendicular to the prescribed wave normal, passing through the origin of the quadric. For an ellipsoidal quadric, this section (known as the indicatrix) is an ellipse. Its major and minor axes are the squares of the refractive indices of the two plane waves that may propagate in the specified direction. Moreover, the orientations of these indicatrix axes are the corresponding directions of the dielectric displacements (D) of the two electromagnetic waves. Note that whereas the dielectric permittivity is a tensor, the refractive index is not [39].

The power series expansion of the dielectric constant (or polarization) is still valid under the electrical field intensities corresponding to normal light levels, but because we look for changes in the index, it is the relative permittivity that is more pertinent in describing optical interactions. All materials are none-the-less weakly nonlinear with the trends in magnitude being $\varepsilon_0\chi^{(3)}|E|^3 << \varepsilon_0\chi^{(2)}|E|^2 << \varepsilon_0\chi^{(1)}|E|$ and with the respective point group symmetry dictating which higher order susceptibilities, through their tensorial components, are nonzero. Isotropic materials and those possessing a center of symmetry are characterized by $\chi^{(2)} \equiv 0$. The quadratic nonlinearity is only possible in materials lacking a center of symmetry, such as all piezoelectrics, which include all ferroelectrics. In the strictest sense, this latter statement precludes $\chi^{(2)}$ optical nonlinearities in glasses since their random structure leads to an effective center of symmetry. However, electrical and thermal poling of glasses also has been used to induce optical nonlinearities. In the thermal case, a region of space charge arises from the diffusion of mobile ions from the anode to the cathode. Under electrical poling, the applied field orients the molecular dipoles in a preferential direction. The result of either form of poling is a net macroscopic anisotropy and the ability for the glass now to generate second harmonic light [76,77]. The cubic nonlinearity is present in all materials regardless of crystal symmetry or lack thereof.

The propagation of monochromatic waves in an optically nonlinear medium has the useful possibility of generating light at other wavelengths. As a simplified example, the change in polarization $\Delta P(t)$, in response to a single incident beam with $|E| = E_0|sin(\omega t)|$, is

$$\left|\Delta P(t)\right| = \varepsilon_o\chi^{(1)}E_o\left|sin(\omega t)\right| + \varepsilon_o\chi^{(2)}E_o^2 sin^2(\omega t) + \varepsilon_o\chi^{(3)}E_o^3\left|sin^3(\omega t)\right| + \cdots \quad (35)$$

The higher order time varying components are equivalently written in terms of higher harmonics of ω since $sin^2(\omega t) \propto 1 - cos(2\omega t)$ and $sin^3(\omega t) \propto [sin(\omega t) - sin(3\omega t)]$. Accordingly, the nonlinear material oscillates and reradiates at frequencies that are integral multiples of ω (i.e.; 2ω, 3ω, etc.). *Second harmonic generation* (SHG) corresponds to the creation of light at double the frequency of the incident radiation. This is significant technologically because it provides an alternative to the direct generation of blue or blue–green light by using instead the conversation of more easily and powerfully generated red or near-infrared light. Similarly, light produced at 3ω is called *third harmonic generation* (THG). However, because the material polarizations at the second and third harmonics are proportional to $\chi^{(2)}$ and $\chi^{(3)}$ respectively, the resultant optical intensities are relatively small in comparison to the incident harmonic wave.

SHG and THG are special cases of nonlinear optical interactions where the incident waves are of the same frequency. The interaction of two monochromatic waves with frequencies ω_1 and ω_2 in a nonlinear optical material can give rise to *sum–difference generation*, i.e., $\omega_{3,4} = \omega_1 \pm \omega_2$ where the plus ($+$) corresponds to the sum frequency (*sum frequency generation*, SFG) and the minus ($-$) to the difference frequency. Frequency combinations of higher complexity are possible through successive SHG and/or sum–difference generation. However, the intensity of light generated at frequencies greater than 2ω, or even 3ω or 4ω (i.e., double-doubled light), typically is very weak. Some studies have shown the possibility of highly efficient third harmonic generation. This can be achieved by directly coupling SHG and SFG through quasi-phase matching in quasi-periodic ferroelectric superlattices [78,79].

The determination of the nonlinear susceptibility or, more specifically, the matrix elements of the nonlinear susceptibility tensor proceed from measurements that quantify a material's *hyperpolarizability*, β_{ijk} (for $\chi^{(2)}$), or either the *n*th *photon absorption coefficient*, $\alpha^{(n)}$, or *nonlinear refractive index*, n_2 (for $\chi^{(3)}$). The hyperpolarizability defines the polarization that oscillates at the second harmonic frequency; i.e., $|P(2\;\omega)| \propto |\beta|$, which in turn, is proportional to $\chi^{(2)}$. The $\chi^{(2)}_{ijk}$ components often are redefined in terms of the *second harmonic coefficients*, d_{ijk} = $1/2\;\chi^{(2)}_{ijk}$ (not to be confused with the piezoelectric d_{ijk} coefficients) and can be measured through either relative or absolute means. Relative measures include the wedge and the Maker fringe techniques. Absolute measures include the second harmonic phase matching, parametric fluorescence, and Raman scattering approaches, where the two former are the more precise and the latter provides simply the sign of the respective coefficient. Measurements of the third-order susceptibility typically utilize the nonlinear dependence of optical transmission with beam intensity. An analogous power series expansion to that in Eq. (31) is used to relate the transmitted intensity, I_1, at a specific wavelength, λ_l, and through a given pathlength, z, so that

$$\frac{dI_1}{dz} = -\alpha^{(1)}I_1 - \alpha^{(2)}I_1I_2 - \alpha^{(3)}I_1I_2I_3 - \cdots, \tag{36}$$

where I_i are the intensities measured at wavelengths λ_i, which typically are different from λ_l, and $\alpha^{(n)}$ are the *n*th photon absorption coefficients; i.e., $\alpha^{(1)}$ corresponds to single-photon absorption (linear absorption), $\alpha^{(2)}$ relates to two-photon absorption, and so forth. Of particular interest to the present discussion is the fact that the $\alpha^{(2)}$ two-photon absorption coefficient is related to the imaginary part of the complex $\chi^{(3)}$ third-order nonlinear susceptibility by

$$\alpha^{(2)} = \frac{3\omega}{2\varepsilon_o c^2 n_o^2}\,\mathrm{Im}\left[\chi^{(3)}\right] \tag{37}$$

where ω is the angular frequency of light ($\omega = 2\pi c/\lambda_1$) traveling through a material of nominal (i.e., low intensity) refractive index, n_0. The nonlinear refractive index, n_2, defined by $n = n_0 + n_2 I$ and related to $\alpha^{(2)}$ through the Kramers–Krönig relations, is a complementary measure and permits the determination of the real part of the complex susceptibility through

$$n_2 = \frac{1}{\varepsilon_o c n_o^2} \operatorname{Re}\left[\chi^{(3)}\right]$$

(38)

Approaches such as electroabsorption or transient gratings methods can be used to measure the complex components of $\chi^{(3)}$ as can single-beam nonlinear transmission, Z scan, eclipsing Z scan (EZ scan), thermal lensing, two-photon luminescence, three-wave mixing [also known as coherent anti-Stokes Raman spectroscopy (CARS), which really is a form of four wave mixing where two of the beams originate from the same source], and four-wave mixing experiments. Each technique has specific conditions, limitations, and nuances that need be considered, and the reader is referred to Ref. 80 for a more complete theoretical discussion and configurational description.

The strength of a nonlinear interaction is governed not only by the relative magnitudes of the operative nonlinear susceptibility-related terms but also by the degree of coherency between the locally generated harmonic fields. This effect is known most commonly as *phase matching* but also may be called *wave number* or *index matching*. For noticeable harmonic generation, phase matching requires $k_3 = k_2 + k_1$ or $k_4 = k_2 - k_1$ where k_i are the wavevectors corresponding to the waves of frequency ω_i (i = 1, 2, 3, 4), $k_i = 2\pi n/\lambda_i$, with n being the refractive index at frequency ω_i and λ_i the wavelength of this wave in the material.

III. MATERIALS

A. Single-Crystal Materials

From a historical perspective, the first synthetic piezoelectric material was Rochelle salt (sodium potassium tartrate tetrahydrate) prepared by Elie Seignette in La Rochelle, France in approximately 1665 [3]. At the time it was discovered by Seignette, the material was developed for medicinal purposes; its piezoelectric response was not discovered until investigations by the Curies in the late 1800s. In fact, it was determined that the piezoelectric response of Rochelle salt was greater than that of quartz. With regard to piezoelectric applications, the use of

Rochelle salt for submarine detection applications was suggested as early as the World War I [3] and the basic concept is pertinent to the date of this writing. However, PZT compositions and PZT-based composites are now the materials of choice for this application. Finally, though the piezoelectric response of Rochelle salt is higher than that of quartz, it is water soluble. Hence, its environmental stability renders it the material of choice for use in resonator and time-keeping device applications. The piezoelectric properties of other single-crystal materials, most notably potassium dihydrogen phosphate (KDP; KH_2PO_4) [81], have also been investigated but today are more important for their critical role in the discovery of the phenomenon of ferroelectricity [3,82] and the realization that ferroelectricity was not an effect related only to a few exceptional cases.

The most widely used single crystal for piezoelectric applications is quartz (SiO_2, point group *32*). Over the past few decades during which great advances in new materials have been made in arguably all genres of technology, quartz remains the classic example of old and simple still being best in many aspects. As early as the 1920s, quartz crystals were used as crystal resonators to stabilize oscillators. These efforts signaled the birth of modern precision frequency control that now encompasses everything from inexpensive wrist watches, to global positioning systems, to mobile communications, to precision munitions, to name just a few. Although silica (SiO_2) crystallizes into several different structures, only the α form is used for time keeping applications. Crystallographically, α-quartz is trigonal with enantiomorphic pairs belonging to space groups $P3_221$ (right handed) and $P3_121$ (left handed). The growth of high-quality quartz using hydrothermal approaches dates back to World War II Germany due to the limited availability of electronic grade natural quartz from Brazil [83]. After the war, growth was continued in the United States reaching commercial production in the late 1950s [84]. In the early 1970s, this synthetic (also called ''cultured'') quartz surpassed natural quartz in use. Today, natural quartz only finds application as jewelry and related aesthetics.

As noted, quartz does not possess the highest piezoelectric coefficients in comparison to other materials. However, it remains a standard for frequency control devices because quartz exhibits the best combination of complementary performance features needed for practicality. This includes machinability into requisite shapes and dimensions; chemical, thermal, and environmental stability; and appropriate crystallography so that resonators/oscillators can be made with resonance frequencies that are tailored with specific temperature and stress dependencies. Particularly noteworthy about quartz is that numerous cuts can be made with respect to one (singly rotated) or two (doubly rotated) crystal axes so that the resultant resonance, be it extensional, flexural, or shear, is temperature independent [85,86].

There are many other single-crystal ferroelectrics used for optical applications:

Lithium niobate: $LiNbO_3$ (LN) and $LaTaO_3$ (LT), both point group 3m at room temperature, represent the most mature material technologies for integrated optic circuits. Their linear electro-optic coefficients are fairly large, and large single crystals of high optical quality can be grown. This latter point, coupled with extensive efforts into diffusion-driven approaches to waveguide fabrication, has led to numerous planar devices that are commercially available based on $LiNbO_3$ [87–89].

Tungsten-bronze niobates: Tungsten-bronze materials are based on the KWO_3 and $NaWO_3$ structures. The more common ferroelectrics of this family include $Ba_2NaNb_5O_{15}$ (BNN), $K_3Li_2Nb_5O_{15}$ (KLN), and $Sr_{1-x}Ba_xNb_2O_6$ (SBN). All crystallize into point group 4mm at room temperature with the exception of BNN, which is mm2. BNN and SBN have been used for EO modulators and deflectors as well as parametric oscillators given their large electro-optic and nonlinear optical coefficients. KLN and its Ta-doped analog, $K_3Li_2(Ta_xNb_{1-x})O_{15}$ (KTLN), not only have higher EO coefficients than BNN but exhibit a higher threshold to laser damage than $LiNbO_3$.

Perovskite family: $BaTiO_3$ (BT), $KNbO_3$, $KTaO_3$, $KTa_{1-x}Nb_xO_3$, point groups 4mm, mm2, mm2, and mm2, respectively, at room temperature, represent the most important materials of this family and possess large electro-optic and second harmonic generation coefficients.

Water solubles: KH_2PO_4 (KDP) and triglycine sulfate (TGS, $\{(NH_2CH_2COOH)_3 \cdot H_2SO_4\}$), point groups 42m and 2, respectively, can be grown to large size with good optical quality and acceptable mechanical properties. The main difficulty with each materials is deliquescence (KDP) and a low Curie temperature (TGS, $T_c = 49°C$). With proper packaging, the large linear electro-optic coefficients of the KDP family (KH_2PO_4, KH_2AsO_4, RbH_2PO_4) have been used for optical shutters, EO modulators, beam steering, and tunable filters.

Potassium titanyl phosphates: $KTiOPO_4$ (KTP), point group mm2 at room temperature, and its analogs $RbTiOPO_4$ (RTP) and $TlTiOPO_4$ (TTP) possess high electro-optic coefficients and low dielectric constants, making them advantageous for EO modulators and switches. KTP also possesses large nonlinear optical coefficients and high conversion efficiencies, and is useful for applications requiring second harmonic conversion of light.

Excellent reviews of properties, structure, and applications of the aforementioned materials, as well as a myriad of others, are available in the literature [89–91].

While the great majority of single-crystal materials are employed in time keeping and electro-optic applications, recent investigations have identified other single-crystal materials that are under development for electromechanical sensor and actuator applications. The single-crystal compositions that have been investigated in greatest depth are $Pb(Zn_{1/3}Nb_{2/3})O_3$-$PbTiO_3$ (lead zinc niobate–lead titanate; PZN-PT) and $Pb(Mg_{1/3}Nb_{2/3})O_3$–$PbTiO_3$ (lead magnesium niobate–lead titanate; PMN-PT) [11]. These materials are relaxor ferroelectric-based compositions that have demonstrated unusually high coupling coefficients ($k_{33} > 0.9$), high piezoelectric strain coefficients ($d_{33} > 2500$ pC/N), and high strain generation capabilities (up to ~2%). These values represent significant improvements compared to even the high performance morphotropic phase boundary PZT compositions. In addition, the preparation of these materials in single-crystalline form has been demonstrated to be substantially easier than the preparation of MPB single-crystal PZT compositions [92–96]. The electromechanical properties of additional single-crystal compositions are reported below in the properties section.

Prior to the fabrication of PMN-PT and PZN-PT in single-crystalline form, these and other relaxor ferroelectric materials had received considerable attention for capacitor and piezoelectric applications as polycrystalline materials. Many of the materials displayed high electrostrictive coefficients with electromechanical strain response that rivaled piezoelectric materials [96–100].

B. Polycrystalline Materials

1. General Aspects

Although single-crystal barium titanate ($BaTiO_3$) is utilized in electro-optic modulator applications, it was first investigated in polycrystalline form for sonar applications during World War II. During these investigations it was also discovered that $BaTiO_3$ possessed advantageous dielectric properties (high dielectric constant and comparatively low dielectric loss), which led to its eventual implementation in capacitor applications [101]. In addition to these two uses, barium titanate is also important historically because it suggested investigation of other perovskite compounds for their dielectric, piezoelectric, pyroelectric, electro-optic, and ferroelectric properties. Today perovskite materials are the most widely used ceramics in each of these applications and the parent crystal structure continues to yield interesting and new results, including, for example, the development of high-temperature superconducting oxides, such as yttrium barium copper oxide, $YBa_2Cu_3O_{7-\delta}$, a layered perovskite compound [10].

To date, piezoelectric ceramics (which are also pyroelectric and possibly ferroelectric) have been formulated from a number of compositions, including barium titanate, barium strontium titanate, PZT, lead niobate, bismuth titanate, sodium potassium niobate, lead magnesium niobate, and other relaxor ferroelectric compositions. Barium titanate continues to be the key constituent in ceramic multilayer capacitors but its use for piezoelectric applications has been largely supplanted by PZT-based materials. This is because PZT compositions (1) possess higher electromechanical coupling coefficients; (2) have a higher Curie point, which permits higher temperature operation or higher temperature processing during fabrication of devices; (3) can easily be poled; (4) possess a wide range of dielectric constants; (5) are relatively easy to sinter; and (6) form solid solutions with many chemical compositions. The similar crystal structure and valences of binary and ternary end members, such as $BaTiO_3$, $SrTiO_3$, $PbTiO_3$, and $PbZrO_3$ result in extensive (usually full) solid solubility, thus allowing a remarkably broad range of achievable properties. The extensive solid solubility that exists in these materials is an important aspect of their use in dielectric, piezoelectric, pyroelectric, and electro-optic applications because it allows for a high level of control of material processing and properties.

2. PZT and PLZT Materials

It is outside the scope of this chapter to provide a detailed discussion of the many different compositional systems that are in use today for transducer applications. However, because of the widespread use of $PbZrO_3$-$PbTiO_3$ (PZT) compositions, attention is focused on these materials and the role of substituents in altering their base properties. The PZT material system was first investigated by Shirane et al. [102] in Japan, and the first piezoelectric compositions from this system were developed by Jaffe [7,103] in the United States. Listed below are some of the major families of materials based on $PbZrO_3$-$PbTiO_3$ compositions:

$Pb(Zr, Ti)O_3$ + additives such as Nb, Sb, Bi, La, Fe, Ta, Cr, Co, and Mn
$Pb(Zr, Ti, Sn)O_3$
$Pb(Mg, Nb)O_3$-$PbZrO_3$-$PbTiO_3$
$PbTiO_3$ + additives of Mn, La, Nd, and In
$(Pb, La) (Zr, Ti)O_3$

As with the compositional modifications that are made to adjust the dielectric response of pure barium titanate for multilayer ceramic capacitors [7], lead zirconate titanate ceramics are almost always used with dopants, modifiers, or other compositions in solid solution. This is commonly done to improve the properties of the basic PZT ceramic material for specific applications. The various substituents are categorized according to whether they are isovalent (the same valence as the host ion) or aliovalent (a different valence than the host ion). The choice of the proper substituent can result in control over the temperature of the Curie transition, a decrease in the temperature dependence of the dielectric constant [7,103], and charge compensation of impurities to improve electrical resistance [104,105]. Substituents can also facilitate poling of the materials or make the materials more resistant to depoling [7]. The role of aliovalent substituents in the defect chemistry of these materials has also been thoroughly investigated by

Smith [106] and is typically described using Kröger–Vink notation [107]. It should be mentioned here that it is common practice in formulating compositions to include more than one type of additive to a given composition in order to achieve a given set of properties. In fact, in following the literature for this area, one can easily be overwhelmed by the large number of additives used to modify the basic properties of BT and PZT-based materials.

Substitution may be on either the large-cation A site, or the small-cation B site in the perovskite structure (Figure 3). Examples of doping practice include the following:

1. *Donor* additives such as Nb^{5+} replacing Zr^{4+} or La^{3+} replacing Pb^{2+} in order to counteract the natural p-type conductivity of PZT [104,105] and thus increase electrical resistivity. Donor additives are usually compensated by A-site vacancies. These additives also enhance domain wall translation and reorientation, and the resulting materials are characterized by square hysteresis loops, low coercivity, high remanent polarization, high dielectric constant, maximum coupling factors, high dielectric loss, high compliance, and higher aging rates [7,108]. Typical applications are in the area of high sensitivity, such as hydrophones, photograph pickups, sounders, and loudspeakers. For electro-optic shutter and modulator applications, La^{3+} additions to PZT enable the fabrication of dense, transparent ceramics [109].

2. *Acceptor* additives such as Fe^{3+} replacing Zr^{4+} are compensated by oxygen vacancies and usually have only limited solubility in the lattice. Domain reorientation is limited; hence, ceramics with acceptor additives are characterized by poorly developed hysteresis loops, lower dielectric constant, low dielectric loss, low compliance, and lower aging rates. Typical applications are in high-power devices such as sonar and ultrasonic transducers.

3. *Isovalent* additives such as Ba^{2+} or Sr^{2+} replacing Pb^{2+} or Sn^{4+} replacing Zr^{4+}, in which the substituting ion is of the same valence, and approximately the same size, as the ion it replaces. Solid-solution ranges with these additives are usually quite high and may result in lower Curie points and increased dielectric constants. Hysteresis loops may be poorly developed without additional additives. Other properties include lower dielectric loss, low compliance, and higher aging rates. These ceramics are used in high-drive applications.

As mentioned earlier, the properties of piezoelectric and electro-optic ceramics are typically optimized through the incorporation of a variety of dopants, with lanthanum being one of the most commonly employed donor additives. Because lanthanum-modified PZT (PLZT) compositions are utilized in both piezoelectric and electro-optic materials, it is selected here as a typical system to demonstrate the impact of the $PbZrO_3$-$PbTiO_3$ ratio on electrical and electromechanical properties, as well as the role of dopants in defect chemistry, processing, and performance.

The solid solution that forms the basis of PLZT materials is a series of compositions resulting from the complete miscibility of lead zirconate and lead titanate in each other (Figure 13), which are modified by the solubility of substantial amounts of lanthanum oxide in the crystalline lattice. A general formula for all compositions in the PLZT system is

$$Pb_{1-x} La_x (Zr_y Ti_{1-y})_{1-x/4} O_3$$

where lanthanum ions replace lead ions on the A site of the ABO_3 perovskite structure (Figure 3). Since La^{3+} (added as La_2O_3) substitutes for Pb^{2+}, electrical neutrality is maintained by the creation of lattice site vacancies [105]. Although this formula presumes that all the vacancies are on the B site, the actual location of these vacancies on either the A^{2+} or B^{4+} sites of the unit cell has not yet been completely resolved despite numerous studies. It is most probable that both A- and B-site vacancies exist [110]. If both A- and B-site vacancies are present in the lattice, it is to be expected that the formulation above would provide excess Pb^{2+} ions, which are expelled from the lattice as PbO vapor during the densification process at elevated temperatures. This behavior does, indeed, occur; in fact, this excess PbO contributes to achieving full density by forming a liquid phase at the grain boundaries and by inhibiting grain growth during the initial stages of densification [109]. Both of these effects are beneficial to the attainment of theoretically dense materials by eliminating residual porosity before it becomes entrapped within the grains.

Ceramics usually formulated from the conventional A-site formula:

$$Pb_{1-\frac{3x}{2}} La_x \left(Zr_y Ti_{1-y} \right) O_3 \tag{38b}$$

must be sintered with excess PbO atmosphere in order to achieve nearly full densification. They characteristically pick up PbO from the atmosphere, indicating the additional likelihood that a combined A-site and B-site vacancy structure is actually present in these materials.

The PZT and PLZT phase diagrams are shown in Figure 13a and 13b, respectively [4,7]. The effect of adding lanthanum to the PZT system is (1) one of maintaining extensive solid solution throughout the PZT system; and (2) one of decreasing the stability of the ferroelectric phases (reducing the Curie points) in favor of the nonferroelectric cubic and antiferroelectric (nonpolar phase possessing adjacent, oppositely aligned polar unit cells) phases. At a 65:35 ratio of $PbZrO_3$ to $PbTiO_3$, a concentration of 9.0% La is sufficient to reduce the stable region of the FE rhombohedral phase to below room temperature. Thus, a material

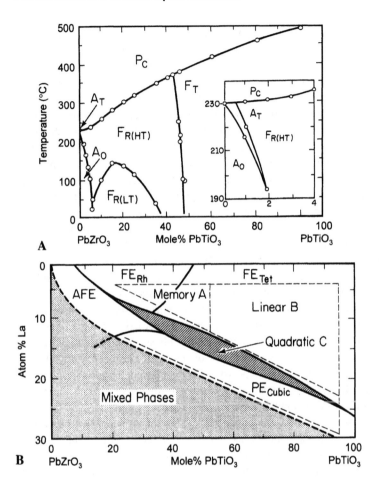

Figure 13 (a) Binary phase diagram for PbZrO$_3$-PbTiO$_3$ showing crystallography and electrical response behavior [7]; (b) PLZT phase diagram illustrating variation in phase boundaries as a function of lanthanum content [4].

of this composition (designated as 9:65:35 or, simply, 9065) is nonferroelectric and cubic in its virgin state. The cross-hatched area existing along the FE-PE (cubic) phase boundary denotes a region of diffuse, metastable ferroelectric phases that can be electrically induced with a sufficiently high electric field. Materials in this region exhibit a quadratic electro-optic response.

Compositions in the PLZT system for piezoelectric applications are usually confined to less than 5% La, whereas electro-optic ceramics possess approximately 6% La or more. Whether within the PZT of the PLZT phase diagrams, compositions located on the tetragonal side of the phase are usually "hard" (high coercive field) materials, whereas those on the rhombohedral side are "soft" (low coercivity) materials. The descriptors '*soft*' and '*hard*' also imply differences in elastic compliance, strain response, and strain–electric field hysteresis.

The most noteworthy base composition for piezoelectric applications is PZT 53:47 (53 mol % PbZrO$_3$ and 47 mol % PbTiO$_3$), which lies at the morphotropic (temperature independent) phase boundary in the PbZrO$_3$–PbTiO$_3$ binary system (Figure 13a). Morphotropic phase boundary (MPB) compositions are known for their high dielectric constants, high piezoelectric coefficients, and high coupling coefficients. The simple explanation that is routinely offered for the high performance of these materials is the increased ease of reorientation during poling [7,43]. The proximity of compositions close to the MPB allow for six polarization directions for the tetragonal state and eight for the rhombohedral state. These 14 available polarization directions for MPB materials lead to high remanent polarizations as well as high piezoelectric response. To further understand the unique performance of these compositions, research into the crystallography of near MPB compositions continues to this day. Recently, researchers have identified a monoclinic phase that is reported to exist at slightly below room temperature [21,22]. It is believed that the existence of this phase may play a role in the greater performance observed performance of the MPB compositions, although this is not yet a universally held perspective. Thermodynamic arguments based on Landau–Gibbs–Devonshire theory have also been utilized to explain the higher dielectric and electromechanical performance of these compositions [111,112].

Lead zirconate titanate–based materials, from soft to hard, may be synthesized for a variety of applications and may be obtained commercially from a variety of vendors. Softer compositions, i.e., those with relatively low E_c values, such as PZT 5A or 5H, are used in applications where low electric fields are utilized and depoling of the piezoelectric is not a concern. These applications include, among others, positioners. In contrast, hard compositions are used in applications where depoling concerns exist, either through temperature excursion or the application of high ac electric field. These applications include sensors, high-power transducers, and motors. Because of the higher coercive field, hard PZT ceramics take less advantage of domain wall translation (or switching) effects, and thus have lower piezoelectric electromechanical coupling coefficients. This sacrifice is required because of the other requirements/conditions associated with these applications.

A variety of descriptors is employed by piezoelectric manufacturers to describe the various PZT-based ceramic compositions suitable for different applications. While it is sometimes frustrating that no compositional information is reported for these materials, many of the properties of the various compositions are disclosed by manufacturers. Some of the original designations, such as PZT-

4 (Navy type I), PZT-5A (Navy type II), PZT-5H (Navy type VI), and PZT-8 (Navy type III), among others, are still in use today. The PZT-4, 5A, 5H, etc, designations are trademarks of Morgan Matroc, but originated with the Clevite Corporation. While the above designators are still commonly employed, there are also a number of other manufacturers, each with their own designators of material performance characteristics. Many manufacturers offer cross-reference charts. A summary of the properties of the more common PZT ceramics is given below in the properties section of this chapter.

Many piezoelectric PZT compositions are based on doped MPB compositions, although the specific application will dictate whether a hard or soft material is used. Applications for compositions along the FE rhombohedral–AFE orthorhombic phase boundary are very few and limited to special devices utilizing a pressure-enforced FE-to-AFE phase transition [113]. Compositions rich in lead titanate are especially useful for surface acoustic wave (SAW) devices [114]. Other work has also indicated that compositions in the AFE region of the phase diagram may find application as high-stability capacitors under high electric fields [115].

Modification of the PZT system by the addition of lanthanum oxide has a marked beneficial effect on several of the basic properties of the material, such as increased squareness of the hysteresis loop, decreased coercive field, increased dielectric constant, maximal electromechanical coupling coefficients, increased mechanical compliance, and enhanced optical transparency. The latter of these properties, optical transparency, was discovered during the 1970s [109,116] but came about as the result of an in-depth study of various additives to the PZT system. Results from this work indicate that La^{3+}, as a chemical modifier, is unique among the off-valent additives in producing transparency. The reason for this behavior is still not fully understood; however, it is known that lanthanum is, to a large extent, effective because of its high solubility in the PZT oxygen octahedral structure, thus producing an extensive series of single-phase solid-solution compositions. The mechanism is believed to be one of lowering the distortion of the unit cell, thereby reducing the optical anisotropy of the crystalline lattice and at the same time promoting uniform grain growth and densification of a single-phase, pore-free structure. Hot pressing is used to obtain optically transparent PLZT materials for shutter and modulator applications [109,116,117].

The solubility of La in the PZT structure is a function of composition and is related directly to the amount of lead titanate present. The compositional dependence of the solubility limit is indicated by the dashed line adjacent to the mixed-phase region (double cross-hatched area) in Figure 13b. For the two end-member compositions, lead zirconate and lead titanate, these limits are 4 and 32 at. % La, respectively. The solubility limits for intermediate compositions are proportional to their Zr/Ti ratios. Compositions in the mixed-phase region are easily identified by their opaque appearance, in contrast to the transparency characteristic of the single-phase, cubic materials.

Electro-optic compositions in the PLZT phase diagram are generally divided into three application areas: quadratic, memory, and linear. As mentioned earlier, the quadratic materials are compositionally located along the phase boundary separating the ferroelectric and paraelectric phases, principally in the cross-hatched area. Memory compositions having stable, electrically switchable optical states are largely located in the FE rhombohedral phase region, and the linear materials possessing nonswitching, linear electro-optic effects are confined to the area encompassing the tetragonal phase. The typical electro-optic response of these compositions, together with the polarization–electric field hysteresis loops, is discussed further below.

3. Piezoelectric Composites

Piezoelectric composites represent the latest in technology trends for literally squeezing the last bit of high performance from a piezoelectric transducer. When one deliberately introduces a second phase in a material, such as in a composite, connectivity of the phases is a critical parameter. Composites are described by a nomenclature that uses two numbers (or three when porosity is also present) to describe the connectivity of the active piezoelectric phase and the inactive (usually polymeric) phase [118–120]. The numbers used in the nomenclature—0, 1, 2, and 3—describe the number of directions in which a given phase is continuous. There are 10 connectivity patterns possible in a two-phase solid, ranging from a 0–0 unconnected three-dimensional checkerboard pattern of a 3–3 interpenetrating pattern in which both phases are three-dimensionally self-connected. Most frequently, the composites that are prepared include PZT rods or fibers and the composites formed are 1–3 in nature, i.e., the piezoelectric (active) phase is connected in one direction while the polymer phase has three-dimensional connectivity. The materials may also include porosity in the inactive 3 phase for further reduction of the dielectric constant of the composites.

Composites with other connectivities have also been developed and their corresponding electromechanical performance has been extensively investigated through both modeling [121,122] and experimental methods [123,124]. Newnham et al. [119] and others [120,125] have demonstrated considerable ingenuity in designing, fabricating, and evaluating the many types of diphasic structures. Excellent treatments of this subject are presented by Pohanka and Smith [120] and Banno [120]. Typical properties of piezoelectric composites compared to standard poled polycrystalline materials are given below in the section on electromechanical properties.

IV. PROCESSING

A. Single-Crystal Growth Methods

1. Quartz

A variety of methods are employed for the growth of single-crystal piezoelectric and electro-optic materials. Today quartz remains the pre-eminent single crystal piezoelectric, although the importance of single-crystal relaxor ferroelectrics is increasing and applications for these materials are rapidly developing. While electronic-grade natural quartz is available, principally from Brazil, today most of the quartz that is utilized in precision time keeping and oscillator applications is synthetic. Interestingly, just as in the development of barium titanate, World War II also accelerated the development of synthetic methods for quartz [83,84]. The synthetic method that was developed at that time, and which is still in use today, is the hydrothermal method. More than 3000 tons of α-quartz is produced annually by this method [83].

   The hydrothermal method involves the use of a high-pressure autoclave that affords the possibility of high-temperature high-pressure processing. While the name of the process is geological in origin, it describes heterogeneous reactions that occur in the presence of water under high-temperature and high-pressure conditions. Under these conditions, dissolution and recrystallization reactions occur that allow for the growth of products that represent the thermodynamically stable phase. Another typical aspect of the process is the use of mineralizers to accelerate the rates of dissolution and recrystallization. Most often, these are strong basic salts such as sodium hydroxide. While the method is useful for both powder and single-crystal preparation, for single-crystal synthesis, including quartz, a seed crystal is used as a site for heterogeneous nucleation for the growth of the large ($\sim$4 cm $\times$ $\sim$12 cm) crystals. For further information on hydrothermal synthesis, the reader is referred to the comprehensive text by Byrappa and Yoshimura [83].

   The hydrothermal method is well suited to the fabrication of α-quartz since it employs temperatures below the α- to β-quartz transition at 573°C. As expected based on its technological significance, while there are a variety of other SiO_2 polymorphs, α-quartz is the only one of these materials to have been grown into large, bulk, single crystals [83]. Typical batch sizes range from 150 kg to 2000 kg per cycle, and process cycle times are between 25 and 90 days to achieve full-size crystals [83]. Typical autoclave conditions include the use of either NaOH or Na_2CO_3 as a mineralizer, nutrient zone temperatures between 355°C and 369°C, a growth zone temperature of 350°C, and a percent fill of about 80% [83]. As expected, the quality of the seed crystal and the growth rate are important considerations for the growth of high-quality materials. Interestingly, the method has also been used for the fabrication of perovskite powders and thin films [126–130].

2. Other Piezoelectric and Electro-optic Single Crystals

Single-crystal materials such as potassium dihydrogen phosphate (KH_2PO_4; KDP) and potassium titanyl phosphate ($KTiOPO_4$; KTP), may be prepared either by flux growth or by hydrothermal synthesis. Preparation of KTP by the flux growth method results in materials that may have high ionic conductivity due to flux incorporation; problems with spurious nucleation must also be handled. However, the flux and hydrothermal methods are utilized as KTP cannot be prepared by melt growth because it decomposes before melting. A review of the preparation of KTP by hydrothermal synthesis is given by Byrappa and Yoshimura [83].

Flux growth is considered to be a subdivision of solution growth, which involves the growth of the solute from a melt of different composition and which takes advantage of the strong solvating power of molten oxides [131,132]. The method is also frequently referred to as molten salt synthesis, and it is widely employed for the preparation of multicomponent oxide crystals [132]. Typically, flux growth involves the dissolution of the component oxides in an oxide flux, followed by the formation of the desired (thermodynamically stable) crystalline compound through a nucleation and growth process. Commonly employed molten salts include KF, PbO, PbF_2, B_2O_3, and combinations of these. A variety of considerations dictate solvent choice, including solute solid-phase stability at the growth temperature, solute solubility and the temperature coefficient of solute solubility, volatility, and reactivity with platinum, among others [132]. The flux growth procedure typically involves the dissolution of the components at temperatures slightly above the saturation temperature followed by slow cooling. Slow growth rates (slow cooling) are maintained to limit the entrapment of solvent inclusions in the crystal. For high-viscosity fluxes, stirring is utilized to shorten the diffusional path close to the growth interface. Most flux growth processes are carried out at temperatures of 1300°C or below using platinum crucibles [132].

Piezoelectric materials that have been prepared by the flux growth method include barium titanate [132,133], lead titanate [134], and single-crystal relaxor (PMN-PT and PZN-PT) ferroelectrics [11]; the growth of single-crystal PZT materials has proven difficult [92–95]. As an example, the flux growth of the lead titanate involves the use of Pb_3O_4, KF, KBF_4, and PbF_2 to form the flux, together with TiO_2 as the titanium precursor [134]. The materials are typically mixed, placed in a lidded platinum crucible, and heated to 900°C for homogenization of the melt. The melt is subsequently cooled to 475°C and held at this temperature for nucleation and growth, followed by slow cooling to 300°C and then to ambient temperature. Materials such as hot acetic acid may then be used to dissolve the solidified flux exposing the crystals [134].

Other investigators have developed methods to prepare single-crystal-like as well as textured relaxor ferroelectrics via templated growth processes [135–138]. These materials are typified by properties that compare favorably with the single-crystal PMN-PT materials, but which are prepared by a simpler method than flux growth. One approach is to use a seed crystal, such as $SrTiO_3$, to induce the transformation of a dense polycrystalline PMN-PT body into a single-crystal-like material via seeded polycrystal conversion [135,136]. The process is carried out at temperatures between 1100°C and 1200°C, and the conversion of the polycrystalline body to the orientation of the seed is verified by a technique such as orientation imaging microscopy.

Seed crystals have also been used effectively in the preparation of textured PMN-PT layers [138] or rapid prototyped actuators of various geometries [139]. For the textured thick films, a typical procedure involves the preparation of a powder precursor based on $(PbCO_3)_2Pb(OH)_2$, $MgNb_2O_6$, and TiO_2 that offered a reactive matrix for sintering. To this mixture, 5 % of {001} $BaTiO_3$ crystals of approximate thickness 50 µm and diameter 100 µm were added as templating agents. This precursor mixture can be used to form a tape cast sheet, which may be densified by hot pressing in oxygen. The result is the oriented growth of the PMN-PT material, resulting in high effective piezoelectric coefficients [138].

Finally, melt or Czochralski growth (crystal pulling) methods have also been used to prepare electro-optic single crystals of materials such as barium titanate and lithium niobate. This method is reviewed briefly from the discussion given by Laudise [132]. Czochralski growth is referred to as a conservative crystal growth method because no material is added to either the liquid or solid phases except by crystallization. In this method, the material to be crystallized is placed in a crucible and is completely melted. Heat transfer and thermal conditions are then arranged such that the melt is isothermal and there is a negative temperature gradient above the melt. A seed is then introduced into the surface of the melt, and a small portion of the surface is melted to ensure a clean crystal surface that is free of spurious nuclei. The seed crystal is then slowly withdrawn, and if the temperature profile and withdrawal rate are properly established a crystal may be pulled from the melt [132]. The method is used extensively in the synthesis of Si single crystals with diameters larger than 300 mm, the current standard in semiconductor manufacturing. With regard to piezoelectric and electro-optic compositions, $BaTiO_3$, $LiNbO_3$, and $KTaO_3$, among others, have been synthesized by this method.

B. Preparation of Polycrystalline Materials by Mixed Oxide and Chemical Methods

As discussed above, a range of electro-optic and piezoelectric single-crystal materials have been synthesized for a variety of applications. However, most piezoelectric devices are based on ferroelectric ceramics that are prepared by a traditional ceramic powder processing route utilizing the individual constituent oxides. In addition to the polycrystalline piezoelectrics prepared via mixed oxides, some of the more optically demanding applications (electro-optic shutters, for example) utilize materials prepared by a chemical coprecipitation technique. The processing method that one selects to prepare the powder depends, to a large extent, on cost; but even more important is the end application (i.e., whether piezoelectric or electro-optic). Understandably, the electro-optic ceramics require a higher purity, more homogeneous, and high-reactivity powder than do the piezoelectrics because inhomogeneities can be detected optically much more easily than electrically [140]. As a result, different powder processing techniques have evolved in the two cases. Piezoelectric ceramics are still prepared from the most economical, mixed-oxide (MO) process, whereas optical PLZT materials utilize a specially developed chemical precipitation (CP) process involving liquid precursors. Although not yet achieved, the trends in this area are toward the development of one unified process that will meet the objectives of both types of materials. There is, in fact, a commonality in these objectives, since the more recent piezoelectric devices demand higher quality material (essentially zero porosity) and the electro-optic devices require a more economical process. The great interest in thin-film devices, including ferroelectric random access memories (FRAMs) [141] and microelectromechanical systems (MEMS) [23], have also resulted in significant developments in preparation methods for ferroelectric thin films. These methods are discussed more extensively in Chapter 8 by Waser and colleagues and include pulsed laser deposition, sputtering, chemical vapor deposition, and chemical solution deposition [142,143]. Because these methods are covered extensively elsewhere in this book, the focus here will remain on bulk polycrystalline materials.

The general processing steps involved in the powder preparation and fabrication of both piezoelectric and electro-optic ceramics are given in Figure 14. A majority of the steps involved in the two processes are quite similar. Essential differences lie in the areas of initial oxide mixing versus coprecipitation and atmosphere sintering versus hot pressing. Calcining (a high-temperature chemical reaction of ingredients to form a final composition) is a critical step in both processes.

The conventional MO process consists of (1) weighing the oxides to the appropriate accuracy; (2) blending in a liquid medium; (3) drying to completeness; (4) calcining the powder at approximately 900°C for 1 h; (5) milling in a liquid medium; (6) drying with binder addition; (7) cold pressing a specific shape; (8) high-temperature firing (sintering) at approximately 1275°C for several hours; (9) shaping; (10) electroding; and (11) final poling. During sintering, covered crucible approaches are typically used in conjunction with PbO-based atmosphere powders to compensate for lead oxide loss. Further details of the process can be found in Refs. 144 and 145.

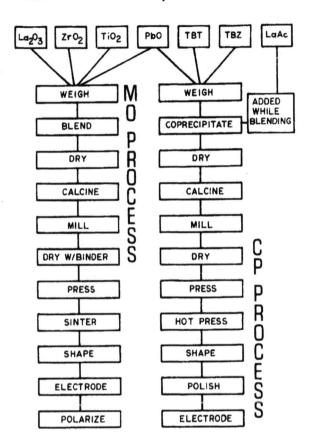

Figure 14 General processing steps for MO and CP synthésis of PZT and PLZT materials.

In the CP process, the high-purity liquid organometallics (tetrabutyl zirconate and tetrabutyl titanate) are first premixed with lead oxide in a high-speed blender and then precipitated by adding the lanthanum acetate, water-soluble solution while blending. The resulting slurry, consisting of a finely divided suspension of mixed oxides and hydroxides in an alcohol-water solution, is dried to completion at approximately 80°C. Subsequent steps in the process involve (1) calcining the powder at 500°C for 16 h; (2) wet milling the powder for several hours to promote additional homogeneity; (3) drying without binder; (4) cold pressing the powder; (5) hot pressing in oxygen; (6) shaping and slicing; (7) polishing; and (8) electroding. Further details may be found in Refs. 116 and 146.

For the preparation of transparent electro-optic materials, hot pressing (Figure 15) is typically utilized to achieve high density and optical transparency [116]. The arrangement shown is typical of many different methods of heating and applying pressure. Experience has shown that a simple, uniaxial, single-ended hot-pressing arrangement is both reliable and economical in producing consistent, optical quality material. A pressed slug is used as the perform for the hot pressing operation. The slug is placed in a silicon carbide mold (which may be alumina-lined) and surrounded by refractory grain (magnesia or zirconia) to prevent high-temperature reactions with surrounding materials. During the heating portion of the hot-press cycle, the chamber is evacuated and back-filled with oxygen. Typical hot pressing conditions are 1250°C for 18 h at 2000 psi. After hot pressing, the slug is extracted from the mold and polished for optical evaluation. This method of vacuum/oxygen hot pressing is used to fabricate optical-quality PLZT slugs up to 6 inches in diameter [147]. An alternative method of hot pressing in flowing oxygen rather than vacuum/oxygen is also known to produce optical quality, but it is generally limited to slug sizes less than 2 inches in diameter.

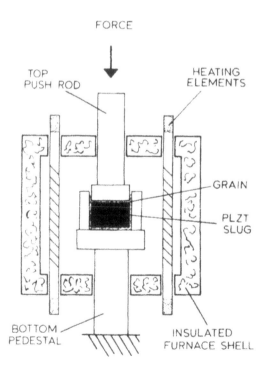

Figure 15   Typical hot pressing setup for the fabrication of transparent PLZT electrooptic ceramics.

The two most critical steps in the fabrication of these ceramics are powder preparation and firing, and of these, the first is most important. Although one can readily determine many measurable characteristics of the powders, such as particle size, surface area, composition, crystalline structure, and morphology, powder reactivity is perhaps the most important—but still the least understood. In fact, a primary objective of the various chemical preparation techniques [148,149] is to increase powder reactivity, thereby achieving full density (zero porosity) and maximal chemical homogeneity. While there have been a number of papers published regarding which phase forms first from the constituent oxides, particle size and mixing procedures also play a role in defining phase evolution and calcination behavior.

C. Poling and the Development of Electromechanical Response

In the as-formed state, upon cooling from the elevated processing temperatures that are typically employed, polar ceramics undergo a transition from a paraelectric cubic state to the polar ferroelectric state as they are cooled through the Curie transition temperature. As noted above, accompanying this transformation is the development of a domain structure within the body that forms in response to the mechanical and electrostatic boundary conditions that are present [41–43]. It simplifies our understanding of the behavior of these materials to view each grain (crystallite) within the ceramic as possessing an overall dipole moment associated with the domain configuration within each grain. For ferroelectric ceramics, the poling process allows us to turn an electromechanically inactive polycrystalline ceramic into a material that acts very like a single-crystal material. When the phenomenon was discovered during the 1950s it was considered a rather amazing revelation, since prior to that time it was believed that only single-crystal materials were capable of displaying piezoelectricity [150,151]. Once poled, the ceramic responds to electrical and mechanical stimuli very much like a single crystal, with the whole body acting as a single entity characterized by the net macroscopic polarization that is induced by poling.

In the poling technique, electrodes are applied to the ceramic, and a sufficiently high electric field is applied to the material such that the dipoles within the individual crystallites are reoriented and aligned in the direction of the electric field. Typically, temperatures slightly above room temperature and applied fields three to four times the coercive field are utilized [152]. During this process, there is a small expansion of the material along the poling axis and a contraction in both directions perpendicular to it, due to the non-180° domain switching processes that occur. An illustration of the effects of this process on ceramic microstructure is shown in Figure 16, and the corresponding polarization and strain are shown as the virgin curves in Figure 7.

The strength of the electric poling field, the temperature utilized, and the grain size and grain size distribution are important factors in determining the extent of alignment and, hence, the resulting properties of the ceramic body. In actuality, alignment is never complete because of the random orientation of the original crystallites and the constraints of domain switching (e.g., 90° or 180° for tetragonal materials or 71°, 109°, and 180° for rhombohedral materials). However, depending on the type of crystal structure involved, the thoroughness of poling can be quite high. For example, the theoretical net polarizations that may be achieved by this process range from 83% (tetragonal) to 91% (orthorhombic

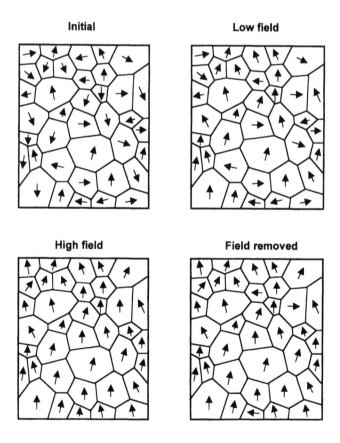

Figure 16 Schematic of microstructural changes in a piezoelectric body during poling and removal of poling field.

phase) of the single crystal [153,154]. It is worth noting briefly that piezoceramics may also be poled mechanically by the application of a stress because of their ferroelastic nature. However, in almost all instances, the electrical poling method is utilized.

D. Composites and Rapid Prototyping Methods

A broad range of techniques has been developed for the fabrication of electrome-chanical composites. One composite fabrication method, which is referred to as the dice-and-fill method, involves the use of a wafering saw to dice or partially dice a PZT sheet into PZT pillars followed by filling of the spacing between the pillars by a polymer [155]. This method has been widely used in the ultrasonics industry for more than 10 years [156]. Pick-and-place techniques have also been used to position PZT rods that are subsequently embedded in an inactive poly-meric host. Another method that has been used to prepare PZT-polymer compos-ites is the polymeric infiltration of a bundle of PZT fibers. The typical process involves extrusion of the fibers from a high-viscosity sol through a multihole spinnerette. The fiber bundle is then placed in a mold that is impregnated with polymer [157]. Yet another method involves the use of injection molding to fabricate PZT performs on a PZT base with hundreds to thousands of elements. This method offers the advantages of precise control of PZT rod alignment and spacing prior to polymer infiltration, as well as low cost, good process reproduc-ibility, and rapid throughput [158].

Another group of methods that has been utilized to fabricate piezoelectric composites is referred to as rapid prototyping, direct fabrication, and three-dimen-sional prototyping [159–162]. An example that is representative of this family of methods developed at Sandia National Laboratories is Robocasting [162]. The method involves the use of PZT inks with pseudoplastic flow characteristics that may be locally deposited via a micropen technique in three-dimensional geometries. The rheological characteristics, particularly the yield stress of the ink, is controlled so that adjacent deposited beads may flow together or remain as individual lines. Typical line widths within the structures that are prepared range from 0.2 to 0.8 mm [159]. The resulting piezoelectric and ferroelectric properties of the materials closely correspond to those of bulk materials prepared by conventional bulk mixed-oxide processes.

V. PROPERTIES

A. Microstructure

Compositions in the PLZT system, whether piezoelectric or electro-optic, are characterized by a highly uniform microstructure consisting of randomly oriented grains (crystallites) intimately bonded together. An example of such a microstructure is given in Figure 17. The sample was polished and thermally etched at 1150°C for 1 h. As is typical for most hot-pressed microstructures, little or no entrapped porosity is evident. This is due to the influence of the external pressure during hot pressing, which aids in pore removal while the material is in a thermo-chemically active state at elevated temperatures. In actuality, some amount of porosity exists in all the materials, whether hot pressed or atmosphere sintered; however, it is of prime importance to the electro-optic ceramics because of their optical transparency. It can also be seen from the microstructure that the average grain size for this sample is approximately 8 μm and is the result of the conditions (temperature, time) selected during fabrication. Piezoelectric ceramics usually possess grain sizes in the range 2–6 μm, whereas the electro-optics cover a wider range, from 2 to 10 μm, depending on the specific electro-optic effect being optimized. A uniform grain size is a highly desirable feature from the standpoint of performance.

Another noteworthy feature of the microstructure of Figure 17 is the complete lack of a second phase, which often precipitates along the grain boundaries or at triple points between grains. Porosity may be considered to be a second phase (air), but here it is not referred to as such. Second phases usually consist of other, unwanted compounds that are present in the material and have developed

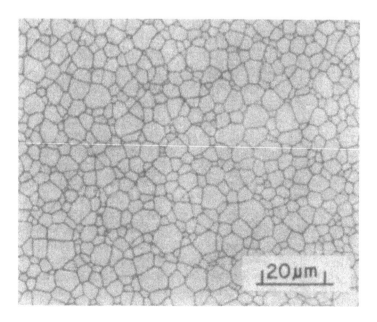

Figure 17   Typical microstructure of a hot-pressed electro-optic PLZT ceramic.

either during calcining or sintering via precipitation or were inadvertently added to the batch during formulation or processing. Such phases are particularly undesirable in electro-optic ceramics and even in minor amounts can produce opacity. Domain (electrical polarization) patterns can also be revealed in the microstructure of ferroelectric ceramics when chemical etching techniques are used and typically appear as a two-level structure because one end of the electric dipole chemically etches faster than the opposite end. Whereas thermal etching may be used to reveal grain boundaries, only chemical etching will reveal domain boundaries. The reason is that at elevated temperatures where thermal etching (partial melting) takes place there are no domains because the material is in the nonpolar, cubic phase.

In the case of piezoelectrics, a somewhat different set of circumstances exist. For piezoelectrics that are utilized as thin disks with thicknesses ranging from 0.002 to 0.005 in., e.g., loudspeakers, second phases (including porosity) must be prevented as much as possible. On the other hand, for larger shapes, second phases can be tolerated if they are adequately controlled. Work by Newnham and coworkers at Penn State on the deliberate introduction of second phases for enhancement of piezoelectric properties has shown that in some cases significant improvements can be made [119,163]. Connectivity is the key word here, i.e., the volume of the second phase or phases and their pattern of dispersion throughout the material are very instrumental in achieving the desired properties.

B. Mechanical Properties

Ferroelectric ceramics, particularly some of the piezoelectrics, are actually structural elements that must withstand reasonably high stress levels in their use condition. Consequently, some knowledge of their mechanical strength is necessary in an optimal design. Although it is always better from a strength standpoint to utilize ferroelectrics (as is true of most ceramics) in compression rather than tension, this is not always possible, and tensile strength becomes a limiting factor. Studies of several hot-pressed ferroelectric compositions in the PZT system have shown that tensile strengths are in the range 65.5– to 96.5 MPa as compositions vary from 52:48 (Zr/Ti) at the MPB to 95:5 near lead zirconate. Additional studies on the fracture processes in ferroelectrics have pointed out that (1) flexure strengths can range from 68.9 MPa to 165.5 MPa for the PZTs; (2) internal stresses significantly affect strength; and (3) ferroelastic twinning, phase transformations, and microcracking are toughening mechanisms preceding failure [164]. The mechanical properties of some of the more common piezoelectric compositions are reported below in Table 3 in the section on electromechanical properties.

The mechanical properties of the PLZT materials have not been investigated as thoroughly as those of the PZTs; however, studies on strain-biased devices using these materials have shown that they have the ability to tolerate unusually

Table 3a Electrical, Electromechanical, and Mechanical Properties of Poled Polycrystalline $BaTiO_3$ and PZT-Based Piezoelectric Ceramics

Property	Symbol	Multiplier	Units	BaTiO₃[a]	DOD-I[b]	DOD-II[b]	DOD-III[b]	DOD-VI[b]	PLZT 2/65/35	PLZT 7/60/40
Dielectric	$\varepsilon_{33}^T/\varepsilon_o$	1	Dimensionless	1900	1300	1700	1000	3400		
	$\varepsilon_{33}^S/\varepsilon_o$	1	Dimensionless	1420	635	830	600	1470		
	$\varepsilon_{11}^T/\varepsilon_o$	1	Dimensionless	1620	1475	1730		3130		
	$\varepsilon_{11}^S/\varepsilon_o$	1	Dimensionless	1260	730	916		1700		
Coupling	k_p	1	Dimensionless	0.354	0.58	0.60	0.50	0.65	0.450	0.720
	k_{33}	1	Dimensionless	0.493	0.70	0.705	0.62	0.75		
	k_{31}	1	Dimensionless	0.208	0.33	0.34	0.295	0.39		
Piezoelectric	d_{33}	10^{-12}	m/V or C/N	191	289	374	218	593	150	710
	d_{31}	10^{-12}	m/V or C/N	-79	-123	-171	-93	-274		
	g_{33}	10^{-3}	Vm/N	11.4	26.1	24.8		19.7	23.0	22.2
	g_{31}	10^{-3}	Vm/N	-4.7						
	e_{33}	1	C/m²	18.6	15.1	15.8		23.3		
	e_{31}	1	C/m²	-4.4	-5.2	-5.4		-6.5		
Mechanical	s_{33}^E	10^{-12}	m²/N	8.93	15.5	18.8	13.9	20.7		
	s_{33}^D	10^{-12}	m²/N	6.76	7.90	9.46	8.5	8.99		
	s_{11}^E	10^{-12}	m²/N	8.55	12.3	16.4	11.1	16.5		
	s_{11}^D	10^{-12}	m²/N	8.18	10.9	14.4	10.1	14.1		
	c_{33}^E	10^{10}	N/m²	16.2	11.5	11.1		11.7		
	c_{33}^D	10^{10}	N/m²	18.9	15.9	14.7		15.7		
	c_{11}^E	10^{10}	N/m²	16.6	13.9	12.1		12.6		
	c_{11}^D	10^{10}	N/m²	16.8	14.5	12.6		13.0		

[a] Ref. 167a.
[b] Ref. 108.

Table 3b   Cross-Reference Table for Various Piezoelectric Manufacturers*

Manufacturer (DOD Designation)	APC International	CTS Wireless	Edo	Morgan Matroc (Morgan Electroceramics)	Sensor Technology Limited	TRS Ceramics
DOD-I	840,841		EC-64	PZT-4	BM400	TRS100
DOD-II	850	3195	EC-65	PZT-5A	BM500	TRS200
DOD-III	880		EC-69	PZT-8	BM800	TRS300
DOD-V			EC-70	PZT-5J	BM527	
DOD-VI	855,856	3203	EC-76	PZT-5H	BM532	TRS600
				PZT-7A	BM740	

* Note DOD is also frequently referred to as Navy. For example, DOD-I and Navy Type 1 represent equivalent materials.

high mechanical strains, approximately 3000 ppm (0.3%), before fracture occurs. A strain of this magnitude is, perhaps, an order of magnitude higher than for most conventional ceramics. This is believed to be due to the ferroelastic properties of these ferroelectrics, which accommodate the strain and relieve the stress by domain reorientation. Such behavior should also result in a change in the polarization state of a poled ferroelectric; this was shown by Maldonado and Meitzler to be the case for PLZT 7:65:35 [165].

C. Thermal Properties

Thermal expansion coefficients can vary considerably in compositions throughout the PZT system, depending on the symmetry of the phases present. Figure 18 illustrates this for various selected Zr/Ti ratio compositions containing 2 at. % niobia. The anomalous behavior below the Curie point is usually a result of subtle changes in the unit cell. For example, in 95:5 the change is between the LT and HT rhombohedral phases; however, above the Curie point the linear expansions are regular and vary in a uniform manner from a value of $6.7 \times 10^{-6}\,°C^{-1}$ at the MPB (53:47 ratio) to $8.8 \times 10^{-6}\,°C^{-1}$ for 95:5. In all cases, the Curie point is marked by a volume decrease in the vicinity of the Curie transition temperature. In contrast to this behavior, PLZT composition tends to demonstrate less inflection in the thermal expansion at the Curie point.

D. Electrical Properties

Since nearly all of the useful properties of ferroelectric ceramics (whether they are dielectric, mechanical, or optical properties) are related in some manner to their response to electrical stimuli, the electrical behavior of these materials is

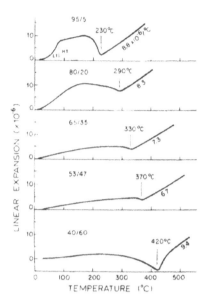

Figure 18   Thermal expansion characteristics of several niobia-doped PZT compositions with varying Zr/Ti ratios.

important to their successful application in piezoelectric or electro-optic devices. Ferroelectrics in general are characterized by (1) higher dielectric constants (ε_r = 500–4000) than ordinary insulating substances (ε_r = 5–100), making them useful as capacitor and energy storage materials; (2) relatively low dielectric loss (0.1–5%); (3) high specific electrical resistivity ($\sim 10^{11}$ to 10^{13} Ω-cm); (4) moderate dielectric breakdown strengths of 250–300 V/mil (100–120 kV/cm) in thin disks; and (5) nonlinear electrical behavior (hysteresis loop), which results in an electrically variable dielectric constant or in electrically variable memory states. In addition, ferroelectrics possess mechanical and optical effects that are interactive with their electrical behavior in producing useful electromechanical and electro-optic properties.

It should be kept in mind that not all of these properties are optimized and realized in a given type of ferroelectric structure or chemical composition, and hence a variety of ceramic materials are manufactured and made available from several different companies throughout the world. As a result, no attempt will be made here to enumerate the specific properties of the various ceramics produced by these manufacturers. In enumerating the electric properties, distinction is usually made between large-signal (>5 V/mil) and small-signal (<1 V/mil) measurements (e.g., dielectric constant is normally a small-signal property, whereas the hysteresis loop is a large-signal property).

1. Dielectric Properties

Small-signal dielectric constant values as a function of temperature for several selected compositions in the PLZT system are given in Figure 19. This figure demonstrates the typical variation in dielectric constant associated with the Curie transition in these materials. For the series with 65:35 Zr/Ti ratios, a consistent and gradual reduction in the Curie point (indicated by the peak in the dielectric curve) occurs up to an La concentration of 8%. Higher La contents reduce the height of the dielectric constant peak, making it more diffuse without significantly changing the temperature of the peak. As a rule, compositions near $PbZrO_3$ and $PbTiO_3$ are characterized by low dielectric constants ($\varepsilon_r = 150-300$) and loss tangents of approximately 1% or less. Maxima in dielectric constants occur along the FE-cubic ($\varepsilon_r = 5700$) and the FE tetragonal–FE rhombohedral ($\varepsilon = 4100$) phase boundaries. Loss tangents in these regions also reach maximal values of 6% and 4%, respectively.

2. Ferroelectric Properties—Hysteresis Loops

The hysteresis loop is the single most important measurement that can be made on a ferroelectric ceramic to characterize its electrical behavior. Indeed, the internal

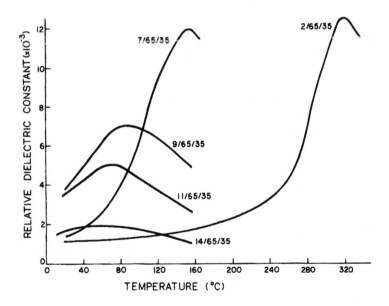

Figure 19 Temperature dependence of the relative dielectric constant for several selected PLZT compositions with a 65:35 Zr/Ti ratio.

polarization of the material must be switchable at some field less than the electrical breakdown for the material to even qualify as ferroelectric. As mentioned previously, piezoelectrics and some electro-optics are routinely subjected to a partial hysteresis loop when they are permanently polarized for device applications.

Hysteresis loops come in all sizes and shapes and, like a fingerprint, identify the material in a special way. Therefore, one should become familiar with such a measurement. Although early workers in the field of ferroelectrics utilized a dynamic (60 Hz) measurement with a Sawyer–Tower circuit and an oscilloscope readout, more recent work has been done with essentially a dc ($\sim 1/10$ Hz) measurement using an X-Y plotter to display the hysteresis loop [7,166]. Hysteresis loop testers are now also commercially available.

Hysteresis loops are usually run on virgin (thermally depoled) electroded disks by varying the electric field ($E = V/t$) across the disk while monitoring the charge collected on a large (approximately 1000 times larger in capacitance than the sample), low-loss capacitor in series with the sample. This charge (actually measured as voltage, which relates to charge by the familiar $Q = CV$) on the low-loss capacitor is proportional to the charge on the sample; thus, one obtains a continuous plot of polarization as a function of electric field. In a strict sense, dielectric displacement rather than polarization is being measured; however, for high dielectric constant materials such as ferroelectrics, the two quantities are very nearly equal and the term "polarization" is usually preferred.

A typical hysteresis loop was shown earlier in Figure 7a. Since a virgin ceramic is macroscopically isotropic (nonoriented crystallites) to begin with, no preferred orientation effects should be evident in the loop; and it should be symmetrical about a point of origin that is selected for the measurement (this, of course, will not be true for previously poled samples). Starting at point 0, the electric field is gradually increased, initially producing very little change in polarization. At some higher field, domain switching occurs fairly abruptly, and a significant increase in effective polarization takes place within the material. This internal polarization (dipole moment) is balanced by externally bound charge that is delivered from the electrical circuit. For square-loop materials, this polarization switching may occur over a relatively narrow field range of 1–5 V/mil; however, a more typical range is 10–20 V/mil.

As the electric field is further increased to point A at approximately $3E_c$, the hysteresis loop saturates at a polarization level that is dependent on composition as well as density and grain size [167]. Reducing the electric field to zero leaves the material in a polarized condition at point C with a remanent polarization value of P_r. Increasing the field to the opposite direction will first reduce the polarization to zero at E_c, the coercive field of the material, and then saturate in the reverse direction at point D. Again, the process of reducing the field to zero traces a path to point F which is equal to $-P_r$. The hysteresis loop is completed by, once again, changing the field direction, increasing it to saturation in the original direction, and finally reducing it to zero at P_r.

Several different types of hysteresis loops are exhibited by compositions in the multicomponent perovskite ferroelectrics, as shown in Figure 20 for the PLZT system. These include an antiferroelectric (AFE orthorhombic phase) loop consisting of a linear portion and a field-enforced FE region; (2) a square (FE rhombohedral phase) loop with a low coercive field; (3) a square (FE tetragonal) loop with a high coercive field; (4) a slim loop exhibiting field-enforced ferroelectric (SFE) behavior; and (5) a linear loop characteristic of normal non-FE dielectrics. Specific applications exist for materials possessing loops of each of these types. Table 3, shown below in the section below on electromechanical properties, presents some typical properties for several selected PLZT compositions.

E. Electromechanical Properties

The electromechanical properties of a piezoelectric (ferroelectric) material are of basic concern to the designer who is planning to utilize these materials in a new or existing application. Since the resulting effects produced by the interaction of

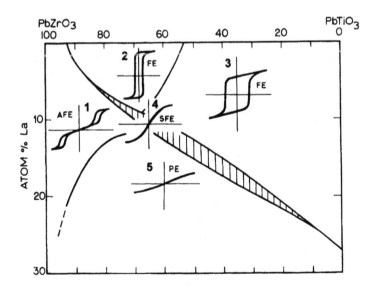

Figure 20   Types of hysteresis loops exhibited by various compositions within the PLZT system: (1) antiferroelectric; (2) square-loop ferroelectric; (3) high coercive field square-loop ferroelectric; (4) slim-loop ferroelectric; and (5) paraelectric. Note: the material compositions are given by the origins.

the material's electrical and mechanical properties are specific and unique, an understanding of these effects is essential to a successfully designed component. In general, one should be familiar with and understand the concepts surrounding (1) the basic types of physical motion involved in material deformation, (2) motor–generator actions, (3) resonant and nonresonant operation, and (4) the equivalent circuit concept of the piezoelectric resonator. The equivalent circuit model of piezoelectrics was introduced above. Each of the other concepts is discussed briefly in this section and the specific electromechanical properties of the various piezoelectric materials are reviewed.

1. Piezoelectric Deformations

The basic deformations of a piezoelectric ceramic are best illustrated by showing a change in physical dimensions as a result of the application of a varying electric field on an electroded and poled plate or disk. For ceramics, it is essential to know the original poling direction and the direction of the applied field (indicated by the electrodes) in order to predict the motion of the material. As we have seen previously, a permanent expansion occurs in the direction of the electric field on initially poling the virgin or thermally depoled ceramic material; and now when considering a dimensional change as a result of the influence of the electric field, it is very important to keep in mind the sign ($+$ or $-$) of the applied field in relation to the sign of the original poling field. If the sign of the applied field is the same as that of the original field, expansion will occur in that direction and contraction will occur in the other orthogonal directions; and when the sign of the applied field is negative with respect to the original field, opposite deformations will take place.

Typical examples of the deformations occurring in plates or disks are given in Figure 21. The deformations are highly exaggerated for illustrative purposes since the actual strains are usually very small, typically on the order of a few micrometers. In Figure 21a thickness expansion or contraction (longitudinal d_{33} effect) in a plate or length expansion in a long rod occurs when the electric field is alternately positive or negative, whereas opposite strains (contraction or expansion; lateral d_{31} effect) take place simultaneously in the lateral dimensions of the plate. Since contraction is considered to be a negative quantity, the lateral electromechanical coefficient that describes this action (d_{31}) is negative and the longitudinal d_{33} coefficient is positive. Similar action occurs in the case of a thin, hollow cylinder or tube (i.e., a continuously curved plate) with electrodes applied to the interior and exterior surfaces. Under the influence of an alternating field, the tube increases and decreases in diameter (hoop mode) as well as in length.

Shear deformations are obtained by initially polarizing the material with temporary electrodes in one direction, removing the electrodes, and then applying permanent electrodes in an orthogonal direction. Subsequent application of an

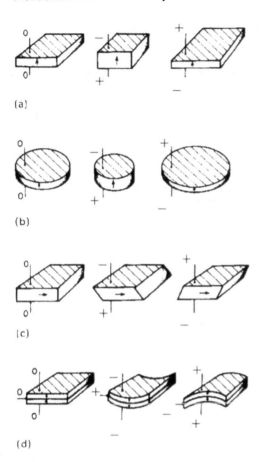

(a)

(b)

(c)

(d)

Figure 21 Typical mechanical deformations of poled piezoelectric plates when subjected to an electric field: (a) thickness and length; (b) radial; (c) thickness shear; and (d) bender.

electric field produces a shearing action, as illustrated for the thickness shear mode in Figure 21c.

The bender action shown in Figure 21d is a special case of the thickness/length extensional mode shown in Figure 21a. Bender devices are discussed below in the section on devices. Briefly, the advantage of this mode over the other modes is the larger mechanical displacements (multiplication factors range from 100 to 1000) achievable through the lever action; however, the higher compliance and lower mechanical impedance limits the amount of force it can generate.

2. Direct and Indirect Piezoelectric Effects: Motor/Generator Actions

A piezoelectric ceramic can operate either as a motor or a generator depending on its configuration and application. As a motor, the material composition and configuration are usually optimized for producing maximal displacements with a minimal input voltage (converse piezoelectric effect); as a generator, the same considerations are made for producing maximum output voltage with a minimum of input stress or strain (direct effect). Common examples of motors are ultrasonic cleaners, sonar, sounders, loudspeakers, pumps, and positioners. Examples of generators include gas ignitors, phonograph pickups, microphones, contact fuses, sensors, and accelerometers. In some instances, both types are utilized in a single device, such as filters, delay lines, and piezoelectric transformers. Specific devices and application issues are discussed below.

3. Resonant and Non-Resonant Operation

In general, any material of a given size and shape has a natural resonance or frequency at which it will vibrate freely. When put into motion with a mechanical force at its resonant frequency, the body vibrates at a higher amplitude than at other frequencies and continues to vibrate until damping forces cause a cessation of movement. This is also true for piezoelectric solids, but due to the coupling between mechanical and electrical properties, these materials can be electrically stimulated to produce mechanical vibrations at many frequencies either below, above, or at resonance. Most piezoelectric devices are operated below resonance, but some notable exceptions are quartz oscillators for frequency control, ceramic IF resonators, and piezoelectric transformers for high-voltage generation. In addition, some must operate efficiently over wide range of frequencies (e.g., tweeters over the range 3–30 kHz). In considering resonant or nonresonant operation, it should be noted that many of the piezoelectric equations are limited to the lower frequencies and do not apply at or near resonance.

4 Summary of Electrical, Mechanical, and Electromechanical Properties

As noted above, one of the most important aspects of perovskite materials for dielectric, piezoelectric, and electro-optic applications is the broad range of properties that may be achieved through control of composition. An illustration of this behavior is shown in Figure 22, where the strain–electric field characteristics are shown for a variety of compositions. As noted in the introduction, tailoring of composition allows for a broad range of attainable properties.

Utilization of piezoelectric materials requires an understanding of their properties as they relate to a given application of interest. Selected compositions

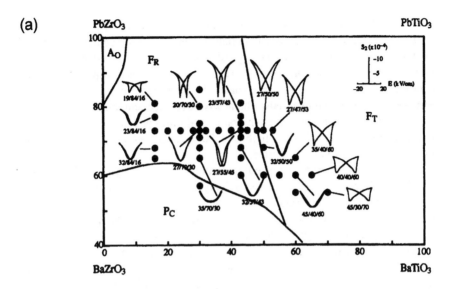

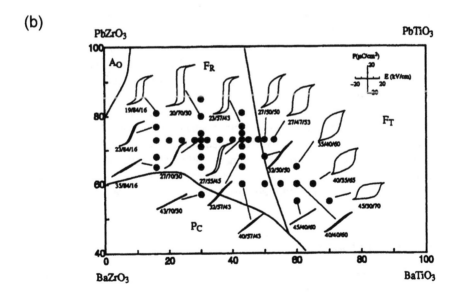

Figure 22 Illustration of the variation in (a) electromechanical response butterfly loops and (b) ferroelectric hysteresis loops for the quaternary system BaTiO$_3$-BaZrO$_3$-PbTiO$_3$-PbZrO$_3$. (After Ref. 173.)

within the PLZT system possess some of the highest electromechanical coupling coefficients attainable in ceramic materials. Other high-strain actuators are based on PMT-PT compositions. Some typical values of k_p, d_{33}, and g_{33} for poled polycrystalline materials are given in Table 3, whereas Table 4 presents similar data for piezoelectric single-crystal materials, including the recently developed PMN-PT single-crystal materials. The newly developed single-crystal piezoelectric materials [11] offer even higher strain capabilities and coupling coefficients. Piezoelectric d_{33} coefficients as high as 2500 pC/N and k_{33} coupling coefficients as high as 0.90 have been reported.

In contrast, the maximum reported values of k_p and d_{33} are 0.72 and 710 $\times 10^{-12}$ C/N, respectively, for poled polycrystalline PZT materials. These values are obtained for ceramic compositions located in the phase boundary region separating the ferroelectric rhombohedral and tetragonal phases. As mentioned in the section on phenomena, there has been considerable speculation concerning the reasons for the fortuitous maximum in coupling at this phase boundary [168–170]. These may be summarized as being due to (1) the existence of a mixture of phases at the boundary, (2) a concurrent maximum in dielectric constant at the boundary, (3) maximal orientational polarization attainable by an electric field in the boundary region, and (4) a maximum in mechanical compliance, permitting maximal domain reorientation without physically cracking. All these factors may contribute to this desirable and important effect; however, more recent work reported by Okazaki [171] and Multani et al. [172] seems to discount the existence of two phases at the boundary when the materials are carefully prepared by chemical means.

A review of the two tables also allows for consideration of the importance of domain wall motion effects on electromechanical response. In reviewing the d coefficients of single-crystal versus poled polycrystalline barium titanate, the contribution of extrinsic (non-180° domain wall motion) effects are seen to more than double the magnitude of the electromechanical response. Finally, the properties of piezoelectric composites are discussed below in the devices and applications sections.

F. Optical Properties

The optical transparency of the PLZT ceramics is highly dependent on the concentration of lanthanum oxide in the material. Neither low nor high concentrations are conducive to producing high optical transparency, but in each case it is for a different reason. At low levels of La, the higher birefringence and the existence of ferroelectric domain walls combine to produce significant light scattering, whereas at the higher levels of La, second phases precipitated at the grain boundaries lead to the high opacity. The specific concentrations of La that yield high optical transparency are dependent on the Zr/Ti ratio of the compositions. For a

Table 4 Electrical, Electromechanical, and Mechanical Properties of Single Crystal Piezoelectric Materials

Property	Symbol	Multiplier	Units	Material				
				BaTiO$_3$[a]	α-SiO$_2$[b]	LiNbO$_3$[c]	LiTaO$_3$[c]	PZN-PT[d] Relaxor
Dielectric	$\varepsilon_{33}^T/\varepsilon_0$	1	Dimensionless	168	4.628	28.5	43.4	5242
	$\varepsilon_{33}^S/\varepsilon_0$	1	Dimensionless	109	4.628	26.7	41.8	869
	$\varepsilon_{11}^T/\varepsilon_0$	1	Dimensionless	2920	4.507	83.3	53.0	3099
	$\varepsilon_{11}^S/\varepsilon_0$	1	Dimensionless	1970	4.420	44.9	41.9	2975
Coupling	k_p	1	Dimensionless	0.560				0.91
	k_{33}	1	Dimensionless	0.315				0.50
	k_{31}	1	Dimensionless					
Piezoelectric	d_{33}	10^{-12}	m/V or C/N	85.6	0	8.69	8.59	2140
	d_{31}	10^{-12}	m/V or C/N	-34.5	0	-1.04	-3.08	-980
	d_{22}	10^{-12}	m/V or C/N	0	0	20.30	8.33	
	d_{11}	10^{-12}	m/V or C/N	0	2.331			
	d_{14}	10^{-12}	m/V or C/N		-0.7763			
	d_{15}	10^{-12}	m/V or C/N			66.38	25.71	130
	g_{33}	10^{-3}	Vm/N	57.5	0	34.46	22.2	
	g_{31}	10^{-3}	Vm/N	-23.0	0	-4.110	-7.970	
	g_{11}	10^{-3}	Vm/N	0	58.41			
	g_{14}	10^{-3}	Vm/N	0	-19.45			
	e_{33}	1	C/m^2		0	1.894	1.894	
	e_{31}	1	C/m^2		0	0.328	0.328	
	e_{11}	1	C/m^2		0.1719			
	e_{14}	1	C/m^2		0.0390			
Mechanical	s_{33}^E	10^{-12}	m^2/N	15.7	9.7329	5.058	4.437	12.0
	s_{33}^D	10^{-12}	m^2/N	10.8	9.7329	4.758	4.246	
	s_{11}^E	10^{-12}	m^2/N	8.05	12.7791	5.854	4.899	8.3
	s_{11}^D	10^{-12}	m^2/N	7.25	12.6429	5.291	4.727	
	c_{33}^E	10^{10}	N/m^2	16.2	10.57816	23.418	27.522	8.9
	c_{33}^D	10^{10}	N/m^2	18.9	10.57816	24.935	28.446	
	c_{11}^E	10^{10}	N/m^2	16.6	8.67997	19.886	23.305	11
	c_{11}^D	10^{10}	N/m^2	16.8	8.75548	21.389	24.214	

[a] Ref. 167b.
[b] Ref. 167c. Note: All values are for right-handed quartz, using the 1978, 1987 IEEE convention.
[c] Ref. 167d. Note: The compliance and open circuit stiffness values included here are calculated from data in this paper.
[d] Ref. 167e; composition: PZN – 4.5%PT

65:35 material, the high-transparency range extends from about 8 at % La to approximately 16 at % La.

Thin (10–15 mil thick) polished plates of PLZT (9:/65:/35 La/Zr/Ti) characteristically transmit about 69% of the incident light and reflect 31%. When broadband antireflection coatings are applied to the major surfaces, transmission is increased to greater than 98%. The surface reflection losses are a function of the index of refraction of the PLZT according to the Fresnel formula (normal incidence, air ambient):

$$R = \frac{(n-1)^2}{(n+1)^2}$$

(39)

where R is the reflection coefficient in air from a single surface at normal incidence of light and n is the index of refraction ($n \cong 2.5$ for PLZT). Since the index of refraction varies as a function of wavelength [174], surface reflection losses may also be expected to vary with the wavelength of light.

Optical absorption in the PLZT materials is wavelength dependent, becoming extremely high in the violet (short wavelength) end of the spectrum near 0.37 μm (370 nm). In the infrared portion of the spectrum, transmittance remains high, to approximately 6.5 μm and gradually decreases in a regular manner to 12 μm, where the material is fully absorbing.

G. Electro-optic Properties

Several distinct electro-optic effects have been found to be operative in the PLZT materials. The three most common types are illustrated in Figure 23: (1) quadratic birefringence, (2) linear birefringence, and (3) memory scattering. Also shown in Figure 23 is a typical setup required for generating each effect together with the observed behavior shown in terms of light-intensity output as a function of electric field.

The first and most widely applied of all the electro-optic responses is the quadratic (Kerr) effect. It is generally displayed by those materials that are essentially cubic, but that are located close to the ferroelectric rhombohedral or tetragonal phases. These materials, by virtue of their natural cubic symmetry, do not possess permanent polarization and are not optically birefringent in their quiescent state. As such, they contribute no optical retardation to an incoming polarized light beam; however, when an electric field is applied to the material, an electric polarization (and consequently, birefringence) is induced in the material and retardation is observed between crossed polarizers (on state). On removing the electric field, the material relaxes again to its cubic state and is in the off state. Relaxation times to the off state vary with composition but generally range from one to

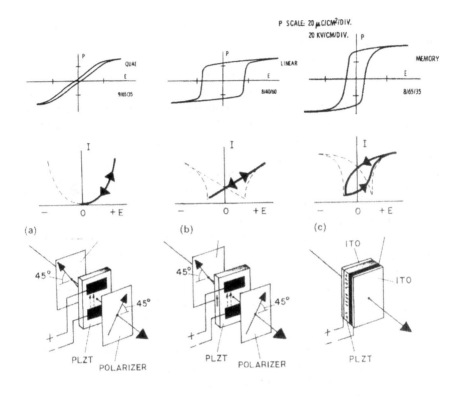

Figure 23 Hysteresis loops, birefringence, and operational configurations for (a) quadratic, (b) linear, and (c) memory PLZT materials. The heavily accented portion of the response curves represents the usable ranges.

$100 \mu s$. Turn-on times are on the same magnitude, and on/off ratios as high as 5000:1 have been measured with high-efficiency polarizers.

The second type of behavior existing in the PLZTs is the linear (Pockels) effect, which is generally found in high coercive field tetragonal materials such as 12:40:60. This truly linear nonhysteretic effect has been found to be intrinsic to the material and not due to domain reorientation processes that occur in the quadratic and memory materials. The linear materials possess permanent remanent polarization; however, in this case the material is switched to its saturation remanence, and it remains in that state. Optical information is extracted from the ceramic by the action of an electric field that causes linear changes in the birefringence, but in no case is there polarization reversal in the material. The experimental setup for observing this effect, as seen in Figure 23b, is identical to that for the quadratic response, except that the PLZT plate is prepoled to saturation remanence before use.

A third type of electro-optic behavior employed almost exclusively for image storage devices is that of electrically controlled light scattering in memory materials. The experimental arrangement involved in observing this effect is also shown in Figure 23c. No polarizers are employed since it is predominantly due to light scattering from domains within the material. The orientation of these domains is electrically alterable; and because light is preferentially scattered along the polar direction of the domains, the light transmitted by the PLZT plate is also electrically controllable. In addition, local areas can be polarized to different levels, leading to an ability for storing images with a gray-scale capability and a resolution of at least 30 line pairs per millimeter. Once a given local area is switched to a specific polarization state, it is permanently locked in until it is electrically switched to a new state or the material is heated above its Curie point (thermally depoled), which erases all the polarization states. The means by which local areas are switched independently of each other is provided by the photoconductor layer sandwiched between one of the transparent (indium-tin oxide, ITO) electrodes and the PLZT. When light impinges on the photoconductor layer, it reduces its resistivity by several orders of magnitude, electrons from the voltage source are transferred from the ITO electrode to the PLZT, and the local polarization is switched to a new state. Erasure of the total image is performed by flooding the plate with light while the voltage is applied in the positive saturation direction. The maximal contrast ratio may be as high as 100:1.

Additional electro-optic effects in PLZT include memory birefringence, depolarization scattering, surface diffraction, and photoinduced changes in refractive index. Although all these effects are more specialized in character than those previously described, they possess reasonable potential for utilization in future applications. Table 5 summarizes all the known electro-optic phenomena occurring in the PLZT materials.

H. Electro-optic Coefficients

A convenient and physically significant comparison of the electro-optic effects in PLZT ceramics with other electro-optic crystals may be made by examining their respective quadratic and linear electro-optic coefficients, R and r_c. These coefficients are listed in Table 6 for various PLZT compositions as well as for several other selected electro-optic crystals. In general, it may be noted that the values for the PLZT ceramics are exceptionally high and exceed those of most other crystalline materials. Indeed, it is primarily these high electro-optic coefficients that make the PLZT materials especially attractive in electro-optic applications.

Table 5 Summary of Electrooptic Effects in PLZT Ceramics

EO Effect	Electrical Address	PLZT Type	Polarized Light	Devices
Birefringence	Transverse	Non-memory, memory	Yes	Shutters, modulators, color filters, goggles, displays, memories
Depolarization scattering	Longitudinal	Non-memory	Yes	Shutters, modulators, displays
Birefringent fringing	Longitudinal	Non-memory	Yes	Displays
Birefringence	Longitudinal (strain bias)	Memory	Yes	FERPIC, image storage, displays, shutters
Scattering	Longitudinal	Memory, non-memory	No	CERAMPIC, image storage, modulators, displays
Diffraction (surface)	Longitudinal	Memory	No	FERICON, image storage
Photoferroelectric (space charge)	Transverse, fringing	Non-memory, memory	Yes	Holographic recording, image storage
Photoelastic (stress-induced $\overline{\Delta n}$)	Transverse	Non-memory	No	Optical waveguides, modulators

I. Aging Effects

Any poled ceramic is theoretically in a constant state of depoling as a function of time. This characteristic, referred to as aging, fortunately does not vary linearly with time. It has been found empirically that most ceramic properties that are dependent on the poling process, such as coupling, dielectric constant, and resonant frequency, exhibit a fairly constant change during each decade. For example, planar coupling may decrease by 0.2% after the first day of poling, an additional 0.2% after the next 10 days, and another 0.2% after 100 days. Extending this to 10,000 days, or 27 years, results in a total decrease in coupling of only 1%. Typical ranges of aging per time decade for planar coupling, relative dielectric constant, and resonant frequency are -0.2% to -2.3%, -0.6% to -5.8%, and $+0.0.8\%$ to $+1.5\%$, respectively. The aging of optical memory states in the

Table 6 Pockels and Kerr Electrooptic Coefficients for Selected Ceramics

Composition La/Zr/Ti	R $(\times 10^{16} \, m^2/V^2)$	r_c $(\times 10^{10} \, m/V)$
8.5/65/35	38.60	—
9/65/35	3.80	—
9.5/65/35	1.50	—
10/65/35	0.80	—
11/65/35	0.32	—
12/65/35	0.16	—
8/40/60	—	1.0
12/40/60	—	1.2
KTN (65/35)	0.17	—
LiNbO$_3$	—	0.17
KDP	—	0.52
SBN	—	2.10
LiTaO$_3$	—	0.22

Source: Ref. 175 and 176

transparent ceramics has not yet been measured but is anticipated to be in a range similar to that for planar coupling.

Since the process of aging begins the moment the ceramic is polarized, it is very difficult to specify exactly a given property that is dependent on its polarization condition. However, one can predict a given property change with some accuracy if the time of poling is known and the aging characteristics for that composition are given. It is also important to know that aging can be interrupted or decelerated by repoling the ceramic, in which case aging begins all over again.

The process of aging may be accelerated by exposing the ceramic to (1) a strong electric depoling field, (2) temperatures that approach the Curie point, (3) high mechanical stresses, and (4) combinations of these factors. Therefore, care must be exercised in selecting a given ceramic for a specific use condition. Ceramic compositions are often formulated to be more resistant to a given depoling environment. A common rule of thumb on temperature environments (the most common depolarization mechanism) is that the ceramic should be subjected to temperatures no higher than one-half of the Curie point.

Aging in ferroelectric ceramics is a complicated process and may occur as a result of several different mechanisms. Jonker [177] summarizes these as (1) a gradual equilibration of the domains with minimization of elastic and dielectric free energy, (2) a segregation of impurities and vacancies on domain walls and crystallite boundaries, and (3) an ordering of impurities and vacancies inside the ferroelectric domains with respect to the polar axis. Okazaki [178], on the other hand, attributes aging primarily to space charge effects in which charge compensation gradually occurs inside every domain because of the effect of the depolarizing field that results from the ordered spontaneous polarization.

VI. DEVICES

A. Piezoelectric Materials

Piezoelectric devices find use in applications including sensing, actuation, and voltage generation. The materials utilized for these applications are typically ferroelectric ceramics, most often the PZT materials that have been discussed. The materials are generally used in a nonswitching mode, although some devices, such as Thunder and Rainbow [16,17,179], take advantage of domain wall motion effects to generate their high electromechanical strain responses. Aside from this application, utilization of the switching response of ferroelectric ceramics has not met with success until the recent development of ferroelectric random access memories (FRAMs) [141]. This is largely due to reliability problems stemming from fatigue (microcracking) and internal biasing (space charge) behavior, which leads to a variable switching threshold and a gradual loss of switched polarization. Piezoelectric devices, some of which are illustrated in Table 7, are discussed in this section. The utilization of these devices in various applications is discussed in the subsequent section.

Table 7 Examples of Piezoelectric Devices and Applications

Device Category	Typical Devices
Ultrasonic generators	Ultrasonic cleaners, degreasers, sonar, pingers, atomizers, ultrasonic welding, intrusion alarms, pest control devices, flaw detectors, flow indicators, medical applications (imaging, therapy, insulin pumps, vaporizers)
Sensors	Phonograph pickups, accelerometers, hydrophones, sonobuoys, depth sounders, auto diagnostic devices (knock sensors, tire pressure indicators, tread wear indicators, wheel balancers, keyless door entry, fuel atomization), flaw detectors
Resonators	IF filters, surface wave filters, delay lines, piezo transformers, TV and radio resonators
Sounders	Loudspeakers, tweeters, tone generators, head sets, buzzers, alarms
Miscellaneous	Relays, pumps, motors, fans, positioners, ink printers, alarm systems, smoke detectors, touch controls, power supplies, ignitors

1. Monolithic Piezoelectrics

The simplest piezoelectric device is a single crystal or poled polycrystalline plate of circular, rectangular, or other geometry that possesses electrodes on the two major parallel faces. The typical response mechanisms that are achievable are shown in Figure 21a–c, and include the axial extensional mode (d_{33} coefficient), the lateral contraction response (d_{31}) and the shear response (d_{15}).

In addition to positioning and relay applications which utilize the deformation response of the material, as calculated above, monolithic piezoelectric devices are often used as resonators. When operated under these (resonance) conditions, device geometry and material properties such as density and modulus define the resonance frequency. The simplest perspective to explain resonance response uses the relations:

$$f_r = v/\lambda = v/2t \tag{40}$$

$$v = (s^E \cdot \rho)^{-1/2} \tag{41}$$

$$\text{and } Z \approx v \cdot \rho \tag{42}$$

where f_r is the resonance frequency of the fundamental thickness mode, v and λ are the velocity and wavelength of the acoustic wave in the material, respectively, t is the thickness, s^E is the closed-circuit compliance, ρ is the density, and Z is the acoustic impedance.

The prototypical resonator material is quartz. Usually an AT cut ($\theta = 35.25°$) is used due to the temperature independence of the resonance frequency $df_R/dT = 0$ for materials of this orientation. Applications include timing circuits for watches, computers, and telecommunications. Typically, for watches, the frequency utilized is 32 kHz, rather than the higher frequencies that are utilized for computer circuitry because of power considerations. A bar or, more commonly, a tuning fork geometry is employed. For computer and telecommunication application frequencies between 24 and 100 MHz are utilized with corresponding resonator thicknesses of 70 to 7 μm, respectively [83]. The preciseness of the cut angle of the crystal is one of the main factors in time keeping accuracy with values for good crystals exceeding 1 second per year.

2. Composites

Composites are utilized to the greatest extent in ultrasonic imaging and sonar applications. The maximal energy output of the composite is related to the product of the g $(\partial E/\partial \sigma)_D$ and d $(\partial D/\partial \sigma)_E$ coefficients. For sonar applications, the hydrostatic stress conditions that are present result in a figure of merit for the performance of these devices that is proportional to $g_h \cdot d_h$. The improvement in performance gained by using a 1–3 piezoelectric composite compared to a simple piezoelectric plate is due to two factors. First, the d_h coefficient is expressed by:

$$d_h = (d_{33} + 2d_{31}) \qquad (43)$$

Because d_{33} and d_{31} have opposite signs, in simple poled polycrystalline piezoe-lectric bodies, d_h has relatively low values. Thus, the use of a 1–3 composite allows for decoupling of the longitudinal and transverse piezoelectric effects, giving a higher d_h value. Second, the g coefficient is inversely related to the relative permittivity of the material through an expression such as:

$$g_{33} = \left(\frac{d_{33}}{\varepsilon_{33}^\sigma} \right) \qquad (44)$$

From the perspective of this relationship, the comparatively high dielectric constant of most PZT materials results in low values of the g coefficient. Therefore, the preparation of 1–3 composites allows for reduced values of d_{31}, which increases d_h, and the low dielectric constant of the polymer matrix phase results in higher g_h values. The incorporation of porosity into the polymer (1–3–0 composites) contributes to still lower dielectric constants and further improvements in the $g_h d_h$ figure of merit that is used to describe hydrophone performance. The composite structure also allows for better impedance matching between water and the hydrophone, which improves energy transfer into the hydrophone. Typical properties are shown in Table 8.

Similar equations to those shown for resonator devices may be used to explain another important aspect of composite properties: impedance and impedance matching with water. The acoustic impedance of PZT is approximately 20–30 Mrayls [180]. This impedance is poorly matched with water ($Z \sim 1.5$ Mrayls), the transmitting medium associated with hydrophone applications. This

Table 8 Typical Properties of Piezoelectric Composites[a]

Material or Composite Device	$d_h g_h$ ($\times 10^{-15}$ m^2/N)
PZT	300
PVDF	1,000
PbNb$_2$O$_5$	2,300
1–3–0 PZT Rod Composite	3,200
1–3–0 Foamed Polymer Composite	10,000
3–2 Perforated Composite	15,000
Diced Encapsulated Composite	20,000
1–3–1 Composite	30,000
Moonie	50,000

[a] Ref. 161.

results in significant reflection ($> 80\%$ in the present case!) of acoustic energy at the transducer surface according to:

$$R = \left(\frac{Z_1 - Z_2}{Z_1 + Z_2} \right)^2$$

(45)

where Z_1 is the acoustic impedance of the transducer and Z_2 the impedance of the medium. This results in a decrease in effective sensitivity since only approximately 20% of the energy of the wave is transferred to the transducer. One approach that has been used to get around this problem is the use of an "antireflection" coating on the surface of the transducer [181], which serves a purpose similar to the use of antireflection coatings in optical applications [182]. This approach can offer significant improvements in energy transfer; however, composite materials offer an alternative to the use of antireflection coatings. Unlike PZT ceramics with their high acoustic impedance, composite materials typically have impedances between 5 and 10 Mrayls, which decreases the reflection losses from 80% to about 30%. This highlights yet another advantage of the 1–3 composites for sonar applications [183].

3. Multilayer Ceramic Actuators

Multilayer ceramic actuators incorporate multiple piezoelectric layers that are driven electrically in parallel but whose displacement response is additive [184]. Therefore, compared to a single-element monolithic actuator, multilayer actuators maintain the comparatively low operational voltages of a monolithic actuator but they can generate high strains due to the number of layers operating in concert with each other [185,186]. The devices are fabricated via a variety of techniques, including the epoxy bonding of piezoelectric (or electrostrictive) sheets to form a multilayer stack, or the use of techniques that resemble those used in the multilayer ceramic capacitor (MLCC) industry, namely, cofired technologies. The fabrication of the cofired devices further resembles that in employed for MLCCs in that Ag-Pd electrodes are utilized. Diffusion bonding methods have also been used with copper and nickel electrodes [187]. While parallel plate geometries are employed most often, multilayer actuators with interdigitated electrodes (IDEs) have also been investigated [188,189] that allow for the use of d_{33} mode actuation for an in-plane response. The IDE actuators exhibited improved lateral displacement response compared to the d_{31} mode parallel plate devices.

A variety of geometries, including devices with various layer thicknesses and number of layers, may be purchased commercially. Due to the comparatively small size of the market for these components, devices are also frequently prepared to customer specification. Operational voltages for the devices are 60, 100, 150, 300, and 600 V, and devices that operate at 200 V are capable of generating strains of 0.2% or higher [190]. Circular actuator dimensions typically range from 10 to 35 mm in diameter and 15 to 60 mm in height. The maximal corresponding strokes, stiffnesses, and loads range from 20 to 80 μm, 20 to 500 N/μm, and 2 to 20 kN, respectively, at an operational voltage of 1000 V. For devices fabricated by epoxy bonding methods, typical layer thicknesses are between 0.2 and 0.5 mm, and the number of layers employed ranges from 20 to 50. For devices prepared by an MLCC-type process, layer thicknesses are typically between 50 and 100 μm and devices of up to 200 layers have been fabricated.

A key consideration in the use of the devices, particularly in automotive applications, is reliability. In this area, multilayer actuators are under investigation as fuel injectors where, because of their quick response times, accurate displacement control, and high force capabilities, they are being considered as a possible replacement for electromagnetic injectors. For devices fabricated through an MLCC process, reliability is frequently crack growth associated with the differential strain at the electroded–unelectroded interface that occurs under drive [191]. This need has resulted in the use of finite element analysis (FEA) techniques for prediction of device performance and estimation of stresses at critical regions within the device [20]. The sophistication of the FEA and other modeling methods has increased greatly in recent years with current models accounting for nonlinearities in material response as well as hysteresis and domain switching.

4. Flextensional Actuators

The term "flextensional" was coined in 1963 and came about as a merging of the words "flexural" and "extensional" into one word. Flextensional devices are typically described as composite devices composed of two different elements: (1) an input driver and (2) a flexing, output member. The devices utilize a bending, or flexing, of the passive member with amplified motion/impedance matching. They also typically employ prestressing of the driving element.

Unimorph, bimorph, Moonie, and cymbal devices are all referred to as flextensional devices [192]. The devices are known for their high-bending deflection responses and find utilization in fans, motors, beam deflectors, accelerometers, loudspeakers, phonograph pickups, noise control, pressure (sound) sensing, pumps, and positioners, among others [193–196]. Newer devices in this general area include multilayer bender actuators, and stress-biased devices, such as Rainbow [16] and Thunder [17]. These later devices take advantage of the same principles as those that dictate the performance of unimorph and bimorph actuators, but utilize fabrication procedures that further enhance the domain switching contribution to response.

(a) *Unimorph and Bimorph Actuators.*   Unimorph (one piezoelectric plate) and bimorph (two piezoelectric plates) devices are typically fabricated by epoxy bonding a piezoelectric plate to a nonactive support at room temperature; schematics of the devices are shown in Figure 24. Circular benders are desirable in certain applications, such as loudspeakers, because they have the highest energy conversion efficiency and possess the largest dielectric capacitance for a given resonant frequency. Bimorph actuators may be prepared with the same or opposite poling directions, although similar displacement response characteristics are achieved through different drive conditions. The center vane allows the plates to be poled in the same direction (parallel type) or in opposite directions (series type) and also permits flexibility of the applied electric signal during operation. The bender shown in Figure 24b (center) is of the parallel type, being initially poled with the positive and negative terminals across the top and bottom surfaces, and subsequently operated with one terminal connected to both major surfaces and the other terminal to the center vane. From an energy standpoint, there is no difference between a series-type bender and a parallel type; however, a choice between the two types is usually made on the basis of voltage and mechanical

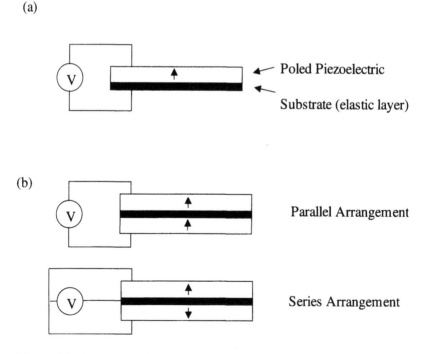

Figure 24 Schematics of (a) unimorph and (b) bimorph actuators. For the bimorph devices, the two poling arrangements are shown together with drive voltage conditions.

displacement considerations. The parallel operation scheme provides the largest output displacement for a given input drive voltage (e.g., loudspeakers), whereas the series operation provides the largest output voltage for a given input displacement (e.g., phonograph pickup).

The deformations are not only large compared to direct extensional devices; they are also large relative to the thickness of the piezoelectric layer(s) employed in the devices. The bending deformation originates from the clamping effect of the elastic substrate (typically a metal layer, although other materials have also been investigated) on the electromechanical response of the piezoelectric layer. The clamping effect results in decreased piezoelectric response at the interface compared to the response at the free surface. Mostly this is associated with differences in non-180° domain wall motion across the thickness of the piezoelectric layer. The greater domain wall motion at the free surface compared to the substrate interface results in the bending type of response that typifies these devices.

The performance of unimorph and bimorph devices has been widely studied, and the effects of the moduli and the relative thicknesses of the two layers on displacement response and forces required for specific deformation have been thoroughly investigated [197–199]. These studies are based on the use of a constitutive equation for the tip deflection of the bender when used in a cantilever geometry:

$$\delta = aF + bV \qquad (46)$$

where δ is the tip deflection, a is a constant, F is the force applied to the tip, b is a constant, and V is the applied voltage. The constants a and b are given by:

$$a = \frac{4s_{11}^p L^3}{wt_p^3} \frac{AB+1}{1+4AB+6AB^2+4AB^3+A^2B^4} \qquad (47)$$

$$b = \frac{3d_{31}L^2}{t_p^2} \frac{AB(B+1)}{1+4AB+6AB^2+4AB^3+A^2B^4} \qquad (48)$$

$$A = \frac{s_{11}^p}{s_{11}^m} = \frac{E_m}{E_p} \qquad B = \frac{t_m}{t_p} \qquad (49ab)$$

where S_{11}^p and S_{11}^m are the (11) mechanical compliances of the piezoelectric and substrate layers, respectively; t_p and t_m are the thicknesses of the piezoelectric and substrate layers, respectively; L and w are the length and width of the cantilever, respectively, and E_p and E_m are the Young's moduli of the piezoelectric and substrate layers, respectively. Wang has shown that the tip deflection response of the actuator may be optimized through control of A and B [199], and for materials such as PZT 5A and stainless steel as the substrate, the highest tip deflections are obtained when the thickness of the piezoelectric is approximately twice as thick as the metal. Unimorph theory has also been applied to the study of Rainbow and Thunder actuators [199,200].

It is difficult to categorize the response of unimorph and bimorph devices due to the range of geometries and sizes that are produced for various applications. Devices 1 cm wide by 4 cm in length with PZT thicknesses of 0.5 mm are capable of generating tip deflections of hundreds of micrometers. However, significantly smaller as well as larger devices, with corresponding changes in electromechanical response, have also been reported. The size scale of the devices ranges from microelectromechanical accelerometers based on PZT cantilevers to friction dampers for large civil structures.

(b) *Stress-Biased Actuators: Thunder and Rainbow.* Stress-biased actuators, commonly referred to as Thunder and Rainbow devices, were first invented by Haertling in 1994 (Rainbow), and have been the subject of intense investigation since that time. The acronyms Rainbow and Thunder represent *R*educed *A*nd *IN*ternally *B*iased *O*xide *W*afer and *TH*in *UN*imorph *D*riv*ER*,[*] respectively. These actuators form a unique family of stress-biased electromechanical devices that display displacement and load-bearing responses that are substantially greater than traditional piezoelectric devices, including unimorph and bimorph actuators [16,17,201–204]. As a result, the devices are of interest in applications where high strain is required and device space is restricted, or where power requirements must also be minimized. Such applications include micropumps, small robotic systems [205], positioners for space-based interferometers [206], needle positioners for textile machinery, propulsion systems for swimming vehicles [207,208], and fuel injectors for heavy-vehicle applications [209]. The performance characteristics of Rainbow actuators have been summarized by Haertling [201], and a comparison of the performance characteristics of the two stress-biased devices has been given by Wise [202]. The reported performance window for Rainbow devices reaches 1 MPa stress and 500% strain (axial displacement of the actuator compared to the thickness of the piezoelectric layer).

Rainbow devices are formed by the partial reduction of lead-based ferroelectric compositions, most typically those based on PZT, to form a composite piezoelectric-cermet structure [16,201,210]. The basic fabrication process involves the partial chemical reduction of a lead-based ferroelectric at temperatures near 1000°C using a graphite block. The process results in the conversion of the piezoelectric oxide to a cermet layer in the bottom portion of the device, forming a piezoelectric-cermet composite structure. During the cooling phase of the pro-

---

* Thunder is a registered trademark of Face International Corporation, Norfolk, VA.

cess, the difference in thermal expansion coefficients of the two layers results in the formation of a domed structure with a superimposed stress profile [211].

Thunder devices are formed by a different process, but again, the devices are composite (piezoelectric-metal) in nature. In this process, a metal substrate is bonded to the piezoelectric layer at elevated temperature ($\sim 300^{\circ}$C) using a curable polyimide layer [17]. After cooling, as for the Rainbow ceramic, the different thermal expansion coefficients of the two principal layers result in a domed device with a superimposed stress profile. Rainbow and Thunder devices thus resemble unimorph actuators, with the most significant differences being the domed nature of the stress-biased devices and the superimposed stress profile. For devices fabricated with layer ratios of about 2:1 (piezoelectric-metal or cermet), the surface region of the piezoelectric is believed to be under significant tensile stress, while the lower portion of the piezoelectric is under compressive stress [211]. Devices fabricated with layers of this thickness ratio have also been shown to generate the highest displacement response [201,202].

Because of their enhanced performance and potential utilization in a variety of applications, study of the underlying mechanisms that contribute to the response of stress-biased devices has received significant attention. Suggested contributing factors include simple mechanical effects, as might be expected based on the resemblance of the devices to unimorphs [199,200], enhanced extrinsic contributions to piezoelectric response due to stress and field effects on domain configuration and switching [55,212,213], and, potentially, mass-loading effects [214,215]. The use of additional mechanical preloads to further improve electromechanical response has also been evaluated [18].

While quantifying the relative contributions of these various mechanisms to the observed response of the devices has proven difficult, the importance of altered domain configuration and enhanced domain wall motion characteristics seem indisputable. The surface region of the device is believed to be particularly important in defining the overall response of the actuator due to the presence of tensile stresses in this region. These lateral stresses result in a higher population of domains with their polarization vectors parallel to the surface. Under the application of an electric field, these domains are switched by 90°, resulting in an enhancement in the extrinsic contribution to electromechanical response. X-ray diffraction studies have been carried out to assess the magnitude of this contribution to device performance [55,212].

B. Electrooptic Shutters and Modulators

Thin polished plates of PLZT, when used in conjunction with polarized light, make excellent wide-aperture electronic shutters. Their advantages over competing technologies such as mechanical shutters liquid crystal light gates include (1) faster response time, in the low-microsecond range; (2) less vibration; (3) lighter weight; (4) thinner profile; and (5) wider operating temperature range, from -40 to $+80°C$. Their main disadvantages are (1) low-on-state transmission of about 15%, primarily a result of the polarizers; and (2) high operating voltages required to reach a full on state. Despite the low on-state transmission characteristics, contrast ratios in excess of 5000:1 are regularly achieved because of the deep off state (0.002%) possible with high-efficiency polarizers. It has also been found that for a large number of applications, the on-state transmission of 15% is quite acceptable. In the case of eye-protective devices, 15% transmission is adequate because of the logarithmic response of the eye to light intensity.

A typical on-state (half-wave voltage) transmission curve of a PLZT shutter as a function of wavelength is given in Figure 25 together with spectral response curves for the Polaroid HN-32 polarizers. As can be seen, the spectral response of the shutter device correlates with that of the polarizers. The PLZT response is flat over the optical range studied. Also, the 15% transmission figure is an average value over the total optical spectrum, which can be expected to increase to approximately 20% with antireflection coatings on the PLZT and polarizers [176].

The polarizer-PLZT configuration usually employed in a shutter device is shown in Figure 26. The interdigital electrodes serve to decrease the required operating voltage by reducing the gap between the positive and negative finger electrodes. These electrodes are produced either by vacuum deposition on the

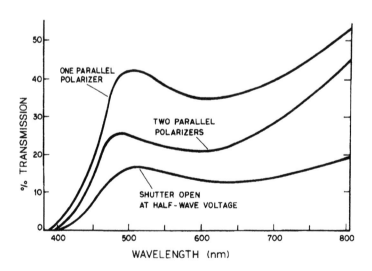

Figure 25 Spectral response of an activated PLZT 9/65/35 electro-optic shutter (25 mm diameter) using crossed HN-32 polarizers.

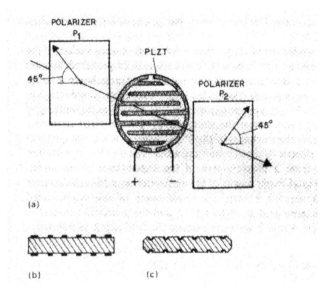

(a)

(b)   (c)

Figure 26   (a) Typical device configuration for an electro-optic shutter using crossed polarizers; (b) vacuum deposited, surface electrodes; and (c) chemically plated, slotted electrodes.

surfaces or by machining and plating small grooves (1.5 mil deep) into the surfaces. Since the width of the fingers is small (1.5 mil or 0.0375 mm) and the gap fairly large (40 mil or 1 mm), the electrodes themselves block only a small percentage of the incident light. These electrodes are thin enough so that they can be optically defocused during viewing without any noticeable effect. To keep the electric fields between the electrodes in a more or less favorable direction (i.e., 90° to the direction of light transmission), it is desirable to maintain an electrode gap-to-plate thickness ratio greater than 1. Reducing this ratio to substantially less than 1 will result in undesirable higher operating voltages. With a 1-mm (40-mil) gap on a 0.375-mm (15-mil)–thick plate as in a normal configuration, the gap-to-thickness ratio is a favorable 2.7.

The slotted electrodes that are partially embedded in the surface of the PLZT by approximately 0.0375 mm (1.5 mil) are usually preferred because they produce a more uniform transverse electric field and hence a more uniform transverse electro-optic effect. This leads to a beneficial reduction in half-wave voltage (On state) of approximately 10% as well as minimizing of residual space charge effects near the electrodes [216]. Since these electrodes do not fully penetrate the thickness of the PLZT wafer, there is the question of just how far the electric field penetrates the wafer. The depth of penetration can be estimated from a measurement of the capacitance of the wafer. Knowing that the total finger length of the electrodes on a 50-mm-diameter wafer with 1-mm gaps is 314 cm, the capacitance of the double-sided electroded wafer is 18.3 × 10^{-9} F and the dielectric constant of 9:65:35 is 5700 (Table 3), we can calculate the field depth as follows:

$$C = \varepsilon \frac{A}{g} \cdot 8.85 \cdot 10^{-14}$$

(50)

where A is the area in cm^2, g the gap length in cm, C the capacitance in farads, ε the relative dielectric constant, and the remaining factor is the permittivity of free space in F/cm. Identifying the area as length of the finger electrodes × depth of penetration d, and rearranging Eq. (50), we have:

$$d = \frac{C \cdot g}{\varepsilon \cdot l} \cdot \frac{1}{8.85 \cdot 10^{-14}}$$

(51)

$$= \frac{18.3 \cdot 10^{-9} F (0.1\, cm)}{5700 \cdot 314\, cm} \cdot \frac{1}{8.85 \cdot 10^{-14}\, F/cm}$$

(52)

$$= 0.0115 \text{ cm or 4.5 mils}$$

(53)

This depth of penetration is, of course, only an estimate, but it does point out that for a 0.375-mm (15-mil)–thick wafer with 0.0375-mm-deep slots, the electric field lines do not extend fully through the thickness of the wafer. For a 0.5-mm (20-mil) electrode gap, this depth of penetration is reduced slightly to 0.1 mm (4 mil), due largely to the reduced gap to thickness ratio of 1.33. Using this value of field penetration, which effectively defines the true optical thickness, and remembering that two surfaces are being activated in series ($t = 2d$), the operating voltage of a PLZT shutter or a color filter may be calculated [217,218].

VII. APPLICATIONS

Since their initial implementation in a range of applications more than 50 years ago, the utilization of piezoelectric and electro-optic materials continues to expand. New uses for piezoelectric thin films are found in sensor and chemical detector applications, and bulk materials are under investigation for friction dampers in buildings to mitigate earthquake damage [219]. Additional new developments include the use of ultrasonic transducers have to accelerate the rate of healing in bone fractures [220,221] and the observation of nonlinear optical response in ferroelectric thin films, which suggests possible applications in optical limiting [222]. Given the breadth of the applications for piezoelectric and electro-optic materials, this section can serve only as an overview of some of the more important applications and recent developments in the field.

A. Piezoelectric Positioners and Relays

Thin electroded piezoelectric plates can be used as mechanical positioners and relays when the required movements are very small (i.e., in the range of a few hundred micrometers or less). Good sources for information, as well as construction details, are given in Refs. 223–225. Sample calculations of the deformations possible are given below.

Consider a poled and electroded rectangular piezoelectric plate as shown in Figure 21a and assume that the plate length is 2.5 cm, the width 1.0 cm, and the thickness 0.25 mm. This plate is to be operated as a micropositioner in a dc mode at 100 V. When voltage is applied, the plate will expand or contract in the thickness direction, depending on the polarity of the voltage with respect to the poled direction of the ceramic. Simultaneously, the length of the plate will contract and expand in similar fashion. The magnitude of these movements may be calculated remembering that the d coefficient is the ratio of resulting strain to applied electric field. For the thickness mode (d_{33}) case, this may be expressed as:

$$d_{33} = \frac{S}{E} = \frac{\Delta t}{t/(V/t)} = \frac{\Delta t}{V} \tag{54}$$

or

$$\Delta t = d_{33} V \tag{55}$$

indicating that the change in thickness of the plate is dependent on the material d coefficient and the voltage and is independent of the plate thickness. Substituting values into Eq. (55) and taking a d_{33} coefficient of 578×10^{-12} m/V from Table 3 for composition 5:55:45 yields $\Delta t = 0.0578$ μm. This is an infinitesimally small movement. This value can be increased but, in a practical sense, only by raising the voltage. There is, of course, a limit on how high the voltage can be raised because either depoling (back switching of the hysteresis loop) or dielectric breakdown will eventually occur. For the length-extensional case, the transverse coefficient is given as

$$d_{31} = \frac{S}{E} = \frac{\Delta l}{l(V/t)} = \frac{\Delta lt}{lV} \tag{56}$$

or

$$\Delta l = d_{31} \cdot \frac{l}{t} \cdot V \tag{57}$$

In this case, the movement is multiplied by the dimensional factor l/t. Again substituting values into Eq. (57) and assuming the appropriate d_{31} coefficient of -268×10^{-12} m/V (the negative sign indicates contraction), we obtain a contraction of 2.7 μm. This is an approximate 50-fold increase in movement for the same voltage, despite the fact that the d_{31} coefficient is only approximately half the value of d_{33}.

Because of the small resulting displacements, other devices, notably multilayer actuators, as well as unimorph and bimorph benders have been developed to amplify these strains. If we consider the same rectangular plate as analyzed for the axial and lateral displacements, and we assume that the plate is bonded to a steel support that is 0.10 mm thick, utilizing Eqs. (46) through (49), the deflection at the end of the cantilever-mounted bender is ~150 μm. This notably larger value represents an additional multiplication factor of 50 over the length-extensional mode and a factor of ~2500 over the thickness mode. It points out that benders are used whenever possible because they provide the highest deflection for the least amount of voltage. Nevertheless, benders are not appropriate in some applications because they are mechanically compliant and incapable of generating significant forces.

1. Gas Ignitors

A poled piezoelectric ceramic disk is a compact, portable power supply if one has the means to extract the electrical power from it. As we have learned, this should be possible by following the example of the Curie brothers and discovering the direct piezoelectric effect all over again! Piezoelectric impact gas ignitors utilize this effect in various applications involving gas appliances, such as stoves, heaters, patio grills, and cigarette lighters. It is a simple and effective device that has gained widespread popularity. A simple example of a gas (spark) ignitor is given in Figure 27, together with a typical construction view.

The ignitor works by applying a force to a properly poled and electroded ceramic whose electrodes are connected to terminals in a preset air gap similar to an automobile spark plug. As the pressure (pressure = force per unit area) builds up on the ceramic from the mechanical input, the voltage across the ceramic increases until the air gap between the terminals breaks down, producing a spark or series of sparks. On releasing the force, voltages of like magnitude but opposite sign will be produced, yielding additional sparks. If the force is applied too slowly

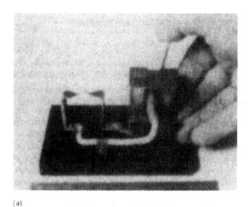

(a)

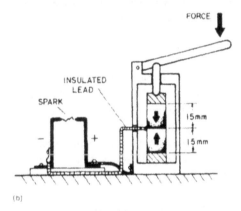

(b)

Figure 27 (a) Laboratory model, squeeze-type spark ignitor; (b) construction view of the device.

no sparking action occurs because of electrical leakage across the electrodes of the ceramic. This is why many ignitors are of the impact (momentary force) variety rather than of the slower squeeze type.

To calculate the magnitude of the voltage generated from a given device, we should first recall that the g coefficient is the ratio of the open-circuit electric field to the applied stress (pressure). From the Heckmann diagram, we may write:

$$-g = \left(\frac{\partial E}{\partial T}\right)_D \tag{58}$$

Rearranging Eq. (58) to solve for the voltage generated, we obtain

$$V = - g \cdot t \cdot T \tag{59}$$

which indicates that the voltage produced is directly proportional to the material g coefficient, the thickness t of the ceramic, and the magnitude of the applied stress T. Assuming an applied stress of 8000 psi (5516×10^4 N/m^2) and obtaining a g_{33} value of $- 23 \times 10^{-3}$ V-m/N from Table 3 for composition 2:65:35, we substitute values into Eq. (59).

$$\begin{aligned} V &= 23 \times 10^{-3} \text{ V·m/N} \times 15 \times 10^{-3} \text{ m} \times 5516 \times 10^4 \text{ N/m}^2 \\ &= 19{,}000 \text{ V} \end{aligned} \tag{60}$$

This voltage of 19 kV is easily high enough to produce a spark across a 6-mm (1/4-in.) gap and possesses sufficient energy to ignite most gases. More thorough analyses of the associated electrical and mechanical work energies associated with the process have also been carried out and are reported in Ref. 30.

2. Speakers and Sounders

One of the more recent successful applications for piezoelectric ceramics has been in the area of piezoelectric loudspeakers, particularly tweeters. Compared with the conventional electrodynamic tweeter, the piezo tweeter is lighter, possesses a shallower profile, is more power efficient, is more reliable, has better transient response, and is cost competitive. These desirable features make piezoelectric tweeters ideally suited to high-power audio applications, stereo systems, portable audio devices, and sound systems for automobiles. Figure 28 illustrates several different types of piezo speakers as well as construction details.

The piezo tweeter development is a notable achievement because it represents the successful application of several technologies to reach a common objective. These involved establishing proper operating principles and design criteria, development of high-quality ceramics, and introducing unique processing techniques for manufacturing. Work was begun by Schafft in the mid-1960s and subsequently reported by his workers [226–228]. The basic principle of operation involves a circular ceramic bender (bimorph) that drives a speaker cone in an unsupported "momentum drive" mode. As an alternating voltage is applied to the disk bender, it will dish in or out, depending on the polarity and the magnitude of the applied signal. The cone, which is lighter than the driver, will thus be moved in a reciprocating in-and-out motion by the momentum of the bender. A circular bender was chosen because (1) it is a high-compliance transducer that is best suited for acoustic devices that radiate into air; (2) it has the highest dielectric capacity, which is required for delivering power from a low-voltage source; (3) it has the highest electromechanical coupling factor of all bender configurations; and (4) it has the highest flexural resonant frequency for a given

(a)

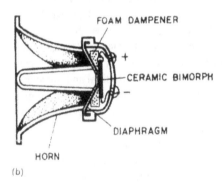

FOAM DAMPENER

+

CERAMIC BIMORPH

−

DIAPHRAGM

HORN

(b)

Figure 28 Examples of piezoelectric tweeters manufactured by Motorola.: (a) models (top left to right), 3.5-in superhorn, satellite tweeter, 3.75-in. exponential horn and (bottom) 2 × 5 in. wide dispersion horn; (b) cross-sectional view of superhorn shown in (a).

dimension or volume. The typical frequency response (100 dB at 0.5 m) for a 3.5-in. piezo superhorn tweeter using a 0.89-in. (22.6-mm)–diameter bender bimorph is fairly flat from 4 kHz to 30 kHz for only 2.8 V applied.

The ceramic drivers (benders) used in the piezo tweeters require the highest quality material available. Since some of the individual ceramic disks may be as thin as 3 mil (0.075 mm), porosity must be absolutely minimal, and grain size should be tightly controlled. Specifications [226] established for the ceramic included a voltage-sensitivity merit factor:

$$\text{Voltage sensitivity} = K \cdot k_p^2 \tag{61}$$

and an energy output merit factor:

$$\text{Energy output} = E_D{}^2 \cdot K \cdot k_p{}^2 \tag{62}$$

where E_D is the magnitude of the electric field (60 Hz applied signal) at which k_p has dropped to 90% of its initial value of 0.63 or greater. Both of these empirical figures of merit should be maximized for the highest tweeter performance. Tweeters today are manufactured in millions of units per year by a number of different vendors.

3. Hydrophones, Sonar, and Other Maritime Applications

Sonar or hydrophone applications represent one of the first uses of piezoelectric materials. This application was first developed during World War I and many advances continue to this day. Older style hydrophones utilized lithium sulfate crystals that were contained within an oil-filled rubber bladder [6]. Since then, a variety of innovations have occurred. Some of these are materials based while others are based on changes in system design. On the materials side, piezoelectric composites aimed at improving the impedance matching between the sensor and the acoustically transmitting medium, water, have been developed. These piezoelectric composites utilize ceramic materials such as PZT in conjunction with polyurethane or other polymers. The decoupling of the longitudinal and transverse piezoelectric coefficients results in a figure of merit for performance that is significantly greater than standard poled polycrystalline ceramics; the composite materials also display better impedance matching.

In addition to the transition from single-crystal-based hydrophones to transducers, there has been significant progress in piezoelectric developments for maritime applications. One advance has been the development of towed array sonar for surface ships and submarines. Moving the sonar to a location remote to the vehicle allows improvements in signal-to-noise ratio due to a reduction in the signal arising from machinery within the vessel. Little information on the actual devices can be found in the available literature [229].

The development of towed arrays is by no means the only recent development in maritime applications of piezoelectrics. Decoys and homing beacons based on piezoelectric materials are widely utilized, and mine counter measures (MCMs) sonar arrays have been developed for forward-looking mine hunting applications [230]. These sonar arrays are constructed using injection-molded composite technology and incorporate 60 1.5° beams scanning a 90° sector in front of the ship. Other developments of military significance include the development of Sonopanels® for underwater applications [231]. The panels use different PZT ceramic elements and types in a 1–3 composite structure to couple more efficiently with water in both transmit and receive modes. The panels may also be used in multielement hydrophone arrays, as well as active surface control applications.

On the commercial front, boat speedometers, fish finders, and ocean floor mapping devices are available. The speedometers make use of a Doppler sonar velocity system mounted in the bottom of the ship that sends ultrasonic pulses to the floor of the body of water and uses the characteristics of the reflected pulses to determine the speed of the ship with great accuracy [6]. Fish finders and ocean floor mapping devices use the reflected signals associated with a more typical type of sonar approach. A final aquatic application of piezoelectrics is in the area of swimming vehicles where investigators have explored the use of stress-biased actuators for propulsion systems [232,233].

4. Medical Applications

One of the most widespread uses of piezoelectric materials is in the medical industry, where ultrasound is used for applications ranging from fetus imaging in obstetrics (see Figure 29) to studies of blood flow through the heart and major associated arteries. Ultrasonic transducers are used to transmit a high-frequency (greater than 20 kHz; typically 1–5 MHz) sound pulse into the body, typically through the use of a gel compound to improve the impedance match between the transducer and human tissue. Once the sound wave has entered the human body,

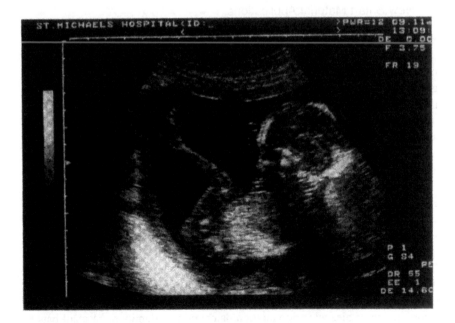

Figure 29 Image of an unborn fetus obtained by ultrasonic imaging. (After Ref. 234.)

the same impedance match/mismatch considerations discussed for hydrophone and sonar applications again apply: the sound waves from the transducer are reflected from interfaces of materials with different impedances, such as bone/ tissue and fluid/soft tissue. The intensity and timing of the reflected waves received by the ultrasonic transducer are processed electronically to generate either two- or three-dimensional images. For cardiological investigations, the Doppler effect is used to determine the direction and rate of blood flow. For a thorough review of ultrasonic imaging, the reader is referred to Ref. 234.

Although the two applications noted above certainly represent the most common use of piezoelectrics in the medical profession, they are by no means the only use. The use of ultrasound transducers for bone growth stimulation for fracture healing has also been investigated [220,221]. Here the transducer is used to produce a pulsed ultrasonic waves that exerts a force at the fracture location of approximately 20 mg. Despite the apparently small size of the force, the fracture healing rate is accelerated by approximately 40% (86 versus 114 days) compared to fractures that were left to heal on their own [235]. It is estimated that the global market for this technology is potentially as large as $200M per year. Other medical applications currently under development include miniature in vivo blood pressure sensors [236] and piezoelectric cochlear implants [237].

5. New Smart Material Applications

Although it is impossible to review all of the applications where piezoelectric ceramics have been used or considered, before concluding it is appropriate to mention some of the newer applications for which piezoelectrics are under investigation. One of the applications that has received significant attention, and which is of particular interest for military applications, is energy harvesting [238]. This technology deals with the recovery of some of the energy expended during normal activities, such as walking or breathing. One investigation focused on the incorporation of piezoelectric devices, such as Thunder actuators, into shoes, which allowed for conversion of some of this mechanical energy into electrical energy.

Another new application of piezoelectrics involves sporting goods. For example, piezoelectrics have been incorporated into skis to reduce chatter when skiing over rough terrain. In this application, piezoelectrics are used to convert some of this unwanted mechanical energy into electrical energy, which is subsequently converted to heat by dumping of electrical energy into an electrical shunt. Developments in this area involved a collaboration between K2, a ski manufacturer, and ACX (Active Control eXperts, a division of Cymer). Other sporting goods applications include smart shock absorbers for mountain bicycles, and the incorporation of piezoelectric vibration dampers into aluminum baseball bats to reduce stinging in the hands when the ball is struck off center.

Other investigators continue to explore additional new applications for piezoelectric materials. Interestingly, one of these is focused at improvements in the processing behavior of traditional ceramics. In this application, piezoelectric sensors are used to monitor acoustic emission associated with cracking of the porcelain bodies during the cooling stages of processing [239]. The use of sensors in this application allows for the design of firing schedules that result in higher production rates.

Another application that has recently been considered is the use of multilayer piezoelectric actuators for the mitigation of earthquake effects in civil structures [219,240]. In this application, the piezoelectric is used as a friction damper, i.e., a device that decreases damage in structures due to vibration through the dissipation of vibrational energy (passive) or actively compensating for the vibration. The main advantage of piezoelectric friction dampers compared to hydraulic actuators is that they offer instantaneous modulation of the clamping force [219].

A final application worthy of mention is the use of composite piezoelectric patch actuators [241] for vibration damping in aircraft [242]. As with most engineering problems, the design of fighter aircraft involves many compromises. In this instance, the design of the FA-18 airframe structure that affords exceptional performance at supersonic speeds creates problems with tail fin buffeting in high-angle-of-attack, low-speed flight conditions. This buffeting can result in significant problems with metal fatigue. Investigators at NASA have developed an active vibration damping circuit that uses small accelerometers together with piezoelectric composite actuators to reduce tail fin buffet, both improving flight control and reducing structural damage to the aircraft.

B. Electro-optic Applications: PLZT Shutters, Modulators, and Color Filters

The ability to modify the refractive index of an optical material through application of an external electric field enables many useful nonlinear optical devices. Modifications to the refractive index affect the light wave's phase velocity, thus permitting tunable phase delays between waves traveling in electro-optically excited regions and those in zero-field regions. Accordingly, the electric-field-induced phase shift promotes intensity tunability when the light traveling these different paths recombines. One example of the technological significance of this is found in electro-optic intensity modulators used in modern telecommunication systems since they enable optical signals to be turned on and off rapidly by a means external to the laser source. Modifications to the refractive index also promote tunability over the angle to which that light is refracted. This is the operable mechanism for electro-optic scanners. Another example of devices utilizing the electro-optic effect is tunable directional couplers, which are of interest for signal routing in wavelength-division-multiplexed optical systems. The coupling of light between adjacent waveguides depends on geometrical factors, wavelength, and refractive index difference between the guides and their cladding. Therefore, modifying the refractive index controls the amount of light routed to different locations. The examples noted above represent only a few applications of electro-optic and nonlinear optical phenomenon. Devices for optical computing, switching, data storage, and telecommunications, to name but a few, utilize such effects making transparent ferroelectric materials additionally useful for current and future photonic applications.

APPENDICES

Appendix A: Relationship Between Reduced and Full Notation for Piezoelectric Response

The full notation description of the electrical, mechanical, thermal, and electromechanical response of piezoelectric ceramics involves using the complete tensors for the material properties of interest. These are shown below for the above material characteristics.

> Mechanical: stress (2nd rank), strain (2nd rank), compliance (4th rank), or stiffness (4th rank)
> Electrical: polarization/dielectric displacement (1st rank), electric field (1st rank), permittivity (2nd rank), or impermeability (2nd rank)
> Piezoelectric: stress (2nd rank), strain (2nd rank), polarization/dielectric displacement (1st rank), electric field (1st rank), piezoelectric coefficient (3rd rank)
> Pyroelectric: polarization/dielectric displacement (1st rank), temperature (0th rank; scalar), pyroelectric coefficient (1st rank)

The relationship in which we are most interested is the piezoelectric response, which will suffice to show the general relationship for the reduced and full notations. Recall that the reduced notation format for the description of dielectric displacement as a function of stress and electric field was:

$$D_i = d_{ij}\sigma_j + \varepsilon_{ii}^{\sigma}E_i \tag{A1}$$

The analogous full notation response for this relationship would be:

$$D_i = d_{ijk}\,\sigma_{jk} + \varepsilon_{ii}^{\sigma}E_i \tag{A2}$$

where d_{ijk} is the full-notation piezoelectric coefficient and σ_{jk} is the full-notation description of stress, which may be normal, such as σ_{11} for a stress perpendicular to the 1 face, or shear, such as σ_{32} for a shear stress in the 3 direction applied to the 2 face [39]. The subscripts i, j, and k may take on values of 1, 2, or 3,

depending on direction. If we arbitrarily set the electric field to zero for both of these equations, we may consider solely the contribution to dielectric displacement arising from the direct piezoelectric effect. We may write the expressions for the reduced and full notation descriptions, respectively, for $E = 0$
as:

$$D_i = d_{ij}\sigma_j \qquad (A3)$$

and

$$D_i = d_{ijk}\sigma_{jk} \qquad (A4)$$

Now consider the associated matrix mathematics required to solve these two expressions. We need to consider the matrices associated with both stress and piezoelectric coefficient. In the full-notation description of stress, there are nine stresses that we must consider: three normal and six shear stresses. The corresponding matrix is a 3×3 matrix of the form:

$$\begin{pmatrix} \sigma_{11} & \sigma_{12} & \sigma_{13} \\ \sigma_{21} & \sigma_{22} & \sigma_{23} \\ \sigma_{31} & \sigma_{32} & \sigma_{33} \end{pmatrix} \qquad (A5)$$

The value of D_i, (a 3×1) matrix (three rows and one column), are obtained by multiplying the above stress matrix by the ($3 \times 3 \times 3$) d_{ijk} matrix. While the matrix mathematics may certainly be carried out, it is burdensome to work with. For homogeneous stresses, the stress matrix is symmetrical about the leading diagonal, i.e., the forces exerted on opposing faces of the cube shown in Figure 5 are equal and opposite. Therefore, $\sigma_{ij} = \sigma_{ji}$, and we may write this matrix as:

$$\begin{pmatrix} \sigma_{11} & \sigma_{12} & \sigma_{13} \\ \sigma_{12} & \sigma_{22} & \sigma_{23} \\ \sigma_{13} & \sigma_{23} & \sigma_{33} \end{pmatrix} \qquad (A6)$$

Consideration of this matrix shows that there are six independent variables, which is the first step in describing the stress state in reduced notation as a (6×1) matrix. Recall our earlier discuss where we defined σ_1, σ_2, and σ_3 as normal stresses and σ_4, σ_5, and σ_6 as shear stresses. We also indicated that for the full-notation description, the first subscript was associated with the stress direction and the second subscript was associated with the face to which the stress was applied. Therefore, in full notation, σ_{11} describes a stress in the 1 direction that is applied to the 1 face, or a normal stress. Thus, σ_1 is equivalent to σ_{11}; similarly, σ_2 is equivalent to σ_{22} and σ_3 is equivalent to σ_{33}. Also, recall that we indicated

that σ_5 was shear stress about the 2 axis, as shown in Figure 5. Further consideration of this figure shows that this stress is applied in the 1 direction to the 3 face.
We may therefore state that $\sigma_5 = \sigma_{31}$, and because of symmetric nature of stress,
we may also write that $\sigma_5 = \sigma_{13}$. Analogous expressions may be written for σ_4,
which is equal to σ_{23}, and σ_6, which is equal to σ_{21}, or from symmetry considerations, σ_{12}. The equivalency of the reduced notation stresses and those used for
the full-notation description are shown below, and the arrow indicates the path
we follow in defining σ_1 through σ_6:

$$= \begin{pmatrix} \sigma_1 & \sigma_6 & \sigma_5 \\ \sigma_6 & \sigma_2 & \sigma_4 \\ \sigma_5 & \sigma_4 & \sigma_3 \end{pmatrix} \tag{A7}$$

We are now prepared to go from a (3×3) matrix description of the stress, σ_{ij},
to a (6×1) matrix description of the stress, σ_j. The symmetry of the stress
tensor has permitted this simplification, and normal and shear stresses may now
be represented as:

$$\begin{pmatrix} \sigma_1 \\ \sigma_2 \\ \sigma_3 \\ \sigma_4 \\ \sigma_5 \\ \sigma_6 \end{pmatrix} \tag{A8}$$

To describe the dielectric displacement resulting from an applied stress, we use
the reduced notation stress tensor together with the reduced notation form of the
piezoelectric coefficient. Equation (A1) is more often used to describe piezoelectric response than Eq. (A2). Although we have clarified the relationship between
the full and reduced notations for stress, we still need to better describe the
relationships between d_{ij} and d_{ijk}. A few examples will serve to clarify these
relationships.

Assume that we want to describe the dielectric displacement in the 3 (poling) direction that results from an applied stress that is normal to the 3 face. We
may define the resulting response of the material using either the full [Eq. (A4)]
or reduced [Eq. (A3)] notation description as:

$$D_3 = d_{333}\sigma_{33} \tag{A9}$$

or

$$D_3 = d_{33}\sigma_3 \tag{A10}$$

where for the piezoelectric d coefficient, i represents the electrical direction (in this case, the charge built up on the faces), and j (reduced and full notation) and k (full notation) represent mechanical direction (in this case, stress). We may thus write:

$$d_{33} = d_{333} \tag{A11}$$

Now assume that we want to describe the dielectric displacement in the 1 direction that results from an applied shear stress about the 3 direction (σ_6 or $\sigma_{12} = \sigma_{21}$). We may write the full- and reduced-notation expressions as:

$$D_1 = d_{112}\sigma_{12} + d_{121}\sigma_{21} \tag{A12}$$

$$D_1 = d_{16}\sigma_6 \tag{A13}$$

And because of the symmetry of stress, we may write that

$$d_{ijk} = d_{ikj} \tag{A14}$$

and thus,

$$d_{16} = 2d_{112} \tag{A15}$$

We could go through additional examples to show that, in general, for the normal stresses, the d coefficients in the reduced notation are equal in magnitude to those of the full notation, while for the shear stresses, the d coefficients in the reduced notation are a factor of 2 greater than those used in the full notation. We may state this mathematically as:

$$d_{ij} = d_{ijk} \quad \text{when, for } \sigma_n, \, n = 1, 2, \text{ or } 3; \text{ and} \tag{A16}$$

$$d_{ij} = 2d_{ijk} \quad \text{when, for } \sigma_n, \, n = 4, 5, \text{ or } 6 \tag{A17}$$

For the relationship between the full- and reduced-notation description of mechanical behavior, the reader is referred to Nye [39].

Appendix B: Crystal Symmetry Effects on Electromechanical Response

While these relations show the general polarization and strain behavior that may be expected from a piezoelectric subjected to an applied stress or electric field, further understanding of the material response and determination of its magnitude dictate that all boundary conditions be more closely considered and that the directions of the applied external stimuli (electrical or mechanical) and the material response be described. This is possible using tensor relations and matrix mathematics.

Nye [39] has presented an elegant approach to show the physical property response of materials to changes in their external environment, including mechanical, electrical, thermal, electromechanical, and pyroelectric behavior. The method that he presents maps well onto the Heckmann diagram (Figure 4) and warrants further comment because it affords a straightforward method to understand, or predict, piezoelectric response. All of the above property information and crystal symmetry aspects are summarized in a concise 10×10 matrix representation for each of the 32 crystal classes. Because of our interest in piezoelectrics, we are most interested in three of these matrices: quartz (32); single-crystal barium titanate because it is a prototypical perovskite piezoelectric (4mm); and materials such as poled polycrystalline lead zirconate titanate. As described in greater detail above, when initially formed, the dipoles within a polycrystalline body posses a random orientation. However, following the poling process, the dipoles possess a preferred orientation, which results in a net macroscopic dipole moment for the ceramic body, and effective 6mm symmetry.

To illustrate the basic aspects of the method, let's consider the change in dielectric displacement (polarization) of a poled polycrystalline PZT body resulting from an applied stress. We will first consider the most general case, where there is no constraint on the applied stress. Because this is a poled PZT ceramic, we use the (10×10) matrix for the equilibrium properties of crystal class 6mm. This matrix is shown in Figure 30. This matrix represents each of the coefficients associated with mechanical, electrical, electromechanical, and thermal properties associated with materials that possess class 6mm symmetry. The (10×10) matrix is subdivided according to these properties into submatrices that represent the mechanical properties (the 6×6 matrix in the upper left that shows the strains S developed for a given stress σ), thermal expansion behavior (the 6×1 matrix at the far right that shows the strains ε that result from a change in temperature ΔT), and so on. The symbols within the matrix represent whether a particular coefficient is either zero or nonzero, and which coefficients are equivalent. The zero coefficients are indicated by the smaller dots and the nonzero coefficients are indicated by the larger dots. A line tying together two coefficients indicates that the two are equal.

With this explanation, we are prepared to describe the electromechanical response of a material. We will start with the most general description possible and will then invoke 6mm symmetry to show how the general response is modified by the constraint of the material's symmetry. Let's assume that we want to describe the dielectric displacement developed for any material when any stress is applied. In other words, at this point we will not assume that any of the coefficients are zero or that we have restricted stress to a particular direction. For the most general case, we will need to consider the dielectric displacement developed in each direction, D_1, D_2, and D_3, and because we have not yet placed any constraint on the applied stress, we need to consider all arbitrary normal stresses, σ_1, σ_2, σ_3, and all arbitrary shear stresses, σ_4, σ_5, and σ_6. We will solve for the ($3 \times$

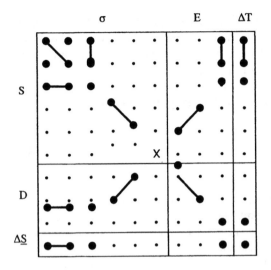

Figure 30 Matrix for equilibrium properties for a 6mm crystal.

1) dielectric displacement matrix in terms of the (6 × 1) stress matrix and the (3 × 6) electromechanical response matrix according to:

$$
\begin{pmatrix} D_1 \\ D_2 \\ D_3 \end{pmatrix} = \begin{pmatrix} d_{11} & d_{12} & d_{13} & d_{14} & d_{15} & d_{16} \\ d_{21} & d_{22} & d_{23} & d_{24} & d_{25} & d_{26} \\ d_{31} & d_{32} & d_{33} & d_{34} & d_{35} & d_{36} \end{pmatrix} \begin{pmatrix} \sigma_1 \\ \sigma_2 \\ \sigma_3 \\ \sigma_4 \\ \sigma_5 \\ \sigma_6 \end{pmatrix}
$$
(B1)

The matrix math above may be described as a (3 × 1) = (3 × 6) × (6 × 1) matrix. We may expand our solutions to solve for D_1, D_2, and D_3 as:

$$D_1 = d_{11}\sigma_1 + d_{12}\sigma_2 + d_{13}\sigma_3 + d_{14}\sigma_4 + d_{15}\sigma_5 + d_{16}\sigma_6 \qquad (B2)$$

$$D_2 = d_{21}\sigma_1 + d_{22}\sigma_2 + d_{23}\sigma_3 + d_{24}\sigma_4 + d_{25}\sigma_5 + d_{26}\sigma_6 \qquad (B3)$$

$$D_3 = d_{31}\sigma_1 + d_{32}\sigma_2 + d_{33}\sigma_3 + d_{34}\sigma_4 + d_{35}\sigma_5 + d_{36}\sigma_6 \qquad (B4)$$

which represent the most general expressions for dielectric displacement and are applicable to materials of any crystal class. Now we consider the above equilibrium property matrix for materials of 6mm symmetry to determine which of the above coefficients (from $d_{11}, d_{12}, \cdots, d_{36}$) are non zero. A review of this matrix shows that there are only five nonzero coefficients (out of a possible 18). These are d_{15}, d_{24}, d_{31}, d_{32}, and d_{33}. Furthermore, the links in the matrix between two pairs of these coefficients tell us that $d_{15} = d_{24}$ and that $d_{31} = d_{32}$; by convention, we represent these coefficients as d_{15} and d_{31}.

Based on the symmetry of the material, the above general expressions may be simplified for the present case (6mm symmetry) to the following:

$$D_1 = d_{15}\sigma_5 \tag{B5}$$

$$D_2 = d_{15}\sigma_4 \tag{B6}$$

$$D_3 = d_{31}\sigma_1 + d_{31}\sigma_2 + d_{33}\sigma_3 \tag{B7}$$

While we might apply a variety of normal or shear stresses to a material, for materials of 6mm symmetry, the only applied stress that will result in dielectric displacement in the 1 direction is a shear stress about the 2 axis ($\sigma_5 = \sigma_{13} = \sigma_{31}$). In contrast, a dielectric displacement in the 3 direction will result from normal stresses applied to the 1 and/or 2 directions (because the d_{31} and d_{32} coefficients are nonzero) or from a normal stress applied to the 3 direction through the nonzero d_{33} coefficient. In contrast, in 6mm materials, a shear stress about the 3 direction ($\sigma_6 = \sigma_{12} = \sigma_{21}$) will never result in a dielectric displacement in a 6mm material. While sometimes the piezoelectric effect is considered as a simple relationship among scalar quantities, this basic method is the approach that should be used to quantitatively describe the piezoelectric response of a material. Two specific examples are given in Appendix C.

Appendix C: Sample Calculations of Piezoelectric Response

Example 1: Determination of charge resulting from an applied stress

Assume that we want to calculate the dielectric displacement (polarization) in the 3 direction for a poled polycrystalline barium titanate body that is subjected to an applied stress of 10 MPa in the poling direction. The basic method to complete this calculation is given below.

First, we need to know the point group for a poled polycrystalline $BaTiO_3$ body. Assuming that we want to predict this response in a material at room temperature, the crystal class for this material is 6mm. This tells us which equilibrium property matrix we should use and the simplified version of the equation that we should use to calculate D_3, the polarization that is developed in the 3 (poling) direction. The equation that results from this approach is:

$$D_3 = d_{33}\sigma_3 \tag{C1}$$

because the only applied stress is σ_3, which is the stress normal to the poling direction that is described in the problem. Because $\sigma_1 = \sigma_2 = 0$ in this instance, we are not concerned with the response that would result from the nonzero d_{31} coefficient. We may determine the resulting polarization with knowledge of d_{33}, which we may find in Jaffe et al. [7] to be reported as 191 C/N. Next we substitute this value and the value for applied stress into Eq. (C1) to determine D_3:

$$D_3 = (191 \times 10^{-12}\ C/N)(50 \times 10^6\ N/m^2) \qquad (C2)$$

$$D_3 = 9.55 \times 10^{-3}\ C/m^2 = 0.955\ \mu C/cm^2 \qquad (C3)$$

This value actually represents a polarization that is slightly more than 10% of the remanent polarization ($P_R = 7.5\ \mu C/cm^2$) of a typical poled polycrystalline barium titanate specimen.

Example 2: Strain resulting from an applied voltage

You have been given a poled lead zirconate titanate 5A plate with lateral dimensions of 1 cm and a thickness of 0.1 cm. One thousand volts is applied to the material in the same direction as the poling field. Calculate the resulting strains in the 1, 2, and 3 directions. Poling was accomplished through the thickness of the plate. As above, we will use an approach based on Nye's method to determine the resulting strain. This problem is an example of the indirect (or converse) piezoelectric effect, so we will use the *d* coefficient with units of m/ V or pm/V (picometers $= 10^{-12}$ m/V). Poled polycrystalline PZT bodies also possess cylindrical symmetry; therefore, we will use the equilibrium property matrix for 6mm, just as we did above for poled barium titanate. For the purposes of this example (we do not need to do this first step because we already know the symmetry of the material), we will show the matrix multiplication that is carried out to solve this problem. It would involve the determination of the strain (6×1) matrix from the (6×3) electromechanical coefficient matrix times the (3×1) electric field matrix. We expect this result because in matrix multiplication, the interior matrix dimensions match, whereas the exterior matrix dimensions give the dimensions of the final matrix, i.e., (6×1) = (6×3) $\times$ (3×1). Above, we saw that the electromechanical tensor that described the polarization resulting from an applied stress was given by a (3×6) matrix. To determine the strain (S) resulting from an applied field (E), we will use the (6×3) electromechanical property matrix:

$$
\begin{pmatrix} S_1 \\ S_2 \\ S_3 \\ S_4 \\ S_5 \\ S_6 \end{pmatrix} =
\begin{pmatrix}
d_{11} & d_{21} & d_{31} \\
d_{12} & d_{22} & d_{32} \\
d_{13} & d_{23} & d_{33} \\
d_{14} & d_{24} & d_{34} \\
d_{15} & d_{25} & d_{35} \\
d_{16} & d_{26} & d_{36}
\end{pmatrix}
\begin{pmatrix} E_1 \\ E_2 \\ E_3 \end{pmatrix}
\qquad (C4)
$$

Before we solve for the resulting strain by looking at the zero and nonzero coefficients, it is worth noting that the subscript notation that is employed for the electromechanical response matrix for the converse piezoelectric effect is reversed from that used above for the direct piezoelectric effect, as well as that employed in standard matrix mathematics. This is due to the fact that in describing the piezoelectric response (i.e., the d coefficient) of a material, we denote the electrical direction by the first subscript and the mechanical direction by the second subscript. It is also important to note that the ε-E matrix is not simply the transpose of the D-σ matrix because we have employed reduced notation to describe the response. Likewise, if we were looking at the mechanical properties of materials, c_{ij} and s_{ij} (stiffness and compliance matrices, respectively) would also not be simple transposes of one another. In the present case, based on a review of Nye, the nonzero coefficients are d_{31}, d_{32} ($= d_{31}$), d_{33}, d_{15}, and d_{24} ($= d_{15}$). We may therefore express the resulting strain(s) for the application of any electric field as:

$$S_1 = d_{31}E_3 \tag{C5}$$

$$S_2 = d_{31}E_3 \tag{C6}$$

$$S_3 = d_{33}E_3 \tag{C7}$$

$$S_4 = d_{15}E_2 \tag{C8}$$

$$S_5 = d_{15}E_1 \tag{C9}$$

$$S_6 = d_{16}E_1 + d_{26}E_2 + d_{36}E_3 = 0 \tag{C10}$$

Based on the above expressions, we are now prepared to solve the resulting strains in the three orthogonal directions. The d_{31} and d_{33} coefficients for PZT 5A are reported to be -119 pm/V and 268 pm/V respectively. The electric field (applied voltage/distance) is 10.0 kV/cm. Using Eqs. (C5) to (C7), we may calculate the resulting strains as:

$$S_1 = d_{31}E_3 = (-119 \times 10^{-12} m/V)(1.0 \times 10^6 V/m)$$
$$= -1.19 \times 10^{-4}$$

$$S_2 = d_{31}E_3 = (-119 \times 10^{-12} m/V)(1.0 \times 10^6 V/m)$$
$$= -1.19 \times 10^{-4}$$

$$S_3 = d_{33}E_3 = (268 \times 10^{-12} m/V)(1.0 \times 10^6 V/m)$$
$$= 2.68 \times 10^{-4}$$

The positive value for S_3 indicates an elongation in the 3 direction while the negative values for S_1 and S_2 indicate lateral contractions.

Appendix D. Thermodynamic Proof of the Equivalency of the Piezoelectric Charge and Strain Coefficients

We want to show that the d coefficient that relates the generated charge to an applied mechanical stress equals the coefficient that relates the strain developed due to an applied voltage. That is, we want to prove that

$$\left(\frac{\partial D_i}{\partial \sigma_j}\right)_E = \left(\frac{\partial \varepsilon_j}{\partial E_i}\right)_\sigma \tag{D1}$$

The d coefficients that represent these relationships appear on the left-hand side of the Heckmann diagram (Figure 4) and connect two apparently different relationships, since different external parameters and resulting material responses are linked. While the same coefficient is used in this figure to describe these relationships, it is not clear from the figure that these must be equal to one another. We will use a thermodynamic approach to demonstrate this.

We begin with an expression for the change in Gibbs energy. There are a variety of expressions that may be written. Some of them include only H, PV, and TS, and others show the effect of changes in other external variables, including, among others, magnetic field. For the case of interest, we start with an expression for G that includes the three principal external variables shown in the Heckmann diagram and describe the change in Gibbs energy as a function of temperature, stress, and electric field. The appropriate expression is:

$$dG = -\underline{S}dT - Sd\sigma - DdE \tag{D2}$$

where $\underline{S}$ is entropy, T is temperature, and all other variables have the same meaning as before. Next, we examine the partial derivatives of G with other external parameters held constant. We may write the following three expressions:

$$\left(\frac{\partial G}{\partial T}\right)_{\sigma,E} = -\underline{S} \tag{D3}$$

$$\left(\frac{\partial G}{\partial \sigma}\right)_{T,E} = -S \tag{D4}$$

and

$$\left(\frac{\partial G}{\partial E}\right)_{T,\sigma} = -D \tag{D5}$$

We next look at the second partial derivatives of G, and because all of the terms in the above expressions are thermodynamic state functions, we may write:

$$\left(\frac{\partial^2 G}{\partial E \partial \sigma} \right)_T = \left(\frac{\partial^2 \, G}{\partial \sigma \, \partial E} \right)_T \tag{D6}$$

which indicates that the resulting function (the second partial of G) must be independent of the order of differentiation. We may now look further at each of the equivalent functions given in (D6). Starting first with the right-hand side of the expression, we may write:

$$\left(\frac{\partial^2 G}{\partial E \partial \sigma} \right)_T = \left(\frac{\partial}{\partial E} \right) \left(\frac{\partial G}{\partial \sigma} \right)_T \tag{D7}$$

But, from Eq. (D4),

$$\left(\frac{\partial G}{\partial \sigma} \right)_T = -S \tag{D8}$$

and therefore,

$$\left(\frac{\partial}{\partial E} \right) \left(\frac{\partial G}{\partial \sigma} \right)_T = -\left(\frac{\partial S}{\partial E} \right)_T = -d \tag{D9}$$

Likewise, we may consider the left-hand side of Eq. (D7) and write:

$$\left(\frac{\partial^2 \, G}{\partial \sigma \, \partial E} \right)_T = \left(\frac{\partial}{\partial \sigma} \right) \left(\frac{\partial G}{\partial E} \right)_T \tag{D10}$$

and from Eq. (D5) we may write that:

$$\left(\frac{\partial G}{\partial E} \right)_T = -D \tag{D11}$$

Combining this expression with Eq. (D10), we can derive an expression for the second partial of the Gibbs energy when the order of differentiation is first with respect to field, then with respect to stress. The resulting expression is:

$$\left(\frac{\partial}{\partial \sigma} \right) \left(\frac{\partial G}{\partial E} \right)_T = \left(\frac{\partial D}{\partial \sigma} \right)_T = -d \tag{D12}$$

Therefore, we may write that

$$\left(\frac{\partial \varepsilon}{\partial E} \right)_T = \left(\frac{\partial D}{\partial \sigma} \right)_T \tag{D13}$$

Stated otherwise, we have shown from a thermodynamic perspective that the coefficients for the direct and converse piezoelectric effects must be equal (although the units will be different). In this scenario, the above result [Eq. (D13)] implies not only that the direct and converse effect piezoelectric coefficients must be equal, but that as we begin to consider crystal symmetry and the directionality and magnitude of the response, the matrices that represent these effects must be equivalent.

REFERENCES

1. Cady W. Piezoelectricity. New York: McGraw-Hill, 1946.
2. IEEE standard on piezoelectricity, ANSI/IEEE Std:176–1987.
3. Busch G. Ferroelectrics 1987; 74:267.
4. Haertling GH. J Am Ceram Soc 1999; 82(4):797.
5. Fukumoto A. Ferroelectrics 1982; 40:217.
6. Richerson DW. The Magic of Ceramics. Westerville, OH: American Ceramic Society, 2000.
7. Jaffe B, Cook WR, Jaffe H. Piezoelectric Ceramics. New York: Academic Press, 1971.
8. Eom CB. JOM–J Min Met Mater Soc 1997; 49(3):47.
9. Balachandran U, Dusek JT, Sweeney SM, Poeppel RB, Mieville RL, Maiya PS, Kleefisch MS, Pei S, Kobylinski TR, Udovich CA. Am Ceram Soc Bull 1995; 74(1):71.
10. Wu MK, Ashburn JR, Torng CJ, Hor PH, Meng RL, Gao L, Huang ZJ, Wang YQ, Chu CW. Phys Rev Lett 1987; 58:908.
11. Park S-E, Shrout TR. J Appl Phys 1997; 82(4):1804.
12. Bokov VA, Myl'nikova IE. Sov Phys-Sol State 1961; 2(11):2428.
13. Smolenskii GA, Isuov VA, Agranovskaya AI, Popov SN. Phys-Sol State 1961; 2(11):2584.
14. Kuwata J, Uchino K, Nomura S. Ferroelectrics 1981; 37:579.
15. Khan A, Meschke FA, Li T, Scotch AM, Chan HM, Harmer MP. J Am Ceram Soc 1999; 82(11):2958.
16. Haertling GH. U.S. Patent 5,471,721, 1995.
17. Fox RL, Hellbaum RF, Bryant RG, Copeland BM. U.S. Patent 6,060,811, 2000.
18. Schwartz RW, Narayanan M. Sens Act A Phys 2002; 101(3):322.
19. Chen W, Lynch CS. Acta Mater 1998; 46(15):5303.
20. Kamlah M, Bohle U. Int J Sol Struct 2001; 38:605.
21. Noheda B, Cox DE, Shirane G, Guo R, Jones B, Cross LE. Phys Rev B 2001; 6301(1):4103.
22. Guo R, Cross LE, Park SE, Noheda B, Cox DE, Shirane G. Phys Rev Lett 2000; 84(23):5423.
23. Polla DL, Schiller PJ. Int Ferro 1995; 7:359.
24. Damjanovic D, Taylor DV, Setter N. Mater Res Soc Symp Proc 2000; 596:529.

25. Megaw HD. Crystal Structures: A Working Approach. Philadelphia: W. B. Saunders, 1973.
26. Bloss FD. Crystallography and Crystal Chemistry: An Introduction. New York: Holt, Rinehart and Winston, 1971.
27. Krause PW. The photon detection process. In Keyes, Ed. RJ, ed. Optical and Infrared Detectors. Heidelberg: Springer-Verlag, 1977.
28. Watton R. Ferroelectrics 1989; 91:87.
29. Hall DA. J Mater Sci 2001; 26:4575.
29b. Mason WP. Quartz crystal applications. Quartz Crystals for Electrical Circuits. New York: Van Nostrand, 1946.
30. Moulson AJ, Herbert JM. Electroceramics. London: Chapman and Hall, 1990.
31. Whatmore RW. Ferroelectrics 1991; 118:241.
32. Callahan JM. Auto Ind Oct., 1991:30.
33. Whatmore RW, Patel A, Shorrocks NM, Ainger FW. Ferroelectrics 1990; 104:269.
34. Fatuzzo E, Merz WJ. Ferroelectricity. New York: American Elsevier, 1967.
35. Hench LL, West JK. Principles of Electronic Ceramics. New York: John Wiley and Sons, 1990.
36. IRE standards on piezoelectric crystals: the electromechanical coupling factor. Proc IRE 1958; 46:764.
37. Kay HF, Vousden P. Phil Mag 1949; 40:1019.
38. Merz WJ. Phys Rev 1949; 76:1221.
39. Nye JF. Physical Properties of Crystals, Clarendon Press. 1989.
40. Ragone DV. Thermodynamics of Materials. Vol. 1. New York: John Wiley and Sons, 1995.
41. Arlt G. J Mater Sci 1990; 25:2655.
42. Merz WJ. Phys Rev 1954; 95(3):690.
43. Arlt G, Sasko P. J Appl Phys 1980; 51:4956.
43b. De Vries RC, Burke JE. J Am Ceram Soc 1957; 40:200.
44. IEEE standard definitions of primary ferroelectric terms, ANSI/IEEE Std, 1986: 180.
45. Hwang SC, McMeeking RM. Int J Sol Struc 1999; 36:1541.
46. Herbiet R, Robels U, Dederichs H, Arlt G. Ferroelectrics 1989; 98:107.
47. Lynch CS. Acta Mater 1996; 44(10):4137.
48. Zhang XL, Chen ZX, Cross LE, Schulze WA. J Mater Sci 1983; 18:968.
49. Damjanovic D. Rep Prog Phys 1998; 61:1267.
50. Hall DA, Stevenson PJ. Ferroelectrics 1999; 228:139.
51. Ganpule CS, Nagarajan V, Li H, Ogale AS, Steinhauer DE, Aggarwal S, Williams E, Ramesh R, De Wolf P. Appl Phys Lett 2000; 77(2):292.
52. Nagarajan V, Ganpule CS, Roytburd A, Ramesh R. Int Ferro 2002; 42:173.
53. Hwang SC, Lynch CS, McMeeking RM. Acta Metall Mater 1995; 43:2073.
54. Zhang QM, Cross LE. Ferroelectrics 1989; 98:137.
55. Schwartz RW, Moon Y-W. Proc SPIE, Smart Struct Mater 2001; 4333:408.
56. Jaffe B, Roth RS, Marzullo S. J Res Nat Bur Stds 1955; 55:239.
57. Butterworth S. Proc Phys Soc 1914; 26:264.
58. van Dyke KS. Phys Rev 1925; 25:895.

59. van Dyke KS. IRE 1928; 16:742.
60. Meeker TR, Shreve WR, Cross PS. Theory and properties of piezoelectric resonators and waves. In Gerber EA, Ballato, Eds. A, eds. Precision Frequency Control. Vol. 1. New York: Academic Press, 1985.
61. Ballato A. Piezoelectric resonators. Design of Crystal and Other Harmonic Oscillators. New York: John Wiley and Sons, 1983.
62. Tiersten HF. J Acoust Soc Am 1963; 35:53.
63. Ballato A. Doubly rotated thickness mode plate vibrators. In Mason WP, Thurston, Eds. RN, eds. Physical Acoustics: Principles and Methods. Vol. 13. New York: Academic Press, 1977.
64. Ballato A, Hatch ER, Mizan M, Lukaszek TJ. IEEE Trans UFFC 1986; UFFC-33: 385.
65. Lamb J, Richter J. Proc Roy Soc (London) 1966; A293:479.
66. Ballato A, Ballato J. Ferroelectrics 1996; 182:29.
67. Marutake M. Proc IRE February 1962; 50:214.
68. Ballato J, Schwartz R, Ballato A. IEEE Trans UFFC 2001; 48:462.
69. Bloss FD. An Introduction to the Methods of Optical Crystallography. Philadelphia: W. B. Saunders, 1961.
70. Neese WD. Introduction to Optical Mineralogy. 2nd ed. New York: Oxford University Press, 1991.
71. Saleh BEA, Teich MC. Fundamentals of Photonics. New York: John Wiley and Sons, 1991.
72. Agulló-López FJ, Cabrera JM, Agulló-Rueda F. Electro-Optics: Phenomena, Materials, and Applications. New York: Academic Press, 1994.
73. Maldonado JR, Meitzler AH. Ferroelectrics 1972; 3:169.
74. Meitzler AH, O'Bryan HM. Appl Phys Lett 1971; 19:106.
75. Wylie CR, Barrett LC. Advanced Engineering Mathematics. 6th ed. New York: McGraw-Hill, 1995.
76. Margulis WF, Garcia F, Hering E, Guedes Valente L, Lesche B, Laurell F, Carvalho I. Mater Res Soc Bull Nov. 1998:31.
77. Myers R, Mukherjee N, Brueck S. Opt Lett 1991; 16:1732.
78. Feng J, Zhu Y-Y, Ming N-B. Phys Rev B 1990; 41:5578.
79. Feng J, Zhu Y-Y, Ming N-B. Science 1997; 278:843.
80. Kuzyk M, Dirk C. Characterization Techniques and Tabulations for Organic Nonlinear Optical Materials. New York: Marcel Dekker, 1998.
81. Busch G. Ferroelectrics 1987; 71:43.
82. Fousek J. Ferroelectrics 1991; 113:2.
83. Byrrapa K, Yoshimura M. Handbook of Hydrothermal Technology. NJ: Park Ridge, 2001.
84. Laudise R, Sullivan R. Chem Eng Prog 1959; 55:55.
85. Meeker T, Shreve W, Cross P. Theory and properties of piezoelectric resonators and waves. In Gerber E, Ballato, Eds. A, eds. Precision Frequency Control. Vol. 1. New York: Academic Press, 1985.
86. Ballato A. Jpn J Appl Phys 1985; 24(Suppl. 24–1):9.
87. Abouelleil M, Leonberger F. J Am Ceram Soc 1989; 72:1311.

88. Alferness R. Titanium-diffused lithium niobate waveguide devices. In Tamir, Ed. T, ed. Guided–Wave Optoelectronics. New York: Springer-Verlag, 1990.
89. Kazovsky L, Benedetto S, Willner A. Optical Fiber Communication Systems. Norwood, MA: Artech House, 1996.
90. Li C, Li Y. Ferroelectric materials. In Gupta, Ed. M, ed. Handbook of Photonics. New York: CRC Press, 1997.
91. Halasyamani P, Poepplemeier K. Chem Mater 1998; 10:2753.
92. Fushimi S, Ikeda T. J Am Ceram Soc 1967; 50(3):129.
93. Kuznetzov VA. J Cryst Grow 1968; 34:405.
94. Clarke R, Whatmore RW. J Cryst Grow 1976; 33:29.
95. Hatanaka T, Hasegawa H. Jpn J Appl Phys 1995; 34:536.
96. Cross LE, Jang S, Newnham R, Nomura S, Uchino K. Ferroelectrics 1980; 23:187.
97. Uchino K. Am Ceram Soc Bull 1986; 65:647.
98. Uchino K. Jpn J Appl Phys 1985; 24(Suppl. 24–2):460.
99. Takahashi S. Jpn J Appl Phys 1985; 24(Suppl. 24–2):41.
100. Takahashi S, Yano T, Fuku I, Sato E. Jpn J Appl Phys 1985; 24(Suppl. 24–2):206.
101. Reynolds TG. Am Ceram Soc Bull 2001; 80(10):29.
102. Shirane G, Suzuki K. J Phys Soc Jpn 1952; 7:333.
103. Jaffe B, Roth R, Marzullo S. J Appl Phys 1954; 25:809.
104. Dimos D, Schwartz RW, Lockwood SJ. J Am Ceram Soc 1994; 77(11):3000.
105. Wu L, Wu T-S, Wei C-C, Liu H-C. J Phys C 1983; 16:2823.
106. Smyth DM. Ann Rev Mater Sci 1985; 15:329.
107. Chiang Y-M, Birnie III D, Kingery WD. Physical Ceramics, Principles for Ceramic Science and Engineering. New York: John Wiley and Sons, 1997:110–111.
108. Jaffe H, Berlincourt D. Proc IEEE 1965; 53:1372.
109. Haertling GH, Land CE. J Am Ceram Soc 1971; 54(1):1.
110. Hardtl K, Hennings D. J Am Ceram Soc 1972; 55:230.
111. Carl K, Hardtl KH. Phys Stat Sol A 1971; 8:87.
112. Mishra SK, Pandey D, Singh AP. Appl Phys Lett 1996; 69:1707.
113. Lysne P, Percival C. Ferroelectrics 1976; 10:129.
114. Ito Y, Takeuchi H, Jyomura S, Nagatsuma K, Ashida S. Appl Phys Lett 1979; 35:595.
115. Biggers J, Schulze W. Am Ceram Soc Bull 1974; 53:809.
116. Haertling GH, Land CE. Ferroelectrics 1972; 3:269.
117. Snow GS. J Am Ceram Soc 1973; 56(9):479.
118. Newnham RE, Skinner DP, Cross LE. Mater Res Bull 1978; 13:525.
119. Newnham R, Skinner D, Klicker K, Rittenmeyer K, Bhalla A, Hardiman B, Gururaja T. Ferroelectrics 1980; 27:49.
120. Pohanka R, Smith P. Electronic Ceramics Levinson, Ed. LM, ed. New York: Marcel Dekker, 1988.
121. Smith WA. IEEE Trans UFFC 1990; 38(1):40.
122. Sigmund G, Torquato S, Aksay IA. J Mater Res 1998; 13(4):1038.
123. Poizat C, Sester M. Comp Mater Sci 1999; 16:89.
124. Randall CA, Miller DV, Adair JH, Bhalla AS. J Mater Res 1993; 8(4):899.
125. Banno H. Ferroelectrics 1983; 50:329.

126. Lencka MM, Riman RE. Chem Mater 1995; 7:18.
127. Rossetti GA, Watson DJ, Newnham RE, Adair JH. J Cryst Growth 1992; 116:251.
128. Wei Z, Yamashita K, Okuyama M. Jpn J Appl Phys 2001; 40:5539.
129. Zeng J, Lin C, Li J, Li K. J Mater Res 1999; 14(7):2712.
130. Cho W-S, Yoshimura M. J Mater Res 1997; 12(3):833.
131. Pamplin BR. Introduction to crystal growth methods. In Pamplin, Ed. BR, ed. Crystal Growth. Oxford: Pergamon Press, 1980.
132. Laudise RA. The Growth of Single Crystals. NJ: Englewood Cliffs, 1970.
133. Remeika JP. J Am Chem Soc 1954; 76:940.
134. Suchicital CTA, Payne DA. 1986:465.
135. Li T, Scotch AM, Chan HM, Harmer MP, Park S-E, Shrout TR, Michael JR. J Am Ceram Soc 1998; 81(1):244.
136. Duran C, Trolier-McKinstry S, Messing GL. J Am Ceram Soc 2000; 83:2203.
137. Seabaugh MM, Kerscht IH, Messing GL. J Am Ceram Soc 1997; 80(5):1181.
138. Sabolsky EM, James AR, Kwon S, Trolier-McKinstry S, Messing GL. Appl Phys Lett 2001; 78(17):2551.
139. Bandyopadhyay A, Panda RK, Janas VF, Agrawala MK, Danforth SC, Safari A. J Am Ceram Soc 1997; 80(6):1366.
140. Dungan RH, Snow GS. Am Ceram Soc Bull 1977; 56(9):781.
141. Taylor DJ, Jones RE, Lii YT, Zurcher P, Chu PY, Gillespie SJ. Mater Res Soc Symp Proc 1996; 433:97.
142. Norga GJ, Fe L. Mater Res Soc Symp Proc 2001; 655:CC9.1.1.
143. Schwartz RW. Chem Mater 1997; 9(11):2325.
144. Okazaki K. Ferroelectrics 1982; 41:77.
145. Matsuo Y, Sasaki H. J Am Ceram Soc 1965; 48(6):289.
146. Brown L, Mazdiyasni K. J Am Ceram Soc 1974; 53:421.
147. McCarthy D, Brooks R. Ferroelectrics 1980; 27:183.
148. Weston TB, Webster AH, McNamara VM. Can Clay Ceram 1967; 36:15.
149. Brinker CJ, Scherrer GW. Sol-Gel Science; The Physics and Chemistry of Sol-Gel Processing. San Diego: Academic Press, 1990.
150. Megaw HD. Nature 1945; 155:484.
151. Wainer E. U.S. Patent 2,467,169, 1949.
152. Berlincourt D. J Acoust Soc Am 1964; 36(3):515.
153. Baerwald HG. Phys Rev 1957; 105:480.
154. Redin R, Marks G, Antoniak C. J Appl Phys 1963; 34:600.
155. Hackenberger W, Helgaland J, Zipparo M, Randall CA, Shrout TR. Proc. 8th U.S.–Japan Seminar on Diel. and Piezo. Ceram, Plymouth. 1997:323.
156. Smith W. Proc SPIE 1992; 1733:3.
157. Meyer RJ, Shrout TR, Yoshikawa S. Proc. 8th U.S.–Japan Seminar on Diel. and Piezo. Ceram. Plymouth, MA, 1997:313.
158. Bowen L, Gentilman R, Fiore D, Pham H, Serwatka W, Near C, Pazol B. Ferroelectrics 1996; 187(1–4):109.
159. Tuttle BA, Smay JE, Cesarano III J, Voigt JA, Scofield TW, Olson WR. J Am Ceram Soc 2001; 84(4):872.
160. Smay JE, Cesarano III J, Tuttle BA, Lewis JA. J Appl Phys 2002; 92(10):6119.

161. Janas VF, Safari A. J Am Ceram Soc 1995; 78(11):2945.
162. Cesarano III J, Segalman R, Calvert P. Ceram Ind 1998; 148:94.
163. Lynn S, Newnham R, Klicker K, Rittenmeyer K, Safari A, Schulze W. Ferroelectrics 1981; 38:955.
164. Pohanka R, Freiman S, Rice R. Ferroelectrics 1980; 28:337.
165. Maldonado J, Meitzler A. Proc IEEE 1971; 59:368.
166. Burfoot JC, Taylor GW. Polar Dielectrics and Their Applications. Berkeley: University of California Press, 1979.
167. Haertling G. IEEE WESCON Proc. Los Angeles, Aug. 1966.
168. Hardtl K. Ferroelectrics 1976; 12:9.
169. Isupov V. Sov Phys-Sol State 1968; 10:989.
170. Benguigui L. Solid State Comm 1972; 11:825.
171. Okazaki K. Ferroelectrics 1981; 35:173.
172. Multani M, Gokarn S, Vijayaraghavan R, Polkar V. Ferroelectrics 1981; 37:652.
173. Li G, Haertling G. Ferroelectrics 1995; 166:31.
174. Thacher P. Appl Opt 1977; 16:3210.
175. Haertling G, Land C. J Am Ceram Soc 1971; 54:1.
176. Cutchen J, Harris J, Laguna G. IEEE WESCON Proc. San Francisco, Sept. 1973.
177. Jonker G. J Am Ceram Soc 1972; 55:57.
178. Okazaki K, Nagata K. J Am Ceram Soc 1973; 56:82.
179. Joon Yoon K, Shin S, Park HC, Goo NS. Smart Mater Struc 2002; 11:163.
180. Smith WA. IEEE Ultrasonics Symp 1989:755.
181. Hinders MK, Rhodes BA, Fang TM. J Sound Vibr 1995; 185(2):219.
182. Hecht E. Vol. Optics. Reading, MA: Addison-Wesley, 1998.
183. Banno H, Ogura K, Sobue H, Ohya K. Jpn J Appl Phys 1987; 26(Suppl. 26–1): 153.
184. Pilgrim SM, Bailey AE, Massuda M, Poppe FC, Ritter AP. Ferroelectrics 1994; 160:305.
185. Takahashi S, Ochi A, Yonezawa M, Yano T, Hamatsuki T, Fukui I. Ferroelectrics 1983; 50:181.
186. Bowen LS, Shrout T, Schulze WA. Ferroelectrics 1980; 27:59.
187. Pritchard J, Bowen CR, Lowrie F. Br Ceram Trans 2001; 100(6):265.
188. Watanabe J, Someji T, Jomura S. Jpn J Appl Phys 1999; 38(Part 1, No. 5B):333.
189. Ohashi J, Fuda Y, Ohno T. Jpn J Appl Phys 1993; 32(Part 1, No. 5B):2412.
190. Adaptronics, Inc., Troy, NY.
191. Winzer SR, Shankar N, Ritter AP. J Am Ceram Soc 1989; 72(12):2246.
192. Shih WY, Shih W-H, Askay IA. J Am Cer Soc 1997; 80(5):1073.
193. Uchino K. Ferroelectrics 1989; 91:281.
194. Jones J, Fuller C. J Sound Vib 1987; 112:389.
195. Steel MR, Harrison F, Harper PG. J Phys D Appl Phys 1978; 11:979.
196. Cheu PL, Muller RS, Jolly RD, Halac GL, White RM, Andrews AP, Lim TC, Motamedi ME. Sens Act 1984; 5:119.
197. Smits JG, Choi W-S. IEEE Trans UFFC 1991; 38(3):256.
198. Wang Q-M, Du X-H, Xu B, Cross LE. IEEE Trans UFFC 1999; 46(3):638.
199. Wang Q-M, Cross LE. J Am Ceram Soc 1999; 82(1):103.
200. Schwartz RW, Cross LE, Wang Q-M. J Am Ceram Soc 2001; 84(11):2563.

201. Haertling GH. Am Ceram Soc Bull 1994; 73:93.
202. Wise SA. Sens Act A Phys 1998; 69(1):33.
203. Mossi KM, Bishop RP, Smith RC. Proc SPIE Smart Struct Mater 1999; 3667:738.
204. Mossi KM, Bishop RP. Proc SPIE Smart Struct Mater 1998; 3875:43.
205. Goldfarb M, Gogola M, Fischer G, Garcia E. ASME International Mechanical Engineering Congress and Exposition. Vol. DSC-Vol. 69–2, November 2000:973.
206. Wise SA, Handy RC, Dausch DE. Proc SPIE Smart Struct Mater 1997; 3044:342.
207. Borgen M, Washington GN, Kinzel G. Adaptive Struct Mater Sys Symp. ASME International Congress and Exposition 2000; AD-Vol. 60:247.
208. Niezrecki C, Balakrishnan S. SPIE Smart Struct Mater Active Mater Behav Mech 2001; 4327:88.
209. Waterfield GR, presented at the 105th Annual Meeting of the American Ceramic Society, April 2003, Nashville, TN.
210. Wang Q-M, Cross LE. Mater Chem Phys 1999; 58(1):20.
211. Li G, Furman E, Haertling GH. J Am Ceram Soc 1997; 80(6):1382.
212. Dausch DE. Ferroelectrics 1998; 210:31.
213. Haertling GH. Proc 9th Int Symp Appl Ferro 1994:65.
214. Mulling J, Usher T, Dessent B, Palmer J, Franzon P, Grant E, Kingon AI. Sens Act A Phys 2001; 94:19.
215. Nothwang WD. Ph.D. dissertation. Clemson, SC: Clemson University, 2001.
216. Cutchen J, Harris J, Laguna G. Proc SPIE 1976; 88:57.
217. Cutchen J. Ferroelectrics 1980; 27:173.
218. Haertling GH. Piezoelectric and electrooptic ceramics. In Buchanan RC, ed. Ceramic Materials for Electronics. 2nd ed. New York: Marcel Dekker, 1991.
219. Chen C, Chen G. Int J Struct Eng Mech 2002; 14(1):21.
220. Talish RJ, Ryaby JP, Scowen KJ, Urgovitch KJ. U.S. Patent 5, 556, 372, 1995.
221. Kristiansen TK, Ryaby JP, McCabe J, Frey JJ, Roe LR. J Bone Joint Surg 1997; 79-A(7):961.
222. Yang PX, Xu JF, Ballato J, Schwartz RW, Carroll DL. Appl Phys Lett 2002; 80(18): 3394.
223. Germano C. Clevite Corp. Tech Pap TP-222 July 1961.
224. Germano C. Clevite Corp. Tech Pap TP-237 March 1969.
225. Piezoelectric Ceramics. London: Mullard, 1974.
226. Schafft H. J Audio Eng Soc 1966; 14:258.
227. Schafft H. Ferroelectrics 1976; 10:121.
228. Bost J. J Audio Eng Soc 1980; 28:244.
229. Scott R. Jane's Defense Weekly, Aug. 16, 2001.
230. Desilets C, Callahan M, Hayward G, Maclean C, Mukjerjee B, Murray V, Nikodym L, Pazol B, Sherrit S, Wojcik G. IEEE Ultrasonics Symp 1997:901.
231. Fiore D, Torri R, Gentilman R. Proc. 8th U.S.–Japan Seminar on Diel. and Piezo. Ceram. Plymouth, MA, 1997:344.
232. Borgen M, Washington GN, Kinzel G, presented at the *7th Mechatronics Forum Int. Conference*, 2000.
233. Balakrishnan S, Niezrecki C. C. J Intell Mater Syst Struct 2002; in press.
234. Wells PNT. Rep Prog Phys 1999; 62:671.

235. Ultrasound speeds bone healing. USA Today 1998; 126(2633):10.
236. MEMS: smaller is the next big thing. NASA Tech Briefs August 2001:16.
237. Mukherjee N, Roseman RD. Mater Res Soc Symp Proc 2000; 604:79.
238. Kendall CJ. B.S. thesis, Massachusetts Institute of Technology. 1998.
239. Ohya Y, Takahashi Y, Murata M, Nakagawa Z, Hamano K. J Am Ceram Soc 1999; 82(2):445.
240. Chen G, Chen C. Proc SPIE Smart Struct Mater 2000; 3988:54.
241. Flexible piezoelectric actuators. NASA Tech Briefs July 2002:27.
242. Nitzshe F, Zimcik DG, Ryall TG, Moses RW, Henderson DA. J Guidance Control Dyn 2001; 24(4):855.

5

Ferrite Ceramics

Rong-Fuh Louh
Feng Chia University
Taichung, Taiwan

Thomas G. Reynolds III
Niceville, Florida, U.S.A.

Relva C. Buchanan
University of Cincinnati
Cincinnati, Ohio, U.S.A.

I. INTRODUCTION

The history of ferrites ("ferro"-oxides) began many centuries ago with the discovery of lodestones consisting of the black ore magnetite (Fe_3O_4), which would attract iron. This is believed to have been discovered in ancient Greece around 800 BC. Magnets found their first application in compasses, which were used by the Vikings in the ninth century, or perhaps even earlier. A first milestone in the history of magnetism was the work done by William Gilbert in 1600 describing the magnetic properties of lodestone up to that point in time. It was not until 200 years later that major developments occurred. The new science of electromagnetism was developed through the work of H. C. Oersted, A. M. Ampere, W. E. Weber, M. Faraday, P. Curie, J. C. Maxwell, and many others. Researchers were getting to know the basis of electromagnetic theory in general and the crystal structures of related materials.

During the 1930s research on soft ferrites was continued by V. Kato, T. Takei, and N. Kawai in Japan and J. Snoeck in Netherlands. By the end of World War II, J. Snoeck of the Phillips Research Laboratories succeeded in describing the basic physics and technology of practical ferrite materials. Work done by Snoeck and his colleagues at Phillips Laboratories in the Netherlands led to magnetic ceramics with strong magnetic properties, high electrical resistivity, and low relaxation losses [1,2]. In 1948, Neel announced his celebrated ferrimagnetism theory, describing the basic phenomenon of "spin–spin interaction" taking place in the magnetic sublattices in ferrites. The stage was now set for the development of microwave ferrite devices. The first practical soft-ferrite application was in inductors used in LC filters in frequency division multiplexer equipment. The combination of high resistivity and good magnetic properties made these ferrites an excellent core material for the filter operating at 50–450 kHz. The major booming of the ferrite industry was led by the large-scale introduction of television in the 1950s. In television sets, ferrite cores were used as high-voltage transformers and electron beam deflection yokes. In 1959, J. Smit and H. P. J. Wijn published a comprehensive book on ferrites. Since then, developments have been made on the magnetic characteristics of ferrite materials that have improved microwave device performances. These involve both compositional and processing modifications. New applications of ferrite materials continue to be realized, such as in the cellular phone, medical, and automotive markets.

All magnetic materials can be grouped into three general categories: (1) diamagnetic materials: having relative permeability (μ_r) slightly below 1; (2) paramagnetic materials having μ_r slightly greater than 1; and (3) ferromagnetic and ferrimagnetic materials having μ_r considerably greater than 1. Ferromagnetic materials are metals whereas ferrimagnetic materials are ceramics. The magnetic materials in category 3 mentioned above exhibit the "hysteresis effect," a nonlinear relationship between an applied magnetic field (H) and the magnetic induction (B) of the material. Consequently, they all have properties associated with hysteresis behavior such as magnetic permeability (μ), saturation magnetization (B_s), remanent magnetization (B_r), and coercivity (H_c). Many materials exhibit these important magnetic properties, including elemental metals, transition metal alloys, rare earth alloys, and ceramics.

Among magnetic ceramics, ferrites are the predominant class. Ferrite has a cubic spinel structure with the chemical formula $MOFe_2O_3$, where Fe_2O_3 is iron oxide and MO refers to a combination of two or more divalent metal oxides, e.g., ZnO, NiO, MnO, and CuO. The addition of such metal oxides in various amounts allows the creation of many different materials with properties that can be tailored for a variety of uses. Ferrites vary from silver gray to black. Cores of magnetic ferrites are used to make inductors, transformers, and proximity switches. The electromagnetic properties of ferrite materials can be affected by operating conditions such as temperature, pressure, field strength, frequency, and time. These materials cover a wide range of crystal structures, compositions, and applications. Therefore, ferrite ceramics exhibit certain common properties:

1. They are all oxides.
2. They are all based on Fe_2O_3 as a major compositional component.
3. They exhibit a spontaneous magnetic induction in the absence of an external magnetic field.

The important properties that make these materials useful are their strong magnetic coupling, high resistivity, and low loss characteristics. Since its commercial beginning in the early 1940s, ferrimagnetic technology has advanced at a rapid pace, due in part to the ease of compositional modifications to meet specific applications.

Ferromagnetism in solids is due to the strong coupling or mutual attraction of magnetic dipole moments, produced by electron spin and electron orbits around the nucleus, on the atoms in the solid. This results in a parallel mutual reorientation of the magnetic moments into a common direction, a condition defined as spontaneous magnetization. Ferrimagnetism, like antiferromagnetism, arises from the antiparallel alignment of the magnetic moments of the ions on different sublattices in the crystal. However, in ferrimagnetism, the oppositely directed magnetic moments do not exactly cancel, so that a net magnetic moment results.

The origin of these coupling behaviors may be related to the outer electron interactions of the neighboring ions. Although the bonding in magnetic ceramics (oxides) is predominantly ionic, secondary covalent bonding also occurs between p orbitals of the filled O^{2-} and the unfilled d orbitals of the transition metal cations. This results in a superexchange interaction or spin alignment between the outer electrons, which is responsible for the coupling. This superexchange interaction is characteristic of coupled ions that both have half-filled or more than half-filled d orbitals, and it accounts for keeping magnetic moments parallel or antiparallel for ferromagnetic, ferrimagnetic, and antiferromagnetic behavior. When the neighboring electrons have antiparallel spins, the superexchange interaction allows these electrons to mutually jump to exchange orbitals. The strength of the coupling due to the superexchange interaction diminishes rapidly as the distance between the ions increase. However, the dumbbell shape of the oxygen $2p$ orbital necessitates a cation-oxygen-cation bond angle of 180° for maximal interaction (or coupling) to occur.

A ferromagnetic or ferrimagnetic oxide exhibits its spontaneous magnetization in the form of Weiss domains, which are microscopic regions of the material within which there are magnetic dipoles aligned in the same direction. Weiss domains are separated by domain walls, within which a gradual transition of magnetic moments occurs. Typically, this transition takes place over a 100-nm distance and has a wall energy of ≈ 0.001 J/m^2 [3]. In the absence of an applied field, the domains orient themselves in a way that minimizes the energy of the system, which results in a net magnetization of zero. The energy of a magnetic ceramic is kept to a minimum when the magnetic flux (due to the domains) is kept mostly within its boundaries. This magnetostatic effect has an important bearing on the design of shapes for optimal magnetic applications.

The effect of permanent ceramic magnet depends on a property of the continuous existence of field withstanding different environments. According to Weiss theory, a selected electron spin within a magnetic material will be subjected to the magnetic field due to the surrounding spins. The magnetic field of these spins enhances the influence of the chosen spin, and the resultant magnetic field comes from the exchange between atomic moments. When the strong magnetic field removed, the material doesn't lose the magnetization completely; it retains the magnetization to some extent, remanence (B_r).

Under an applied magnetic field, the Weiss domains of the ferrimagnetic material are responsible for the hysteresis behavior exhibited, the degree of magnetization being dependent on the strength and direction of the applied field. The hysteresis loop characteristic of a ferrite is obtained when it is subjected to an external field (Figure 1). Irreversible changes can result from exposure to very low temperatures, and the magnetic quality is restored only by remagnetization. Starting from a state of zero magnetic induction (usually obtained by heating the material to a temperature above its Curie temperature, T_c), as the material is exposed to an increasing magnetic field of strength (H), the magnetic induction (B) increases linearly. The rate at which the induction increases defines the initial permeability (μ_i).

$$\mu_i = \frac{1}{\mu_0} \lim_{H \to 0} \frac{B}{H} \tag{1}$$

where $\mu_0 = 4\pi \times 10^{-7}$ H/m,

B = magnetic induction, Wb/m^2 (1 Wb/m^2 = 1 tesla = 10^4 gauss), and
H = magnetic field, A/m (1 A/m = 0.126 oersted)

As the field (H) continues to increase, magnetic domains within the ceramic grains rotate to align with it. When the rotational alignment is complete, near the knee of the curve, the domains that are aligned with the field now grow to maintain their lowest free-energy state. At point 2, the material has become "saturated"; any further increase in the magnetic field will not increase the magnetic induction. The value of induction obtained at the maximal field is the saturation magnetization (B_s).

Reducing the magnetic field from this value does not cause a drastic reduction in the induction initially because the energy stored in the domain walls must be overcome. Consequently, even when the external field has been reduced to zero, there remains a significant remanent magnetic induction (B_r), point 3. Applying a magnetic field in the opposite sense causes the domains to switch their magnetic polarity to align themselves with the field. The magnetic field that causes this switching to take place is called the coercive field (H_c). Increasing the field further in the negative direction causes the domains to grow, as they

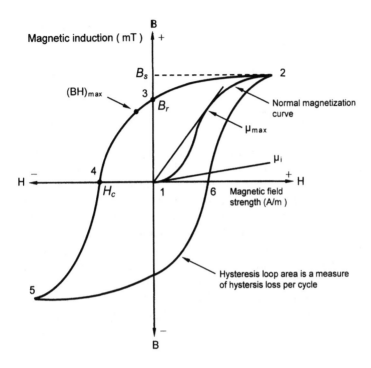

Figure 1 Hysteresis loop behavior of typical ferrites showing important properties. B_r, Permanent polarization; B_s, saturation polarization; H_c, coercive field; μ_i, initial permeability.

store an increasing amount of magnetic energy, until the material is once again saturated. At this point in the third quadrant of the hysteresis loop, the material is in a state completely analogous to the first quadrant. Reversing the field causes the magnetic properties to continue around the hysteresis loop in a counterclockwise direction.

The magnetic properties described by the hysteresis loop (and these are by no means all the magnetic parameters of interest) are highly dependent on both the intrinsic properties of the materials themselves (crystal structure and composition) and on such extrinsic processing parameters as grain size and density. The shape of the hysteresis loop can be defined by the squareness of the B_r/B_s ratio as the ratio of residual flux density to the saturation flux density. The coercive field varies with grain size; therefore, controlling the grain size within certain limits can control the width of the hysteresis loop. Furthermore, the hysteresis loop can be explained by a combination of bowing, unpinning, and displacement processes on the grain boundaries. It is for this reason that the ferrimagnetic oxides commonly known as ferrites play such a key role as electronic materials.

For example, typical magnetic values for the cubic spinel ferrites and the hexagonal magnetoplumbite ferrites are significantly different, as illustrated below:

	H_c (Oe)	μ_i	B_r (Gauss)
Spinel	0.1	100–20,000	400–1200
Magnetoplumbite	4000	1–10	4000

Consequently, these materials find uses in considerably different applications. Spinel ferrites, used as transformer cores in high-frequency power supplies, are materials engineered to have low hysteresis losses (high resistivity) and high saturation magnetization in order to make the transformer as small and efficient as possible. The hexagonal ferrites, on the other hand, are widely used as ceramic magnets, which have three basic sublattices combined in different numbers in a hexagonal structure. In this application the material must have very high coercivity to resist the demagnetizing forces encountered in motor applications. Both of these uses can be accomplished by ferrites, using relatively cheap and abundant Fe_2O_3 as the major raw material. For four decades ferrite components have been used in an ever-expanding range of applications and in steadily increasing quantities. Table 1 is a brief list of design advantages and major applications of soft ferrites.

Table 1   Design Advantages and Major Applications of Soft Ferrites

Design advantages	Major applications
• High resistivity	• Power transformers and chocks for high-frequency power supplies
• Low loss combined with high permeability	• Inductors and tuning transformers for frequency tuners
• High operating frequency range	• Pulse and boardband transformers for matching circuits
• Good temperature and time stability	• Magnetic deflectors for TV sets and monitors
• Wide materials selection	• Recording heads for audit/video/data storage devices
• Versatility of core shapes	• Rotating transformers for video camera recorders
• Low material cost	• Shield beads and chocks for EMI suppression
	• Magnetic transducers for vending machines

Table 2 Summary of Ferrite Structure Types, Typified by Changes in the Fe_2O_3-MeO (or Me_2O_3) Modifier Oxide Ratios

Spinel	1 Fe_2O_3-1 MeO	(where MeO = transition metal oxide)
Garnet	5 Fe_2O_3-3 Me_2O_3	(where Me_2O_3 = rare earth metal oxide)
Magnetoplumbite	6 Fe_2O_3-1 MeO	(where MeO = divalent metal oxide from group IIA—BaO, CaO, SrO)

Ferrites crystallize in a large variety of structures; including spinel, magnetoplumbite, and garnet (Table 2). These three also account for a majority of the technological applications. Common to all these materials is the simple ratio of Fe_2O_3 to the other oxide component (Table 2). The crystal structure of these materials holds the key to their diverse magnetic properties because in each case it defines the interactions between the various ions on an atomic level.

II. SPINEL FERRITES

A. Composition

The spinel ferrites are isostructural with the naturally occurring spinel $MgAl_2O_4$ and conform to the general formula AB_2O_4. The structure was first described by Bragg [4] and Nishikawa [5] in 1915. It is cubic (Figure 2), consisting of eight AB_2O_4 units. The relatively large oxygen anions are arranged in cubic close packing, with the octahedral and tetrahedral interstitial sites occupied by transition metal cations. Of the 64 available tetrahedral sites (A sites), 8 are occupied, and 16 of the 32 octahedral (B sites) are occupied.

Although to a first approximation the anions are considered to be rigid spheres, the presence of cations in the interstitial sites causes the oxygen sublattice to be distorted and the tetrahedrally coordinated sites to be slightly expanded, causing a slight displacement of the oxygen anions along the cell diagonals [3]. Consequently, the spherical radius of the tetrahedral and octahedral sites can be calculated in terms of the oxygen displacement parameter u:

$$r_{tet} = \left(u - \frac{1}{4} \right) a_o \cdot 3 - R_o \quad \text{and}$$

$$r_{oct} = \left(\frac{5}{8} - u \right) a_o - R_o \tag{2}$$

where R_0 is the ionic radius of oxygen and a_0 is the lattice parameter.

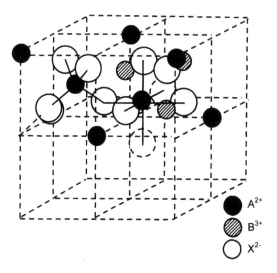

Figure 2 Unit cell for spinel structure with alternating tetrahedral (AO_4) and octahedral (BO_6) coordinated units (four each per unit cell).

For materials typical of the spinel ferrites, R_0 amounts to:

$$0.58 < r_{tet} < 0.67 \text{ Å}$$

$$0.70 < r_{oct} < 0.75 \text{ Å} \tag{3}$$

Within the transition metal series the d-electron structure of the cations changes in a regular and systematic way, whereas the ionic radius remains relatively constant (Table 3). This offers an excellent opportunity to engineer specific magnetic interactions in the crystal lattice by substituting various transition metal cations into the crystal structure, thereby altering the macroscopic magnetic properties.

The AB_2O_4 structure offers many possible combinations of cations, which could balance the -8 charge contribution of the oxygen ions. The most important of these are the ($+2$, $+3$) ferrospinels, in which Fe_2O_3 is a major component. In order to define the location of the cations on the octahedral and tetrahedral sites, the following convention has been adopted:

$$Me^{2+}[Fe_2^{3+}]O_4 \qquad \text{normal spinel} \tag{4}$$

where the ions on the octahedral sites are enclosed in brackets. The ferric ion can occupy either tetrahedral or octahedral sites, depending on what other cations

Table 3 Comparison of the d-Electron Structure, Coordination, and Cation Radii for Transition Series Elements

Element	d^0	d^1	d^2	d^3	d^4	d^5	d^6	d^7	d^8	d^9	d^{10}
Ti	Ti^{4+} (6)[a] 0.0605[b]	Ti^{3+} (6) 0.067	Ti^{2+} (6) 0.086								
V		V^{4+} (6) 0.059	V^{3+} (6) 0.0640	V^{2+} (6) 0.079							
Ta		Ta^{4+} (6) 0.066	Ta^{3+} (6) 0.067								
Nb		Nb^{4+} (6) 0.069	Nb^{5+} (6) 0.070	Nb^{2+} (6) 0.071							
Cr		Cr^{3+} (4) 0.0350	Cr^{4+} (4) 0.044 Cr^{4+} (6) 0.055	Cr^{3+} (6) 0.0615							
Mo		Mo^{5+} (6) 0.063	Mo^{4+} (6) 0.0650	Mo^{3+} (6) 0.067							
Mn				Mn^{4+} (6) 0.0540 HS^d 0.065	Mn^{3+} (6) LS^c 0.058 HS 0.0820	Mn^{2+} (6) LS 0.067					
W		W^{4+} (6) 0.0650									
Fe						Fe^{3+} (4) HS 0.049 Fe^{3+} (6) LS 0.055 HS 0.0645	Fe^{2+} (4) HS 0.063 Fe^{2+} (6) LS 0.061 HS 0.0770				
Co							Co^{3+} (6) LS 0.0525 HS 0.061	Co^{2+} (6) LS 0.065 HS 0.0735			

(Continued)

Table 3 Continued

Element	d^0	d^1	d^2	d^3	d^4	d^5	d^6	d^7	d^8	d^9	d^{10}
Ni								Ni^{3+} (6) LS 0.056 HS 0.060	Ni^{2+} (6) 0.0700		
Cu										Cu^{2+} (6) 0.073	
Cd											Cd^{2+} (4) 0.084 Cd^{2+} (6) 0.095
Zn											Zn^{2+} (4) 0.060 Zn^{2+} (6) 0.0745
Ga											Ga^{3+} (6) 0.0620

[a] Number in parentheses represents the coordination number.
[b] The number is the cation radius in nm for a given valence state.
[c] LS stands for low spin.
[d] HS stands for high spin.
Source: R. D. Shannon and C. T. Prewitt, *Acta Crystallogr.*, 925–946 (1969).

are present, and this results in the inverse spinel:

$$Fe^{3+}[Me^{2+}Fe^{3+}]O_4 \qquad \text{inverse spinel} \qquad (5)$$

As pointed out by Economos [6] and Broese Van Groenou [7], normal spinels are formed when Me^{2+} = Zn or Cd, and inverse spinels when Me^{2+} = Ni, Co, Fe, Mn, or Cu.

The crystal structure can be considered as two magnetic sublattices: an A-site and a B-site lattice. Magnetic superexchange interactions between the octahedrally coordinated B sites and the A sites, through the intervening oxygen anions, cause these two sublattices to be antiferromagnetically coupled. Depending on the type and site preference of the cations, a variety of magnetic properties are possible. These site preferences, related to the octahedral site preference energy (OSPE), are indicated for the common spinel cations in Table 4. The OSPEs have been estimated from a variety of thermodynamic data [8–14]. A relatively large negative OSPE for a given cation in Table 4 indicates a strong octahedral (B) site preference for that cation. Conversely, tetrahedral (A) site preferences are indicated by positive OSPEs. Note the strong A-site preference for Zn^{2+} and Cd^{2+} ions as well as the no-preference indications for Fe^{3+} and Mn^{2+} ions. These site preferences have a major role in the development of ferrites for particular applications, as described below.

The saturation magnetization of the spinel ferrites is readily explained on this basis. For example, magnetite or natural mineral "lodestone," Fe_3O_4, is an

Table 4 Octahedral Site Preference Energy (OSPE) for Cations in Ferrite Spinel

Cation	OSPE (kcal/mole)	Site preference
Al^{3+}	−18.6	Strong B
Cd^{2+}	10	Strong A
Co^{2+}	−7.06	B
Cr^{3+}	−37.7	Strong B
Cu^{2+}	−15.2	Strong B
Fe^{2+}	−4.0	B
Fe^{3+}	∼0.0	A or B
Mg^{2+}	−1.5	B
Mn^{2+}	∼0.0	A or B
Ni^{2+}	−2.06	B
Ti^{4+}	1 ∼ 2	A
V^{3+}	−12.8	B
Zn^{2+}	4	Strong A

Source: Data from Refs. 8–14.

inverse spinel, and using the previous notation,

$$Fe^{3+}[Fe^{2+}Fe^{3+}]O_4 \qquad (6)$$

The d-electron spins between A and B sites are antiferromagnetically ordered,

$$d^5[d^4d^5] \qquad (7)$$

and would indicate a net magnetic moment of four Bohr magnetons; there is good agreement between this value and the value of 4.1 observed at 0 K. The Bohr magneton values (μ_B) for the cations are determined from the net spin of the d-orbital electrons on the different cations, given in Table 3.

Each unpaired electron spin produces a magnetic moment, measured and designated as one Bohr magneton (μ_B), which counts for a value of 9.27 $\times$ 10^{-24} A-m^2. The discrepancies between the calculated and the measured μ_B values of the net magnetic moments are partly due to the fact that the exact locations of the cations are unknown, as in mixed spinel, which have unequal distributions of ions between the A and B sites and partly due to imperfect coupling between electron spins.

Considering the electronic configuration of the outer shells of the elements most important in the structures of ferrites (Fe, Co, Ni, Mn, Zn, Cu, Mg), Fe^{3+}, Mn, and Mn^{2+} have five unpaired electrons in $3d$ orbital; Fe and Fe^{2+} have four unpaired electrons in $3d$ orbital; Co and Co^{2+} have three unpaired electrons in $3d$ orbital; Ni and Ni^{2+} have two unpaired electrons in $3d$ orbital; Cu and Cu^{2+} have one unpaired electron in $4s$ and $3d$ orbital, respectively; while Zn, Zn^{2+}, Mg, and Mg^{2+} have no unpaired electrons. In pure ferromagnetic materials such as metallic Fe, Ni, and Co, the spontaneous alignment of spins results for neighboring ions, with cooperative interaction of electrons necessary for the ferromagnetic effect to be realized.

In spite of the metal cations being separated by oxygen anions and the lack of direct exchange interactions, the spin interactions taking place in the ferrite ceramic of spinel structure (AB_2O_4) arise from cations in spinel interact toward antiparallel alignment of unpaired electron spins. The superexchange interaction model has been proposed between the $3d$ orbitals of metal ions and the $2p$ orbitals of oxygen ions. The interaction is critically dependent on the distribution of Fe ions or other ions with unpaired outer electrons, such as Ni, Co, Mn, and Cu between the tetrahedral (A) and octahedral (B) sites of the oxygen sublattices in the spinel structure. The potential for interaction heavily depends on the type of site in which the cation sits.

For a given ionic separation, the shape of the p orbital of the oxygen ion suggests that the interaction is most prominent when the metal–oxygen bond angle is close to 180°. In the spinel crystal, the order of bond angle increase with the interaction A-O-A (79°), B-O-B (90° and 125°), and A-O-B (126° and 154°); therefore, the superexchange interaction between the tetrahedral and octahedral sites (A-O-B) is strong and spins align, whereas interactions between the A-O-A or B-O-B sites are weak, and no spin alignment results.

In inverse spinel containing iron ions, an equal number of Fe^{3+} occupies both tetrahedral A sites and octahedral B sites to give a strong interaction (A-O-B). This results in antiparallel spin alignment with no net resultant magnetic moment. However, inverse spinels can be doped with divalent oxides (MeO), where Me prefers the octahedral sites such that some of the octahedral (B) sites are occupied by Me^{2+}. The strong (A-O-B) interaction of the Me^{2+} in the octahedral sites with the Fe^{3+} in the tetrahedral sites results in a net uncompensated moment of magnetization. This moment is proportional to the number of unpaired electrons in the Me^{2+}. That is, all the Fe^{3+} alignment is still antiparallel.

The manganese zinc (MnZn) ferrites and nickel zinc (NiZn) ferrites appear to have anomalous behavior of the saturation magnetization—the addition of nonmagnetic zinc ferrite *raises* the saturation magnetization. This behavior can be understood using the same model as for Fe_3O_4.

Zinc ferrite, $ZnFe_2O_4$, is a normal spinel, and as such the unit cell has no net magnetic moment:

$$ZnFe_2O_4$$

$$Zn^{2+}[Fe^{3+}Fe^{3+}]O_4 \tag{8}$$

$$d^0[d^5d^5] \tag{9}$$

In normal spinel containing iron, all the Fe^{2+} ions occupy the tetrahedral A sites, e.g., $FeAl_2O_4$, and all the Fe^{3+} ions occupy the octahedral B sites, e.g., $ZnFe_2O_4$. In both cases, the Al^{3+} and Zn^{2+} have paired spins with $\mu_B = 0$ and the A-B interaction cannot give any spin alignment. In addition, the Fe-Fe interaction through A-A or B-B is too weak to offer any alignment, so that no net magnetic moment results in normal spinels containing nonparamagnetic ions associated with iron ions. Manganese ferrite is approximately 20% an inverse spinel and consequently the two magnetic sublattices are antiferromagnetically aligned.

When the nonmagnetic zinc ion (d^{10}) is substituted into the manganese ferrite lattice, it has a stronger preference for the tetrahedral site than does the ferric ion and thus reduces the amount of Fe^{3+} on the A site. Because of the antiferromagnetic coupling, the net result is an increase in magnetic moment on the B lattice and an increase in saturation magnetization [Figure 3 and Eq. (10)]:

$$Zn_x^{2+}Fe_{1-x}^{3+}[Mn_{1-x}^{2+}Fe_{1+x}^{3+}]O_4 \tag{10}$$

$$(d_x^0 d_{1-x}^5)\,[d^5d^5] = (5 - 5x) + [(5 + 5x) - (5 - 5x)] = 5 + 5x \tag{11}$$

where x is molar fraction of Zn^{2+} ions. However, at high levels of zinc substitution, the A-site magnetic ion becomes so diluted that the coupling between the two lattices is lost and the saturation magnetization drops. An additional consequence of the weakening magnetic interactions is that the spin couplings can more easily be moved out of alignment by thermal energy vibrations, so that the

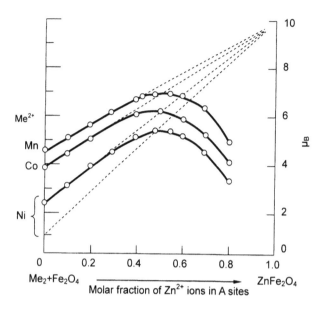

Figure 3 Plot of magnetization versus Zn additions to Ni, Co, and Mn ferrites showing increased magnetization and saturation effects.

temperature at which the magnetic properties disappear (a characteristic temperature T_c called the Curie temperature) decreases with increasing zinc content (Figure 4). Thus, the Curie temperature for manganese ferrites is 450°C decreasing to 250°C with 0.2 mols of zinc ferrite added. The limit of ferrite application is determined by the Curie temperature. With increasing temperature up to approximately 250°C there is little significant loss, but thereafter increasing demagnetization has to be reckoned with. As described above, magnetite (Fe_3O_4) has low permeability and very low resistivity; therefore, better magnetic properties will be found with the "mixed ferrites," i.e., NiZn and MnZn ferrites. In general, NiZn ferrites have low permeability and high resistivity, and are used at higher frequency (5~500 MHz), whereas MnZn ferrites have high permeability and low resistivity, and are used at lower frequency (from 10 kHz to 5 MHz). The stoichiometric ratio of Mn/Zn not only influences the Curie point but also changes the magnetic properties. Figure 5 shows the composition diagram for MnZn ferrites in mol % for MnO, ZnO, and Fe_2O_3, indicating the optimal regions for high-saturation flux density (B_s), high quality factor (Q), and high initial permeability (μ_i), which are in accordance with the three common types of MnZn ferrites, i.e., high-permeability ferrites, low power loss ferrites, and low-drive and low-loss ferrites [16].

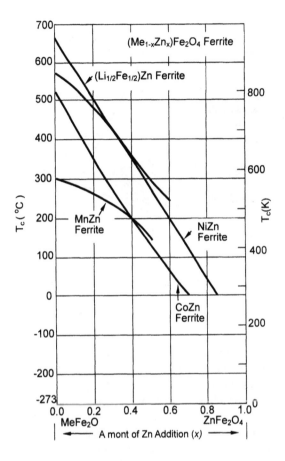

Figure 4 Depression of magnetic Curie temperature (T_c) with Zn additions to Mn, Co, Ni, and Li/Fe ferrites.

B. Structural Anisotropy

Magnets exist in two basic forms: isotropic and anisotropic. Anisotropic magnets have a preferred orientation with a substantially higher energy density, whereas isotropic magnets can be magnetized and set along all axes. To achieve optimal magnetic properties in the spinel ferrites it is necessary to consider the anisotropy energies, which are related to the crystal structure and microstructure of the materials. These anisotropy energies control most of the useful macroscopic properties of magnetic materials. They are, respectively, (1) the magnetocrystalline anisotropy energy, which favors spin alignment in an easy crystallographic direc-

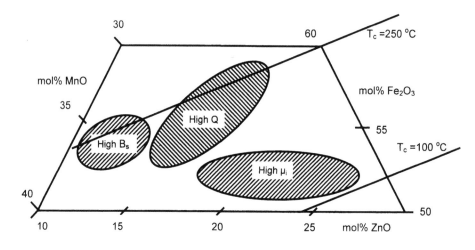

Figure 5   Composition diagram for MnZn ferrites in mol % for MnO, ZnO, and Fe_2O_3.

tion; (2) the magnetostrictive energy, related to strain anisotropy; and (3) the magnetostatic energy, related to shape anisotropy. Toroidal shapes, for example, are often selected for magnetic components because of their ability to minimize the demagnetizing or fringing fields, which would otherwise represent an energy loss to the system. When a direct current flows in an inductor a fixed magnetizing force is maintained in the core (H_{dc}). For a small to medium gap, 85% of the core flux is confined to the cross section of the core face adjoining the gap, and the remaining 15% of the core flux is fringing flux caused by shunting of the gap. The fringing flux decreases the total reluctance of the magnetic path and makes the observed inductance greater than the calculated value. Thus, very large gaps are sometimes broken up into many smaller gaps to reduce fringing.

The magnetocrystalline anisotropy is that part of the crystal energy that is dependent on the direction of magnetization in the lattice. For a cubic material it can be represented by an equation of the type [17–19]:

$$E_A = K_0 + K_1(\alpha_1^2\alpha_2^2 + \alpha_1^2\alpha_3^2 + \alpha_2^2\alpha_3^2) + K_2(\alpha_1^2\alpha_2^2\alpha_3^2) + \cdots \qquad (12)$$

where α_1, α_2, and α_3 are the direction cosines of the magnetization direction with respect to the cubic axes of the crystal, and the K_n are the anisotropy constants, which decrease rapidly with the order n. K_1 indicates the value of minimal energy position, an arbitrary constant, K_2 and K_3 are the anisotropy constants of the [110] and [111] directions, respectively. In cubic systems the easy directions are along the cube edges, if K_1 is positive and K_2 is not more negative than $-9K_1/4$. When

K_1 is negative, the case of nickel, the easy directions are along the cube diagonals. When the easy direction of magnetization is parallel to a cube edge, the anisotropy energy is $E_A = K_0 = 0$, since $\alpha_1 = \alpha_2 = \alpha_3 = 0$, and thus $K_2 = K_3 = 0$. The anisotropy energy for the easy direction if along a face diagonal is $E_A = K_0/4$ since $\alpha_1 = 0$ and $\alpha_2 = \alpha_3 = 1/\sqrt{2}$. The anisotropy energy for the easy direction along a cube diagonal is then equal to $E_A = K_1/3 + K_2/27$ since $\alpha_1 = \alpha_2 = \alpha_3 = 1/\sqrt{3}$. In ferrites the $\langle 100 \rangle$ directions of the cube are easy for magnetization or the directions of minimal anisotropy energy.

The anisotropy constants are different from one material to another and are also temperature dependent. The magnetocrystalline anisotropy is predominantly dependent on K_1, the first-order anisotropy constant, and is strongly influenced by the cations present in the crystal structure [20]. In the manganese zinc ferrite system, for example, there is very little contribution to K_1 by Mn^{2+}, while the Fe^{3+} ion has a slight positive contribution to the anisotropy constant and Fe^{2+} has a strong positive contribution. Because of the temperature dependence of these effects, it is possible to have a net zero anisotropy value at some temperature. That temperature coincides with the well-known secondary maximum in the magnetic permeability (see Figure 15). Hence, the temperature dependence of K value and its magnitude have important consequences with respect to permeability.

A majority of the applications of the spinel ferrites depend on achieving high permeability, low coercive force, and low hysteresis losses. Consequently, the strong effect of divalent iron places stringent requirements on the composition and processing of the spinel ferrites; these are discussed in more detail in Section V.

Ohata [21] has investigated the relationship between magnetocrystalline anisotropy and magnetic permeability in the MnO-ZnO-Fe_2O_3 system and effectively mapped out the regions where highest magnetic permeability can be expected to occur (Figure 6). The work of Stoppels et al. [22,23] extended these investigations to well-characterized single-crystal materials; this work has helped to explain the role of the second-order constant, K_2, on permeability.

Magnetostriction is another anisotropy effect that determines the properties of ferrites. The application of a stress to a magnetostrictive material causes a change in the energy of the crystal lattice and hence in the magnetic properties of that lattice [24,25]. The inverse effect is also possible whereby the application of a magnetic field causes a measurable strain to be induced in the material. Ferrites usually display a small volume change, of the order of 10^{-9}, under the influence of a magnetic field. This is caused by the dependence of the exchange energy on atomic spacing. The length of the crystal changes along the direction of the applied field, increasing with the applied field until saturation. This phenomenon is called magnetostriction and is due to the gradual orientation of a domains under the influence of the applied field. Thus, the magnetomechanical

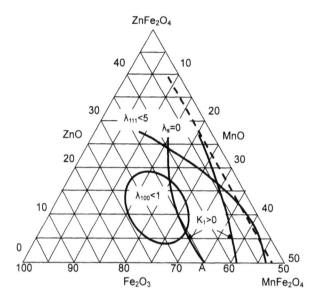

Figure 6 Composition field showing magnetocrystalline anisotropy, magnetostriction, and magnetic permeability regions in the MnO-ZnO-Fe$_2$O$_3$ ferrite system. (λ_{hkl} values are the magnetostrictive coefficients along direction [hkl].)

relationship of a crystal can be expressed in terms of strain tensor coefficients U_{ij} and magnetostriction coefficients λ_{khl}, along the direction [hkl]. For instance:

$$U_{11} = \frac{3}{2}\alpha_1^2 \lambda_{100} \qquad U_{22} = \frac{3}{2}\alpha_2^2 \lambda_{100} \qquad U_{33} = \frac{3}{2}\alpha_3^2 \lambda_{100}$$

$$U_{12} = \frac{3}{2}\alpha_1\alpha_2 \lambda_{111} \qquad U_{23} = \frac{3}{2}\alpha_2\alpha_3 \lambda_{111} \qquad U_{13} = \frac{3}{2}\alpha_1\alpha_3 \lambda_{111} \qquad (13)$$

For polycrystalline materials, average saturation magnetostriction, λ_s, is equal to

$$\lambda_s = \frac{2}{5}\lambda_{100} + \frac{3}{5}\lambda_{111} \qquad (14)$$

This magnetostrictive property can be directly applied to transducers by transferring the electrical energy into mechanical energy or vice versa, for the many common applications such as ultrasonic machining (drilling, welding, and cleaning), strain gauge, speedometer, mechanical filter, etc. The magnetostrictive or "piezomagnetic" ferrites are nickel ferrite-based, such as NiZn, NiCu, NiMg, and NiCo systems. However, NiZn ferrite is typically used because it has a high resistivity and is suitable for high-efficiency components at high frequency. The main consequence of magnetostriction in ferrites is to render them sensitive to stress. The observed anisotropy behavior in highly textured hexaferrites with a high-frequency magnetic field applied in the direction where fixed magnetic moments exist, can be explained by the presence of typical Bloch walls, which contribute to the increase in initial magnetic permeability. This behavior is similar to that of soft ferrite structures with positive magnetostriction.

Prior to magnetization, each grain of ferrite is divided into two domains separated by a Bloch wall. The domain walls become unpinned and displaced within the grains for $H > H_{cr}$ and cause a visible increase in magnetization [26]. To attain a single domain oriented along the field direction, a rotation mechanism to form single-domain grains in the permanent ferrite magnets must occur.

The magnetic permeability may drop by 20–50% if they are handled roughly or dropped while being wound or assembled. Many ferrite components are coated to improve insulation resistance, and the shrinking of this coating can seriously alter the magnetic performance. Finally, the use of ferrite recording heads in such applications as audiotape recorders and video cameras requires a material with low magnetostriction to eliminate stress-induced noise in the signal. Fortunately, those areas where the magnetocrystalline anisotropy is low also correspond to low magnetostriction.

III. HEXAGONAL FERRITES

Most commonly, the M-type ferrite, e.g., $BaFe_{12}O_{19}$, is widely used as permanent magnets and as the magnetic strips on credit and debit cards. A number of hexagonal compounds exist in the general compositional field BaO-MeO-Fe_2O_3 (Figure 7). These are designated M, W, Y, and Z and correspond to $(BaO + MeO)/Fe_2O_3$ ratios of 1:6, 3:8, 4:6, and 5:12, respectively [27]. The compound of greatest technological interest is the M-hexagonal ferrite, $BaFe_{12}O_{19}$, which is isostructural with the naturally occurring magnetoplumbite, $PbFe_{7.5}Mn_{3.5}Al_{0.5}Ti_{0.5}O_{19}$ [28]. The structure is represented in Figure 8 and is considerably more complex than spinel, being built up of oxygen layers that have both cubic and hexagonal close packing [29]. The coordination environment of the cations is complicated and the competing, magnetically opposing moments of the Fe^{3+} ions result in a saturation magnetization of ≈ 4000 gauss [30]. Bloch wall displacement in multidomain structures dominates the magnetic properties. Domain wall movement also limits the quality of magnetic materials. The most important hexagonal ferrite (magnetoplumbite) is based on $BaO: 6Fe_2O_3$. The unit cell is elongated along the mechanically active c axis, where Ba^{2+} and O^{2-} ions, of similar size,

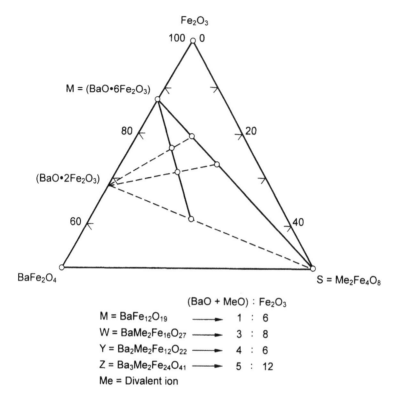

Figure 7 Hexagonal ferrite compositions in the field of the BaO-MeO-Fe_2O_3 system.

to form a closed-packed lattice with interstitial sites occupied by Fe^{3+} ions. The lattice site of Ba^{2+} can be replaced by Sr, Ca, and Pb ions. [31].

Hexaferrite is one of the three families of microwave ferrite materials. Hexaferrites exhibit very broad line widths and use is generally limited to millimetric applications. Strontium or barium hexaferrite exhibits good balance of magnetic strength as well as resistance to demagnetization. The raw materials used are strontium or barium carbonate and iron oxide, which are readily available and in expensive. As ceramic is very hard and brittle, these magnets are formed to near-net shape with final critical dimensions achieved by wet grinding using diamond abrasive wheels. They are the most widely used permanent magnets. As a result of the hard ferrite's high degree of brittleness, when screwing down or adding reinforcing impact stress should be avoided. When integrating hard ferrite with other materials, different types of expansion have to be taken into consideration. Magnetized magnets attract iron swarf and dust. The magnets,

(a)

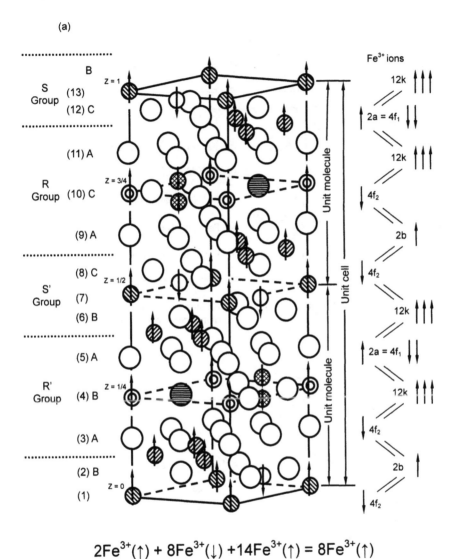

$$2Fe^{3+}(\uparrow) + 8Fe^{3+}(\downarrow) + 14Fe^{3+}(\uparrow) = 8Fe^{3+}(\uparrow)$$

Figure 8 Unit cell structure of M-hexagonal ferrite ($M^{2+}Fe_{12}O_{19}$). Stacking of oxygen layers in the C (magnetically active) direction shows both cubic and hexagonal close packing [32].

(b)

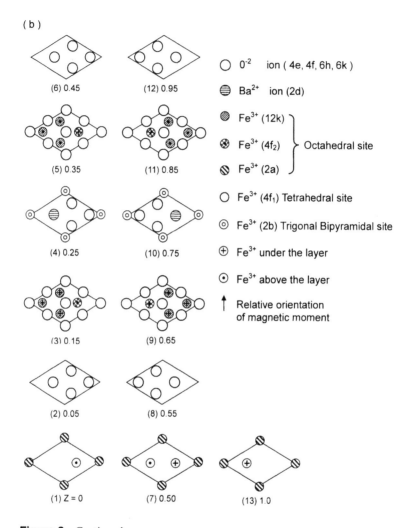

Figure 8 Continued.

therefore, should preferably be magnetized after mounting. The hard ferrites are often used to produce "plastic magnets" by embedding the ferrite powder in a flexible plastic matrix.

The magnetocrystalline anisotropy of these materials is high, making them ideally suited as permanent magnets [32]. The performance of a hard magnet depends on its microstructure. Grain size and the degree of orientation of the c axis determine the remanent magnetization and coercive field. It is critical for permanent magnet materials to have the final grain size of the order of 1 μm. This critical domain size prevents formation of domain walls that could cause demagnetization by domain wall movement. Thus, the grain size of hexaferrites should be kept sufficiently small (<1.6 μm for barium ferrite) to ensure a microstructure with single domain in order to make demagnetization difficult.

IV. GARNET

In 1952, C. L. Hogan from Bell Labs made the first nonreciprocal microwave device at 9 GHz that was based on the Faraday rotation effect. Gyromagnetic resonance line width has been a measure of magnetic losses in a ferrite material. This is the fundamental property that enables microwave ferrites to exhibit nonreciprocal properties at microwave frequencies and thereby enable isolation to be achieved in circulators and isolators. Various cation substitutions were employed to improve the properties of the spinel ferrite materials. This modified the magnetic properties for different frequency ranges, power requirements, and phase shift applications. Soft ferrite can be further classified into "nonmicrowave ferrites" for frequencies from audio to 500 MHz and "microwave ferrites" for frequencies from 100 MHz to 500 GHz. Garnet is one of the three "families" of microwave ferrite materials.

Based on $Y_3Fe_5O_{12}$ (YIG) the magnetic and electric properties can be tailored by substitutions of metallic ions for either Fe or Y. Garnet exhibits the lowest losses of the ferrite families. Yttrium iron garnets are used as electromagnetic radiation waveguides and phase shifters. The high-anisotropy fields have been utilized in microwave ferrite devices in the millimeter range. In 1956, Neel, Bertaut, Forrat, and Pauthenet discovered the garnet ferrite class of materials. YIG is prototypical of the rare earth ferromagnetic insulators. The material is cubic, like spinel, and is isostructural with the mineral garnet, $Ca_3Al_2(SiO_4)_3$, the structure having been determined by Bertaut and Forrat [33] and Geller and Gilleo [34] (Figure 9). The lattice constant (12.376 Å) is significantly larger than that of spinel (8.41 Å is typical for nickel zinc ferrite) and the unit cell contains eight formula units or 160 atoms. The coordination of the cations is considerably more complex, with the metal cation located in tetrahedral, octahedral, and dodecahedral sites. The structure has been investigated in detail, and a number of workers have reviewed known compounds and the crystal chemistry [35,36]. The garnet structure has three sublattices and is also referred to as rare earth iron garnets. These materials, although having a magnetization lower than that of spinel ferrite, possess extremely low ferromagnetic line widths. The tetrahedral sites, of which there are 24, are the smallest and are normally occupied by Fe^{3+} ions; the octahedral sites number 16 per unit cell and are occupied by Fe^{3+}; the 24 dodecahedral sites in the structure have the largest space and are occupied by Y^{3+} ions:

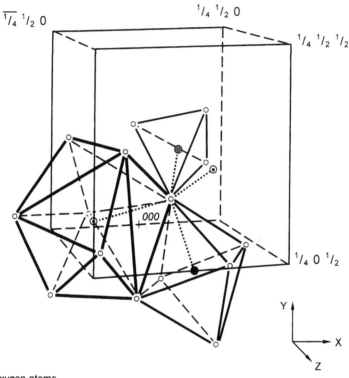

○ Oxygen atoms
● Octahedral sites Fe^{3+}(a) at $(0\ 0\ ^1/_2)$
◉ Tetrahedral sites Fe^{3+}(d) at $(0\ ^1/_4\ ^3/_8)$
◉ Dodecahedral sites Y^{3+} at $(\overline{^1/_4}\ ^1/_8\ ^1/_2)$ and at $(0\ ^1/_4\ ^5/_8)$

Figure 9   Unit cell structure of the garnet ferrite (YIG; $Y_3Fe_5O_{12}$) showing cation coordination structures [29].

$$\{Re_3^{3+}\}\quad [Fe_2^{3+}]\quad (Fe_3^{3+})O_{12}$$

$$\begin{array}{cccc} (c) & (a) & (d) & \text{Crystal site positions} \\ CN=12 & CN=6 & CN=4 & \hfill (15) \end{array}$$

Dodecahedral   Octahedral   Tetrahedral

As is the case with the spinels, a wide range of transition metal cation substitutions are possible, and rare earth ions may be substituted on the octahedral and dodecahedral sites [37]. Consequently, the saturation magnetization, magnetocrystalline anisotropy, and other magnetic properties may be altered over a wide range.

V. PROCESSING

A. Powder Preparation

A wide variety of processing methods can be used to prepare ferrite powders, Economos [6], Takada [38], Robbins [39], and Goldman [40] have described a number of chemical precipitation methods from low-cost chloride and carbonate precursors. The usual synthetic approach is to start with a solution containing the appropriate metal ion sulfates and to precipitate a ferrite intermediate using hydroxide or carbonate as the precipitating agent. The precipitation is carried out at pH $10\sim12$ at an elevated solution temperature; often air is bubbled through the solution and the precipitate to encourage oxidation. Repeated washing of the precipitate may be necessary to remove trace amounts of sulfate, which can have detrimental effects on the sinterability and magnetic properties of the powder. The resulting powder is finely divided, $0.05\sim1.0$ μm, and chemically homogeneous. However, this method has found application only in certain specialized areas because of higher manufacturing costs and the necessity of disposing of large quantities of aqueous solutions.

Schnettler and Johnson [41] investigated the possibility of preparing finely divided, highly reactive powders by a freeze-drying technique, whereby an aqueous salt solution containing the proper metal cations is sprayed into chilled hexane. The frozen pellets are then freeze dried, and the resultant anhydrous salt is reacted in a furnace to yield a ferrite powder. The anhydrous powder is highly reactive and forms the ferrite phase at temperatures substantially below those required for solid-state reaction. The method, while yielding high-purity reactive material, is not readily adaptable for scale-up to commercial quantities. These and other preparative methods are quite useful in the preparation of very pure materials, which are instrumental in the study of the structure and magnetic properties of ferrites; however, standard ceramic processing (Figure 10) is the predominant technology for most of the ferrite industry.

Initially, the raw materials are assayed for purity. For the spinel ferrites specific attention is paid to the silica content of the Fe_2O_3 (it should be less than 200 ppm and, for the highest-grade materials, less than 100 ppm) because Fe_2O_3 constitutes approximately 65% of the composition. The usual source of iron oxide is chemically precipitated powder prepared from sulfate solutions or spray-roasted iron oxides prepared from HCl-based pickle liquors. These spray-roasted oxides are considerably less expensive but contain a higher level of SiO_2 (approximately 200 ppm) because of the silica in the steel strip being pickled. Typically, spray-roasted oxides find application in entertainment ferrites, in ferrites for trans-

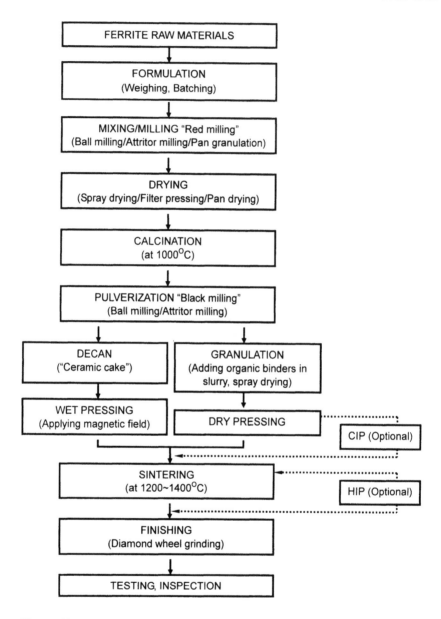

Figure 10 Flowing chart of standard ceramic processing of ferrite components.

formers and power supplies, and in the hexagonal ferrites, also known as hexaferrites, used for ceramic magnets.

Around the 1950s, Takei and Sugimoto discovered that small amounts of dopants could significantly improve the magnetic properties of ferrites by affecting the grain growth and grain density [42]. Since then, numerous studies have shown that such properties as power loss, electric resistance, grain growth, and permeability can be improved by the addition of dopants, such as CaO, CoO, B_2O_3, SiO_2, TiO_2, V_2O_5, Ta_2O_5, and Nb_2O_5. For example, gel processing has been used to make controlled, homogeneous oxide additives to the ferrites [43]. Niobium doping has been used to alter electrical properties in nonferrite applications as well, such as facilitating domain switching in lead-zirconate-titanate (PZT) films, or enhancing the permittivity of Nb-doped bismuth titanate. The role of oxygen mobility in connection with niobium is important. Adding niobium to ferrites is effective in reducing power loss at high frequencies, and in increasing the permeability over a wide range of frequencies. Accordingly, small amounts of Nb_2O_5 (0.01–0.08 wt%) in transformer MnZn ferrites reduces power loss in the high-frequency range. The addition of Niobium oxide likely impacts the magnetic properties by altering the grain boundaries [44]; however, only high-purity Nb compounds can be used to achieve such an effect. Nb oxide–containing ferrites, which are used as cores for line filters, have shown an improved permeability and superior frequency dependence at frequencies of 100–500 kHz [45]. Improved power loss and high-frequency performance have been realized in Nb-containing ferrites in the MnZn-ferrite system [46]. As a result, the size of power supply transformers can be reduced, and they can be used over a wider temperature range.

Since the full effectiveness of Nb doping usually takes place within a narrow composition window, it is important to distribute the Nb additive evenly in the ceramic slurry. Otherwise, islands of various niobium concentrations may evolve, thus distorting the magnetic properties of the device. One way to improve the Nb dispersion within the ceramic slurry is by using a soluble aqueous solution of niobium oxalate, which gives superior blending results. Substantial precipitation of niobium hydroxide occurs by adding ammonia in the solution to adjust pH to > 6. Heating niobium oxalate in an oxidizing atmosphere leads to its decomposition and to the formation of niobium pentoxide around 600°C [47]. The aqueous route offers advantages such as no use of organic solvents, acceptable solution stability against hydrolysis, and easy handling. Due to its acidic and oxidizing properties, niobium oxide can enhance various catalytic reactions such as selective oxidation, hydrocarbon conversion, and carbon monoxide hydrogenation. In the case of ferrites, the use of Nb oxalate allows the preparation of homogeneous mixtures on an atomic scale.

Conventional synthesis of ferrites involves multiple grinding, heating and cooling cycles of suitable precursor materials, in which reactions require a longer time mainly because interdiffusion in solids is slow, even at high temperatures. In contrast, the kinetics of self-propagating, high-temperature synthesis (SHS) reaction, in the combustion process demonstrated by Q. A. Pankhurst et al. results in the formation of various kinds of ferrite powders in a very short time [48].

The SHS reactions proceed by a combustion wave that moves away from the source of ignition with uniform velocity. Most of the chemical reaction occurs in the short-lived molten reaction zone created by the propagating shock wave. By applying a magnetic field during the reaction, the wavefront of molten zone travels faster through the mixture, effectively promoting the formation of short-lived intermediate phases. The final products can have very different properties compared to those made either by zero magnetic field SHS or by conventional methods. For example, a field of 1.1 T produces a 20% decrease in coercivity in $BaFe_{12}O_{19}$ and magnetization increases of 15% in $MgFe_2O_4$ and 35% in $Mg_{0.5}Zn_{0.5}Fe_2O_4$. With the SHS method, changes in the crystal structure of the ferrite can occur, due to changes in the site distribution of iron and lithium atoms in $Li_{0.5}Fe_{2.5}O_4$, in the ratio of tetragonal to cubic phase in $CuFe_2O_4$, and to the spinel inversion parameter in $MgFe_2O_4$. A field-induced flow of ions and electrons at the wave front or field-induced texturing of the starting mixtures are two possible key factors in the process.

Following weighing, the mixture of oxides, chlorides, or carbonates is mixed to yield a homogeneous powder. Mixing is either done dry or, more usually, by wet milling in a steel mill with steel balls. This process is commonly referred to as red milling because the oxide mixture has a characteristic red color. Two types of ball mills are used in ferrite processing: ball mills and attritor mills. Ball mills are large cylindrical mills, which are filled with steel balls, powder, and water and are rotated about their cylindrical axis so that the balls fall by gravity and impact on the powder particles. These mills are inexpensive to operate and maintain, and are available in a wide range of sizes, from 1-kg mills that rest on rotating rollers to mills that will accommodate a 2500-kg charge of powder.

Attritor mills are filled with smaller steel balls. The milling action is caused by a steel rod with "lifters" that stirs the balls, causing them to lift up and fall back. The milling action is much more intensive because the force restoring the media downward is the weight of all the media above it. Because of the higher impact energy it is possible to use smaller diameter media, and it is usual that attritor mills produce a powder with a smaller particle size distribution. Also, because the mill itself does not rotate, it is possible to use the attritor in a semicontinuous fashion by pumping slurry and an oxide mixture through it. However, attritor mills, require more power and are more complicated to run and maintain. Following this first (red) milling, the water is removed from the slurry by a high-capacity filter press or by spray drying.

Because of the energy costs associated with red milling and the subsequent removal of the water, for less demanding technical grades of ferrite the red milling step is omitted completely. In that case the raw materials are weighed and mixed dry (or with a very low percentage of water). Mixing can be done by simple pan granulation, or if it is desired that the material be compacted slightly to enhance the solid-state reaction, it can be passed between rollers to form thin sheets or extruded to form small pellets. The hexagonal ferrites are weighed on a semicontinuous basis and dry mixed, granulated, and fed by conveyor belt or bucket belt to direct-fired rotary kilns.

Following this first mixing step, the raw materials are reacted to form the ferrite phase; this operation is called calcining or presintering. The reaction temperature is typically 700–1000°C in air. As is the case with milling, there are various ways in which calcining can take place. The most common is for the semidry powder to be fed into a gas-fired rotating cylindrical kiln. Smaller quantities can be reacted in ceramic boxes in batch kilns. The calcining reaction itself is critical to the future properties of the ferrite because it determines the reactivity of the powder and the final shrinkage in the subsequent sintering step. While NiZn ferrites are reacted almost 100% to the spinel phase during this step, the MnZn ferrites are not fully converted to spinel (only 40~65% spinel).

The presintering of hexagonal materials takes place at a higher temperature (100~1250°C) than that for the spinels. In this case it is important to convert the oxides 100% to the hexagonal phase so that there is little, if any, reaction sintering with subsequent grain growth taking place when the parts are fired. It is necessary to keep the high-heat soaking time of the kiln short (≤ 30 min) so that the hexagonal particles remain as small as possible, thus minimizing the energy required to mill them. To further enhance millability, the hot pellets exiting the kiln are quenched to induce stresses and microcracks in them.

The result of the calcining process is a powder that is lightly sintered into friable lumps. This material must be milled again to further homogenize its composition and improve reactivity by having a uniform small particle size. Milling is done again in ball or attritor mills with water as the carrier. The so-called black-milling step is monitored or controlled by measuring the particle size or surface area of the powder. Extended milling beyond the desired particle size should be avoided because it introduces a small amount of ultrafine particles, which can nucleate nonuniform grain growth during the sintering stage and cause a degradation of the magnetic properties. This is especially important in the M-hexagonal materials, where the final grain size target is 1–2 μm.

During the final milling the chemical composition is carefully analyzed and additions are made to bring it to specified values. For example, in microwave ferrites of the type $Y_3Fe_{5(1-\Delta)}O_{12}$, it has been shown that Δ must be controlled to ± 0.002 to achieve optimal properties [49]. At this time, permanent-magnet materials are doped with small amounts of oxide mainly CaO, and SiO_2. These additives combine to form a secondary phase during sintering, which retards grain growth [50]. Near the end of the black-milling cycle of the soft ferrites, one or several organic binders are added to the mill [51]. Additives function as binders, as lubricants; or, as in the case of niobium, to enhance certain magnetic and electrical properties. Prior to forming, the mixture is granulated by spray drying, improving the powder's flowability. This multicomponent binder system provides several functions as (1) a dispersant that enhances milling and improves the solids-to-liquid ratio, thereby reducing costs and increasing throughput; (2) a binder and plasticizer to increase green strength in the pressed parts; and (3) a lubricant to help reduce die wall friction during pressing. It has been pointed out by Ghate [50] that the binder system itself can play a major role in determining the final magnetic properties, especially of high-permeability materials. One such system in use is the following [52]:

0.4 wt% polyethylene imine (Polymin)
0.4–1.6 wt% polyvinyl alcohol (PVA)
10.2–1.0 wt% polyvinylpyrrolidone (PVP)
0.03–0.1 wt% methylcellulose (MC)

This binder offers the advantage of being clean burn-off, leaving no residue in the ceramic, and allowing high permeability levels to be obtained. The ceramic slurry, including binders, is spray dried to yield a free-flowing powder that can be dry pressed into a wide variety of popular shapes such as toroids, rods, tubes, stripes, pot cores, E cores, U cores, ETD cores, RM cores, EP cores, balun cores, EMI cores, and planar cores, among others (Figure 11), of which typical applications include inductors, filters, delay lines, and transformers.

The design of core shape selection should take the following factors into consideration: type of application, power level, temperature rise and heat dissipation, safety and shielding requirements, space availability, ease of winding and assembly, and inductance requirement. Temperature variation can result in both reversible and irreversible changes in magnetization. As temperature rises above ambient, induction (B) will decrease at a rate of about $-0.2\%/°C$. As temperature rises, a ferrite magnet will increase in coercivity at a rate of about $+0.27\%/°C$. Currently major shape development trends relate to low-profile and surface mount device applications. Low-profile cores containing windings can be inserted through the printed circuit board (PCB), while surface mount design involves simple as well as complex approaches. Small beads can be connected with a simple formed conductor. Conventional through-hole bobbins for mated cores can be tooled with surface mount terminations so that the magnetic design can be the same.

Isotropic ferrite is produced by a dry pressing technique. Ferrite magnets are formed in a dry press with dedicated, multicavity dies followed by high-temperature sintering. Hard ferrite must be magnetized in the direction of orientation, which is the same as the direction of pressing. As a result of the arbitrary alignment of ferrite crystals, magnetization in three dimensions is possible. Although some ceramic magnets are dry pressed in an orienting magnetic field, their magnetic properties are typically low. So as to optimize the shape and magnetocrystalline anisotropy, most ceramic magnets are wet pressed in the pres-

Ferrite Geometry	Type	Description	Uses
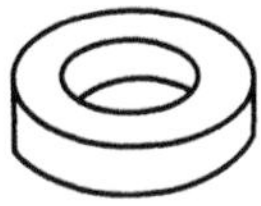	Toroid	Very good magnetic shielding, low material cost but very high assembly cost	Boardband and pulse transformers, common mode chokes
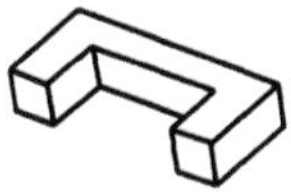	U Core	External magnetic field generated close to the exposing winding, low cost	Power transformers and inductors
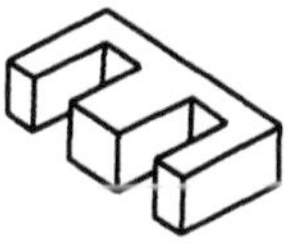	E Core	External magnetic field generated close to the exposing winding, de facto standard, low cost	Power transformers and inductors
	RM Core	Good magnetic shielding, IEC core standard, high cost	High Q inductors and tuned transformers
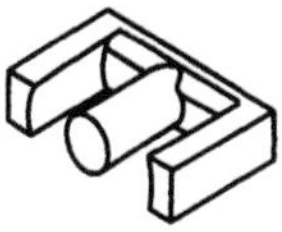	ETD Core	External magnetic field generated close to the exposing winding, IEC core standard, very low cost	Power transformers and inductors
	Pot Core	Very good magnetic shielding, IEC core standard, high cost	High Q inductors and tuned transformers

Figure 11 Various types of ferrite cores with different shapes.

Ferrite Geometry	Type	Description	Uses
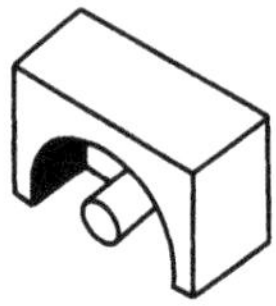	EP Core	Very good magnetic shielding, de facto standard, high cost	Power transformers
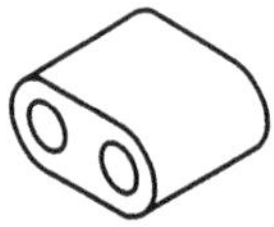	Balun Core	Excellent magnetic shielding, high cost	Boardband and pulse transformers, common mode chokes
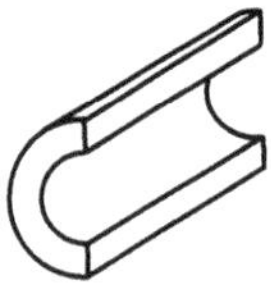	EMI Core	Good magnetic shielding, average cost	Cable EMI suppression
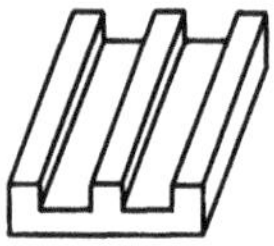	Planar Core	Good magnetic shielding, average cost	Low profile SMD of power transformers
	Rods	Very poor magnetic shielding, very low cost	Antennas and high frequency welding
	Tubes		
	Stripes		

Figure 11 Continued.

ence of a field. Ferrite crystals are then aligned parallel to one another under the influence of a strong magnetic field. The green density of dry-pressed bodies may be increased by dry-bag cold isostatic pressing (CIP), where the green body is placed in thick rubber mold, with an external pressure of 10~30 ksi subsequently applied. Rather than having a binder added and then be spray dried, the black-milled ceramic magnet slurry is allowed to stand in decanting tanks for several days and the excess water is drawn off. The resultant slurry, with approximately the consistency of yogurt, is pumped directly to the presses and injected into the die cavity. Once the die cavity is closed, a magnetic field is applied and the particles align themselves with the field because of the water present in the slurry. After alignment is complete the water is removed by vacuum filtration through holes in the die itself (the inside of the die cavity is lined with filter paper) and the parts are removed. Since there is no binder to impart strength to these wet-pressed parts, they must be handled and stored gently. The parts must be dried for a period ranging from 2–3 days to several weeks to reduce the moisture content from 10–12% to 2–4% before they can be sintered. The ferrite magnets with long, small cross-section such as rods or tubes may be formed by extrusion method.

B. Firing

The firing of the ferrites can be divided into two cases:

1. Materials that are prepared in the fully oxidized state, e.g., nickel zinc ferrites, hexagonal (magnetoplumbite) ferrites, and garnets.
2. Materials in which it is necessary to control the valence state of the cations, typically the MnZn ferrites

The nickel zinc (NiZn) ferrite, $Ni_{1-x}Zn_xFe_{2-y}O_4$, is formulated with a slight deficiency of iron ($0 < y < 0.025$), and the firing is done in air or oxygen at a temperature of 1100–1350°C to ensure that all the cations are in their highest oxidation state. These compounds are characterized by very high resistivity ($>$ 10^7 Ω-cm), relatively low permeability, and low saturation magnetization. Because of their high resistivity, they are used primarily in high-frequency applications. One additional feature of NiZn ferrite is that the material can be sintered to a dense ($>$99.8%), pore-free ceramic. This makes it ideally suited for the fabrication of magnetic recording heads for computer and audio applications. During the sintering cycle the initial ceramic grains grow from 0.5–1.0 μm to 10–25 μm.

The ferromagnetic garnets must also be sintered in air or oxygen to ensure that all iron is in the trivalent state; otherwise, operating at frequencies in the range 100 MHz to 500 GHz they would exhibit excessive dielectric losses.

The hexagonal M-type ferrites, $SrFe_{12}O_{19}$, are also fired in an oxidizing atmosphere. However, in this case the sintering cycle is optimized to yield a dense material with minimal grain growth. For use as a ceramic magnet it is necessary to keep the grain size as small as possible ($<1{\sim}2$ μm) so that the grains are single domain and the high coercive force necessary to prevent demagnetization is not degraded. The firing of hexagonal ferrites usually takes place in large gas-fired car kilns, with the ceramic parts stacked on ceramic tiles or placed in ceramic boxes. Although the sintering cycle may be as long as 24 h, the time at peak temperature (1350°C) may be as short as 30 min to minimize grain growth. The structure of crystal lattices and sublattices, defining the intrinsic properties, is often interrupted by grain boundaries, which in turn strongly influence the bulk properties. The grain boundary mobility during sintering and grain growth must be adapted to the desired microstructure and the resistivity must be as high as possible in order to decrease the eddy current loss. On route to improve the magnetic properties of MnZn ferrite, close investigation of grain boundaries on the nano scale is of most concern.

The MnZn ferrites, $Mn_{1-x}Zn_xFe_{2+y}O_4$, are formulated with an excess of iron ($0.05 < y < 0.20$). This additional iron permits compensation of the magneto-crystalline anisotropy to a net value of zero by controlling the amount of Fe^{2+} present in the spinel structure. As a result, the resistivity of MnZn ferrite is much lower due to the simultaneous presence of Fe^{2+} and Fe^{3+} ions, but the permeability and saturation magnetization are much higher than in the NiZn series (see Table 6). Consequently, the majority of the applications of MnZn ferrites occur in the frequency range 100 kHz to 3 MHz.

The precise control of the Fe^{2+}/Fe^{3+} ratio (typically 0.05: 2.00) during the sintering of large quantities of ceramic parts represents a major technological challenge in the production of ferrites. A number of workers have studied the gas–solid equilibrium [53,54] between ferrites and the atmosphere in which they are fired [55–58] and the existence boundaries of the various phases are well defined (Figure 12). Another way to visualize the interaction with the atmosphere is through an Arrhenius-type plot of the relationship between temperature, atmosphere, and the boundaries between the various phases (Figure 13). In this instance the phase boundaries approximate straight lines with slopes defined by $-C_1$; hence the atmosphere–temperature relationship can be expressed as

$$\log Po_2 = \frac{-C_1}{T} + C_2 \qquad (16)$$

In theory, then, to control the amount of divalent iron, and hence the temperature at which there is zero magnetocrystalline anisotropy, it is necessary for the sintering to follow a specific equilibrium line (defined by C_1 and C_2) throughout the firing and cooling down of the product. In practice, however, ferrites are conveyed and fired in tunnel kilns. These kilns are rendered relatively gas tight by doors or gates at each end, which open and close to permit one ceramic tile load of parts to enter at a time. Atmosphere regulation is accomplished by introducing nitrogen gas into the cooling portion of the kiln through entrance ports. The result is a stepwise reduction in the amount of oxygen from the 5–21% present at the sintering temperature (1250–1450°C) to 100–50 ppm at an annealing temperature of 1050–1200°C (Figure 14). Poorly controlled initial stage of heating or the last stage of cooling may introduce thermal stress in the fired materials. In addition, improper control, to the extent of 10–100 ppm O_2 during the very end of cooling stage for MnZn ferrite, will be deleterious to the properties.

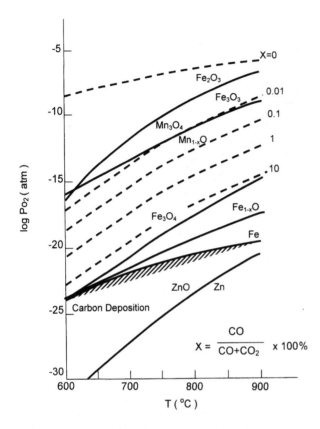

Figure 12 Gas–solid equilibrium phases as a function of oxygen partial pressure (P_{O_2}) and temperature for ferrite oxides.

At this elevated temperature, equilibrium with the ambient atmosphere occurs quickly, which means that the annealing zone can be relatively short before the ferrite parts are pushed out through the cooling portion of the kiln. As the temperature drops, the reaction kinetics between the ferrite and the atmosphere also decline rapidly because the equilibrium atmosphere required in this low-temperature portion of the kiln cannot be achieved in practice.

Small laboratory and production batch kilns with working volumes of 0.7–1.0 m^3 have made it possible to approximate atmosphere control along a C_1–C_2 curve using computer-controlled mixing of O_2–N_2 gases. With a total cycle time of 8–24 h, these curves offer the potential for firing parts individually tailored for specific properties. It is also possible, as pointed out by Plaskett et al. [58], to use a dynamically varying gas–gas buffer system (CO–CO_2) to control

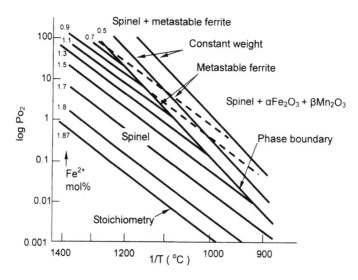

Figure 13 Relationships between oxygen partial pressure P_{O_2}, temperature, and phase stability in spinel ferrites.

the atmosphere; however, this has not found an application in production equipment. It is also possible to "pack" the ferrite parts to be fired in a coarse, fully reacted powder of the same composition, the high specific surface area powder acting as a shield or atmosphere source for the parts under sintering process. For most of magnetic toroids, the sintered pieces after tumbling need the coating process to enhance dielectric strength, reduce edge chips, and provide a smooth winding surface. Inappropriate coating steps and poor material selection will degrade the core performance. The available coating materials include nylon, epoxy paint, and parylene.

Depending on the specific ferrite, the linear shrinkage after sintering can range from 10%–17%. Maintaining correct dimensional tolerances as well as the prevention of cracking and warpage related to this shrinkage are essential concerns of the manufacturing process. The mechanical and electromagnetic properties of the ferrite are heavily affected by the sintering process, which is time–temperature–atmosphere dependent. Magnets are usually ground by diamond wheels and magnetized with produce dimensions with close tolerances; customer-specific tolerances can also be accommodated. The end product is then subject to visual inspection and random sample testing to check the physical and functional specifications. One has to be aware that these ceramic magnets should

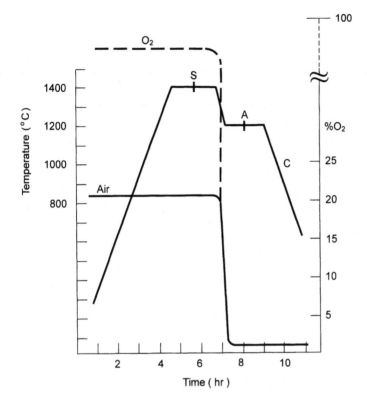

Figure 14 Sintering profile as a function of temperature and atmosphere for MnZn ferrite. S, Soak; A, anneal; C, cooling temperatures.

not still be considered a structural member in an assembly because they are brittle and should be handled so as to avoid chipping and cracking.

C. Hot Pressing

For most applications the densities obtained by conventional sintering are adequate. However, applications such as magnetic recording require dense, pore-free materials that have the highest saturation and magnetic permeability. The MnZn ferrites are ideal, having permeabilities of 5000–6000 (NiZn ferrite = 2000) and saturation magnetization 5000–6000 gauss (NiZn ferrite = 3500). However, the sintered density is in the range 94–97%, and some other densification technique must be used before the MnZn ferrites are suitable.

Uniaxial hot pressing was the first method employed to obtain fully dense MnZn ferrite material. The usual method is to press a preform in the shape of a flat circular disk and then to place that disk in a die constructed of graphite, silicon carbide, or silicon nitride [59]. Densification takes place by heating the die assembly at 1000–1380°C for periods of 1/2–2 hr. Following the hot pressing step it is usual to anneal at a lower temperature in order to optimize magnetic properties but not induce further grain growth [60]. This method, while yielding dense parts, has the disadvantages that (1) pressure is limited by die wall strength at elevated temperatures, (2) thermal shock characteristics of the die place a lower limit on cycle time, (3) the pressed volume of each disk is relatively small, and (4) each individual piece sees a separate hot pressing cycle and thus piece-to-piece reproducibility may be poor. For these reasons the hot isotatic pressing (HIP) technique has been developed for ferrites.

Hardtl [61] demonstrated that ceramic materials could be densified in an HIP using inert gas under high pressure (N_2, 1250–1350°C, 2000 psi, 2–10 hr) without the need for an impervious metal envelope to transmit the forces. The key to this method is to sinter the pressed part at a low temperature prior to hipping [62]. This sintering step must be carefully controlled to minimize grain growth yet assure that there is no interconnected porosity through which the gas could permeate during hipping.

VI. SINGLE-CRYSTAL FERRITE

Initial attempts to fabricate single-crystal ferrite during the late 1960s concentrated on the Verneuil (flame fusion) growth method, in which the ferrite powder was melted in a flame and fell on a cooler rod of ferrite, causing the single crystal to grow upward. The results of this method were less than suitable because the crystal was highly stressed and deficient in ZnO due to volatilization in the flame. The magnetic properties of these materials were greatly reduced in consequence.

Currently, MnZn ferrite single crystals are grown for fabrication into video recording heads by the Bridgman method. In this technique, a small single-crystal seed of ferrite is placed in a capillary–shaped extension at the bottom of a cylindrical platinum-iridium crucible. The area above the single crystal is filled with ferrite powder or a preform of sintered ferrite. A narrow high-temperature (1500°C) zone passes from the seed to the upper end of the crucible as it is lowered slowly through the furnace (~1 mm/h), causing the ceramic to melt and resolidify as a single crystal. The platinum crucible is then cut off the crystal ingot, to be reclaimed, and the single-crystal rod, 100 × 800 mm, is fabricated into video recording heads.

The use of the nonmagnetic gadolinium gallium garnet ($Gd_3Ga_5O_{12}$; GGG) as a substrate for the epitaxial growth of substituted yttrium iron garnet ($Y_3Fe_5O_{12}$;

YIG) films in bubble memory device fabrication has resulted in the necessity of growing large, well-characterized single crystals of GGG. Fortunately, the rare earth gallium garnets melt without decomposition (congruently melting) and can be pulled from the melt as single-crystal boules or ingots by means by Czochralski technique, which is a common practice for electronic grade silicon wafer production [63]. Because of the high melting temperature of the materials (1500–1700°C), iridium crucibles are used and heating is done by radio frequency (RF) coupling to the crucible. A suitable seed crystal is attached to the end of a cooled rod, then slowly brought down to the molten interface and raised at a rate of 0.5–5.0 mm/h. To enhance stoichiometry, the crucible and the growing crystal boules are rotated in opposite directions while growth is taking place. After growth is complete, the crystal is sawed into thin disks and polished. It then becomes the host lattice on which a thin layer of magnetic garnet is grown for the fabrication of magnetic bubble memories [64].

Yttrium iron garnet single crystal is shaped into spheres (dia. = 0.4~1.0 mm), which can have a role as ferrimagnetic resonator and band pass filter. The YIG sphere is placed at the intersection of two orthogonal coils (loops, strip lines, or cavity modes) on x-y plane and is easily oriented in the magnetic field. When it is not magnetized, no power is transferred between the coils. When a dc field is applied along the z axis, the coils are coupled through the transverse components of the dipolar field of the ferrite resonator [65].

VII. APPLICATIONS OF MAGNETIC FERRITES

Ferrites are especially useful in the electronics industry due to two key characteristics: (f) high magnetic permeability, which concentrates and reinforces the magnetic field, and (2) high electrical resistivity which limits the amount of electric current flow in the ferrite. Ferrites exhibit low energy losses and are highly efficient at high frequencies (1 kHz to 1 GHz). Hard ferrites have desirable properties such as higher magnetic flux density, higher coercive force, and higher resistance to demagnetization, and they have low tendency to become oxidized, compared to other non–rare earth permanent magnets. The low-cost attractiveness of this material make ceramic magnets useful for many permanent magnet applications. Along with substantial increase of flux density, there is a significant reduction in the size of the magnets as well. Permanent magnet ceramics are used in many applications where a constant magnetic field without electric current is needed. With the advent of high-frequency communications, the standards techniques for reducing eddy current losses, using lamination or iron cores, became no longer efficient or cost effective. The high electrical resistivity of ferrites, combined with desired magnetic characteristics, make them suitable for high-frequency operations. Numerous daily applications for using permanent ceramic magnets include information storage media, communications, electrical power generation and distribution, dc brushless motors, lawn mowers, speakers, refrigerator doors, separators to isolate ferrous materials from nonferrous ones, repulsive suspension of levitated railway, and magnetic resonance imaging [66–68].

Inductance is defined as the rate of change of flux linkages with respect to the current producing the flux. Inductor design is, to a large extent, the proportioning of values of air gap and magnetic pathlength divided by either ac or dc permeability. The higher the induction, the higher the permeability. The permeability is inversely proportional to the saturation flux density of the core material used. Saturation should be avoided if possible; it yields nonlinear results in inductors. The permeability of a magnetic core is decreased by the insertion of an air gap. Inserting an air gap causes the magnetization curve to "shear over" to make a hysteresis loop become wider due to greater core loss. Establishing the proper air gap in an inductor carrying direct current is quite crucial. It is a common practice to build such inductors with a gap to prevent dc saturation. The air gap size, core size, and number of turns depends on three interrelated factors: inductance desired, dc current in the winding, and ac voltage across the winding. If the air gap is large, the inductor is not much affected by changes in permeability; it is thus called a linear inductor. For a small air gap, changes in permeability due to current or voltage variations would alter inductance in the inductor, which is then classified as a nonlinear inductor.

In inductor design, the quality factor (Q) of a coil is the ratio of inductive reactance ($X_L = 2\pi f L$) and the serial resistance (R_s). The larger Q is, the better the inductor performs. R_s is determined by multiplying the length of the wire used to wind the coil by the dc resistance per unit length for the wire gage used. This quality factor is dramatically effected by the frequency of operation. At lower frequencies Q is very good because the dc resistance of the winding has the only effect. As the frequency increases, Q increases up to a point where the skin effect and the combined distributed capacitances begin to dominate until the self-resonance frequency of the inductor (f_r) is reached. We can enlarge Q in the inductor by spreading the winding out, using a bigger wire size or air gap between the windings. To determine the optimal condition for core material in linear operation the frequency should be considered as well as the maximal flux density the core material can support. At lower freqencies selecting a flux density below the saturation level for ferrite is essential. In higher frequencies the flux density is selected where the loss is limited to approximately 50 mW/cm^3 (or 5 W/lb) for ferrite before losses start to limit flux density. MnZn ferrite toroids have a low resistivity and shorting the wire to the core will reduce its Q value; the high resistivity of NiZn ferrite cores often minimizes this kind of loss in Q.

Hard ferrites were developed in the 1960s as a low-cost alternative to metallic magnets. Even though they exhibit low energy storage (compared to other permanent magnet materials) and are relatively brittle and hard, ferrite magnets have won wide acceptance due to their good resistance to demagnetization, excellent corrosion resistance, and low price per pound. In fact, measured by weight, ferrite represents more than 75% of the world magnet consumption. It is the first choice for most types of dc motors, magnetic separators, magnetic resonance imaging, and automotive sensors. The hard ferrite magnets are manufactured to rigid magnetic and physical standards, which normally exceed the Magnetic Materials Producers Association (MMPA) standards. Ferrites are used predominately in three areas of the electronic field: (1) low-level applications, (2) power applications, and (3) electromagnetic interference (EMI) suppression. As a part of the strategy for controlling EMI, ferrites have a crucial role as absorptive filters. As their excitation frequency approaches ferrimagnetic resonance, loss properties increase drastically. Common-mode noise is electrical interference that is common to both lines in relation to earth ground. In contrast, differential-mode noise, also known as normal-mode noise, is electrical interference that is not common to both lines but is present between lines. Hence, the ferrite can appear as a low-loss inductor at low frequency but as a lossy impedance device at high frequency, say above 25 MHz. NiZn ferrites are normally used for EMI suppression in the end use of circuit boards, cables, connectors, common-mode or differential-mode chocks, as it is possible to adjust the frequency–loss curve by tailoring its composition. Though MnZn ferrites may be employed, their resonance frequencies are rather low and would hinder the attenuation range. Ferrites are widely used as attenuators of unwanted high-frequency signals. These ferrites are known as EMI suppressors. They are typically available as beads, split cores, flat ribbon core and toroidal cores. Ferrite tiles are also available for use in anechoic chambers. The breadth of application of ferrites in the field of customer electronics, computers, and communications (3C) continues to grow. The wide range of possible geometries, continuing improvements in material characteristics, and their relative cost effectiveness make ferrite components the choice for both conventional and innovative applications.

Slick [69] has detailed a wide variety of applications for which ferrites are used. These cover the frequency range from 60 Hz to 500 MHz, with the majority of the applications being served by two materials: MnZn ferrite and NiZn ferrite ferrites. Within these systems a variety of compositions have been developed with desired properties as detailed in Table 5, for a series of proprietary formulations by Ferroxcube (Phillips) Corporation. Formulations 4C4 and 3D3 are NiZn ferrites, whereas 3B7, 3B9, 3C8, and 3E2A are Mn-Ni ferrites. The initial permeability (μ_i) changes with temperature for these formulations are shown in Figure 15, while typical values for the initial permeability at 298 K are given in Table 7. Within the temperature range shown in Figure 15, the MnZn ferrite samples show a shift in Curie point and an increase in permeability consistent with the data presented in Figures 3 and 4. The temperature factor in Table 6 represents the

Table 5 Types of Ferrites and Properties

Composition	Lattice parameter a_0(Å)	Curie point (°C)	Saturation magnetization B_s, at 20°C	Bohr magneton (μ_B)	Crystal structure
$MnFe_2O_4$	8.50	300	5200	4.6	Spinel
$FeFe_2O_4$	8.39	585	6000	4.1	Spinel
$NiFe_2O_4$	8.32	590	3400	2.3	Spinel
$CoFe_2O_4$	8.38	520	5000	3.7	Spinel
$CuFe_2O_4$	8.71	455	1700	2.3	Spinel
$BaFe_{12}O_{19}$	—	450	4000	—	Hexagonal
$SrFe_{12}O_{19}$	—	453	4000	—	Hexagonal
$MgFe_2O_4$	8.36	440	1400	1.1	Spinel
$Li_{0.5}Fe_{2.5}O_4$	8.33	670	3900	2.6	Spinel
γ-Fe_2O_3	—	575	5200	—	Spinel
$Y_3Fe_5O_{12}$	—	287	1700	—	Spinel
$ZnFe_2O_4$	8.44	60	—	0	Spinel

Source: Ref. 15.

incremental change in the initial permeability with unit of temperature. The phenomenon commonly known as disaccommodation (D), listed in Table 6, is a magnetic aftereffect, which is a proportional decrease of permeability after a disturbance of a magnetic material, measured at constant temperature at 10~100 kHz over a given time interval, i.e. D = $[L(t_1) - L(t_2)]/L(t_1)$. Normally, the initial permeability (μ_I) decreases with increasing frequency and time. Such slow relaxation processes are associated with the relaxation or diffusion of impurity atoms to equilibrium positions. The atoms assemble at domain walls and dislocations, with consequent pinning of the domain walls, thereby reducing domain wall motion and initial magnetic permeability. This relaxation effect becomes more noticeable for materials with impurity atoms of low solubility. Disaccommodation is a measure, as a function of time, of this decrease in permeability due to the gradual demagnetization of a magnetic sample [70].

Table 6 also lists applications for the different ferrite formulations, and these are developed in greater detail in Table 7, where the various devices, their functions, frequency of operations (<500 MHz), and desired properties are summarized. For instance, a choke is an inductor, which is intended to filter or choke out signals. One of the common applications for ferrite is as a transformer, which transforms one voltage into another without changing the frequency and is employed in power supplies for satellites, mobile phones, VCRs, cable TV, and in

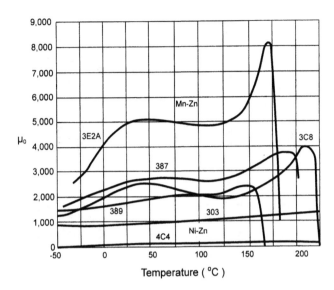

Figure 15 Initial permeability versus temperature for ferrite compositions in the NiZn ferrite (3D3, 4C4) and MnZn ferrite (3B9, 3B7, 3E2A, 3C8). The latter show Curie point shifts due to compositional changes as well as secondary maxima attributed to magneto-crystalline anisotropy effects.

broad-band digital transmission technologies such as ADSL and high-frequency cellular phone systems.

Microwave ferrites are very dense ceramics, which are dark gray or black and of polycrystalline microstructure. Ferrites are electrically nonconductive but highly magnetic so they are widely used as transformer cores. Microwave ferrites have an unique property called gyro magnetic resonance. If microwave electromagnetic waves pass through a microwave ferrite material, there is an interaction between the electrons in the material and the electromagnetic radiation. This causes the ferrite material to act in a nonreciprocal manner. In effect, the material only allows radio waves to pass in one direction but not in the reverse direction, which is an extremely useful property in microwave component design for components such as circulators and isolators, extensively employed in microwave cellular and satellite communication systems and in radar. Summarized also in Table 8 are typical ferrite and garnet formulations (systems) for microwave applications, as detailed, along with pertinent properties at the frequencies indicated.

The general applications can be divided into two types: signal handling (low level) and power handling. The low-level uses are those in which the applied field (or excitation current) and the resulting magnetic flux in the ferrite are linear;

Table 6 Properties of Commercial (Ferroxcube) Ferrite Formulations

Characteristic	Symbol and unit	Ferroxcube grade					
		4C4	3D3	3B9	3B7	3C8	3E2A
Initial permeability at 298 K	μ_i	125	750	1800	2300	2700	5000
Saturation flux density at 298 K	B_s (mT)	300	380	320	380	440	360
Coercive force	H_c (mA/m)	37.7	12.6	3.8	2.5	2.5	1.0
Residual flux density	B_r (mT)	—	—	—	—	100	—
Loss factor at:							
4 kHz	$\tan\delta/\mu_i$	35×10^{-6}	—	$\leq 1 \times 10^{-6}$	$\leq 1 \times 10^{-6}$		
100 kHz	$\tan\delta/\mu_i$	35×10^{-6}	—	$\leq 5 \times 10^{-6}$	$\leq 5 \times 10^{-6}$	$\leq 10 \times 10^{-6}$	$\leq 10 \times 10^{-6}$
500 kHz	$\tan\delta/\mu_i$	$\leq 40 \times 10^{-6}$	$\leq 14 \times 10^{-6}$	25×10^{-6}	25×10^{-6}		
1 MHz	$\tan\delta/\mu_i$	60×10^{-6}	$\leq 30 \times 10^{-6}$	120×10^{-6}	120×10^{-6}		
5 MHz	$\tan\delta/\mu_i$						
Temperature coefficient of inductance (TCI)	$\dfrac{\Delta\mu_i}{\mu_i\,\Delta T}$ (°C^{-1})	-6.0×10^{-6} min. -6.0×10^{-6} min.	$+1.0 \times 10^{-6}$ min. $+3.0 \times 10^{-6}$ max.	$+0.9 \times 10^{-6}$ min. $+1.9 \times 10^{-6}$ max.	-0.6×10^{-6} min. $+0.6 \times 10^{-6}$ max.	—	-2×10^{-6} min. $+2 \times 10^{-6}$ max.
Power losses at 16 kHz and 200 mT	mW/cm³	278–328 K	243–343 K	243–343 K	253–343 K	$\leq$110 (25°C) $\leq$100 (100°C)	293–343 K
Disaccommodation (D)	$\dfrac{\Delta\mu_i}{\Delta\mu_i^2\, t_1\, \log_2/t_1}$	$<15 \times 10^{-6}$	12×10^{-6}	$\leq 2.5 \times 10^{-6}$	$\leq 3.5 \times 10^{-6}$	—	—
Hysteresis losses at kHz	ηB (t^{-1})	—	$\leq 1.8 \times 10^{-3}$	$\leq 1.1 \times 10^{-3}$	$\leq 1.1 \times 10^{-3}$	—	—
Curie temperature	T_C (K) (°C)	<573 (>300)	>423 (>150)	>418 (>145)	>443 (>170)	>483 (>210)	>443 (>170)
Uses		1–20 MHz, filter coil applications, high-frequency, wide-band, and pulse transformers	200 kHz – 2.5 MHz, matches polystyrene capacitors inductors, pulse transformers	Audio to 300 MHz, inductors and tuned transformers	Audio to 300 kHz, inductors used with low-temperature-coefficient capacitors, pulse and power transformers	High-flux density applications, switched-mode power supplies	Wide-band and pulse transformers, inductrs

Table 7 Summary of Nonmicrowave Ferrite Compositions and Applications

Ferrite chemistry	Device	Device function	Frequencies	Desired ferrite properties
		Linear B/H, low-flux density		
MnZn, NiZn	Inductor	Frequency selection network	≤1 MHz (MnZn ferrite)	High μ, high μQ, high stability of μ with temperature and time
MnZn, NiZn	Transformer pulse (and wide band)	Filtering and resonant circuits	~1–100 MHz (NiZn)	High μ, low hysteresis losses
		V and I transformation	Up to 500 MHz	
NiZn	Antenna rod	Impedance matching	Up to 15 MHz	High μQ, high resistivity
MnZn	Loading coil	Electromagnetic wave receiver	Audio	High μ, high B_s, high stability of μ with temp., time, and dc bias
		Impedance loading		
		Nonlinear B/H, medium to high-flux density		
MnZn, NiZn	Flyback transformer	Power converter	<100 kHz	High μ, high B_s, low hysteresis losses
				High μ, high B_s
MnZn	Deflection yoke	Electron beam deflection	<100 kHz	Mod. high μ, high B_s, high hysteresis losses
MnZn, NiZn	Suppression bead	Block unwanted ac signals	Up to 250 MHz	Mod. high μ, high B_s, high hysteresis losses
MnZn, NiZn	Choke coil	Separate ac from dc signals	Up to 250 MHz	High μ, high density, high μQ, high wear resistant
MnZn, NiZn	Recording head	Information recording	Up to 10 MHz	
MnZn	Power transformer	Power converter	<60 kHz	High B_s, low hysteresis losses
		Nonlinear B/H, rectangular loop		
MnMg, MnMg-Zn, MnCu, MnLi, etc.	Memory cores	Information storage	Pulse	High squareness, low switching coefficient, and controlled coercive force
MnMgZn, MnMgCd	Switch cores	Memory access transformer	Pulse	High squareness, controlled coercive force
MnZn	Magnetic amplifiers			

Table 8 Ferrites and Garnets for Microwave Frequency Range

Material	T_c, (°C)	ϵ_r	B_s (V·s/m^2)	Frequency range (GHz)	Applications
MgMn ferrites	290–320	12–14	0.18–0.29	>7	Phase shifter, circulator Resonator isolator
MgMnAl ferrites	95–295	8–14	0.06–0.21	>2	Phase shifter, isolator Field displacement
NiZn ferrites	300–590	13–16	0.32–0.50	>5	Faraday rotation
NiCo ferrites	500–590	10–12.5	0.18–0.30	>0.5	Faraday rotation isolator Resonance isolator
NiAl ferrites	160	9.0	0.05	>1	Phase shifter
YFe garnets	110–265	14–16	0.06–0.16	>1	Phase shifter, circulator
YFeAl garnets	130–280	14–15	0.04–0.17	>1	Phase shifter, circulator Field displacement
YFeGd garnets	280	15	0.10		Temperature stabilization

ϵ_r represents the relative dielectric constant at frequencies given [71].

the permeability of the material is thus constant (see Figure 16). This means that driving fields are low and that core and winding losses can be neglected. The application of linear design principles can be found in low-level transformers, inductors, telecommunications filters, and the like, and are usually limited to a magnetic induction level of less than 250 gauss [72]. These are special types of transformers. Among them, baluns are generic devices for converting a balanced signal to an unbalanced signal. High-frequency baluns can be constructed with transmission lines and be driven from the unbalanced side to give a balanced output of some multiple of the input impedance. The typical shape of a balun is shown in Figure 11. Ferrite rods are used as antennae for audio receivers in the telecommunication systems. In pulse-compression systems, soft ferrites serve the function of wave guide and wave shaper. The permeability of ferrites does not change much with frequency up to a critical frequency range (10–100 MHz) but then decays rapidly with increasing frequency.

Manganese zinc ferrites having high saturation magnetization and low power losses (e.g., Ferroxcube 3C8, Siemens N27, TDK H7Cl) offer considerable advantages over laminated transformer steels at frequencies above 1 kHz. At

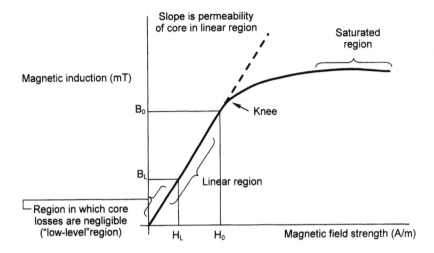

Figure 16 Plot of permeability versus magnetic field strength for a linear ferrite.

these frequencies the low resistance of metal laminates ($\sim$50 $\mu\Omega$-cm) causes eddy current losses and transformer heating to be significant.

Although ferrite has a lower operating flux density (3300 gauss) than that of transformer steel (18,000 gauss), the higher operating frequency reduces the maximum magnetic induction in the transformer:

$$B = \frac{E \times 10^8}{4.44 \times f \times N_p \times A_e} \tag{17}$$

where B = flux density (gauss), E = volts (rms), f = frequency (Hz), A_e = effective area of transformer (cm^2), and N_P = primary turns. Because of the inverse relationship between flux density and frequency, high-frequency operation reduces the flux density to within acceptable limits. It is also apparent from this relationship that the transformer can be made smaller (reduced A_e) and the number of turns (N_P) reduced as the frequency increases. Ferrites do, however, have a much lower thermal conductivity than metallic transformer cores, so for applications greater than several kilowatts it becomes difficult to cool the transformer effectively. Core loss increases exponentially with increasing flux density or frequency; however, a number of material grades have been designed so that their minimal core loss occurs at a specific temperature. Ferrites can be manufactured to permeability of over 15,000 with little eddy current loss. However, the high permeability of the ferrite makes it unstable at high temperatures and easily saturated. In a frequency range between 5 and 100 kHz, ferrite material is suitable for applications such as DC-to-DC converters, magnetic amplifiers and inverter power supplies, because it is cheaper than tape wound cores and is used in applications where high flux density and high temperature stability are not critical. It must be noted that driving ferrites with excessive current may cause permanent damage to the core.

A ferrite core has a high electrical resistivity ($\sim 10^6$ Ω-cm) and exhibits low core loss hence it can be used at higher frequencies than metallic alloys. The performance of the ferrite is determined by its composition. Ferrites are limited by their low Curie temperature and low operating flux density. The maximal flux density decreases as the temperature increases. Despite such limitations ferrite cores are used extensively at higher frequencies. A major application of high-frequency power components is in the switched-mode power supply (SMPS), as used in DC power supply voltage protector circuits [73]. These devices rectify and smooth the primary voltage and switch it through the transformer at high frequencies (15–100 kHz). The output of the transformer is then a highly regulated, filtered dc voltage. Offering high efficiencies (60–90%) and better performance in a smaller package, SMPS are finding wide use in the instrument and computer power supply markets. Many of these principles have been applied to fluorescent lighting ballasts. Rather than operating at line frequencies using metal transformers, new more efficient ballasts and self-contained fluorescent bulbs are used in incandescent fixtures.

Except for certain military and other specialized applications, the use of ferrite memory cores ($LiFe_5O_8$) in computer mainframes has all but disappeared because of advances in semiconductor memory technology. However, spinel ferrites are used extensively in the recording of both analog (audio and video recording) and digital signals. In 1966, IBM introduced the 2314 series of peripheral memories using an NiZn ferrite recording head. This ceramic recording head was easier to construct than one using thin metal laminations because it could be machined to close tolerances from bulk ceramic and glass bonded in air to form a precise, durable gap. The track density was low (100 tracks per inch) and the system storage density was less than 1/4 megabit per square inch. Storage densities of digital systems have increased steadily at 27% per year since that time, and current systems offer storage densities in excess of 12 megabits per square inch. With improvements in the processing of MnZn ferrites (e.g., hot pressing), it is now possible to use these materials with their higher permeability and saturation in applications where the track width is less than 0.001 inch (25 μm).

Ferrite substrates are used with thin-film magnetic heads for hard disk drives in high-capacity, high-density magnetic storage systems. Sumitomo Special Metals have the market leader for these systems. The soft ferrites are sintered at 1200–1400°C, followed by polishing procedure to the desired shape and dimensions. After sintering, hot isostatic pressing (HIP) is done to obtain higher density materials that enhance the performance of the magnetic heads and other electronic components (Figure 11). To meet the requirements of device miniaturization, thin-film magnetic heads are made by magnetron sputtering technique onto Al_2O_3 and TiC ceramic substrates, which are hot pressed and mirror-finish polished. These structures are designed for high-capacity floppy disk drives, tape drives, and a wide range of other drives. The substrates must have smooth surfaces and be free of residual strain, especially resistance to wear caused by the magnetic storage medium, to assure proper operation of the hard disk drive.

In order to increase the resistivity of NiZn ferrites, CoO and MnO are added by replacing some portion of Ni^{2+} ions in the spinel structure. In particular, the Co^{2+} dopant in the NiZn ferrite enhances fixation of the domain walls, to effectively reduce magnetic loss. Because of the large contribution to magnetic anisotropy by Co^{2+} ions in the spinel, the K_1 value, as shown in Eq. 12, of CoNiZn ferrites is almost zero and makes the temperature coefficient of permeability adjustable. Recently developed is a low-fire type CoNiZn ferrite with sintering temperature at 900°C or less, to be used exclusively for chip coils. The attractiveness of CoNiZn ferrite is in allowing the use of high-silver content or pure silver paste for the inner electrodes used to produce multilayer chip inductors. The multilayer LC chips in LCD modules are the integration coil chips and chip capacitors, for which additional care must be expended to avoid occurrence of defects in the chips. Cracks perpendicular to the interface can result from thermal mismatch between the inductor and capacitor. Delamination at the interface of the multilayer chip is due to a difference in shrinkage rate of inductor and capacitor. Silver migration from the inner electrode in the LC chip can cause undesirable reactions between the silver and oxide, or promote densification of the winding region.

The co-ferrite thin film is a well-known candidate for high-density magnetic recordings because of its high coercivity, mechanical hardness, and chemical stability. Co-ferrite fine powders and thin films have been grown on thermally oxidized silicon substrates using a sol-gel method. The results show that most Co-ferrite grains on the silicon substrates are oriented randomly and the magnetic properties of Co-ferrite powders are transited from a ferrimagnetic to a paramagnetic phase on decreasing the annealing temperature. It is observed that thin films annealed at 650°C have grain-like features, typically 25 nm. The saturation magnetization of the 600°C-annealed powder is about 60 emu/g and the coercivity of the film is about 1900 oersted.

The disappearance of the magnetic properties of ferrites at the Curie temperature can be used to construct a fixed-point temperature sensor. The Curie temperature can be controlled precisely by substituting varying amounts of zinc ferrite into manganese ferrite. After it has been sintered the Curie temperature is fixed and it is possible to fabricate a miniature fixed-point temperature sensor either by using the material as a signal transformer (above the Curie temperature a signal on the primary coil will not appear on the secondary coil) or as the inductive component in a tuned circuit.

The high-frequency properties have been used by Tanka and Reynolds [74] to fabricate a very sensitive magnetic field detector by gating the flux through a sense coil (Figure 17). A controlling coil is wound onto a narrowed portion of the magnetic path and then energized with a square wave driving that portion of the circuit into magnetic saturation. Saturation of this portion of the magnetic patch prevents magnetic flux from circulating through the sense coil. As the square wave decays, the narrowed portion desaturates abruptly, allowing the flux to circulate. The high-frequency flux gating through the sense coil induces a voltage in the sense coil. Of particular advantage in such applications as the sensing of ground fault currents is the fact that the output is essentially independent of the frequency of the signal being measured and the device is active for a very brief period, making it immune to spurious noise effects.

The research efforts for commercial ferrites today are almost dedicated to electronic circuit design needs. For example, as computers are designed for increased clock speeds and decreased currents, switch-mode power suppliers have become essential for the supply of pure currents to electronic chips. Device digitalization has produced greater EMI drawback, which must be reduced. Commercial communication and vehicular radar systems call for reconfiguration of standard ferrite devices to meet size, weight, and integration specifications. Future space and satellite communications will require broad-band operation at millimeter and

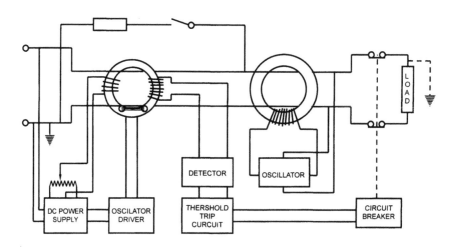

Figure 17 Circuit block diagram for magnetic field detector utilizing magnetic ferrite core for flux control and sensing.

submillimeter wavelengths. Many technologies, such as tape casting, low-temperature co fired ceramics, hexagonal ferrites, and component integration, will continue to be widely employed for manufacturing and material development. However, one must overcome the over-riding challenges of performance and cost-down requirements.

REFERENCES

1. Okutani K. Magnetic Materials in Japan: Research, Application, and Potential. Oxford: Japan Technical Information Service, Elsevier, 1991:213–251.
2. Snoeek JL. Ferromagnetic Materials, with Introductory Chapters on the Statics and the Dynamics of Ferromagnetism. 2nd ed. New York: Elsevier, 1949.
3. Bragg WH. Philos. Mag. 1915; 30:315.
4. Nishikawa S. Proc. Tokyo Math. Phys. Soc. 1915; 8:199.
5. Smit J, Wijn HP. J. Ferrites. New York: John Wiley and Sons, 1959:139.
6. Economos G. J. Am. Ceram. Soc. 1955; 38(9):335.
7. Broese A, Groenou Van. Mater. Sci. Eng. 1968–1969; 3:317.
8. Numerical Data and Function Relations in Science and Technology, Group 3, Crystal and Solid State Physics. 1969.
9. Petric A, Jacob KT. J. Am. Ceram. Soc. 1982; 65(2):117–123.
10. Petric A, Jacob KT, Alcock CB. J. Am. Ceram. Soc. 1981; 14(11):632–9.
11. Pelton AD, Schmalzried H, Sticher J. J. Phys. Chem. Solids. 1979; 40(12):1103–22.
12. Katsura T, Wakihara M, Hara S-I, Sugihara T. J. Phys. Chem. Solids. 1975; 13: 107–113.
13. Charette GG, Flengas SN. J. Electrochem. Soc. 1968; 115(8):796–804.
14. Jacob KT, Jeffes JHE. High Temp. High Pressure. 1972; 4:177–182.
15. Somiya S, ed. Advanced Technical Ceramics. Tokyo: Academic Press, 1984.
16. Soft Ferrite: A User's Guide, Magnetic Materials Producers Association. Chicago, 1992.
17. Broese A, Groenou Van. Mater. Sci. Eng. 1968 to 1969; 3:317.
18. Crangle J. The Magnetic Properties of Solids. London: Edward Arnold, 1977, Chapter 6.
19. Bradley FN. Materials for Magnetic Functions. New York: Hayden, 1971, Chapter 1.
20. Watanabe Y. Phys. Status Solidi (b). 1978; 90:697.
21. Ohata K. J. Phys. Soc. J. 1963; 18:685.
22. Stoppels D. J. Appl. Phys. 1980; 51:2789.
23. Stoppels D, Enz U, Damen JPM. J. Magn. Magn. Mater. 1980; 20:231.
24. Jones RV. IEEE Trans. Son. Ultrason, SU-13. 1966:86.
25. O'Connor M, Belson H. J. Appl. Phys. 1970; 41:1028.
26. Valenzuela R. Magnetic Ceramics. London: Cambridge University Press, 1994: 156.
27. Smit J, Wijn HP. J. Ferrites. New York: John Wiley and Sons, 1959:177.
28. Adelslöld V. Ark. Kem. Min. Geol. 1938; 12A(29):1.

29. Albanese G. J. Phys. Colloque C1. 1977; 38(Suppl. 4):C1-85.
30. Smit J, Wijn HP. J. Ferrites. New York: John Wiley and Sons, 1959:194.
31. Janasi SR. IEEE Trans Magnets. 2000; 36:3327.
32. Stuijts AL. Philips Tech. Rev. 1954; 16:141.
33. Bertaut AL, Forrat AL. Compt. Rendu. 1956; 242:383.
34. Geller SZ. Kristallogr. 1967; 125.
35. Langley RH, Sturgeon GD. J. Solid State Chem. 1979; 30:79.
36. Hawthorne FC. J. Solid State Chem. 1981; 37:157.
37. Gilleo MA. In: Wohlfarth EP, ed. Ferromagnetic Materials, Vol. 2. New York: North-Holland, 1980:1.
38. Takada T. Ferrites: ICF 3 19:3.
39. Robbins H. Ferrites: ICF 3 19:7.
40. Yu Bu-Fan B, Goldman A. Ferrites: ICF 3 19:52.
41. Schnettler FJ, Johnson DW. Ferrites: ICF 3 19:105.
42. Sugimoto M. J. Am. Ceram. Soc. 1999; 82:269–280.
43. Fan J, Sale FR. IEEE Trans. Magnetics. 1996; 32:4854.
44. Yamazaki M, Nayutani T, Kobiki H. US Patent 5,368,763, 1994.
45. Inoue S. U.S. Patent 5,779,930, 1998.
46. Yasuhara K, Ito T, Inoue S. US Patent 5,846,448, 1998.
47. Margalit J. Ceram Ind 150 [13]. 2000:20.
48. Parkin IP, Kuznetsov MV, Pankhurst QA. J. Mater. Chem. 1999; 9:273–81.
49. Nicolas J. In Ferromagnetic Materials, Vol. 2. Wohlfarth EP, ed. New York: North-Holland, 1980:243.
50. Ghate BB. In Grain Boundary Phenomena in Electronic Ceramics, American Ceramic Society. 1981:477.
51. Onoda GY, Jr. In Ceramic Processing Before Firing Onoda GY, Jr, Hench LL, eds. New York: John Wiley and Sons, 1978:235.
52. Dixon M. U.S. Patent 4,247,500, 1979.
53. Ernst WG. J. Am. Sci. 1966; 264:37.
54. Eugster HP, Skippen GB. Res. Geochem. 1967; 2:492.
55. Blank JM. J. Appl. Phys. 1961; 32(Suppl. 3):378S.
56. Macklen ED. J. Appl. Phys. 1965; 36:1022.
57. Slick PI. In Ferrites. Hoshino Y, Iida S, Sugimoto M, eds. Proceedings of the International Conference held in July 1970, Japan. Baltimore: University Park Press, 1971:81.
58. Plaskett TS. J. Appl. Phys. 1982; 53:2428.
59. Ikeda A. In Ferrites. In: Hoshino Y, Iida S, Sugimoto M, eds. Proceedings of the International Conference held in July 1970, Japan. Baltimore: University Park Press, 1971:337.
60. Hirota E, et al. In Ferrites. In: Watanabe H, et al, eds. Proceedings of the International Conference, Sept.–Oct. 1980, Japan. Japan: Center for Academic Publications, 1981:667.
61. Hardtl KH. Am. Ceram. Soc. Bull. 1975; 54:201.
62. Bunthker C, Berben Th. J. Phys., Colloque C1. 1977; 38(Suppl. 4):C1–341.
63. Brandle CD, Valentino AJ. J. Cryst. Growth. 1972; 12:3.

64. Chang Hsu. Magnetic Bubble Technology: Integrated-Circuit Magnetics for Digital Storage and Processing. New York: IEEE Press, 1975.

65. Helzaju J. Microwave Engineering: Passive, Active, and Non-Reciprocal Circuits. London: McGraw-Hill, 1992:310.

66. Levinson LM. Electronics Ceramics, Properties, Devices, and Applications. New York: Marcel Dekker, 1988:170.

67. Valenzuela Raul. Magnetic Ceramics. London: Cambridge University Press, 1994: 167.

68. Hench LL, West JK. Principles of Electronic Ceramics. New York: Wiley-Interscience, 1990:311.

69. Slick PI. In Ferromagnetic Materials Wohlfarth EP, ed. Vol. 2. New York: North-Holland, 1980:189.

70. Bradley FN. Materials for Magnetic Functions. New York: Hayden, 1971, Chapter 1.

71. Heck Carl. Magnetic Materials and Their Applications, 1974.

72. Linnear Ferrite Magnetic Design Manual. Saugerties, NY: Ferroxcube Division of Amperex Electronic Corp..

73. Kit Sum K. Switch Mode Power Conversion: Basic Theory and Design. New York: Marcel Dekker, 1984.

74. Tanka D, Reynolds T. Ground fault interrupter, U.S. Patent 4,280,162, 1981.

6

Ceramic Sensors

**Bernard M. Kulwicki,* Stanley J. Lukasiewicz, and
Savithri Subramanyam**
*Texas Instruments Incorporated
Attleboro, Massachusetts, U.S.A.*

Ahmed Amin
*Naval Sea Systems Command
Newport, Rhode Island, U.S.A.*

Harry L. Tuller
*Massachusetts Institute of Technology
Cambridge, Massachusetts, U.S.A.*

I. INTRODUCTION

The operation of our industrialized world has depended, in large part, on our ability to measure and control a variety of physical and chemical parameters, including temperature, pressure, and the chemistry of our environment. In recent decades, development of electrical measurement and control systems that couple input sensors and output transducers via ever more sophisticated signal processing units has continued at a rapid pace. Furthermore, the scope of potential applications of such systems has expanded significantly. However, comparable improvements in the input and output functions provided by the sensors and actuators have not kept pace and now often limit the feasibility of the overall system.

---

* Retired.

Increasing attention to developments in these areas can, therefore, be expected to continue.

In this chapter the authors focus on ceramic sensors, which, because of their stability, low cost, and, in many cases, unique properties, find application in areas unsuitable for other types of sensors. To provide a basis for understanding and predicting the operation of the most common sensors, the fundamental relationships between the electrical properties of ceramics and important variables, such as composition, temperature, atmosphere, and microstructure, are considered. Attention is also focused on practical difficulties that must be addressed during the development and application of new and improved sensors.

Particular sensors that are discussed here include positive temperature coefficient (PTC) thermistors, negative temperature coefficient (NTC) thermistors, gas sensors, humidity sensors, capacitive pressure sensors, and detectors for "uncooled" thermal imaging applications. The first three device categories were covered in some detail by Hill and Tuller [1] who emphasized bulk ceramic thermistors and limited the discussion of gas sensors to oxygen sensors for automotive air-to-fuel ratio control. In this chapter, these subjects are updated, and new sections are added on pyroelectric-ferroelectric sensors for infrared (IR) imaging, combustible and toxic gas sensors, humidity sensors, and pressure sensors. Because of the additional significant new material, much of the detail in the original work is not repeated but rather included by reference [1].

For each type of sensor, the following features will be addressed: theory of operation, device manufacture, applications, and trends and directions. In this manner it is hoped that this chapter will serve the needs of the applications engineer, the processing specialist, and the research scientist concerned with developing new or improved materials and sensor devices. In order to address these subjects in a logical flow, the order chosen for presentation is as follows:

1. Pyroelectric IR sensors
2. NTC thermistors
3. PTC resistors
4. Gas sensors
5. Humidity sensors
6. Pressure sensors

In this manner, the theory and application of now mature pyroelectric devices for uncooled thermal imaging are covered before newer developments in NTC bolometry. In addition, the grain boundary barrier layers originally proposed to account for the electrical behavior of PTC resistors are discussed before similar barrier layers key to the performance of gas and humidity sensors. Capacitive pressure sensors have become quite important in the last 10 years, but are independent of the other sensor types and therefore discussed last.

Uncooled IR imaging is a very interesting application because it represents perhaps the first successful attempt to form hybrid structures with a ceramic detector material, bump bonded to a silicon readout integrated circuit (ROIC), and it involves a new type of sensor material in devices currently manufactured in quantity. The pyroelectric-ferroelectric detector did not exist (except in the laboratory) when this volume was last revised [2]. Thermal imaging is also of interest here because it represents a new application for NTC thermistor materials, formed as thin films in "monolithic" microbolometer arrays, which promise to be lower in cost than hybrid devices. Also, thin-film pyroelectric sensors are being actively developed to address certain performance deficiencies of microbolometers. This considerable ferment in recent years has arisen because of the tremendous interest in both military and commercial applications for medium-wavelength IR cameras and the desire to achieve their maximal performance at the lowest achievable cost.

II. PYROELECTRIC INFRARED SENSOR ARRAYS

Traditionally, focal plane arrays for high-resolution IR imaging have been produced using photoconductive detector materials, such as mercury cadmium telluride. The latter devices require cooling with liquid nitrogen or comparable refrigeration in order to achieve sufficiently low thermal noise so that the signal is not overwhelmed. These "cooled" detector arrays continue to be useful in applications requiring high resolution and rapid response time. However, they are very expensive (more than $10,000 per system). Uncooled detectors may require temperature control, but not significant refrigeration, and operate in the neighborhood of room temperature. They have the advantage of considerably lower cost. Several reviews of alternative technologies for producing high-quality uncooled thermal imaging systems provide significantly more detail than will be covered in this chapter [3–6].

A. Infrared Imaging

Infrared imaging in the medium-wavelength range (7.5–14 μm), corresponding to the thermal black body maximum near room temperature and to the atmospheric window, is desirable for a wide range of applications. Night vision, target recognition, reconnaissance, driving and navigation aid in foggy and poor visibility weather are among the applications of this technology. Densely populated arrays fabricated from ceramic barium strontium titanate (BST) have made it possible to produce compact, low-cost, uncooled IR cameras that are capable of delivering television-quality images.

Thermal detectors are generally based on two different modes of operation. The technologies considered most seriously are pyroelectric detectors and thermistor bolometers [6]. Pyroelectric materials undergo a change in polarization due to change in temperature, which makes them excellent candidates for IR detectors. A thermistor bolometer is a sensitive means to detect the thermal effect of IR radiation [7,8]. In the latter instance, the rapid change of resistance with temperature of semiconducting NTC or PTC thermistors is utilized. This topic will be covered in the section on NTC thermistors. Both technologies offer the capability to operate near room temperature without the need for extensive cooling.

The basic problem of processing thermal images stems from low-level signals of interest. They represent a small component of the total detectable radiation flux. In addition, the dark current level (in the absence of incident flux) further limits the sensitivity of thermal detectors. This effect is relatively inconsequential for point detectors. For scanned linear arrays this problem has been solved by ac coupling the output of each pixel. This is not an acceptable solution for densely populated two-dimensional staring arrays (which view the scene directly without scanning) for two reasons: (1) the limited real estate on the readout circuitry, and (2) the low operating frequency, which requires large ac coupling capacitors.

Pyroelectric detectors, on the other hand, are ac-coupled devices, thereby reducing system complexities by providing a practical solution to the problem of detecting small signals in a fairly large background thermal scene. A hybrid approach with thermally isolated detector array, solder bump bonded, pixel by pixel to the silicon ROIC was pioneered independently by Texas Instruments [9,10], Honeywell [11], and Plessey [12]. In this section we will briefly describe pyroelectric operation modes, thermal detection and figures of merit, noise and sensitivity limits, signal enhancement, imaging array, and system description.

B. Pyroelectric Operation Modes

There are certain requirements imposed by the crystal symmetry group for pyroelectricity to occur: (1) the absence of an inversion operation and (2) the existence of a unique polar axis. Therefore, pyroelectricity is limited to the following 10 crystallographic point groups: $1(C_1)$, $2(C_2)$, $3(C_3)$, $4(C_4)$, $6(C_6)$, $m(C_s)$, $mm2(C_{2v})$, $3m(C_{3v})$, $4mm(C_{4v})$, $6mm(C_{6v})$, in addition to the limiting group $\infty mm(C_{\infty v})$ which represents that of an electrically poled ferroelectric ceramic. The direction of the pyroelectric vector is along the unique polar axis, i.e., 2, 3, 4, 6, or ∞ in the preceding groups. It is arbitrarily set for point group $1(C_1)$ and lies in the mirror plane for $m(C_s)$.

The short circuit (current mode) pyroelectric coefficient p is defined as:

$$p = (\partial P/\partial T)_E \qquad (1)$$

In the MKS system of units, the polarization vector P (*i.e.* the charge per unit area taken perpendicular to P) has the units $C\ m^{-2}$. Therefore, the pyroelectric coefficient is expressed as $C\ m^{-2}\ K^{-1}$, or more often in the literature as $\mu C\ cm^{-2}\ K^{-1}$. If a pyroelectric crystal with a metallized surface area A (perpendicular to the polar axis) is subjected to a uniform heating rate $(\partial T/\partial t)$, the pyroelectric current is expressed as:

$$I = pA\ (\partial T/\partial t) \tag{2}$$

The open circuit (voltage mode) pyroelectric coefficient $(V\ m^{-1}\ K^{-1})$ is given by the following identity:

$$(\partial E/\partial T)_P = (\partial P/\partial T)_E\ (\partial E/\partial P)_T = p/\varepsilon \tag{3}$$

where ε is the dielectric permittivity $(F\ m^{-1})$. Noting that $P = \varepsilon_0 \chi E$ where χ is the dielectric susceptibility, we get $(\partial P/\partial E)_T = \varepsilon_0 \chi$. The relative permittivity (dielectric constant) is given by $\varepsilon_r = \varepsilon/\varepsilon_0 = (1 + \chi)$; therefore, we may write for high dielectric constant materials that $\varepsilon = \varepsilon_0 \chi$ and $(\partial E/\partial P)_T = 1/\varepsilon$.

Consider electric field–biased E_b operation of a pyroelectric-ferroelectric material with a spontaneous polarization P_s. There are significant performance gains from this mode of operation, which will become evident from the analysis that follows. The dielectric displacement D takes the form:

$$D = \varepsilon E_b + P_s$$
$$(\partial D/\partial T)_E = p' = (\partial \varepsilon/\partial T)_E\ E_b + (\partial P_s/\partial T)_E \tag{4}$$

where p' is the total pyroelectric coefficient. The first term on the right-hand side in Eq (4) is the dielectric contribution (dielectric bolometer mode), whereas the second term is the spontaneous polarization contribution (pyroelectric mode). Rigorously, the dielectric contribution term should be written as a field integral to account for dielectric nonlinearities of the ferroelectric material. We also have:

$$(\partial E/\partial T)_D = (\partial D/\partial T)_E/(\partial D/\partial E)_T = p'/\varepsilon = (1/\varepsilon)[(\partial \varepsilon/\partial T)_E\ E_b + (\partial P_s/\partial T)_E] \tag{5}$$

It must be emphasized that p' and $(\partial E/\partial T)_D$ do not vary linearly with bias field E_b as implied by the equations above. This is because of the nonlinear dielectric response to E_b. The derived quantities p' and p'/ε are related to the material figures of merit, as will be described next. The principle of detector operation near the transition temperature T_t, where properties are maximal, is schematically illustrated in Figure 1. A dc bias field (typically $\sim 10^6\ Vm^{-1}$) is applied to prevent depoling and boost signal. Properties of some pyroelectric-ferroelectric barium strontium titanate compositions [13] are listed in Table 1.

The material properties that are necessary for analyzing the performance (signal and noise) of a pyroelectric bolometer have been presented in scalar form. Rigorously, the tensor representation should be used for completeness and deeper

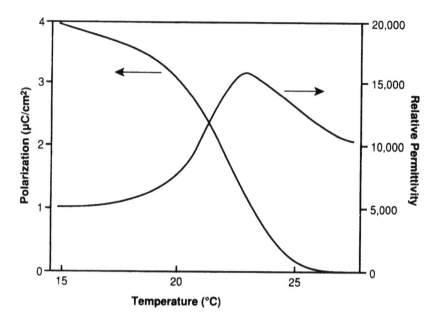

Figure 1   Schematic depiction of the polarization and dielectric responses of a ferroelectric bolometer [10]. (Courtesy of Texas Instruments.)

Table 1   Properties of dc-Biased Barium Strontium Titanate [13]

Ba/Sr ratio	Bias (kV/cm)	T_t (°C)	p'_{max} (μC cm^{-2} K^{-1})	ϵ_r	tan δ	F_j (cm^3 J^{-1})$^{1/2}$	F_v (cm^2 C^{-1})
67 : 33	1	21	23	31000	0.028	0.84	2700
67 : 33	2	22	6.3	33000	0.021	0.25	670
67 : 33	6	24	0.70	8800	0.004	0.12	280

$F_v = p'/C_p \epsilon \epsilon_0$
$F_J = p'/C_p \, (\epsilon \epsilon_0 \tan \delta)$
$C_p = 3.2$ (J cm^{-3} K^{-1})

insights. In polarizable deformable nonlinear dielectrics, such as bonded ferroe-lectric BST arrays, the effect of elastic, thermal, and electrical boundaries on the pyroelectric signal must be underscored [14]. For instance, the pyroelectric effect in a mechanically "free" and "clamped" crystal could differ greatly (both in magnitude and sign). Nonuniform heating causes undesirable signals due to ter-tiary pyroelectricity. On the other hand, the effect of bias field on the pyroelectric signal is influenced by the order of the ferroelectric phase transition.

C. Thermal Detection and Figures of Merit

The pyroelectric detector equivalent circuit can be represented by a current source $I = Ap(dT/dt)$ driving a parallel detector–load impedance. If C_d and R_d are the capacitance and leakage resistance of the detector and R_L and C_L are load resis-tance and capacitance (deliberate or parasitic), then the parallel detector–load capacitance C_E and resistance R_E are $C_E = (C_d + C_L)$ and $R_E^{-1} = G_E = R_d^{-1} + R_L^{-1}$. Here G_E is the electrical conductance and $C_d = \varepsilon_0\varepsilon A/z$ with z being the detector thickness. The circuit admittance is $Y = G_E + i\omega C_E$.

The detector (BST pixel) shown in Figure 2 is represented by a thermal capacitance H connected by a thermal conductance G_T to the surroundings (mesa, silicon ROIC, etc.). Solution of the heat transfer equation for a sinusoidally modu-lated incident thermal power $W = We^{i\omega t}$ yields the following expression for the detector voltage responsivity R_v:

$$R_v = |I/YW| = \eta\omega pA/[G_TG_E(1 + \omega^2\,\tau_r^2)^{1/2}\,(1 + \omega^2\tau_E^2)^{1/2}] \tag{6}$$

where η is the fraction of incident power absorbed by the detector, A is the detector area, and $\tau_T\,(= H/G_T)$ and $\tau_E\,(= C_E/G_E)$ are the thermal and electrical time constants, respectively. The voltage responsivity R_v has the following charac-teristics:

1. A maximum at a frequency $f = 1/2\pi(\tau_T\tau_E)^{1/2}$, given by $R_v\,(\text{max}) = \eta pA/G_TG_E(\tau_T + \tau_E)$,
2. A flat response within 3 dB of $R_v\,(\text{max})$, in the region bounded by the thermal and electrical relaxation frequencies,
3. For frequencies higher than the characteristic frequency $f\,(= 1/2\pi\tau_E)$, the response is attenuated by C_L, i.e., $R_v = \eta p/z\omega c_p(C_d + C_L)$, where c_p is the heat capacity per unit volume. Nonetheless, R_v exhibits a $1/f$ behavior.

Detector material figures of merit can be readily defined from the responsivity equation. For instance, if $C_d \gg C_L$ then $R_v = \eta p/A\omega c_p\varepsilon$, and a voltage respon-sivity figure of merit F_v may be taken as $p/c_p\varepsilon_0\varepsilon$. On the other hand, if $C_d \ll C_L$, a current responsivity figure of merit F_i may be taken as p/c_p. An additional figure of merit F_J, related to noise-limited sensitivity, will be introduced in the next section.

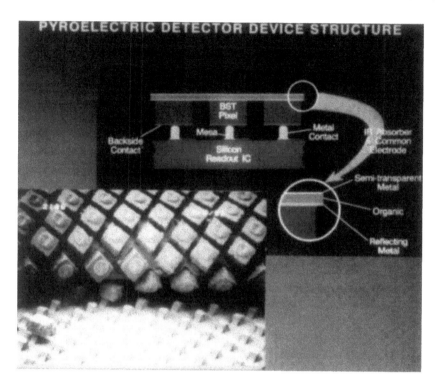

Figure 2   Array structure (a peel-off) showing the BST pixels, and the mesas for thermal isolation. The inset depicts vertical integration (hybrid process) of optical coating, BST, interconnect, and polymer mesa on the Si substrate [10]. Note the 1:3 connectivity pattern of BST pixels. According to this notation, the BST elements are connected in one dimension, *e.g.*, along the *z* axis in the elevation plane and totally disconnected in the target, *i.e. xy* plane. The second phase (xenon or air) surrounding the pixels is three dimensionally connected. (Courtesy of Texas Instruments.)

D. Noise and Sensitivity Limits

The pyroelectric detector sensitivity is limited by three noise sources: Johnson, temperature fluctuation, and amplifier current and voltage. Johnson noise is also known as the dielectric loss noise. There are two sources by which Johnson noise is generated in a capacitive detector: (1) leakage and (2) poor electrical contacts. A leaky capacitor is represented by an ideal capacitance C_d in parallel with a resistance R_d. The impedance is $Z_p = R_d/(1 + i\omega R_d C_d)$, where $R_d = 1/(\omega C_d \tan \delta)$. For a high-quality ferroelectric, $\tan \delta$ is very small (<0.01), hence dielectric loss contributions are insignificant. Poor electrical contacts can have a profound effect on losses. A series resistance R_s can contribute to dielectric losses. The impedance of a series capacitance–resistance circuit is $Z_s = R_s + (1/i\omega C_d)$, where $\tan \delta = \omega R_s C_d$. The series resistance has to be in the $\sim 10^6\ \Omega$ range to significantly contribute to losses. The spectral density of Johnson noise is $v_J^2 = 4kT\ [R_E/(1 + \omega^2 R_E^2 C^2)]$, where $R_E = R_d R_L/(R_d + R_L)$ is the total parallel detector–load resistance, and $C_E = C_d + C_L$ is the total parallel detector–load capacitance. For high-frequency or large R_L, we have:

$$v_J^2 = 4kT\ (\tan \delta/\omega)[C_d/(C_E^2 + C_d^2 \tan^2 \delta)]$$

or

$$\sim 4kT\ (\tan \delta/\omega)[C_d/(C_d + C_L)^2]$$

for low-loss ferroelectrics.

From the expression above, the dielectric loss noise has the following characteristics: (1) it is attenuated by the load capacitance C_L; (2) it behaves as $1/f$ noise at higher frequencies; (3) at lower frequencies R_L dominates the impedance and the response is flat. From the definition of detector sensitivity $D^* = (R_v/v_J)$, a figure of merit F_J for detectors limited by Johnson noise is taken as $F_J = p/c_p(\varepsilon\varepsilon_0 \tan \delta)^{1/2}$.

Temperature fluctuation noise is the limiting factor of the sensitivity of thermal detectors. It is the signature of the random exchange heat transfer by photons (radiation), phonons (conduction), or convection between the detector and its surroundings. The voltage fluctuation spectrum for a band-limited detector is:

$$v_T^2 = (4kT/H)[pA/(C_d + C_L)]^2\ [\tau_T/(1 + \omega^2\ \tau_T^2)]\ [\omega^2\ \tau_E^2/(1 + \omega^2\ \tau_E^2)] \tag{7}$$

If the characteristic frequency $f_E\ (= 1/2\pi\tau_E)$ of the high-pass filter is comparable with the thermal bandwidth, then the noise is significantly suppressed. Otherwise, if τ_E is large, then the high-pass filter has very little effect on the noise. As it turns out, $D_T^* = (R_v/v_T) = \eta(A/4kT^2\ G_T)^{1/2}$ is independent of materials parameters.

E. Signal Enhancement

Dielectric bolometers fabricated from polycrystalline BST are characterized by high electrical resistivity. Therefore, no signal degradation will occur as a result of charge spreading. On the other hand, they suffer from a large lateral thermal diffusion coefficient D_T. As a result, the thermal distribution in the target (detector) plane will be degraded, and so will the signal levels. The spatial resolution on the target plane is determined in accordance with the thermal modulation transfer function (MTF):

$$\text{MTF} = (f_c/\pi^2 D_T n^2) \tanh(\pi^2 D_T n^2/f_c) \tag{8}$$

where D_T is the lateral thermal diffusion coefficient ($m^2 s^{-1}$), f_c is the chopper frequency (Hz), and n is the spatial frequency (line pairs/cm). Note that D_T is defined as the ratio of thermal conductivity κ and the heat capacity per unit volume, $D_T = \kappa/\rho C_p$, where ρ is the density and C_p the specific heat. Therefore, the voltage responsivity R_v as defined by Eq. (6) is modified according to:

$$R_v(n) = [\text{MTF}][R_v(0)] \tag{9}$$

The lateral thermal diffusion coefficient for polycrystalline barium titanate $BaTiO_3$ is $\sim 2 \times 10^{-7} m^2 s^{-1}$. This is 2 orders of magnitude larger than that of polymer PVF_2 ($\sim 2 \times 10^{-9}$). To compare the MTF for $BaTiO_3$ and PVF_2, consider a spatial line frequency $n = 80$ line pairs/cm (corresponding for instance to 50 μm thermal line width on 75-μm centers) and let the chopper frequency f_c be 50 Hz. The MTF values for polycrystalline $BaTiO_3$ and polymer PVF_2 will be 0.39 and 0.99, respectively. To minimize thermal spreading, the BST detector is reticulated (sliced) to form a 1:3 connectivity pattern (pixels). According to this notation, the BST elements are connected in one dimension, e.g., along the z axis in the elevation plane and totally disconnected in the target, i.e., the xy plane (Figure 2). The second phase (xenon or vacuum) that surrounds the detector is three-dimensionally connected.

F. Detector Material, Imaging Array Fabrication, and System Description

The detector material is doped ceramic barium strontium titanate: $(Ba,Sr)TiO_3$. BST was selected because of its high pyroelectric figure of merit in the neighborhood of the phase transition and because of its high dielectric constant [13,15]. The Sr content is approximately 33 mol %, which shifts the phase transition to the vicinity of ambient temperature. Dopants include a donor ion (e.g., Dy) on the A site of the perovskite lattice (ABO_3) and a transition metal ion (typically Mn) on the B site [15]. The donor ion serves to limit grain growth during sintering [16], and the acceptor ion increases the dc resistivity to $> 10^{13}$ Ω-cm, limiting leakage current under bias [15].

The overall manufacturing process flow has three branches, shown in Figure 3: the sensor element flow, the thermal isolation flow, and the assembly flow. The detector element fabrication begins with stoichiometric, doped BST powder, prepared by solution pyrolysis [15,17] and milled to a fine particle size. The powder is isostatically compacted to form a green blank approximately 200 mm in diameter. Sintering at 1450°C yields a dense ceramic wafer having a diameter of 150 mm and a grain size in the range 1–2 μm. Typical ceramic microstructures appear in Figure 4.

NaN# Ceramic Sensors

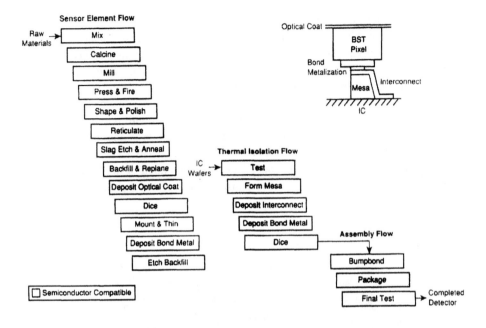

Figure 3   Process outline for manufacturing hybrid infrared detector arrays [99].

$(Ba_{0.67}Sr_{0.33})_{1-x}Dy_xTiO_3$

0.5% Dy   0.7% Dy

Figure 4   SEM microstructures of BST for two Dy-doping levels, 0.5 mol % (left) and 0.7 mol % (right). The polymer calibration spheres are 1.1 μm in diameter. The linear intercept grain sizes are 2.1 μm and 1.2 μm, respectively.

The wafers are ground and polished. A glass substrate provides mechanical integrity during thinning. The final thickness of the BST sensor layer is 15–25 μm. After laser reticulation, the wafers are etched to remove the slag and subsequently annealed in oxygen to reoxidize the surface. Several finished wafers, each containing 41 reticulated sensor arrays of 245 × 328 pixels, are shown in Figure 5.

In a separate flow, the silicon ROICs are provided with polyimide mesas for thermal isolation, and the desired metallization pattern is applied. Then the reticulated BST pixels are directly bump bonded to the ROIC. The pixels are on 48.5-μm centers and the kerf is about 10 μm. Reticulation of the BST detector material to produce a 1:3 connectivity array enhances the thermal modulation

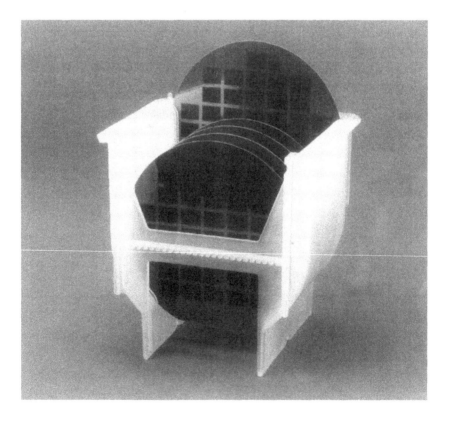

Figure 5 Reticulated 150-mm-diameter and 18-μm-thick BST wafers (on glass), each containing 41 pixel arrays [100].

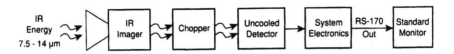

Figure 6 A block diagram of the thermal imaging system [101]. (Courtesy of Texas Instruments.)

transfer function (MTF) as described in the previous section. A resonant cavity IR absorber is formed on the IR-sensitive side of the detector. It provides better than 90% average absorption over the 7.5 to 14-μm spectral region. The finished detector is an array of pixels with a tight property distribution. Pixel-to-pixel uniformity is an important requirement for an imaging array. Variations in properties from pixel to pixel can result in undesirable fixed-pattern noise.

The readout IC is a complementary metal oxide semiconductor (CMOS) device. The unit cell consists of CMOS inverter preamplifier, a high-pass filter, a gain stage, a tunable low-pass filter, a buffer, and an address switch. The high-pass filter has a characteristic frequency $f_{3db} \sim 10$ Hz. It consists of the pixel capacitance and a large-feedback resistor ($\sim 10^{12}$ Ω). The low-pass resistance is a diode whose effective impedance is controlled by an in-cell current source. An off-chip voltage source determines the current level and the resistance. Therefore, the filter is tunable. The operation of the system is briefly discussed with the aid of the block diagram in Figure 6. An antireflection Ge window covers the package and allows IR transmission in the 7.5 to 14-μm spectral region. The IR lens (typically $f/1.0$) with 100 mm focal length projects the IR image on the focal plane BST array. The array is mounted on a thermoelectric cooler for optimal operation near the ferroelectric-paraelectric transition temperature T_c. The chopper provides the thermal modulation (30 Hz/line). The ROIC filters, amplifies, samples, and multiplexes the detector signal one row at a time and delivers the output at a standard RS170 rate.

A useful system performance figure of merit is the noise equivalent temperature difference (NETD). This is the temperature difference at the scene, ΔT_{scene}, that is required to generate an rms signal-to-noise ratio of 1. A remarkable NETD of less than 0.04 K has been achieved with recent pyroelectric imaging cameras fabricated from hybrid BST detectors that operate near room temperature.

G. Future Trends

The forces driving product evolution are cost reduction, performance improvement, and miniaturization. These goals are being addressed by reducing process complexity using thin-film detector materials that can be deposited directly on the ROIC, thus eliminating two branches in the process flow for hybrids. In one approach. NTC thermistor materials, principally vanadium oxide (VO_x) or amorphous silicon (a-Si), form the thermal detector in microbolometer arrays (see next section). The latter represent dc-coupled systems having certain disadvantages, however, and therefore efforts have also continued toward developing thin-film pyroelectric arrays [18–20]. Perhaps the most serious difficulty in using thin-film pyroelectrics is in achieving the necessary response with processing temperatures sufficiently low as not to adversely affect the silicon-integrated circuitry.

Initially, thin-film lead zirconate titanate (PZT) sensors were developed [19], but their performance was not competitive with that of hybrid devices. Sensor materials with improved properties were investigated, including $PbTiO_3$ [21], $PbSc_{0.5}Ta_{0.5}O_3$ [22,23], and $(Pb,La)TiO_3$ [20,24]. Recent pyroelectric detectors using $(Pb,Ca)TiO_3$-based thin films have been developed that produce imagery comparable to that of hybrid devices, with excellent promise for even better performance, at lower cost [18,25]. Figure 7 illustrates the pixel design for the latter device. An example of a still pyroelectric image produced by a corresponding experimental thin-film focal plane array and $f/1.0$ optics is shown in Figure 8.

Impressive progress in thermal imaging using the pyroelectric effect in ferroelectric materials has occurred. It is possible to produce compact, low-cost IR cameras that operate near room temperature and are capable of delivering television-quality images. Detector (BST) reticulation to produce a 1:3 array

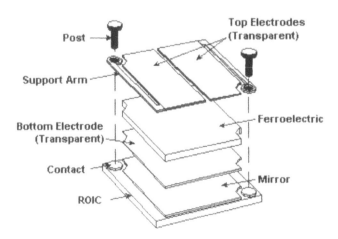

Figure 7   Exploded view of a thin-film ferroelectric pixel [25].

Figure 8   Single-frame thermal image extracted from a driving video, produced by a thin-film pyroelectric detector with $f/1.0$ optics [18].

enhances the thermal MTF, thereby reducing signal attenuation due to in-plane thermal diffusion. Applications of this technology are varied and include night vision, target recognition, driving and navigation aids in foggy and poor visibility weather, and locating of victims in heavy smoke–fire fighting. Equally impressive progress, from the viewpoint of ceramic processing, has resulted from its intersection with semiconductor processing technology. This has led to a capability for routine manufacturing of devices, each containing an assemblage of about 80,000 individual ceramic elements on a scale of $50 \times 50 \times 20$ μm per pixel, and doing so in reasonably high volume; with excellent uniformity and reliability.

III. NTC THERMISTORS

Faraday discovered and reported on the semiconducting behavior of Ag_2S in 1833. For the next 100 years the negative temperature coefficient behavior of the resistivity of semiconductors remained a curiosity, due to the difficulty in reproducibly making useful devices. Commercial utilization of ceramic sensors began in the 1930s and early 1940s with pioneering work at Siemens, Philips, and later Bell Telephone Laboratories. Major applications of these early devices were in telecommunications, circuit temperature compensation, and temperature measurement.

More sophisticated applications were introduced, and sensor manufacturing expansion occurred, in the postwar era (1950–1960) when aerospace and cryogenic devices were introduced. Greater emphasis on sensor stability and performance resulted in better sensor housing designs and materials capabilities. From 1960 to 1970, accuracy and interchangeability were emphasized for the newer applications. Significant development efforts for medical applications were also made as a result of government support programs. Thermistor manufacturers began to focus on volume production, and high-accuracy "chip" and bead thermistors were introduced.

This branch of ceramic sensing is thus quite mature and need not be discussed at great length here. Rather, newer technology utilizing ceramic thin films will be emphasized, and emerging applications of microbolometers for thermal imaging will be highlighted. A rich literature adequately covers preexisting technologies [1,26–28].

A. Semiconductors

The electrical conductivity (σ) of a semiconductor is the sum of several terms:

$$\sigma = Z_1 e \mu_1 n_1 + Z_2 e \mu_2 n_2 + \cdots \tag{10}$$

where e is electron charge, n_i represent the concentrations of various charge carriers (electrons, holes, ions) and μ_i represent their respective mobilities. Ions may have multiple charges, i.e., $Z_i > 1$. In practical devices one type of charge carrier tends to dominate, so the resistivity (ρ) becomes:

$$\rho = \sigma^{-1} \sim (Z_1 e \mu_1 n_1)^{-1} \tag{11}$$

or for a p-type semiconductor $\rho = (e \mu_h p)^{-1}$ (for example) where p is the hole concentration and μ_h the hole mobility. Both the carrier concentration and mobility are temperature dependent and can be thermally activated. At low temperature, extrinsic carriers are generated from impurity levels in the band gap:

$$n = BT^{3/4} \exp\left(-\frac{E_d}{2kT}\right) \tag{12}$$

$$p = CT^{3/4} \exp\left(-\frac{E_a}{2kT}\right) \tag{13}$$

in which E_d and E_a are the donor and acceptor ionization energies and B and C are constants proportional to the square root of the donor and acceptor densities, respectively, and k is Boltzman's constant.

At sufficiently high temperatures (*i.e.*, $1/2\ kT \sim E_d$, E_a), full ionization of the impurities is achieved and the carrier densities become independent of temperature. This region, called the "exhaustion" or "saturation" region in semiconductor terminology, separates the thermally activated intrinsic region at elevated temperatures (both electron and hole excitation across the band gap, E_g) from the much weaker thermally activated region at low temperature.

In ionic compounds such as oxides, the electron mobility is determined in large part by its interaction with the polar modes of the crystal. Here the electron induces a deformation in the surrounding lattice, thus tending to trap itself in a potential well of its own making. The electron and its surrounding polarization cloud are commonly referred to as a polaron.

Two types of polarons are distinguished. When the radius of the well is many atomic spacings, the electron continues to move in a band but with an enhanced effective mass. The mobility of a large polaron above the Debye temperature is given by:

$$\mu = \mu_0 T^{-1/2} \tag{14}$$

and is expected to be about $1-100\ \text{cm}^2/\text{V-s}$. If, instead, the well radius approaches one atom spacing, the electron becomes self-trapped at a given lattice site and can move to an adjacent site only by an activated hopping process similar to that exhibited by ionic diffusion. Small polaron mobility is described by:

$$\mu = \left[\frac{(1-c)ea^2 v_0}{kT} \right] \exp\left(-\frac{E_H}{kT} \right) \tag{15}$$

in which E_H is the electron hopping energy, v_0 the attempt frequency, and $(1 - c)$ the fraction of sites not already occupied by other electrons. Small polaron mobilities are generally reported to be on the order of 10^{-4} to $10^{-2}\ \text{cm}^2/\text{V-s}$ at elevated temperatures. Since many thermistor formulations include large fractions of transition metal cations, small-polaron transport can often be expected given the narrow $3d$ bands characteristic of these elements. Hopping energies typically add $0.1-0.4$ eV to the overall activation energy of conduction.

Ionic conduction may also occur, particularly at elevated temperature, but it is not a significant contributor in thermistors. As a matter of fact, precautions are always taken to avoid ionic conduction because it contributes to instability. A full discussion of ionic conduction relevant to ceramic sensors is given by Tuller [1,29–31].

To summarize this section, the temperature coefficient of resistance (TCR) value of material in which the bulk transport properties are controlling is related to (1) carrier generation and (2) carrier mobility. In general, at least one of these factors is thermally activated, and so the resistivity takes the form:

$$\rho = \rho_0 \exp\left(\frac{E}{kT}\right) \tag{16}$$

When discussing practical devices, resistance R is generally used and Eq. (16) is conventionally rewritten in the form:

$$R = A \exp\left(\frac{B}{T}\right) \tag{17}$$

where the material constant, B value or β factor, is simply E/k, and A is a constant related to device dimensions and material resistivity. Typical values of B fall around 3500 K, which corresponds to an activation energy of about 0.3 eV. It should be noted that B does not always remain constant over the full range of interest but may vary due to a transition between transport regimes, as discussed above.

Another constant already mentioned (i.e., TCR) is often designated by the symbol α and is defined by:

$$\alpha = \frac{1}{R}\frac{dR}{dT} \tag{18}$$

It may easily be shown by differentiating Eq. (17) with respect to temperature that α and B are related to each other by:

$$\alpha = \left(-\frac{B}{T^2}\right) \tag{19}$$

B. Thermistor Materials

Traditional NTC thermistors are formed from oxides with the spinel crystal structure based on $NiMn_2O_4$. They are bulk effect devices possessing very stable temperature-dependent electrical resistivity. Principal applications include precise temperature measurement and control, and temperature compensation of other nonlinear components and circuits. Other, more specialized NTC materials include Ti-doped Fe_2O_3 and Li-doped $(Ni,Co)O$. More recently, thin-film thermistors fabricated from vanadium oxide (VO_x) and amorphous silicon (a-Si) have found use in microbolometers for IR imaging. The latter application will be highlighted below.

Materials in the oxide system $(Mn,Ni,Fe,Co,Cu)_3O_4$ possess high temperature coefficients (B = 3000–4500 K) and excellent stability. The conduction mechanism requires that ions of different valence exist on the octahedral B site (AB_2O_4). Alternate ion distributions are possible, even for fixed chemistry [27]:

$$(Ni_{1-x}^{2+}Mn_x^{2+})\ (Ni_x^{2+}Mn_{2-2x}^{3+}Mn_x^{4+})O_4 \tag{20}$$

and

$$(Mn_{1-x}^{2+}Mn_x^{3+})\ (Ni^{2+}Mn_x^{3+}Mn_{1-x}^{4+})O_4$$

In both cases charge transport can occur because ions with different valence are present on the closely spaced B sites. The loosely bound electrons ($Mn^{4+} + e^-$ $\leftrightarrow Mn^{3+}$) become mobile by virtue of the availability of nearby unoccupied sites. The carrier density is

$$N_d = 16P_d/a^2c \tag{21}$$

where P_d is the probability that a B cation is an electron donor, and a and c are the lattice parameters. The mobility is related to the probability that lattice vibrations will induce hopping, and is thermally activated [Eq. (15)]. The high activation energy for hopping gives rise to a correspondingly high dependence of conductivity on temperature, making such materials especially suitable for application as temperature sensors. Figure 9 illustrates how the resistivity can be adjusted by varying the composition in the Mn-Ni-Fe spinel system.

For the intended applications, stability is an essential requirement. Potential sources of instability include:

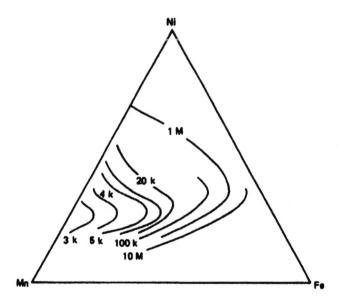

Figure 9 NTC thermistor resistivity contours at 25°C for the compositional system Ni-Fe-Mn oxides [1].

Changes in composition due to oxidation–reduction reactions, point defect migration, or diffusion of ions from the contacts

Ionic rearrangements due to more favorable free energy at the temperature of use in comparison with the temperature of preparation, and

Possible changes in surface states at the grain boundaries (*e.g.*, by adsorption of electron donor molecules)

The first and second sources of instability can be mitigated by proper choice of composition, and by annealing or preaging the device. Further, it is desirable that the operating range corresponds to the saturation region where carrier generation is no longer thermally activated [1,32]. The last may require encapsulation. Highly stable thermistors are commercially available. However, a trade-off always exists between stability and cost that must be considered in the selection of a sensor for a given application.

Although the $NiMn_2O_4$-based spinels used for conventional thermistors can be prepared in thin-film form, with the TCRs comparable to bulk values (B as high as 6500 K) [33], the materials of choice for IR sensor films are presently limited to VO_x and a-Si. The spinels are not welcome in semiconductor front ends. The same would seem to be true for VO_x, but serendipitously the technology for processing VO_x had already been developed for other applications [34]. Figure 10 depicts the relationship between the TCR (α) of sputtered VO_x thin films as a function of electrical resistivity [35]. It should be noted that $x < 2$, since it is desirable to have low electrical resistivity and a smooth change in resistivity with temperature (i.e., absence of any phase transition such as occurs for $x = 2$). The α coefficient for typical IR detector films is -0.02 to -0.03 K^{-1}. The relationship of $\alpha(\rho)$ typifies semiconductors; higher resistivity corresponds to higher TCR, other things being equal.

Because of excellent front-end compatibility, silicon has also been considered for thermal sensors. The temperature coefficient of resistivity for polysilicon is just -0.005 K^{-1}, but processing simplicity makes it a potential sensor candidate [36]. Amorphous Si (a-Si), more precisely a-Si:H, has a higher TCR of about -0.027 K^{-1}, i.e., similar to that of VO_x, and this makes it a strong candidate as well [37]. Additional materials with higher temperature coefficients exist, *e.g.*, a-SiC with $\alpha \sim -0.05$ K^{-1} [36], $NiMn_2O_4$ with similar α (mentioned above), and GaAs ($\alpha \sim -0.09$ K^{-1} [34]. Phase transition materials such as YBCO and VO_2 have even higher temperature coefficients, but until now such materials suffered from process incompatibilities, complexities or stability problems.

C. Microbolometer Sensor Arrays

As noted in the discussion of pyroelectric sensor arrays, several drivers have prompted the investigation of alternate thermal sensor technologies, and micro-

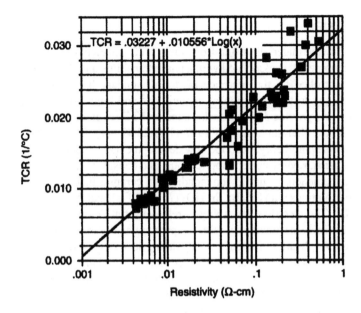

Figure 10 TCR *versus* resistivity for VO$_x$ [35]. (© 1995 IEEE.)

bolometer arrays offer potentially significant cost benefits. Efforts to produce practical devices using nickel manganite films [38] and CMOS n-well polysilicon membranes [36,39] have demonstrated some success in the laboratory. However, the only devices on the market that can compete with the hybrid pyroelectric arrays are systems using thin films of VO$_x$ [34,35] or a-Si [40,41].

Pixel structures, including thermal isolation gaps, are formed using standard micromachining techniques. The sensor membranes are usually deposited by sputtering, and the properties are controlled by the process conditions and doping. Although a-Si can also be deposited by plasma-enhanced chemical vapor deposition (CVD), the method normally used for manufacturing solar cells and thin film transistor (TFT) liquid crystal displays, such films exhibit high excess noise, attributed to uncontrolled dangling bonds. This appears to be less of a problem with sputtered films.

In theory, microbolometers should be less costly than hybrid systems, but this is not yet the case because temperature control and the use of a chopper cannot be eliminated without degrading performance. In addition, the detector is only one part of the system, and each camera requires similar electronics, packaging, and costly germanium optics. So the earlier commercialization, installed manufacturing base, economy of scale, and production experience in hybrid pyroelectric detectors offset the cost savings derived from the simpler array fabrication of microbolometers. All three types of device currently share the market.

IV. PTC RESISTORS

Ceramic PTC thermistor materials differ from their NTC counterparts in two ways. The PTC phenomenon is associated with a phase transition, and the electrical resistivity is controlled by the presence of grain boundary barrier layers. Similar barriers are key to the operation of ceramic gas and humidity sensors. Several general reviews provide detailed background on PTC resistor technology and summarize the key literature through 1987 [1,27,28,42–44]. This chapter will summarize the relevant theory, describe new developments in understanding operational mechanisms, review current applications, and consider future trends.

The PTC phenomenon occurs in semiconducting, ceramic $BaTiO_3$ ($T_c \sim$ 120°C) and its solid solutions with Sr: $(Ba,Sr)TiO_3$ ($T_c < 120$°C), and Pb: (Ba,Pb)-TiO_3 ($T_c > 120$°C). The electrical resistivity is relatively flat below the Curie temperature (T_c) and increases markedly with temperature as the transition temperature is traversed. Applications include self-regulating heaters, current limiters, over-current protectors, and temperature sensors. Devices are used in home appliances, automobiles, telecommunication equipment, and industrial processes. Polymeric composite PTC resistors, based on carbon-filled, crystalline polyethylene, also exist and are useful in certain protector applications (see Ref. 43). Figure 11 illustrates the small-signal resistivity versus temperature for several compositions exhibiting transition temperatures from -25°C to $+300$°C.

The PTC in semiconducting titanates arises from the change in resistance of grain boundary Schottky barriers during the phase transition from the low temperature, ferroelectric state to the high-temperature paraelectric state. Below T_c the grains are polarized, and the spontaneous polarization compensates trapped interfacial charge. Above T_c the spontaneous polarization vanishes; this, along

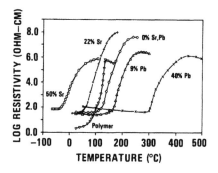

Figure 11 *R/T* behavior of ceramic and polymer composite PTC resistors [43]. [Reprinted by permission from the Society for the Advancement of Material and Process Engineering (SAMPE).]

with the changing dielectric permittivity, affects the shapes, heights, and distributions of barrier potentials, giving rise to the observed dependencies of resistivity on temperature and applied voltage.

A. Barrier Layers

By their nature, grain boundaries represent inhomogeneities in structure and composition. Compositional inhomogeneities can be marked due to processes such as impurity segregation and/or kinetically limited reactions in the vicinity of the boundaries. These grain boundary defects and impurities give rise to a distribution in energy of a planar array of localized interface states. If we assign different individual Fermi energies E_{FB} and E_{FG} to the boundary regions and to the grains prior to their effective contact and equilibration with each other, we can expect the formation of space charge regions and potential barriers subsequent to such equilibrium. For two semiconducting grains and their boundary, one now has, in effect, two back-to-back diodes with barrier heights ϕ_{BO}, space charge layer widths $d_l = d_r = d$, and boundary width w, which for real materials is typically much smaller than d_l and d_r. The symbol ζ represents the energy difference between the Fermi energy and the conduction band edge.

Application of the one-dimensional form of Poisson's equation to such junctions gives the following relationship between the built-in barrier height ϕ_{BO} and the depletion layer width d:

$$\phi_{BO} = \frac{qN_D d^2}{2\varepsilon_r \varepsilon_0} \tag{22}$$

where N_D is the donor density in the bulk and ε the dielectric constant. If we take N_b to be the two-dimensional charge trapped in the boundary states, then

$$N_b = 2dN_D \tag{23}$$

to satisfy charge neutrality for the junction. Therefore, ϕ_{BO} may be rewritten as:

$$\phi_{BO} = \frac{qN_b^2}{8\varepsilon_r \varepsilon_0 N_D} \tag{24}$$

Note the inverse relationship between the barrier height and the dielectric constant (ε_r). This relationship plays a crucial role in ferroelectrics in which ε_r is a strong function of temperature in the vicinity of the Curie temperature (at least above T_c where $BaTiO_3$ can be considered to be a linear dielectric).

When a bias voltage V is applied across the boundary, the spatial energy distribution changes, with the left side going into forward bias and the right side into reverse bias. Beginning with the assumption that the only important mechanism for current flow across the barrier is thermionic emission, Pike and Seager [45] derived the following current–voltage relationship:

$$J = A \exp\left(\frac{\zeta + \phi_B}{kT}\right)\left[1 - \exp\left(-\frac{eV}{kT}\right)\right]$$

(25)

in which J is the current density, A the Schottky–Richardson prefactor, ϕ_B the band bending, and V the applied voltage.

For a large percentage of polycrystalline oxide semiconductors, space charge barriers at the grain boundaries result from segregation of dopants, more rapid redox equilibrium at the grain boundaries due to enhanced grain boundary diffusion, electron exchange with adsorbed molecules, and/or intrinsic interface states. For NTC thermistors a B value roughly equal to ϕ_{BO}/k would result, in addition to any temperature dependence resulting from bulk effects already discussed.

Metal electrode–semiconductor contacts are often also rectifying. Analysis of such Schottky barriers proceeds in a fashion similar to that of p-n junctions except for the fact that the space charge is now entirely localized on the semiconductor side. In developing thermistors and related devices, one is often concerned with identifying electrodes which are "ohmic" and of low resistance, thereby adding little to the temperature dependence of the device. In general, however, the TCR of a material is made up of bulk, grain boundary, and electrode contributions. In developing a device with well-defined characteristics, each of these factors must be taken into account and controlled.

B. Ferroelectric Implications

The importance of ferroelectric polarization in controlling the resistivity below T_c was recognized qualitatively by Jonker [46]. Quantitatively, Poisson's equation is not strictly applicable to ferroelectrics, since a linear relationship between field and displacement (i.e., ε, = constant) is assumed. Instead of Poisson's equation, one must begin with Gauss' law [47]. The dielectric nonlinearity can be represented via Devonshire's thermodynamic formalism:

$$E = \chi P + \xi P^3 + \zeta P^5$$

(26)

where E is electric field, P is polarization, χ is the dielectric susceptibility $(\varepsilon\varepsilon_o)^{-1}$, and ξ and ζ are third- and fifth-order coefficients, respectively. The temperature dependence of the susceptibility is given by the Curie–Weiss law:

$$\chi = (T - \Theta)/C\varepsilon_0$$

(27)

where Θ is the Curie–Weiss temperature and C is the Curie constant. The remaining coefficients and their temperature dependences are known for barium titanate as well (see Refs. 47–48).

The condition wherein the spontaneous polarization is normal to the interface leads to:

$$qN_D\phi = 1/2\chi(P - P_s)^2 + 1/4\xi(P - P_s)^4 + 1/6\zeta(P - P_s)^6 \qquad (28)$$

The width of the depletion layer becomes:

$$2d = (qN_b - 2P_s)/qN_D \qquad (29)$$

Since $P_s = (\xi/2\zeta)^{1/2} [1 + (1 - 4\chi\zeta/\xi^2)^{1/2}]^{1/2}$ for $T \le T_c$, and $P_s = 0$ for $T > T_c$, ϕ can be calculated as a function of N_D, N_b, and T [47]. The resistivity versus temperature curves obtained in this manner resemble real behavior, although incompletely, since piezoelectric charges due to grain clamping are ignored, and a distribution of ferroelectric domain orientations exists in actual ceramic microstructures such that only a variable fraction of the polarization can screen the trapped interfacial charge.

Still, a picture has emerged to suggest that low resistivity of PTC resistors below the Curie temperature occurs because of polarization compensation, and the resistivity increases as the spontaneous polarization vanishes above T_c. In addition, one would expect to observe considerable variation in R-T behavior as a function of grain orientation, and this is in fact seen experimentally [49–51].

C. Grain Boundary Chemistry

The grain boundaries appear to be more complex than generally appreciated. Alternate proposals to account for the interfacial trapping sites include Jonker's suggestion that the interfacial charges arise from electron transfer to chemisorbed oxygen molecules [46,52] and Daniels' strong theoretical case in favor of metal ion vacancy enrichment in the neighborhood of the interface [53–56]. In addition, trapping sites could be affected by concentration of transition metal ions near the grain boundaries. The interfacial chemistry develops during sintering.

A typical PTC thermistor composition (with $T_c = 180°C$) can be represented by the formula:

$$(Ba_{0.9}Pb_{0.1})_{0.996}Y_{0.004}Ti_{0.999}Mn_{0.001}O_3 \qquad (30)$$

Several mol % of liquid phase forming compound may be added to the formula. The donor dopant could be a different trivalent ion residing on the A site (e.g., La), or a pentavalent ion (e.g., Nb) on the B site. The liquid phase former could be excess TiO_2 (eutectic $\sim$ 1320°C), SiO_2 (eutectic $\sim$ 1260°C), a combination of TiO_2 and SiO_2, or "AST," a mixture of Al_2O_3, SiO_2, and TiO_2 in the mole ratio 4:9:3. Sintering occurs above the eutectic temperature (e.g., at 1350°C), and a certain amount of liquid phase forms. The various oxides dissolve in the liquid, and grain growth of the doped titanate proceeds by recrystallization from the liquid. Oxygen evolution accompanies donor incorporation during grain growth

[57]. In addition, it is likely that segregation occurs when component oxides become partitioned between the liquid and crystalline phases [58].

The resistivity minimum shifts to higher donor concentration as the amount of liquid phase increases [59]. For the composition above, the resistivity minimum occurs near 0.4 mol % Y. The minimum also shifts to higher donor concentration as the acceptor concentration increases [60,61]. The latter effect is shown in Figure 12. As the amount of liquid is increased, the dopant concentrations in the liquid (and thus incorporated in the solid) decrease, unless compensated by a higher overall concentration in the formula. Likewise, acceptor ions compensate an equivalent number of donors. The latter effect is crucial in manufacturing, since native impurities in the raw materials are predominantly acceptors. Variable impurity levels can thus cause severe problems in maintaining satisfactory control of the manufacturing process.

After the grain growth portion of the sintering process completes (e.g., at 1350°C), the material is cooled. The liquid tends to collect at triple points where most of it solidifies. Transmission electron microscopy (TEM) micrographs generally do not detect significant amounts of secondary phase within the grain boundary region itself. However, it is conceivable that acceptor-enriched remnants could still be present in the surface layer in the immediate vicinity of the boundary. Titanate "particles" have been observed using high-resolution TEM [58]. Also, theoretical considerations suggest that acceptors do tend to become enriched at the grain surfaces [62].

In order to develop optimal PTC properties, the cooling process must be conducted in an oxidizing atmosphere, and the initial cooling rate must be reason-

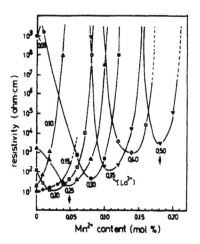

Figure 12 Room temperature resistivity of La-Mn codoped BaTiO$_3$ [61].

ably slow, e.g., 100°C per hour. Instead of slow cooling, the ceramic can be cooled more rapidly but annealed at about 1150°C. This annealing step is thought to foster grain boundary oxidation, either through chemisorption of O_2 or by a diffusion process that produces a thin layer of metal ion vacancies in the vicinity of the interface. Further support that oxygen penetration along the grain boundaries is crucial for good PTC property has been provided by means of O^{18} tracer studies [63]. The oxidant need not be O_2; adsorption of halogen gases (electron acceptors like O_2) also enhances the PTC effect [64,65]. The latter observation lends support to the significance of chemisorbed O_2 as the primary source of interfacial traps. STEM studies confirm that excess oxygen is present at the interface and that the acceptor-rich space charge formed at high temperature is retained after cooling [66]. Similar conclusions have been reached based on analysis of impedance spectra [67].

Additional support arises from the ease of chemical degradation when the PTC resistor is exposed to reducing molecules at fairly low temperatures. Significant resistivity degradation can occur under very mild conditions [68]. An example is displayed in Figure 13. In this example, PTC resistors were plated with electroless nickel and electroplated with gold over the Ni. The PTC virtually disappeared after gold plating (50°C, 5 mA/cm^2) and annealing at 350°C in air. Samples that were not electroplated did not degrade. It is surmised that the degradants originate in the bath chemistry and their diffusion into the PTC ceramic is promoted by reactions occurring during the electroplating process. Substantial recovery of the PTC occurred after further annealing the degraded samples in air at 500°C. Similar results were observed with unplated ceramics, either annealed at 600°C in vacuum, or exposed to NH_3 gas.

Mass spectral analysis results are reproduced in Table 2. Clearly, the degraded ceramics were characterized by absorption of H_2, CO, and CO_2 (or their precursors). The principal degradant was hydrogen, since the other molecules were largely retained following recovery. The key point here is not that H_2 molecules, acting as electron donors, serve to neutralize the interfacial acceptor states and release the trapped electrons to the bulk; this behavior is analogous to the operational mechanism of ceramic gas sensors. Rather, it illustrates the ease with which adsorption and accompanying electron transfer reactions occur at the grain boundaries in ceramic $BaTiO_3$. This, together the evidence cited above, enhances the plausibility of the chemisorbed oxygen model.

D. Role of Manganese

PTC resistors exhibit their best properties when the material is codoped with Mn [69]. Mn is incorporated in donor-doped single crystals on the lattice B site as Mn^{2+} and as Mn^{3+}. For otherwise undoped BST the Mn^{2+} exists above the Fermi level in the band gap, and Mn^{3+} occupies an energy level below E_F [70].

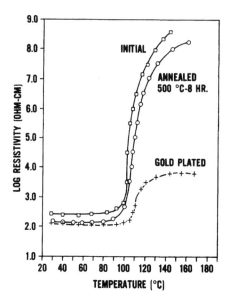

Figure 13 Degradation of PTC characteristics during gold plating and recovery after moderate temperature annealing in air [68]. (© 1986 IEEE.)

Table 2 Mass Spectral Analysis of Gases Evolved (cm^3/kg, 800°C, 1 hr) from PTC Ceramics During Vacuum Heating of Virgin, Plated, and Annealed Samples [68]

Gas evolved	As fired	Plated	Contact removed	Annealed
CO_2	1.1	29.8	14.7	33.2
CO	1.6	30.0	18.8	26.8
H_2	0.1	48.3	22.8	2.9
SO_2	—	0.04	—	0.5
HC	1.4	1.9	1.7	1.6
TOTAL	4.2	110.0	58.0	65.0

Note: For uniform 10-μm grains, one monolayer represents approximately 23 cm^3/kg. Sample thickness = 1.25 mm.

In ceramics, native acceptors including Mn, if present, exist within the grains but tend to preferentially segregate at the grain boundaries. Chemisorbed oxygen interacts with the segregated acceptor ions to form trapping levels in the gap [67]. The presence of Mn results in a wider and deeper distribution of trapping levels than would be the case if only native acceptors were present because Mn^{2+} can be oxidized to Mn^{3+}. (This argument is somewhat speculative but seems plausible.) Other multivalent acceptors, such as Ru, appear to function similarly [71].

E. Resistivity Versus Temperature

Considering the current evidence, the preferred model for the PTC effect may be described as follows. Donor doping produces free electrons within the grains. A depletion layer forms near the grain boundaries due to the existence of a distribution of interfacial electron traps in the band gap. The trapped charges give rise to potential barriers at the grain boundaries. Below the Curie temperature the trapped charges are compensated by ferroelectric polarization [Eqs. (28) and (29)], the barriers are small, and the resistivity is low. Above the Curie temperature, the spontaneous polarization within the grains vanishes and the barrier potential and resistivity grow due to rapidly decreasing permittivity [Eq. (24)]. Eventually the trapping states are pulled up to the Fermi level where the barrier potential and resistivity reach maxima, and thereafter decrease, as the trapped electrons are released. The NTC effect at high temperature is in large part due to thermal ionization of electrons across the grain boundary barriers by thermionic emission.

F. Applications and Future Trends

Specific applications for PTC resistors have not changed greatly since publication of the preceding edition of this volume. Several automotive applications related to efficient carburetion no longer exist. Other applications in refrigeration (motor starting), air conditioning (compressor crankcase heaters), self-regulating heaters, overcurrent protectors, and thermistors continue in wide use. Device design issues, including electrical contacts, voltage sensitivity, and stability of electrical properties, have been adequately discussed in the general references cited above. This is mature technology, like bulk ceramic NTC thermistors, and future progress is likely to be evolutionary (barring unforeseen circumstance, of course!). It seems unlikely that thin-film technology will apply to PTC resistors as we know them today. But progress in increasing our understanding of the physical mechanisms that determine the properties and behavior of PTC resistors is expected to continue as new instrumentation and measurement techniques become available for probing the microstructures and microchemistries of these materials on finer scales.

These semiconducting ceramics remain a very interesting and challenging subject for research.

V. GAS SENSORS

Several changing physical properties can be used for sensing gases. These include changes in resistance, current, potential, capacitance, work function, mass, temperature, or optical absorption. These variables determine the nature of the sensor design. In terms of ceramic sensors, ZnO, TiO_2, SnO_2, and ZrO_2 have been the most widely investigated for detecting gases such as CO, CO_2, O_2, NO_x, H_2, CH_4, and other hydrocarbons.

Several low-temperature ceramic sensors have been commercially successful. At the low-cost end, the most common ones are the SMO (semiconducting metal oxide), catalytic, and electrochemical devices. At the high-cost end are the infrared sensors. These are capable of detecting only nonlinear molecules such as CO_2. Chemical or gas sensors have applications in a wide range of industries such as the automotive, aerospace, environmental, food, safety, transportation, process, utility, and power industries.

In this section, we will examine the basic operating characteristics of the three ceramic-based gas sensors, namely, electrochemical sensors and solid-state devices, which include both the SMO and the FET (field effect transistor) sensors. Each type of sensor will be discussed using a specific example. The intent in this section is to provide only a brief overview of the three types mentioned above with citation to some general papers. In the interest of conciseness, excessive literature citations will not be reproduced here.

Several general references provide broader reviews, describing processing technologies, sensor materials, and detailed behavior of a variety of applications. Schwank [72] covers an overview of oxygen sensor technology, both past and current; Janata [73] gives an overview of potentiometric sensors, and Hill [1], Annino [74], Heisig [75], and Yamazoe [76] cover some details with respect to a few specific sensors. Maskell [77] provides detailed information on the physics involved in gas sensing using zirconia. Barsan and Weimar provide an in-depth analysis of the operation of SMO sensors [78]. Takahata offers a comprehensive review of tin dioxide sensors after 20 years on the market [79]. Lastly, the reader may be interested in the book of Moseley et al. [80] and the special issue on sensors edited by Wang et al. [81] in which a wide variety of gas-sensing technologies are compared.

A. Solid Electrolytes

In 1965, the first zirconium oxide–based O_2 sensor was introduced. Over the years, this type of sensor has become one of the leading oxygen sensors to monitor the level of oxygen in the exhaust system in the automotive industry. Figure 14 shows a schematic view of an oxygen sensor. The main function of the oxygen sensor in the exhaust system is to monitor and feedback the information on the level of oxygen in the system. This is crucial to maintain the stoichiometric air/fuel ratio centered around 14.5. By maintaining the air/fuel ratio in close vicinity of the so-called stoichiometric ratio, the three-way catalyst, placed in the exhaust stream, is able to simultaneously reduce NO_x, CO, and hydrocarbon emissions with high efficiency. The requirement arose from the increasing concern with the environmental impact of these pollutants. An adequate level of oxygen is necessary to ensure complete combustion of the fuel and minimize undesirable emissions.

In the oxygen sensor shown, a potential is developed when the partial pressure of oxygen on one electrode is different from that on the other electrode. The cell voltage depends on the ratio of oxygen concentration (related to partial pressure) as well as the temperature. The electromotive force (EMF) is given by Nernst equation:

$$EMF = RT/4F \ln [p(O_2)_{reference}/p(O_2)_{sample}] \tag{31}$$

where R is the gas constant, T is the absolute temperature in K, F is Faraday's constant, ln is the natural log, and p is the partial pressure.

In Figure 14, the reference gas used is air $(+)$ and the exhaust gas stream on the outside $(-)$ of the ceramic tube is separated by the electrolyte ZrO_2. To make the electrolyte more stable and the ions $(O_2{}^-)$ more mobile, ZrO_2 is doped with yttrium oxide (Y_2O_3). Y_2O_3 is usually chosen over the less expensive CaO (calcium oxide) due to the higher ionic conductivity induced in zirconia as well as improved electrode kinetics.

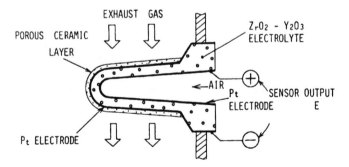

Figure 14 Schematic view of a practical zirconia auto exhaust sensor [1]. (Courtesy of Toyota Motors.)

The outer surface of the tube in Figure 14 shows a porous layer of platinum that serves as a catalyst for the reaction in addition to acting as a metallic conducting layer to measure the EMF. The inert oxide layer outside the platinum acts as a filter to keep away some of the gases that poison the sensor. Gases containing lead, halogens, sulfur, or phosphorus constituents are detrimental to the life of platinum, particularly in certain nonautomotive applications.

The amount of excess oxygen or fuel in the exhaust can also be determined by measuring the limiting electrochemical current of amperometric oxygen sensors. These become particularly important in so-called lean-burn engines. Here the partial pressure of oxygen does not strongly vary with the air-to-fuel ratio, in contrast to engines operating at or near the stoichiometric air-to-fuel ratio. Under these circumstances, sensors are needed that have a stronger than logarithmic sensitivity to oxygen partial pressure variations. Amperometric sensors rely on current limiting due to diffusion or interfacial phenomena at the electrode, which are linearly dependent on the partial pressure of the gas constituent [82]. Sensors based on this principle are also being developed to detect other gases including NO_x.

Badwal [83] has suggested that the electrodes (rather than the electrolytes) limit the sensor operation at low temperatures. The electrode impedance and the response times increase at low temperature and are more sensitive to CO. It has been suggested that using a mixed conductor such as CeO_2-Sm_2O_3 improves performance at lower temperature. Doping zirconia with CeO_2 was found to improve the kinetics of water vapor reduction.

The advantage of the electrochemical principle is a stable, linear, repeatable output over long periods at a reasonable cost. In addition to oxygen applications, electrochemical cell sensors are usually designed to be used as toxic gas monitors. However, most of the sensors based on the electrochemical principle are affected by changes in the environment such as temperature, pressure, and humidity. Where higher resolution is required, the effects due to these variables are compensated.

B. Solid-State Devices

Under solid-state ceramic sensors, there are two common types: the gas-sensitive SMO devices and the chemically sensitive FETs. A comprehensive treatment on the physics and various aspects of design of these sensors can be found in Ref. 73. First, a brief overview of the SMO sensors will be given followed by a review of FET sensors. Finally, a summary on the focus of future technology for gas sensing using ceramic materials will be provided.

1. SMO Sensors

SMO devices rely on the properties of semiconducting metal oxides. They typically operate at elevated temperatures (300–500°C), to avoid water adsorption, as well as to decrease response time. The practical performance of the SMO sensors is governed by three properties: sensitivity, selectivity, and stability. Depending on the target gas, the material, the surface structure, and the temperature are fine tuned to improve sensitivity and selectivity.

The most widely used oxide for gas sensing is SnO_2 (tin dioxide) with dopants added to improve specificity (TiO_2-based sensors and perovskite-type sensors are some others). The response of an SnO_2 sensor to various gases is shown in Figure 15.

The accepted model of an SnO_2 sensor is shown in Figure 16 [84]. The negatively charged oxygen adsorbates play an important role in detecting flammable gases such as CO and H_2 among others. Around 300–500°C, the O^- ion is most reactive with the flammable gases. With the n-type semiconductor, the oxygen adsorbate builds a space charge region on the surface of the metal oxide grains resulting in an electron-depleted surface layer. Electrons are transferred from the grain surface to the adsorbate. The depth of the space charge layer is

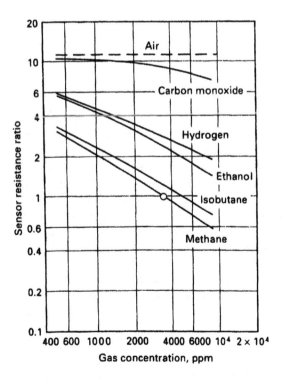

Figure 15 Response of a commercial SnO_2 sensor to various reducing gas concentrations in air. (Courtesy of Figaro Engineering Inc.)

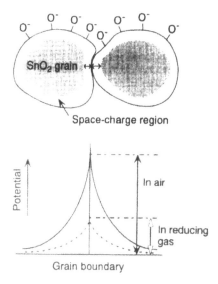

Figure 16 A model of a potential barrier to electronic conduction at a grain boundary [84].

a function of the surface coverage of oxygen adsorbates and the electron concentration in the bulk. In air, the sensor has a high resistance due to the potential barrier developed at each grain boundary or at the semiconductor–contact interface. When exposed to flammable gases at elevated temperature, the oxygen adsorbate takes part in the reaction. The electrons return to the oxide grain resulting in a decrease in the resistance. This change is measured and is used as an indicator to detect the presence of gases. With the p-type semiconductor, the resistance in air is low because the charge carriers are positive holes. In this case, the reaction of the oxygen adsorbates with the gases leads to an increase in resistance.

SMO sensors typically operate at elevated temperature. Sensors have SnO_2-based thick films deposited on a substrate (usually alumina). The substrate has a heater made of platinum that maintains the sensor at a desired temperature (Figure 17).

SMO sensors are used in several applications today. Many toxic and flammable gases are detected using this technology. Figaro Engineering Inc. [85] has successfully deployed this technology to detect a wide range of toxic and flammable gases. Many of the commercially available sensors have a small quantity of palladium (Pd) added to promote reducing-gas response. It also lowers the optimal

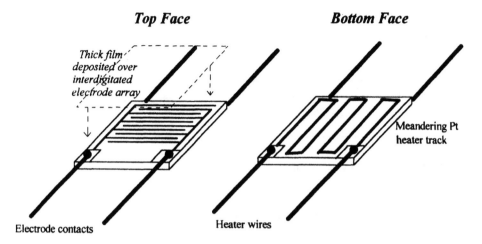

Figure 17 Schematic diagram showing the structure of a typical gas-sensor substrate used to support tin dioxide–based thick films [102].

detection temperature. Other noble metals such as Ag and Pt have also been studied to enhance the sensitivity of the film.

2. FET Sensors

Lundstrom [86] first reported gas-sensitive metal oxide semiconductor (MOS) devices to sense hydrogen. They are based on the field effect generated by gases in the (metal oxide semiconductor field-effect transistor (MOSFET) using catalytic metals. Gas molecules adsorbed on MOS structures with gas-sensitive gates induce a change in the threshold voltage, detectable as a shift in the capacitance–voltage or current–voltage characteristics. Field effect gas sensors have evolved from, firstly, simple Pd-gate devices to those with a variety of catalytic metals (Pd, Pt, Ir, Ru, Rh); secondly, thicknesses ranging from "thick" gates of >100 nm to "thin" gates of <10 nm; thirdly, silicon-to-silicon-carbide–based devices operating at much higher temperatures; and fourthly, from individual devices to arrays [87]. The practical sensitivity of the sensor is usually given by the practical detectable voltage change. Noise and drift can limit this value. A typical Si-SiO$_2$ system used in MOSFETs to detect hydrogen-based compounds is shown in Figure 18 [73].

Pd-gate devices are typically sensitive to hydrogen, hydrogen sulfide, and ammonia. At low concentrations of hydrogen, palladium (or a Pd alloy) Schottky diodes on silicon substrates have been used in the semiconductor electronic indus-

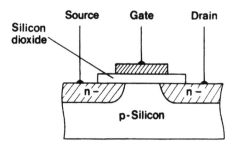

Figure 18   The basic MOS transistor [103].

try. The advantage of the Schottky diode is its high sensitivity in gas sensing. When the concentration of gases is high, then a resistor-based approach is used. Devices have been fabricated in which both approaches are integrated in a single device to cover the entire range. Hydrogen sensors are used primarily in leak detection, process monitoring, fuel cells, and electrical transformer monitoring applications.

The gate metal is most likely to be an alloy of palladium such as Pd-Ag or Pd-Cr. Other metals have also been investigated such as Pt, Au, Ni, Al, Cu, Mg, Zn. "Oxide" (dielectric) layers investigated include materials such as TiO_2, ZnO, GaP, and CdS. The gate metal in these devices has three functions: it dissociates the gas through catalytic action, it transports hydrogen atoms to the metal-oxide interface, and it adsorbs the hydrogen atoms at this interface as detectable dipoles [73]. The amount of hydrogen adsorbed on the Pd–semiconductor interface depends on the chemical reaction on the metal surface as well as the hydrogen pressure. In an inert atmosphere, these devices are extremely sensitive to hydrogen.

The properties of the Si-SiO_2 (or Si-Si_3N_4) system are sensitive to the oxide growth process, which introduces defects as well as chemical impurities. Alkali ions, mostly sodium, are detrimental to the device. Due to its high mobility even at room temperature, ionic charge will cause instability if present. Despite the stability and sensitivity issues, FET devices have been in use to detect hydrogen for various applications.

For low-temperature application ($<200°C$) the device is Si based. For higher temperature application such as process flue stacks and automotive exhaust systems, the device is SiC based. SiC is also chemically inert, making it attractive in corrosive environments.

C. Future Trends

The trend toward improving the performance of SMO-based gas sensors has been focused primarily on improving the material system/structure. One of the major areas of focus has been in developing nano-sized metal oxides. The main advantages are reduced grain size, tighter control of grain growth, and increased surface area. Increased sensitivity also results from the fact that the space charge layers formed during gas adsorption deplete a larger fraction of the volume of each of the nano-sized grains. Much effort is also being expended toward the production of SnO_2 and other SMO films by laser ablation, gas-phase condensation, sputtering, and so forth. Another area of research is using mixed-metal oxides to improve the stability over long periods of operation as well as sensitivity. Specificity is still one of the major concerns of SMO gas sensors. Much effort has been devoted to the development of sophisticated pattern recognition techniques to remove cross-sensitivity of devices. Figure 19 shows a sensor array fabricated on a silicon wafer which includes an integrated heater and temperature sensor. To reduce interference as well as to protect the sensor from poisoning, many sensors use special filters in the actual device.

Another major area of focus has been in miniaturization of these sensors. Using bulk micro-machining processing as well as CMOS technology, the sensing and heating elements are integrated on a thermally isolated membrane to reduce power consumption as well as reduce thermal time constant. Silicon-based substrates are being studied to replace the conventional alumina substrate. As the technology matures, there is reason to believe that mass production and integration of control functions on the same chip will make the price of these sensors even more attractive.

In addition, research laboratories around the world are working toward microelectromechanical systems (MEMS) – based approaches for sensor element

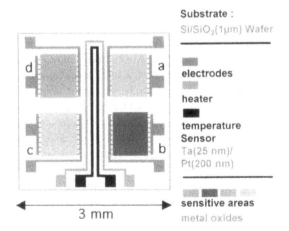

Figure 19   Design of a 2 × 2 multioxide gas sensor array [104].

fabrication, using nanomaterials for oxides and/or sensing elements to improve the overall performance of gas sensors. The MEMS approach also reduces power requirements quite dramatically. In addition, a multiple microsensor approach is gaining ground. In this approach, several types of sensors, such as resistors, diodes, and electrochemical cells, are integrated on a single device. Sophisticated algorithms will interpret the combined output. The response from such a sensor array will overcome some of the shortcomings of individual sensors, such as cross-sensitivity and selectivity. With the development of a combined microfabricated gas sensor array to detect multiple gases with a single system and by proper choice of material systems, the detection capabilities could be vastly improved. A review of advances related to the use of MEMS and microelectronic technologies in sensor technology was recently published by Tuller and Mlcak [87].

VI. HUMIDITY SENSORS

Humidity sensors represent a special category of gas sensor, limited to detection of H_2O molecules. They are important in control systems for industrial processes, agricultural processes, and human comfort. Ceramic humidity sensors represent one of several approaches. Other types of humidity sensor in widespread use include devices based on polymer dielectrics (capacitive devices) and polyelectrolytes (resistive devices). Ceramic devices are especially useful when heat cleaning can be used for restoration of proper sensing action after exposure of the sensing element to contamination. General reviews are available that describe the various approaches and the technical trade-offs among them [88–90]. A comprehensive review of ceramic humidity sensors has been given by Nitta [91].

A. Materials and Stability

A large number of oxides and inorganic salts, including Al_2O_3, SiO_2, TiO_2, ZrO_2, Fe_2O_3, ZnO, as well as various titanates (perovskite structure) and chromites (spinel structure), possess useful sensitivity to water vapor. However, the selection of a specific material is less related to its sensitivity and more related to its stability. Reproducible device performance, over long periods of time, in the presence of degradants and contaminants can be very difficult to achieve. But this requirement cannot be overstressed. It is of critical importance for gas sensors in general.

Stability can be inherent, *i.e.*, by design, or it can be achieved through regeneration. An interesting example of regeneration is represented by the $MgCr_2O_4$-TiO_2 spinel sensor developed for application in microwave ovens [92]. In this application, the sensor ascertains the onset of cooking by detecting the evolution of water vapor from heating food, and the sensor must function repeatably with exposure to a wide range of molecules present in the kitchen, many emanating during the cooking process. The problem was addressed by mounting the sensor chip within a concentric heater, and periodically energizing the heater to burn out the contaminants and return its condition to the clean state.

Inherent specificity can be enhanced by controlling the chemstry and pore structure of the ceramic sensor material, or by protecting the sensor surface with filters that favor passage of the desired molecule (H_2O) but limit the passage of anticipated contaminants. The latter strategy has been implemented very successfully for one of the better resistive polymeric sensors on the market, but it is less common in ceramic devices for which inherent stability and regeneration have been favored (see Ref. 88). The importance of pore distribution and control has been discussed by Shimizu [93] and Sun [94], respectively. Inherent stability may also be enhanced by introduction of a thin second phase on the grain surfaces which isolates the active sensor material from condensate and contaminants. This approach appears to succeed for sensors based on $ZnCr_2O_4$-$LiZnVO_4$ [95].

The change in humidity response of a commercial SiO_2-ZnO sensor, which was aged in the presence of a 50:50 vol % solution of methanol and ethanol, at room temperature, is displayed in Figure 20. This figure illustrates both the response characteristic for this resistive sensor and its aging behavior. It should be noted that exposure to saturated alcohol vapor is a very severe test for humidity sensors, employed just for comparison purposes, and that this sensor was among the most stable of those designed to operate without heat cleaning or protective coating. Periodic regeneration would be beneficial but is not necessarily required for the intended application. In all applications there is always a trade-off between cost and performance. Ideally one would prefer maximal performance at minimal cost, and these desires are the drivers for progress in developing improvements in technology.

B. Operating Mechanism

Many ceramic humidity sensors are thought to share a common physical mechanism, involving chemisorbed water molecules residing at grain boundaries, bound to favorable cation sites, and physically adsorbed water occupying additional monolayers more loosely bound to the chemisorbed layer [96]. Their interfacial chemistry, involving adsorbed molecules at the grain boundaries, in this sense resembles the microchemistry of both PTC resistors and gas sensors. The physisorbed H_2O molecules become dissociated by the high electrostatic fields within the chemisorbed layer:

$$2H_2O \Leftrightarrow H_3O^+ + OH^- \tag{32}$$

It has been estimated that the fraction dissociated is on the order of 1%, or a factor of 10^6 greater than that in liquid water (see [96]). At relatively low humidity,

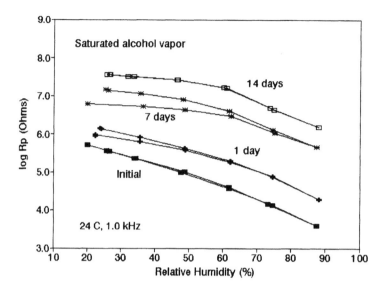

Figure 20 Aging of a commercial SiO_2-ZnO humidity sensor during exposure to saturated alcohol vapor [88].

electrical conduction occurs when H_3O^+ releases a proton to a neighboring water molecule that accepts it, releasing another proton, etc. This process is referred to as the Grotthuss chain reaction, and it is thought to represent the conduction mechanism in liquid water as well as in water adsorbed on surface layers of solids. An illustration of this for a Li-doped material can be viewed in Figure 21. At high humidity (*e.g.*, >40% relative humidity), water also condenses within the pores of the sensor, giving rise to electrolytic conduction in addition to the protonic conduction within the adsorbed layer.

Similar activity is thought to occur in other oxides, e.g., $MgCr_2O_4$-TiO_2, a p-type semiconductor. Exposure of $MgCr_2O_4$-TiO_2 to humidity at low temperature (<150°C) results in adsorption of water. Molecules attach at active Cr^{3+} sites forming hydroxyl groups that dissociate to form protons and Cr^{4+} [91]. Additional adsorption of water results in formation of hydroxyl multilayers and protonic conduction. At intermediate temperatures (300–500°C) chemisorption of water, an electron acceptor, leads to increasing conductivity since the hole concentration increases due to electron transfer to the adsorbed molecules. Complete desorption occurs above 550°C, enabling the virgin surfaces to be regenerated by heat cleaning.

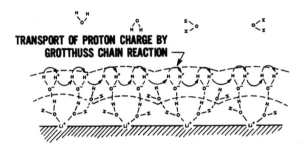

TRANSPORT OF PROTON CHARGE BY
GROTTHUSS CHAIN REACTION

LAYER I. UNDERLYING LAYER OF WATER MOLECULES CHEMISORBED TO
Li⁺ CATIONS.
(THIS LAYER IS COMPOSED OF CATION-HYDROXYL-WATER
COMPLEXES OF INDETERMINATE STRUCTURE).

LAYER II. WATER MOLECULES PHYSISORBED ONTO LAYER I.

Figure 21 Simplified physical model of water adsorption and proton charge conduction on the surface of a solid-state humidity sensor. A salt-doped type of resistive sensor is depicted where Li^+ cation is assumed to be the active constituent [96]. (Reprinted with permission from SAE Paper No. 810432 © 1981 Society of Automotive Engineers, Inc.)

C. Applications and Future Trends

Applications for humidity sensing and control are extremely varied and range from industrial air conditioning systems (where handling of fine powders is humidity dependent) and semiconductor processing, to the rather mundane microwave oven application mentioned above. Additional applications exist in agriculture (cereal stocking, greenhouses), food production (drying), medicine (respiratory equipment), high-energy combustion processes, and home appliances (clothes dryers). The overall range of operation is from 1 ppm to saturation at temperatures up to 100°C.

The market is accordingly highly fragmented for both humidity and gas sensors, with no one preferred sensor material or design approach. Part of the problem is that different solutions are advantageous for operation at very low humidity, at very high humidity, or over a wide range of humidity. Polymeric devices have lower sensor element cost than ceramics and serve many applications quite well, although they lack a certain robustness that characterizes cermics and are not generally applicable at elevated temperatures. The market is likely to remain fragmented until such time that a truly large-scale application surfaces to justify investment of resources adequate for demonstrating sufficient accuracy, operating range, and stability to serve many applications.

As with other sensor types, efforts are being undertaken to develop integrated devices, particularly systems that utilize thin films of SiO_2 or porous Si for the active sensing element. The principal challenge is the same as it is for discrete devices: to achieve adequate sensitivity with reproducible properties over the lifetime of the device, during exposure to various thermal excursions and exposure to contaminants.

VII. CAPACITIVE PRESSURE SENSORS

Ceramic capacitive pressure transducers are electromechanical devices that are widely used in vehicular and industrial applications when accurate and reliable electronic pressure sensing is required. They are used to monitor or control the fluid pressure in air conditioning, power steering, transmission, oil, fuel, and brake systems and to measure manifold air pressure for engine control applications. Their specified operating temperature range in vehicular systems is from $-40°C$ to $+135°C$ [97].

A. Description

A capacitive pressure transducer contains a ceramic capacitive sensing element that converts an applied pressure to an output capacitance in the picofarad range. Since this capacitance output is nonlinear with pressure and is susceptible to electrical interference, electronic signal conditioners are used to convert it to a voltage output that is linear with pressure and relatively immune to electrical interference (Figure 22).

The capacitive sensing element and conditioning electronics are typically placed in a protective package that isolates them from corrosive conditions in the external environment. This package also provides a means of connecting the

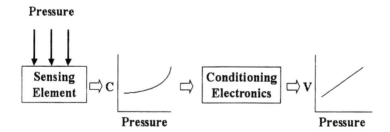

Figure 22 A schematic representation of how a capacitive pressure transducer converts an applied pressure to a linear voltage output.

sensing element to the pressure source and contains an electrical connector for the output signal. The pressure port fitting and the electrical connector are adaptable to various mounting configurations and plug requirements. Also, the package is designed to maintain pressure integrity in overpressure conditions. The general location of the sensing element and conditioning electronics within the package is shown in Figure 23.

B. Design and Construction

The capacitive sensing element is essentially a parallel plate capacitor with a defined electrode geometry and separation distance. It consists of two support plates held at a fixed separation distance by a glass seal. Ninety-six percent alumina is generally used as the structural plate material because it has a high flexural strength, behaves in a purely elastic manner over the operating temperature range of the sensor, is gas tight, can be readily formed in a variety of shapes, and can be matched in thermal expansion by commercially available sealing glasses.

Each alumina plate has a metal electrode on its surface that forms the two electrodes of the sensing capacitor. These electrodes are often gold but can be any electrically conductive material. Standard deposition techniques, such as printing, plating, or evaporation, can be used to apply the metal to the alumina plates. One of the alumina plates in the sensing element, generally referred to as the dia-

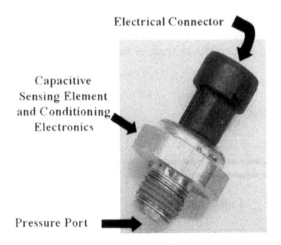

Figure 23 Location of the pressure sensing element and conditioning electronics with a pressure transducer package. (Courtesy of Texas Instruments.)

phragm, is relatively thin and is designed to deflect approximately 20–50% of the initial gap at the maximal operating pressure. The other plate is relatively thick and is designed not to deflect with pressure. This plate is generally referred to as the substrate. A simple capacitive sensing element is schematically illustrated in Figure 24.

The sealing glass that rigidly bonds the alumina plates at a fixed separation distance is selected so that it has a slightly lower coefficient of thermal expansion than alumina. This will place the glass in a compressive state after the sealing process. Glass is used as the bonding medium rather than organic adhesives because of its purely elastic response to stress and because it can be closely matched in thermal expansion to the alumina support plate. This minimizes the effects of temperature change on output capacitance. The brittle nature of glass and its lack of plastic flow over the operating temperature range results in a constant electrode separation gap over the life of the sensor. Proper selection of the sealing glass and design of the sensing element and packaging results in a transducer that can maintain an output error less than ± 2.5% without temperature compensation.

C. Device Operation

The diaphragm of the sensing element deflects when there is a pressure imbalance between its top side, which is exposed to the pressurized fluid, and its bottom side, which is inaccessible to the pressurized fluid. The pressure exerted on the bottom side of the diaphragm is generally referred to as the internal reference pressure. The sensor operates in the gauge mode when the space between the two electrodes is vented to the external atmosphere. In this case the sensor is always measuring the pressurized fluid with reference to ambient atmospheric pressure. Conversely, the sensing element can be made so that the space between

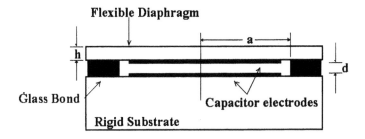

Figure 24 A schematic representation of a simple pressure-sensitive capacitive sensing element.

the electrodes is hermetically isolated from the ambient atmosphere. When this internal reference pressure is a vacuum, the sensing element operates as an absolute device. When the sensing element is sealed in an air atmosphere, the internal reference pressure will be several psi in magnitude. Although it may seem desirable to always use a vacuum as the internal reference pressure, there is a cost penalty associated with this option because more expensive processing is required. In a practical sense, air sealing does not significantly degrade output accuracy in many applications because the small internal reference pressure is accounted for during device calibration.

The capacitance of an ideal parallel plate capacitor is given by:

$$C = \frac{\varepsilon_r \varepsilon_0 A}{d} \tag{33}$$

where C = capacitance (F), ε_o = permittivity of a vacuum (F m^{-1}), d = plate separation distance (m), A = electrode area (m^2), and ε_r = dielectric constant of the medium between the plates.

It can be seen from this equation that d, plate separation distance, is the only parameter that changes when the diaphragm plate deflects under pressure. Initial plate separation is typically in the 15 to 50-μm range for most sensing elements. This ideal equation does not strictly hold during pressurization because the deflection of the diaphragm plate with pressure, as shown in Figure 25, will not be uniform across the radius of the diaphragm. The deflection will be greatest at the center and least near the glass seal where the diaphragm plate is firmly bonded to the substrate. A more accurate approximation of the output capacitance of the sensing element is given by the equation [97]:

$$C = 2\pi \varepsilon_r \varepsilon o \int_0^{r_d} \frac{r_d \partial r}{d - \Delta d} \tag{34}$$

where Δd is the change in the initial electrode separation distance at any given r_d during pressurization.

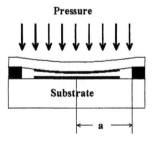

Pressure

Substrate

a

Figure 25 Non-linear diaphragm deflection in a pressurized capacitive sensing element.

If the glass seal in the sensing element has a circular geometry and only the diaphragm deflects with pressure, the change in the original separation distance between the plates, Δd, as a function of radius can be approximated from a general analysis given by Timoshenko [98]:

$$\Delta d = \frac{12}{64}\left[\frac{Pa^4(1-v^2)}{Eh^3}\right]\left[1-\frac{r^2}{a^2}\right] \tag{35}$$

where a = inside radius of the glass seal, P = applied pressure, v = Poisson's ratio, E = Young's modulus, and h = thickness. At $r = a$, there is no deflection under pressure and the original electrode separation is unchanged. At $r = 0$, deflection is a maximum and the electrode separation distance under pressure is approximated by $d - \Delta d$.

It can be seen from Eq. (35) that the operating pressure range of the sensing element can be adjusted by changing the radius of the glass seal (a) or the thickness of the diaphragm (h) to maintain a maximal deflection between 20% and 50% of the gap at maximal pressure. Adjustment of either or both of these parameters provides a means for a given manufacturing process to cover a wide range of operating pressures with minimal processing changes.

D. Comparative Performance

Slide potentiometers and strain gauges that are bonded to a ceramic or metal diaphragm or diffused into a silicon diaphragm and arranged in a Wheatstone bridge configuration have been sometimes used in place of capacitive pressure transducers in vehicular applications. The capacitive approach, however, is generally considered superior because of its overall accuracy, cycle life, and ease of packaging. A comparison of the three pressure sensing methods is shown in Table 3.

VIII. SUMMARY AND CONCLUSION

The principal ceramic sensors have been reviewed. The sections on NTC thermistors, PTC resistors, and gas sensors have been brought up to date. New sections on thermal imaging, humidity sensors, and pressure transducers have been added.

Hybrid pyroelectric imaging arrays based on electrically biased barium strontium titanate have enabled the proliferation of low-cost, uncooled IR cameras for all manner of night vision applications. Microbolometers utilizing integrated NTC thermal sensors based on thin films of vanadium oxide or amorphous silicon are becoming competitive in some of these applications. These devices exemplify a growing confluence of ceramic and semiconductor processing technologies.

Table 3 A Performance Comparison of Various Pressure Transducer Technologies [97]

Attribute	Capacitive	Potentiometric	Strain gage
Accuracy			
Temp stability	1	3	2
Linearity	2	3	1
Hysteresis	1	3	1
Reliability			
Stability	1	3	1
Cycle life	1	3	1
Electrical	2	1	2
Packaging			
Ruggedness	1	1	2
Scalability	1	2	2
Cost			
Tooling	2	1	3
Unit cost	2	1	2
Average	1.4	2.1	1.7

1 = favorable.
Source: Reprinted with permission from SAE Paper No. 880413 © 1988, Society of Automotive Engineers, Inc.

PTC resistors, gas sensors, and humidity sensors all rely on adsorbed molecular species bound within the grain boundaries of the ceramic. These internal surface layers not only account for useful sensor activity in these devices but also have implications with respect to their stability and aging behavior. The various adsorption processes have not been well characterized, are not well understood in detail, and can provide an interesting as well as challenging area for new research. Additional emphasis in gas and humidity sensor development will be required to improve device stability and life.

Capacitive pressure transducers have become commonplace as well as commercially important in automotive and some industrial applications. They are manufactured at low cost, in high volume, and demonstrate that some applications will continue to be served by highly robust, discrete sensors.

A final point relates to device cost. Although the cost of individual sensor elements is important, it may be only incidental to their eventual application. The decisive factor is the overall system cost, which generally includes electronics, interconnections, and packaging, in addition to the sensor element. This is at least in part what is driving the development of integrated sensors, including ancillary thin film and micromachining (*e.g.*, MEMS) technologies. Thin films can offer performance benefits too, especially with respect to faster response times for thermal, gas, and humidity sensing.

In pondering the future of ceramic materials for sensing, Hill and Tuller concluded the first edition of this chapter in 1986 with the expectation that "as the need for different types of transducers expands, ceramics, often with compositions or structures unlike those present in today's sensors, will provide the most cost-effective solutions" [1]. We have seen this to be evident during the ensuing interval and expect that it will characterize progress in the coming years as well.

REFERENCES

1. Hill DC, Tuller HL. Ceramic sensors: theory and practice. In Buchanan RC, ed. Ceramic Materials for Electronics. Vol. 31. New York: Marcel Dekker, 1986: 265–374.
2. Buchanan R. Ceramic materials for electronics. In Thurston MO, Middendorf W, eds. Electrical Engineering and Electronics. 2nd ed.. New York: Marcel Dekker, 1992.
3. Shorrocks NM, Porter SG, Whatmore RW, Parsons AD, Gooding JN, Pedder DJ. Uncooled infrared thermal detector arrays. Proc SPIE, 1320, Infrared Technology and Applications 1990:88–94.
4. Kruse PW. Uncooled infrared focal plane arrays, Proc. 9th Int. Symp. Appl. Ferroelectrics. State College, PA, IEEE, 1995:643–646.
5. Tidrow MZ, Clark WW, Tipton W, Hoffman R, Beck W, Tidrow SC, Robertson DN, Pollehn H, Udayakumar KR, Beratan HR, Soch K, Hanson CM, Wigdor M. Uncooled infrared detectors and focal plane arrays. Proc SPIE, 3553 Detectors, Focal Plane Arrays and Imaging Devices II 1998:178–187.
6. Kruse PW, Skatrud DD. Uncooled infrared imaging arrays and systems. In Willardson RK, Weber ER, eds. Semiconductors and Semimetals. Vol. 47. Boston: Academic Press, 1997.
7. Brandao GB, deAlmeida AL, Deep GS, Lima N, Neff H. Stability conditions, nonlinear dynamics, and thermal runaway in microbolometers. J Appl Phys 2001; 90:1999–2008.
8. Keyes RJ. Topics in Applied Physics: Optical and Infrared Detectors. Vol. 19. New York: Springer-Verlag, 1980.
9. Hopper GS. Ferroelectric imaging system U.S. Patent 4,080,532, 1978.
10. Hanson CM, Sweetser KN, Frank SN. Uncooled thermal imaging. TI Tech Journal 1994; 11(5):2–10.
11. Butler N, McClelland J, Iwasa S. Ambient temperature solid state pyroelectric IR imaging arrays. Proc SPIE, 930 Infrared Detectors and Arrays 1988:151–163.
12. Putley EH. The possibility of background limited pyroelectric detectors. Infrared Phys 1980; 20:149–156.
13. Kulwicki BM, Amin A, Beratan HR, Hanson CM. Pyroelectric imaging, Proc. 8th Int. Symp. Appl. Ferroelectrics. Greenville, SC, IEEE, 1993:1–10.

14. Amin A. Texas Instruments Technical Report, 1991.
15. Kulwicki BM. Fine-grain Pyroelectric Detector Material and Method U.S. Patent 5,314,651, 1994.
16. Yamaji A, Enomoto Y, Kinoshita K, Murakami T. Preparation, characterization, and properties of Dy-doped small-grained $BaTiO_3$ ceramics. J Am Ceram Soc 1977; 60:97–101.
17. Faxon RC, McGovern RT. Method for making ceramic titanates and materials therefor U.S. Patent 3,637,531, 1972.
18. Hanson CM, Beratan HR, Belcher JF. Uncooled thermal imaging using thin-film ferroelectrics. Proc SPIE, 4288, Photodetectors: Materials and Devices VI 2001: 298–303.
19. Shorrocks NM, Patel A, Walker MJ, Parsons AD. Integrated thin-film PZT pyroelectric detector arrays. Microelect Eng 1995; 29:59–66.
20. Takayama R, Tomita Y, Iijima K, Ueda I. Pyroelectric properties and application to infrared sensors of $PbTiO_3$, $PbLaTiO_3$ and $PbZrO_3$ ferroelectric thin films. Ferroelectrics 1991; 118:325–342.
21. Polla DL, Ye C, Tamagawa T. Surface-micromachined $PbTiO_3$ pyroelectric detectors. Appl Phys Lett 1991; 59:3539–3541.
22. Watton R, Todd MA. Induced pyroelectricity in sputtered lead scandium tantalate films and their merit for IR detector arrays. Ferroelectrics 1991; 118:279–295.
23. Patel A, Shorrocks NM, Whatmore RW. Lead scandium tantalate thin films for thermal detectors. Mater Res Soc Proc 1992; 243:67–72.
24. Ye C, Tamagawa T, Lin Y, Polla DL. Pyroelectric microsensors by sol-gel derived $PbTiO_3$ and La-$PbTiO_3$ thin films. Mater Res Soc Proc 1992; 243:61–66.
25. Hanson CM, Beratan HR. Thin-film ferroelectrics: breakthrough. Proc SPIE, 4721, Infrared Detectors and Focal Plane Arrays VII 2002:91–98.
26. Kulwicki BM. Thermistors and related sensors. In Engineered Materials Handbook. Ceramics and Glasses: ASM International. Materials Park, OH. Vol. 4, 1991: 1145–1149.
27. Macklen ED. Thermistors. Ayr, Scotland: Electrochemical Publications, 1979.
28. Sachse HB. Semiconducting Temperature Sensors and their Applications. New York: John Wiley and Sons, 1975.
29. Tuller HL. Highly conductive ceramics. In Buchanan RC, ed. Ceramic Materials for Electronics. Vol. 31. New York: Marcel Dekker, 1986:425–473.
30. Knauth P, Tuller HL. Solid state ionics: roots, status and future prospects. J Am Ceram Soc 2002; 85:1654–1680.
31. Tuller HL. Chapter 2 in this volume, 2004.
32. Tuller HL. Review of electrical properties of metal oxides as applied to temperature and chemical sensing. Sensors and Actuators 1983; 4:679–688.
33. Parlak M, Hashemi T, Hogan MJ, Brinkman AW. Effect of heat treatment on nickel manganite thin film thermistors deposited by electron beam evaporation. Thin Solid Films 1999; 345:307–311.
34. Wood RA. Monolithic silicon microbolometer arrays. In Uncooled Infrared Imaging Arrays and Systems, Semiconductors and Semimetals. Kruse PW, Skatrud DD, eds. Vol. 47. Boston: Academic Press, 1997:45–121.

35. Cole B, Horning R, Johnson B, Nguyen K, Kruse PW, Foote MC. High performance infrared detector arrays using thin film microstructures, Proc. 9th Int. Symp. Appl. Ferroelectrics, State College, PA, IEEE, 1995:653–656.

36. Tezcan DS, Kocer F, Akin T. An uncooled microbolometer infrared detector In any standard CMOS technology, Proc. 10th Int. Conf. Solid State Sensors and Actuators. Sendai. Japan, 1999:610–613.

37. Ropson S, Syllaios AJ, Weathers J. Sensor system for environmental applications, InTech Online, 2000.

38. Lee M, Yoo M, Bae S, Shin H. Detectivity of thin-film NTC infrared sensors. Proc SPIE, 4288, Photodetectors: Materials and Devices VI 2001:422–429.

39. Tezcan DS, Eminoglu S, Akar OS, Akin T. A low cost uncooled infrared bolometer focal plane array using the CMOS N-well layer, Proc. 14th Int. MicroElectroMech. Syst. Conf., Interlaken, Switzerland, 2001.

40. Unewisse MH, Passmore SJ, Liddiard KC, Watson RJ. Performance of uncooled semiconductor film bolometer infrared detectors. Proc. SPIE. Infrared Technology XX. Vol. 2269, 1994:43–52.

41. Chatard J, Tribolet P. Uncooled infrared detector technology, from research to production within 6 months. Proc. SPIE. Photodetectors: Materials and Devices VI. 2001; 4288:100–111.

42. Herbert JM. Ferroelectric Transducers and Sensors. New York: Gordon and Breach, 1982.

43. Kulwicki BM. Trends in PTC resistor technology. SAMPE 1987; 23:34–38.

44. Kulwicki BM. PTC materials technology 1955–1980. Adv Ceram 1981; 1:138–154.

45. Pike GE, Seager CH. The dc voltage dependence of semiconductor grain-boundary resistance. J Appl Phys 1979; 50:3414–3422.

46. Jonker GH. Some aspects of semiconducting barium titanate. Sol State Electr 1964; 7:895–903.

47. Kulwicki BM, Purdes AJ. Diffusion potentials in $BaTiO_3$ and the theory of PTC materials. Ferroelectrics 1970; 1:255–263.

48. Kulwicki BM. Critical electric field in $BaTiO_3$. J Appl Phys 1969; 40:3118–3120.

49. Nemoto H, Oda I. Direct examinations of PTC action of single grain boundaries in semiconducting $BaTiO_3$ ceramics. J Am Ceram Soc 1980; 63:398–401.

50. Ogawa H, Demura M, Yamamoto T, Sakuma T. Estimation of PTCR effect in single grain boundary of Nb-doped $BaTiO_3$. J Mater Sci Lett 1995; 14:537–538.

51. Kuwabara M, Morimo K, Matsunaga T. Single-grain boundaries in PTC resistors. J Am Ceram Soc 1996; 79:997–1001.

52. Jonker GH. Equilibrium barriers in PTC thermistors. Adv Ceram 1981; 1:155–166.

53. Daniels J, Hardtl KB. Electrical conductivity at high temperatures of donor-doped barium titanate ceramics. Philips Res Rep 1976; 31:489–504.

54. Daniels J. Defect equilibria in acceptor-doped barium titanate. Philips Res Rep 1976; 31:505–515.

55. Daniels J, Wernicke R. New aspects of an improved PTC model. Philips Res Rep 1976; 31:544–559.

56. Daniels J, Hardtl KB, Hennings D, Wernicke R. Defect chemistry and electrical conductivity doped barium titanate ceramics. Philips Res Rep 1976; 31:487–559.

57. Drofenik M, Popovic A, Irmancik L, Kolar D, Krasevec V. Release of oxygen during the sintering of doped $BaTiO_3$ ceramics. J Am Ceram Soc 1982; 65:C203–C204.
58. Lee JK, Park JS, Hong KS, Ko KH, Lee BC. Role of liquid phase in PTCR characteristics of $(Ba_{0.7}Sr_{0.3})TiO_3$ ceramics. J Am Ceram Soc 2002; 85:1173–1179.
59. Fukami T, Tsuchiya H. Dependence of resistivity on donor dopant content in barium titanate ceramics. Jpn J Appl Phys 1979; 18:735–738.
60. Peng C, Lu H. Compensation effect in semiconducting barium titanate. J Am Ceram Soc 1988; 71:C44–C46.
61. Ting C, Peng C, Lu H, Wu S. Lanthanum-magnesium and lanthanum-manganese donor-acceptor-codoped semiconducting barium titanate. J Am Ceram Soc 1990; 73:329–334.
62. Lewis GV, Catlow CRA, Casselton REW. PTCR effect in $BaTiO_3$. J Am Ceram Soc 1985; 68:555–558.
63. Hasegawa A, Fujitsu S, Koumoto K, Yanagida H. The enhanced penetration of oxygen along the grain boundary in semiconducting barium titanate. Jpn J Appl Phys 1991; 30:1252–1255.
64. Jonker GH. Halogen treatment of barium titanate semiconductors. Mater Res Bull 1967; 2:401–407.
65. Alles AB, Amarakoon VRW, Burdick VL. Positive temperature coefficient of resistivity effect in undoped, atmospherically reduced barium titanate. J Am Ceram Soc 1989; 72:148–151.
66. Chiang YM, Takagi T. Grain-boundary chemistry of barium titanate and strontium titanate II, origin of electrical barriers in positive-temperature-coefficient thermistors. J Am Ceram Soc 1990; 73:3286–3291.
67. Alles AB, Burdick VL. Grain boundary oxidation in PTCR barium titanate thermistors. J Am Ceram Soc 1993; 76:401–408.
68. Kulwicki BM. Instabilities in PTC resistors, Proc. 6th Int. Symp. Appl. Ferroelectrics Lehigh University, Bethlehem, PA, IEEE, 1986:656–664.
69. Ueoka H. The doping effects of transition elements on the PTC anomaly of semiconductive ferroelectric ceramics. Ferroelectrics 1974; 7:351–353.
70. Nakahara M, Murakami T. Electronic states of Mn ions in $Ba_{0.97}Sr_{0.03}TiO_3$ single crystals. J Appl Phys 1974; 45:3795–3800.
71. Kulwicki BM, St.Martin NP. Ceramic semiconductors, U.S. Patent 4,101,454, 1978.
72. Schwank JW, DiBattista M. Oxygen sensors: materials and applications. MRS Bull 1999; 24:44–48.
73. Janata J, Huber RJ. Solid State Chemical Sensors. Orlando: Academic Press, 1985.
74. Annino R, Cheney MC, Connelly JP. A stack probe for the analysis of carbon monoxide and oxygen. ISA Trans 1986; 25:7–17.
75. Heisig CG. Process oxygen analysis: more than just picking an instrument. InTech 1984; 31:73–77.
76. Yamazoe N, Miura N. Gas sensors using solid electrolytes. MRS Bull 1999; 24: 37–43.
77. Maskell WC. Advanced zirconia gas sensors. In Mosley PT, Norris J, Williams DE, eds, Chapman & Hall. 1991.
78. Barsan N, Weimar U. Conduction model of metal oxide gas sensors. J Electroceram 2000; 7:143–167.

79. Takahata K. Tin dioxide sensors—development and applications. In Seiyama T, ed. Chemical Sensor Technology. Vol. 1. New York: Kodansha-Elsevier, 1988: 39–55.

80. Moseley PT, Norris JOW, Williams DE. Techniques and Mechanisms in Gas Sensing. New York: Adam Hilger, 1991.

81. Wang DY, Logothetis E, Ichinose N, Traversa E. Special issue on sensors. J Electroceram 1998; 2:213–308.

82. Takeuchi T. Oxygen sensor. Sensors and Actuators 1988; 14:109–124.

83. Badwal SPS, Bannister MJ, Garrett WG. Low temperature behaviour of ZrO_2 oxygen sensors. Adv Ceram 1984; 12:598–606.

84. Shimizu Y, Egashira M. Basic aspects and challenges of semiconductor gas sensors. MRS Bull 1999; 24:18–24.

85. Figaro Engineering, General information for TGS sensors, 1998.

86. Lundstrom I, Shivaraman MS, Svensson C, Lundkvist L. A hydrogen sensitive Pd-gate MOS transistor. J Appl Phys 1975; 46:3876–3881.

87. Tuller HL, Mlcak R. Inorganic sensors utilizing MEMS and microelectronic technologies. Curr Opin Solid-State Mater Sci 1998; 3:501–504.

88. Kulwicki BM. Humidity sensors. J Am Ceram Soc 1991; 74:697–707.

89. Hubert T. Humidity sensing materials. MRS Bull 1999; 24:49–54.

90. Michell AK. Humidity sensing, In Moseley PT, Norris JOW, Williams DE, eds. Techniques and Mechanisms in Gas Sensing. New York: Adam Hilger, 1991: 189–197.

91. Nitta T. Development and application of ceramic humidity sensors, In Seiyama T, ed. Chemical Sensor Technology. Vol. 1. New York: Kodansha-Elsevier, 1988: 57–78.

92. Nitta T, Terada Z, Hayakawa S. Humidity-sensitive electrical conduction of $MgCr_2O_4$-TiO_2 porous ceramics. J Am Ceram Soc 1980; 63:386–391.

93. Shimizu Y, Arai H, Seiyama T. Theoretical studies on the impedance-humidity characteristics of ceramic humidity sensors. Sensors and Actuators 1985; 7:11–22.

94. Sun H, Wu M, Yao X. Porosity control of humidity-sensitive ceramics and theoretical model of humidity-sensitive characteristics. Sensors and Actuators 1989; 19: 61–70.

95. Yokomizo Y, Uno S, Harata M, Hiraki H. Microstructure and humidity-sensitive properties of $ZnCr_2O_4$-$LiZnVO_4$ ceramic sensors. Sensors and Actuators 1983; 4: 599–606.

96. Fleming WJ. A physical understanding of solid state humidity sensors, SAE Tech. Paper Ser. 810432, Proc. SAE Int. Congress and Exposition. Detroit, MI, 1981: 51–62.

97. Sabetti AJ, Bishop RP, Charboneau TJ, Wiecek TE. Automotive pressure transducer for underhood applications, SAE Tech. Paper Ser. 880413, Proc. SAE Int. Congress and Exposition. Detroit, MI, 1988:53–59.

98. Timoshenko SB, Woinowsky-Krieger S. Theory of Plates and Shells. New York: McGraw-Hill, 1959.

99. Owen R, Belcher HBJ, Frank S. Producibility advances in hybrid uncooled infrared detectors. Proc SPIE, 2225, Infrared Detectors and Focal Plane Arrays III 1994: 79–86.

100. Owen R, Belcher HBJ, Frank S. Producibility advances in hybrid uncooled infrared detectors II. Proc SPIE, 2746, Infrared Detectors and Focal Plane Arrays IV 1996: 101–112.

101. Neal HW, Kyle RJS. Texas Instruments uncooled infrared systems. TI Tech 1994; 11(no. 5):11–18.

102. Williams G, Coles GSV. The gas-sensing potential of nanocrystalline tin dioxide produced by a laser ablation technique. MRS Bull 1999; 24:25–29.

103. Lundstrom I, Svensson C. Gas-sensitive metal gate semiconductor devices. In Janata J, Huber RJ, eds. Solid State Chemical Sensors. Orlando: Academic Press, 1985: 1–63.

104. Wollenstein J, Bottner H, Plaza JA, Cane C, Min Y, Tuller HL. Materials and technologies for metal oxide-based gas sensor arrays. Presented at International Meeting on Chemical Sensors, Boston, MA, 2002.

7

ZnO Varistor Technology

Lionel M. Levinson
Vartek Associates LLC
Schenectady, New York, U.S.A.

I. INTRODUCTION

Metal oxide varistors are novel ZnO-based ceramic semiconductor devices with highly nonlinear current–voltage characteristics similar to back-to-back Zener diodes. Over the past few decades these varistor devices have become the preferred approach to protecting electronic, electrical, and power distribution and transmission circuits from the destructive voltage levels induced by lightning impulses or switching surges.

The varistors are produced by a ceramic sintering process that provides a structure largely composed of conductive ZnO grains surrounded by thin electrically insulating barriers. In this chapter the physical properties of the resulting three-dimensional series-parallel junction network are discussed, and the novel electrical properties are interpreted in terms of the material microstructure and grain-to-grain conduction mechanism.

The applications of ZnO varistors are predominantly in the field of circuit overvoltage protection, although some uses of the material as an active circuit element have been described [1,2]. Overvoltage protection is a necessity for both electronic circuits and in the electrical power distribution and transmission industries. As a specific example of the need for this type of device one need only consider the fact that circuits and systems containing low-cost, high-reliability solid-state components are typically designed for use on 120-V ac power lines. However, it is generally accepted that proper design practices must take into account the fact that solid-state components cannot, in general, withstand the amount of overvoltage imposed on the circuitry by typical power system transients. This drives the requirement for voltage surge protection. The magnitude of this type of problem is graphically illustrated in Table 1, which gives surge data from a survey on the nature of surge voltage transients in residential power circuits [3]. The data, collected from a variety of homes in urban and rural locations in New York, Florida, and South Carolina, indicate that surges in the 1000-V range and above are prevalent with a frequency of the order of several surges per day. Such transients usually result from switching surges or lightning storms and can play havoc with unprotected solid-state components connected to a power line. Product reliability therefore requires that the solid-state component be provided with adequate protection from these transient overvoltages.

In this chapter the status of ZnO varistors is assessed. The ZnO varistor, often called a metal oxide varistor, was initially developed in Japan [5] and became available in the United States in the early 1970s under the trade name GE-MOV varistor, largely for the protection of consumer and industrial equipment, designed to work in the voltage range below 1000 V. As an interesting aside, it

Table 1 Surge Voltages Observed in a Variety of Urban and Rural Homes in New York, Florida, and South Carolina

Location number	Most frequent surge crest (V)	Duration (μs or cycles)	Average number of surges per hour	Highest crest (V)	Most severe surge duration
1	300	10 μs	0.07	700	10 μs
2	500	20 μs	0.14	750	20 μs
3	300	1 cycle	0.05	600	1 cycle
4	300	2 cycle	0.2	400	2 cycles
5	250	1 cycle	0.01	400	1 cycle
6	800	1 cycle	0.03	1800	1 cycle
7	300	10 μs	0.1	1200	4 cycles
8	1500	1 cycle	0.2	1500	1 cycle
9	2000	1 cycle	0.4	2500	1 cycle
10	1500	1 cycle	0.15	1500	1 cycle
11	1400	1 cycle	0.06	1700	1 cycle
12	600	3 cycles	0.05	800	3 cycles
13	200	15 μs	0.1	400	30 μs
14	1000	4 cycles	0.1	5600	1 cycle

Source: Ref. 3.

may be noted that work on the nonlinear electrical properties of ZnO ceramics had been underway in Russia in the late 1950s [6,7] but the application potential of these materials was first recognized by Matsuoka and his coworkers in Japan [5] about 10 years later. An intensive effort was then launched to develop the understanding and application capabilities of ZnO varistor materials, and this established the basis for the use of this material as the predominant overvoltage surge protection device.

Varistor ceramics and their applications have been the subject of numerous reviews and the reader wishing to obtain more detailed information is referred to these detailed discussions [8–13]. The most recent review by Clarke [13] provides a thorough description of the current state of understanding. Gupta's review [10] provides a discussion of the criteria important to the use of varistors in high-voltage power distribution and transmission applications.

This chapter is structured as follows: following a description of the electrical characteristics of the varistor the general microstructure and fabrication of the devices are discussed. Varistors have an interesting and rich behavior as circuit elements, and the effect of frequency and voltage on the varistor impedance is briefly described. A review of our current understanding of the mechanism of the highly nonlinear current–voltage characteristic of ZnO ceramic varistors follows. This understanding requires that we address both the behavior of a single grain–grain junction and the complexities introduced by variability in the grain structure. Both issues are discussed.

As with any electronic device, industrial application requires that one takes into account possible degradation and failure effects, and material related to these issues is presented. Finally, some illustrative examples of ZnO applications are featured.

II. VARISTOR ELECTRICAL CHARACTERISTICS

The electrical characteristics of ZnO varistors are similar to back-to-back Zener diodes but with much greater voltage-, current-, and energy-handling capabilities. Figure 1 shows an oscillograph of the symmetrical current–voltage characteristics of this material. The device can be regarded as "insulating" up to a certain field, the breakdown field F_{BR}, above which it becomes conducting.

A more complete examination of the device behavior requires the display of current and voltage on a logarithmic scale. Figure 2 gives current I versus applied voltage V for a typical varistor measured at 77 K and for a range of temperatures near 300 K. Note that there is an enormous variation in current (a factor of 10^{11}), while the applied voltage varies about a factor of 3. From Figure 2 it is also evident that there is wide range of current (10^{-6} to $\approx 10^2$ A) over

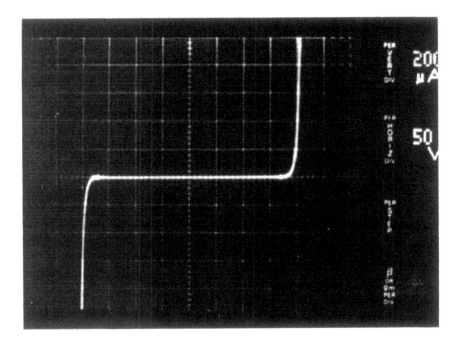

Figure 1   Current–voltage trace (actual photo) of a typical ZnO varistor.

which the log current–voltage relation is approximately linear; that is

$$\frac{J_1}{J_2} = \left(\frac{F_1}{F_2}\right)^{\alpha} \tag{1}$$

where J is the current density passing through the varistor, α is the measure of non-linearity, and F is the field applied to the device. Equation (1) is the so-called empirical varistor power law equation. If $\alpha = 1$, we would have an ohmic device. As $\alpha \rightarrow \infty$, we obtain a perfect varistor (i.e., the current density varies infinitely for small changes in the applied field). The exponent α is 25 to 50 or more in a typical ZnO varistor. Actual values of α, breakdown field, and off-state resistance in typical devices are available elsewhere [8–13] and will not be specified here.

At very low and very high currents, Eq. (1) is no longer a good representation of the varistor characteristic. The low-current region has been denoted "pre-breakdown" and corresponds physically to the transport of very low-level current through the varistor at voltages below the breakdown region. The very high

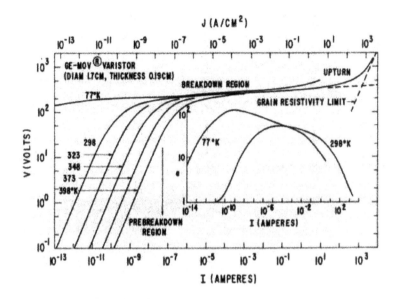

Figure 2 Current–voltage characteristics of a metal oxide varistor at 77 K and for a small range of temperatures near 300 K. The exponent α equals the inverse slope of the curve and is a measure of device nonlinearity.

current deviations from the power law Eq. (1) are believed to result, at least in part, from an ohmic series resistance derived from the ZnO grains in the varistor. This region has been labeled "upturn" in Figure 2.

The use of ZnO varistors to protect sensitive equipment from transient overvoltages is in principle extremely simple; the varistor is directly connected across the power line in parallel with the load to be protected. This is indicated schematically in Figure 3. Care is taken to choose a varistor having a breakdown voltage slightly greater than the maximal design voltage applied to the system to be protected. In normal operation the varistor is insulating (i.e., it operates in the prebreakdown region of Figure 2). If a transient is incident so that the total voltage applied to the load rises above the varistor breakdown voltage, the varistor current rapidly increases along its characteristic current–voltage curve (Figure 2), becoming a conducting shunt path for the incident transient pulse.

The varistor response [14] to ≈500 ps rise-time pulses is shown in Figure 4. In 4a, the applied pulse voltage is 50 V and only the capacitive charging current is observed. The decay of the varistor voltage at the end of the pulse is associated with a negative varistor discharge current (not plotted in the figure). In 4b, the pulse voltage is 150 V, which is above the breakdown voltage of the varistor. In this case, the capacitive charging current is almost hidden by the increased con-

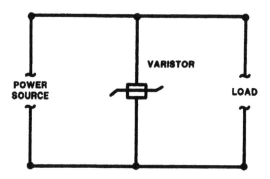

Figure 3   Typical application of ZnO varistor as a transient protective element.

duction current and the varistor voltage shows a peak that then decreases with increasing time. There is no indication in this curve of any delay in the initiation of the conduction current once the varistor voltage has risen above some minimal value. Since the varistor has the ability to "clamp" the incoming pulse, the varistor "turn on" time is less than the rise time of the pulsing equipment, $\approx 5 \times 10^{-10}$ s. The initial peaking of the varistor voltage and its subsequent decrease with time has been denoted the "overshoot" effect and represents a time-dependent modification of the varistor conduction process that is inherently very fast.

The existence of the overshoot effect has an important device implication. The protective voltage quoted is usually that determined using standard measuring procedures (e.g., pulses of 8 μs rise time) where the voltage overshoot, V_{max} −

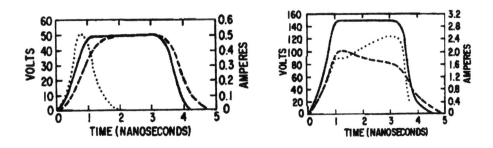

Figure 4   Voltage response (dashed line) and current response (dotted line) of a varistor chip in a 50-circuit to $\approx$500-ps rise-time pulses of magnitude (a) 50 V and (b) 150 V. The solid line indicates the response without the chip in the circuit.

V_{dc} is small. However, for faster rise time, high-current pulses, the protective level may be appreciably poorer. The magnitude of the overshoot varies from device to device and depends on the varistor formulation.

III. VARISTOR MICROSTRUCTURE AND FABRICATION

ZnO varistors are highly complex multicomponent oxide ceramics whose electrical behavior depends both on the ceramic microstructure of the device and on detailed processes occurring at the ZnO grain boundaries. The primary constituent of the varistor is ZnO, typically 80 mol % or more. As a result of the varistor processing, the ZnO in varistors is semiconducting, with $\rho_{ZnO} \leq 1$ Ω-cm at room temperature. In addition to the ZnO, the varistor contains smaller amounts of a number of other metal oxide constituents (thus the name "metal oxide varistor"). A typical early composition given by Matsuoka [5] contains 97 mol % ZnO, 1 mol % Sb_2O_3, and 1/2 mol % each of Bi_2O_3, CoO, MnO, and Cr_2O_3. The varistor constituents are often indicated by the cations alone, and thus a formulation containing Bi_2O_3 is said to contain Bi. Current commercial varistor mixes may contain as many as eight or more metal oxide additives. The reasons for this are predicated on the empirical approach taken to maximize varistor performance in *all* of the areas required for product acceptance. For example, if the coefficient of nonlinearity, or exponent α, was the only important device parameter, acceptably high coefficients of nonlinearity could easily be achieved in the simplest of varistor systems composed of a single varistor-forming ingredient, such as Bi or Pr, and one or two varistor performance ingredients, such as Co, Mn, or Ni. However, a number of other considerations, such as stability, conduction uniformity, power dissipation, maximal surge current, energy absorption capability, grain resistivity, and so forth, in addition to exponent, must be optimized to obtain overall long-term reliability as well as electrical performance. Unfortunately, this concert of requirements necessitates a proliferation of varistor ingredients. Thus, while we have developed a reasonably precise theoretical model for varistor conduction, commercial improvements proceed mainly by "educated" trial-and-error activity in mix selection and processing schedule development. There are, of course, good reasons for this approach. The varistor behavior depends both on the detailed ceramic microstructure of the device and on the related chemical and physical processes occurring at the ZnO grain boundaries during high-temperature sintering. Most conduction models ignore this chemical and microstructural complexity, and while they can indeed account for a wide variety of observed varistor phenomena, they are usually not specific in the role of the individual chemical additives. Hence, they cannot be used to optimize overall performance by the a priori selection of process and ingredients. The above notwithstanding, it is useful to group various additives to ZnO varistor devices by the predominant effect they have on the device behavior. A nonexhaustive list is as follows:

1. Nonlinearity inducers—Bi, Pr.
2. Nonlinearity enhancers—Co, Mn, Sb, Cr
3. Grain growth retardants—Sb, Si
4. Grain growth enhancers—Ti, Al, Ba
5. Stability enhancers—Ba, B, Cr, Ag, Li, K
6. ZnO conductivity enhancers—Al, Ga, In

The above list should be treated with caution since some additives may have more than one effect depending on other additives present. In addition, the amount of additive (or inadvertent impurity present) can be critical. For example, additions of Li in the 10- to 100-ppm level can enhance stability whereas larger amounts of lithium can dramatically reduce the degree of nonlinearity, especially in the high-current region (Li is known to make ZnO resistive).

Fabrication of ZnO varistors follows standard ceramic techniques. A simplified flow diagram used in varistors for electronic component protection is given in Figure 5. The ZnO and other constituents are mixed, for example, by milling in a ball mill. The mixed powder is dried and pressed to the desired shape. The resulting pellets are sintered at high temperature (typically, 1100–1400°C). The sintered devices are then electroded, often (but not invariably) with a fired silver contact. Leads are attached by solder and the finished device may be potted in epoxy and tested to meet required specifications.

For larger single-disk volumes—e.g., the 3-inch-diameter, 1-inch-thick disk used in high-voltage arrester applications—the basic flow diagram of Figure 5 is also followed but some of the steps are more complex. For example, the milling procedure is carried out in a series of steps to ensure proper additive dispersal throughout the disk volume. The additives are milled separately before addition of the ZnO. The resulting powder is mixed in a shear blender and spray dried. The mix is then calcined at temperatures in the range 800–900°C to prereact the ingredients. The calcined mix is again milled and spray dried prior to pressing. The electroding procedure is also composed of a series of steps since in practice the disks are stacked one on the other for arrester insertion on lines of up to 10^6 V or more. The disks must be lapped prior to electroding to achieve maximal disk–disk contact. The electrodes are often plasma-sprayed aluminum for maximal disk face conductance and to allow high-temperature operation. The disk rim may also be passivated with an insulating glass or ceramic layer at this time. The final step is an anneal at about 600°C to improve electrical stability. Disk assembly in the arrester or other protective device, which may have a large number of disk columns in parallel, will not be discussed here.

The microstructure of fired varistor is depicted schematically in Figure 6. The varistor is idealized as comprised of a set of conducting grains, size d,

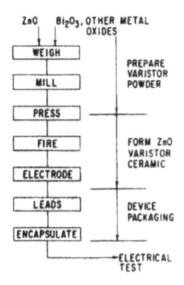

Figure 5 Simplified flow diagram for the fabrication of ZnO varistors used to protect electronic equipment.

surrounded by an extremely thin (few nm) adsorbed layer. Transmission electron microscopy [15–18] and Auger electron spectroscopy [19] indicate that bismuth is segregated to the ZnO grain boundaries. The form of the segregation and whether it exists as a film, as isolated segregated ions, or perhaps in some other form has been the focus of various high-resolution electron microscopy studies [15,18]. At this point it appears likely that the form of segregation depends on the particular formulation and processing used for the varistor device under study. Current flows between the electrodes as indicated. Typical grain sizes are $d \sim$ 10 μm, and the grain resistivity ρ_{ZnO} is about 1 Ω-cm.

The actual microstructure of a ZnO varistor is considerably more complex than the idealized depiction of Figure 6. In Figure 7, a photomicrograph of a polished and etched section of a typical varistor is shown. Three phases—grains, intergranular phase, and particles—are evident. While the microstructures of varistors exhibit considerable variation from one formulation to another, they typically exhibit the characteristics of a ceramic prepared by liquid-phase sintering. In this case, the ceramic consists of large ZnO grains with a bismuth-rich second phase at the triple junctions [15,20–22].

The grains are the predominant phase in Figure 7 and consist of small ($\sim$20 μm) conducting ZnO microcrystals. The straight lines evident in some of the ZnO grains correspond to twin boundaries delineating different ZnO crystal planes

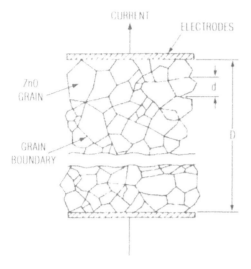

Figure 6 Schematic depiction of the microstructure of a ZnO varistor with grains of conducting ZnO, average side d. Bismuth is segregated to the ZnO grain boundaries. Electrodes are attached and current flows as indicated.

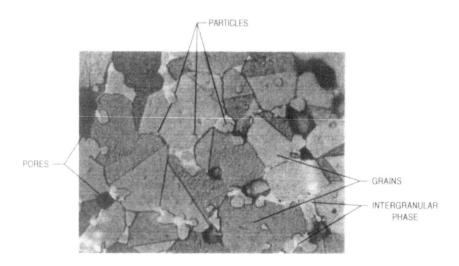

Figure 7 Optical photomicrograph of a polished and etched section of a typical varistor. The various features are discussed in the text.

having different etch properties. The whitish areas lying largely between the ZnO phase at triple points have been labeled "intergranular material" and are bismuth rich. Actually, the true intergranular barrier relevant to varistor action is not visible in Figure 7. Varistor action is controlled by depletion layers lying within the ZnO grains at grain–grain interfaces. These potential barriers are formed at the grain boundaries when the varistor is cooled from its sintering temperature. Barrier formation is believed to occur in the temperature range of 450–700°C (or on subsequent annealing). Indeed, heat treatment or annealing somewhere in this temperature range is often a key process step in the manufacture of varistors for high-voltage applications. The necessary annealing step was discovered serendipitously at General Electric in the mid-1970s and is often an essential requirement to obtain the high degree of stability necessary for high-voltage varistor applications in the utility industry. The electronic processes occurring during the annealing stage are presumably related to dopant or trap rearrangement at the ZnO grain boundary. However, the details of these processes remain essentially unknown.

In addition to the ZnO grains and intergranular material, a third phase, labeled "particles," is visible in Figure 7, X-ray studies [20] have shown that this phase has a spinel-type structure with the approximate formula $Zn_7Sb_2O_{12}$. These spinel particles are insulating and have only a secondary role in determining device properties. Their presence is a result of the particular composition and processing of the varistor in Figure 7. Other varistor formulations have other (or no) such secondary phases. For example, depending on the composition, processing, and amount of additives, the microstructure also can contain particles of pyrochlore.

As will be discussed later, an electrically insulating boundary region, thickness $t \sim 1000$ Å, is also formed at each ZnO grain boundary. This boundary or "depletion" region is wholly within the ZnO grain and is similar to the Schottky barrier commonly associated with semiconductor interfaces.

To analyze varistor behavior, it is useful (though an oversimplification) to represent the varistor microstructure by the block model shown in Figure 8. This model presumes that the device is assembled of conducting ZnO cubes of size d separated from each other by an insulating barrier region of thickness t. It should be emphasized that the insulating barrier is not a separate phase but is a representation of the back-to-back depletion layers at the ZnO grain boundaries.

Referring to Figure 7, we note that in this sample the varistor grain size d is about 20 μm. For this varistor material the macroscopic average breakdown field F_{BR} specified at a current density of 1 mA/cm^2 is around 130 V/mm. It follows that the macroscopic breakdown voltage per intergranular barrier, v_g, is

$$v_g = F_{BR}d = 2.6 \text{ V per barrier} \qquad (2)$$

The foregoing value of breakdown voltage per grain barrier is characteristic of

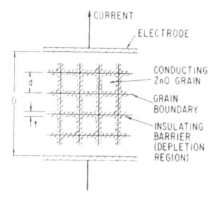

Figure 8   Block model of a ZnO varistor having grain size d ($\sim$20 μm) and intergranular depletion barrier thickness t ($\sim$1000 Å). D is the electrode separation.

ZnO varistors when the grain breakdown voltage is evaluated by measurements on bulk samples. This bulk value is expected to be lower than the true breakdown voltage per grain because the current always seeks the easiest path (i.e., the path with fewest barriers between the electrodes). The number of grains for the actual current path is lower than the average number of grains between the electrodes; equivalently, we should increase the value of d used to estimate v_g. Indeed, microscopic data on varistor behavior of single-grain junctions (presented below) give $v_g \approx 3.5$ V per grain for varistor materials.

It should also be emphasized that an important and significant feature of the behavior of ZnO varistors is that macroscopic breakdown voltages of 2 or 3 V per barrier are observed for a wide variety of ZnO varistor materials. Substantial variations in device processing and composition have relatively minor effects on v_g.

From the schematic of Figure 8 it is clear that the electrical characteristics of ZnO varistors are related to the bulk of material (i.e., the device is inherently multijunction with varistor action shared between the various ZnO grain boundaries). This implies that tailoring the device breakdown voltage V_{BR} is simply a matter of fabricating a varistor with the appropriate number of grains, n, in series between the electrodes. Thus, to achieve a given breakdown voltage, one can change the varistor thickness (for fixed grain size) or one can vary the grain size to increase the number of barriers, n, keeping the device thickness constant. In either case,

$$V_{BR} = nv_g = \frac{Dv_g}{d} \tag{3}$$

where D is the electrode spacing and d is the ZnO grain size. Typical varistor values for protection of equipment on 120-V ac power lines are $V_{BR} = 200$ V, $d = 20$ μm, $D = 1.6$ mm, and $n = 80$.

The block model (Figure 8) approach to understanding the behavior of ZnO varistors is equally useful in developing a simple equivalent circuit representation of this material. How this comes about is outlined in the following section.

IV. VARISTOR EQUIVALENT CIRCUIT

Measurement of the dielectric constant ε_r of ZnO varistors typically gives values $\varepsilon_r \sim 1000$. This high value, while perhaps initially surprising, is quite easily understood in terms of the varistor model of Figure 8 and the equivalent circuit of Figure 9. Since $t << d$ (Figure 8), it is clear that the volume between the varistor electrodes is largely occupied by conducting ZnO grain material. In fact, the thickness of insulating dielectric lying between the sample electrodes is not D but Dt/d. Therefore, if ε_g is the dielectric constant of the ZnO in the depletion layers, we expect the varistor capacitance C to be given by

$$C = \varepsilon_g \varepsilon_0 \frac{A}{Dt/d}$$
$$= \frac{d}{t} \varepsilon_g \varepsilon_0 \frac{A}{D} \tag{4}$$

That is, the effective dielectric constant is increased by a factor of d/t.

Measured values of the effective dielectric constant are ~ 1000 [23], implying $d/t \sim 100$ since $\varepsilon_g \sim 10$ for ZnO. Taking $d \sim 10$ μm [24], we find that $t \sim 1000$ Å. This value represents the thickness of the depletion layer at the ZnO grain boundary [25–27]. The depletion layer largely controls the varistor low-voltage capacitance. The thickness of any non-ZnO material between the grains is about 2 orders of magnitude less [15–19].

Returning to Figures 8 and 9, it is reasonable to describe the varistor intergranular barriers by a parallel capacitor (C_p) and resistor (R_p) pair. R_p is clearly voltage dependent. In addition, both R_p and C_p vary with frequency and temperature (see below). In series with R_p and C_p is the small resistance r_{ZnO} of the ZnO grains. One can usually ignore r_{ZnO} except for very high varistor currents or very high varistor frequencies [28].

In Figures 10–12 we plot the measured dielectric constant ε_r, dissipation factor $D = \tan \delta$, and parallel resistivity ρ_p of a commercial varistor. The high dielectric constant evident in Figure 10 is consistent with our previous discussion

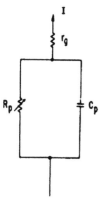

Figure 9 Simple equivalent circuit representing a metal oxide varistor as a capacitance in parallel with a voltage-dependent resistor. C_p and R_p are the capacitance and resistance of the intergranular layer, respectively; r_g is the ZnO grain resistance. For low applied voltages R_p behaves as an ohmic loss.

and reflects the two-phase nature of ZnO varistors. ε_r decreases somewhat, from 30 to 10^5 Hz, with a sharper dispersive drop evident in the range 10^5–10^7 Hz.

In Figure 11 the dissipation factor D is peaked in the vicinity of 300 kHz. This peak is reminiscent of a (broadened) Debye resonance, and some discussion of the origin of the resonance peak can be found in Ref. 23. It is interesting to

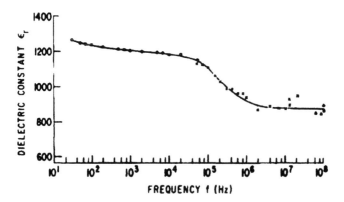

Figure 10 Variation with frequency of the room temperature dielectric constant of a ZnO varistor.

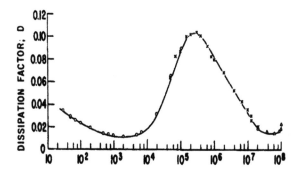

Figure 11 Dissipation factor $D = \tan \delta$ versus frequency at room temperature.

note that in the range $30-10^8$ Hz, the dissipation factor varies by only a factor of 10. This has immediate consequences for the frequency dependence of the parallel resistivity, since

$$\rho_p = \frac{1}{\varepsilon_r \varepsilon_0 \omega D}$$

(5)

Thus, noting from Figures 10 and 11 that to a first approximation D and ε_r are constant, we expect $\rho_p \sim 1/\omega$ (i.e., the varistor resistivity should drop as the inverse frequency).

This behavior is illustrated in Figure 12, where we plot the parallel resistivity ρ_p versus frequency. From the figure we have $\rho_p \approx 10^9$ Ω-cm at 30 Hz and $\rho_p \approx 10^3$ Ω-cm at $f = 10^8$ Hz, implying that the varistor resistivity drops 6 decades when the frequency increases by 6.5 decades. Clearly, R_p of Figure 9 cannot be regarded as a simple frequency-independent leakage path. Superimposed on the general drop in R_p with frequency we observe a dispersive anomaly in the region 10^5-10^7 Hz, as expected from Figures 10 and 11, and Eq. (5).

In addition to the frequency dependence of C_p, $D = \tan \delta$, and R_p, measurement indicates that these quantities also vary with temperature. Relevant data and an analysis of the results are given in Refs. 23 and 28.

Finally, it is important to emphasize that the data of Figures 10–12 were obtained from measurements at very low voltages compared with the varistor breakdown voltage V_{BR}. In this voltage region the varistor is an essentially linear circuit element. As the applied voltage approaches V_{BR}, more complex phenomena can be anticipated, and an example of such behavior is given by the voltage dependence of the capacitance described in Ref. 29.

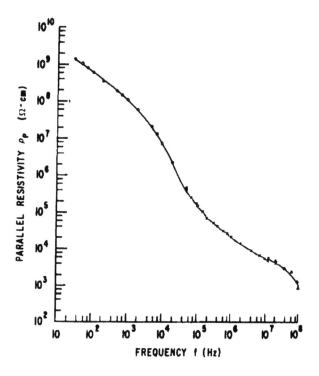

Figure 12   Parallel resistivity ρ_P of a ZnO varistor at room temperature.

V. MECHANISM OF VARISTOR BEHAVIOR

Any complete theory of the electrical behavior of ZnO varistors must account for features that depend on the varistor ceramic microstructure and also account for the detailed behavior of a single ZnO grain–intergranular barrier heterojunction. Generally, the simplified block model of Figure 8 has proved adequate for understanding the implications of varying varistor microstructures, although the idealized representation of the ceramic of Figure 8 should certainly be used with caution. Certainly the disorder associated with different arrangements of grain structures must affect the varistor behavior, and this will be discussed in greater detail below.

Current understanding of the phenomena associated with a single ZnO grain–intergranular barrier junction is incomplete, and perhaps this imperfect understanding is not unexpected since it is inordinately difficult to obtain well-controlled electrical or chemical data on an irregular grain interface imbedded between two grains in a ceramic. We shall discuss in turn

a. Single grain boundary phenomena
b. Effects of microstructure
c. ZnO grain resistivity
d. Degradation and failure

A. Single Grain Boundary Phenomena

An appreciation of the behavior of a single grain boundary can be obtained by
a more detailed examination of Figure 2. Three regions can be distinguished in
these curves. In the breakdown region, the current is a highly nonlinear function
of the applied voltage for many decades of current and is essentially temperature
independent. The dependence of current on voltage in this region of the current–voltage curve is often described by the empirical relation given in Eq. (1),
where α can achieve values in the range 50–100 near 10^{-3} A/cm^2. At very large
currents, the curve exhibits an upturn. This feature is not an intrinsic property of
the breakdown mechanism but is associated with the finite resistivity of the ZnO
grains themselves; the upturn represents the voltage drop in the grains. In the
prebreakdown or leakage region at low voltages, the $I–V$ characteristic is close
to linear. In many cases, the temperature dependence of the current can best be
described [30] in terms of an activation energy.

$$I = I_0 e^{-e\phi/kT} \tag{6}$$

where $e\,\phi \approx 0.6$–0.8 eV near 300 K.

An interesting and important feature of the breakdown characteristic is that
it is relatively insensitive to the details of chemical composition and processing
within reasonable limits. It is, of course, possible to prepare samples using somewhat arbitrary ingredients in which the nonlinearity is minimal ($\alpha < 10$) and a
clear, sharp breakdown characteristic is not observable. However, for compositions that contain a varistor-forming ingredient (heavy elements such as Bi, Pr,
and Ba, with large ionic radii) and at least one or two varistor performance
ingredients (generally transition metal elements such as Co, Mn, and Ni), and
for sintering temperatures in the range 1100–1350°C, the breakdown behavior
illustrated in Figure 2 with $\alpha \approx 50$ can be achieved.

The data of Figure 2 relate to measurements on a bulk varistor, i.e., the
data are averaged over many thousands of grain boundaries. However, for typical
commercial varistors, the gross ceramic microstructure does not dominate the
conduction process. In Figure 13, $I–V$ data are shown for bulk varistor material,
≈ 0.15 cm thick, for the as-sintered varistor surface obtained using surface electrodes spaced ≈ 0.10 cm apart, and for a single grain–boundary junction [29].
The single grain junction measurements were obtained using microelectrodes
evaporated and photolithographically defined on carefully prepared varistor surfaces. A pair of electrodes separated by a junction is shown in Figure 14. The

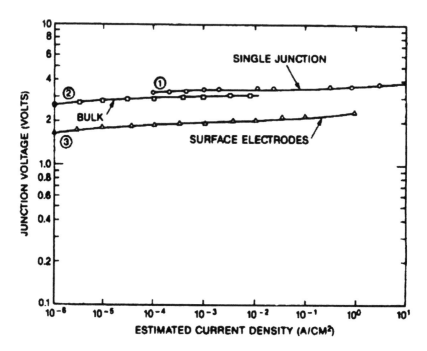

Figure 13 Current–voltage characteristics of commercial varistor material obtained using (1) microelectrodes placed across a single-grain junction, (2) electrodes placed 0.1 cm apart on the as-sintered varistor surface, and (3) electrodes placed on the opposite faces of a varistor disk ≈0.15 cm thick. Voltages measured in (2) and (3) are 100 times that indicated on the ordinate scale.

measured voltage is 100 times larger for the bulk and surface electrode data than that given on the ordinate scale. The shapes of these curves are virtually identical, showing that the electrical properties are determined solely by the behavior of the individual grain–grain junctions. The varistor voltage itself is determined by the number of grain boundaries between electrodes. However, we do not mean to imply here that the uniformity of the microstructure is not an important device consideration. We shall discuss this in more detail in Section V.B below.

A number of theoretical models have been proposed that explain grain–grain conduction and other varistor phenomena with reasonable credibility [25–27,29,31–36]. These models are based on the double depletion layer concept for the region of closest grain–grain contact, and this is the generally accepted starting point for any physical model of the varistor junction. The origin of the double depletion layer can be understood by considering the formation of a grain boundary when two identical semiconducting grains are joined. The grain bound-

Figure 14   Photomicrograph of evaporated Al electrode configuration on the surface of a ZnO varistor. The electrodes have ohmic contact to two ZnO grains separated by a single-grain barrier. The measured current – voltage curve is given in Figure 13.

ary is assumed to comprise the same semiconducting material as the grain but also contains defects and dopants. As a result, its Fermi level is different from that of the two grains, and it also has additional electronic states because of the defects and dopants within the band gap. Electrons flow to the grain boundary, where they are trapped by the defects and dopants. The result of this electron flow is that the electrons trapped at the grain boundary act as a sheet of negative charge at the boundary, leaving behind a layer of positively charged donor sites on either side of the boundary thereby creating an electrostatic field with a barrier at the boundary. The magnitude of the potential barrier can be calculated by solving the Poisson equation for the potential. A schematic picture of the grain boundary potential barrier is given in Figure 15.

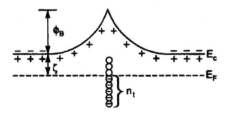

Figure 15   Double depletion layer at a ZnO grain boundary. Electrons move from the grain into traps at the grain boundary, inducing the electron barrier shown. E_F and E_C denote the Fermi level and conduction band, respectively. (Adapted from Ref. 21.)

This double depletion layer model is adequate to describe the leakage region of the varistor characteristic and its temperature dependence. However, it was realized at an early date [29] that this simple approach based only on electron participation in current transport cannot account for the very large nonlinearities (nonlinearity exponent α much greater than 50) exhibited by varistors in the breakdown region. The participation of minority carriers in the conduction process was first postulated by Mahan et al. [29] and validated by Pike et al. [37] who observed band gap electroluminescence at 3.2 eV in some varistor compositions. The greater the luminescent intensity, the higher the varistor nonlinearity. Electroluminescence was observed in three different varistor compositions, containing Bi, Bi + Mn, and Bi + Mn + Sb additives, respectively. The observed luminescent intensity is proportional to the square of the varistor current, which is consistent with a model of hole creation by impact ionization near the ZnO grain boundaries. Impact-ionized holes have also been invoked to explain the unusual phenomenon of negative capacitance observed in ZnO varistors when subjected to bias voltages in the vicinity of breakdown [34].

This effect of holes is shown schematically in Figure 16. Since the depletion layer is very thin (about 100 nm), the electric field in this region can reach very high values (around 10 MV/cm) when the applied voltage is in the breakdown range. Under such high electric fields, some of the electrons crossing the barrier gain sufficient kinetic energy that they can produce minority carriers (holes) by impact ionization of the valence states and acceptor states within the depletion region. The hole produced by this process diffuses back to the grain boundary and compensates for part of the trapped negative interface charge. This lowers the potential barrier at the grain boundary. This impact ionization hole creation process can result in very large nonlinearity coefficients. A mathematical model of this process is given in Ref. 38.

Most of the available theoretical models are phenomenological in their construction in that the physical origin of the surface state and bulk donors describing the varistor barriers are not considered. For example, in one phenomenological model [33], the surface state density and doping profile are extracted from $C-V$ and $I-V$ data. While it may be possible in this model to assign a set of such parameters to each and every varistor sample and thus account for materials and processing variations in terms of these parameters, they are not of themselves independently reasoned or justified and are thus of limited predictive utility.

B. Effects of Microstructure

As with all ceramics we expect the microstructure of ZnO varistors to be irregular. The idealized schematic of Figure 8 depicts conduction between the electrodes taking place through a set of identical parallel paths. If in fact all the grain boundaries were identical electrically and the shape and size of all the grains the

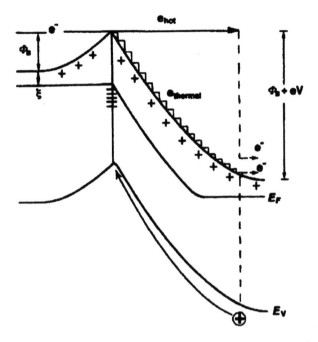

Figure 16 Band diagram for a grain barrier under an applied voltage in the breakdown region. Hot electrons create holes by impact ionization. These holes move to the grain boundary where they decrease the barrier height. The hole creation process and consequent barrier height drop produces the extremely high nonlinearity observed in ZnO varistors. (Adapted from Ref. 12.)

same, then the electrical characteristics of the bulk varistor would be same as that of an individual grain boundary. In reality, small perturbations in the ceramic microstructure—for example, in the number of grain boundaries in the various parallel conduction paths—can cause the current flow to vary considerably from place to place over the electroded faces of the varistor disk. This behavior can drastically affect varistor performance, especially in the high-nonlinearity region, even in cases where the individual grain–grain barrier electrical characteristics are "optimal."

In general, we can expect variations in the nature of the grain boundary as well as variations in the arrangement of the grain boundaries. For example, by measurements on large numbers of individual grain boundaries, Einzinger [39] has shown that there can be considerable variation in the breakdown voltage from one boundary to another (Figure 17). His measurements also showed that some boundaries were ohmic, some nonlinear, whereas others were insulating. This

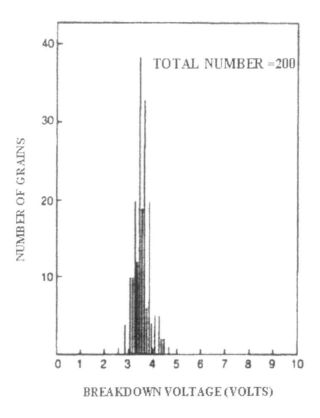

Figure 17 Variation of breakdown voltage across single grain boundaries. The data were obtained using microcontacts. (Adapted from Ref. 39.)

type of informations, together with observations regarding the complexity of varistor microstructures, has led to the realization that varistors are far from uniform.

The need to account for microstructural disorder has been apparent from the very early days of varistor technology. For example, low-voltage varistors operate in the 10-V range. Since the breakdown voltage of an individual grain boundary in ZnO varistors is ~3.5 V, low-voltage varistors are typically only a few grains thick. This places severe constraints on microstructural uniformity in the device. Clearly, modest variations in grain size will lead to significant preferential channeling of current in the regions where the grains are larger (fewer grains boundaries between the device electrodes). Intuitively one can convince oneself that the result of microstructural nonuniformity will be to degrade the varistor performance by increasing the leakage and decreasing the nonlinearity.

Modeling the detailed effects of microstructural nonuniformity on varistor behavior is difficult due to the extremely nonlinear nature of the device. However, the advent of more powerful computers has permitted the varistor behavior to be simulated, and the results of extensive computations have been published by the groups working with Clarke [40–43] and Mahan [44–46]. In modeling the current flow the microstructure is described in terms of an equivalent electrical network of nonlinear resistors; then disorder is introduced into the network. The spatial distribution of current flow is then determined in response to an applied voltage.

As expected intuitively, the simulations reveal that disorder in the form of variations in grain size, grain–grain contact area, or grain boundary barrier height decrease the attainable nonlinearity, cause a rounding of the current–voltage characteristic in the vicinity of the breakdown voltage turn-on, and decrease the breakdown voltage itself. The simulations provide insight on the spatial distribution in current flow and current localization in particular. This is described in detail in Ref. 13. As might be expected, localization is most pronounced in the nonlinear region. The individual grain boundaries adjust their conductivities in response to the applied voltage, and we observe a spatial redistribution of the principal current paths through the varistor as the voltage is increased or decreased. Experimental evidence for inhomogeneous current flow through a varistor comes from thermal imaging of the cross section of a varistor under electrical load [46]. The high-current paths are visualized by the local increase in temperature caused by Joule heating. The data show that the electrical energy is primarily dissipated at the grain boundaries and that the current localization paths are typically only one or two grains wide.

The ability to model and observe the effects of microstructural nonuniformity in varistors has enabled manufacturers to enhance the energy handing capability of their material by tailoring manufacturing processes to produce material with superior uniformity.

C. ZnO Grain Resistivity

In the upturn region (Figure 2) the ZnO grain resistivity is an important determinant of device performance. Conduction within the grain is ohmic and in its simplest representation (Figure 9) adds a voltage (per unit length) of value

$$V_g = J\rho_g \tag{7}$$

to the breakdown voltage, where ρ_g is the grain resistivity. This voltage produces the upturn shown in Figure 2 at high currents and is the feature that limits the performance of metal oxide varistors in high-current surge suppression applications. The circuit voltage protective level is not simply the grain boundary breakdown voltage but is higher because of the voltage drop in the grains themselves.

While pure ZnO is an insulating semiconductor with an ≈ 3-eV band gap, sintered ZnO is reasonably conducting. As can be estimated from Figure 2, at room temperature ρ_g is around 1 Ω-cm, whereas ρ_g is 10–100 times higher at 77 K. The grain resistivity can be controlled to a certain extent by dopants such as Al or Li [47–49]. However, the simple concept that grain resistivity, ρ_g, solely determines the upturn is only approximate. For example, one can attempt to evaluate ρ_g by linearly extrapolating the varistor breakdown characteristic (in a log-log plot) into the upturn region. The difference between observed varistor voltage and extrapolated voltage serves to define a voltage drop ΔV_g, which should be associated with the voltage drop across the ZnO grain interior. Unfortunately, if we then compute $\Delta V_g/I$, we find that this quantity is not independent of current I. In fact, $\Delta V_g/I$ decreases with increasing current and is not a very good measure of r_g.

Other measurements of the grain resistivity are also of interest and provide additional insight into the interpretation of upturn data. At high frequencies [28], $f > 10^8$ Hz, the intergranular capacitance becomes an effective short circuit, and the equivalent circuit for the varistor shown in Figure 9 can be represented as a resistor r_g whose resistivity is the ZnO grain resistivity. The value of r_g can then be determined by high-frequency impedance measurements. Infrared optical techniques [50] also provide a method for measuring the ZnO grain resistivity since the reflectance of ZnO has a sharp minimum whose wavelength position depends on the free-electron density and thus on the ZnO grain resistivity. Both of these techniques give values for the grain resistivity ρ_g, which are lower than those obtained from upturn measurements by perhaps a factor of 5 or more [28].

The grain resistivity can be increased by doping with Li, which presumably introduces deep levels in ZnO compensating the shallow donors responsible for grain conduction. Doping with Al, Ga, or In, on the other hand, reduces the grain resistivity. Doping varistors with Al is now common practice, and grain resistivities of ≈ 0.1 Ω-cm are achieved in commercial varistor materials. Although ZnO resistivities below this value are achieved in single-crystal studies [51], efforts to reduce the resistivity below ≈ 0.1 Ω-cm by conventional means in varistor materials have not met with practical success. One of the problems encountered in varistor grain doping is the fact that leakage conduction is also affected by the use of grain dopants such as Al. Thus, for example, the voltage clamp ratio, $V_{1KA}/V_{0.1mA}$, which is one figure of merit for varistor surge protection capability, may have its optimal (lowest) value at an Al doping concentration that does not minimize the grain resistivity. The connection between Al doping and leakage conduction is made somewhat clearer by secondary ion mass spectroscopy (SIMS) depth profiles showing that the element Al tends to segregate at grain boundaries rather than just within the ZnO grains themselves [52]. Thus, the grain boundary is also doped by Al, and apparently such doping increases leakage conduction.

D. Degradation and Failure

Device degradation has been an issue affecting the application of ZnO varistors from the advent of the use of the technology as a surge protection device. Degradation usually refers to the steady increase in leakage current when the varistor is subjected to a constant dc or ac voltage or to a series of pulses [53–59]. It is important to note that the magnitude and temperature dependence of the leakage conduction can be very sensitive to the materials and processing schedules used in the sample preparation. However, degradation as used here refers to the fact that some electrical, mechanical, chemical, and environmental stresses can tend to increase leakage conduction [59]. An example of the effect of dc electrical overstress on the $I-V$ characteristics of a ZnO varistor is shown in Figure 18. The leakage conduction after stress has increased by more than 4 orders of magnitude and is now polarity dependent.

Degradation phenomena in ZnO varistors exhibit several general features [59]. First, the rate of degradation (rate of increase of leakage conduction) can be strongly influenced by very small changes in the chemical makeup and/or processing of the varistor, as well as by heat treatments well below the sintering temperature. Second, degradation associated with a diverse set of stresses (electrical, mechanical, environmental, etc.) can often be rapidly reversed by annealing at ≈300°C or lower. Third, and perhaps most striking, the degradation process affects mainly the varistor leakage and prebreakdown conduction and not the varistor behavior for voltages above the breakdown voltage. This behavior is clearly indicated in Figure 18. Therefore, changes in barrier height, depletion

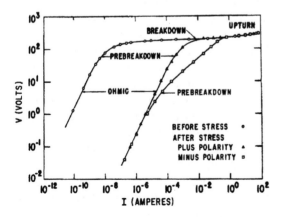

Figure 18 The effect of de electrical overstress on the current – voltage characteristics of a ZnO varistor.

layer thickness, intergranular layer trap density, distribution, etc., which are associated with the degradation process should be such as to leave unchanged the highly nonlinear breakdown conduction process.

In the early 1970s it was found by serendipitous experimentation that in some commercial varistor formulations the extent of continuous ac voltage–induced degradation can be minimized by low-temperature (400–600°C) annealing. Also, some varistor formulations are less susceptible to degradation than others. Moreover, after degradation, the original current–voltage characteristics can often be restored by annealing the degraded varistor at relatively low temperatures (~200°C). Annealing is accompanied by the generation of current (thermally stimulated currents), indicating that charge is trapped during the degradation process and can be released from relatively shallow traps in the degraded material by thermal activation [60].

Two principal mechanisms for degradation have been proposed. Gupta and Carlson [61] proposed a form of electromigration, in which interstitial ions (e.g., Zn) in the depletion layer diffuse preferentially in the direction of current flow. On reaching the grain boundary, the ions combine with the defects defining the electrostatic barrier, thereby lowering the potential barrier and correspondingly increasing the leakage current. A second mechanism proposes that the grain boundary interface states are associated with chemisorption of oxygen and that during degradation oxygen desorption occurs. Although the mechanism primarily responsible for degradation has not yet been definitively established, it has been shown that small additions of monovalent ions, such as sodium and potassium, to the varistor formulation can enhance the stability to degradation [62]. The beneficial effect of these ions has been attributed to their occupying interstitial sites and thereby blocking the migration of zinc interstitials.

In addition to issues related to long-term degradation, system application of ZnO varistor devices requires that the material be used under conditions that do not lead to device failure induced by excessive pulse energy or thermal energy input. In practice, two main types of device failure are commonly observed. One is a so-called puncture mode in which a hole penetrates through the varistor with distinct signs of material melting, blackening, and vaporization. In the other form of failure the varistor fractures into two or more pieces but with no melting. Both failure modes occur above a threshold energy, as shown in Figure 19. It is generally found that fracture mode failures are associated with short pulse widths whereas long pulses failures result in punctures [63].

The puncture mode of failure is generally assumed to occur as a result of thermal runaway. The current becomes so high locally that the ZnO heats sufficiently to cause even more current to channel through the local hot spot. Since the grain boundary leakage increases rapidly with temperature, the increased current caused by local heating along the localization path results in further current

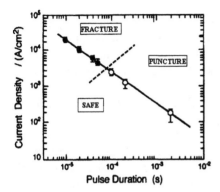

Figure 19   Current and pulse duration map for varistor failure modes. (Adapted from Ref. 63.)

localization and heating. This positive-feedback loop continues with melting and grain boundary electrical shorting being the ultimate result.

When varistors are subjected to extremely short (less than 10 μs) high current density pulses (Figure 19), varistor failure tends to occur by fracture of the material. This short pulse behavior is generally regarded as adiabatic in nature since the associated thermal diffusion length is less than the ZnO grain size. Material fracture morphology and fracture mode observed depends on the shape and size of the varistor. Tall varistors used for lightning arresters often fracture near the varistor block center with the fracture surface parallel to the electrode. It is fairly clear that this type of fracture is induced by a shock wave generated as a result of the very high-current pulse.

Both puncture and fracture failures can be reduced if the varistor ceramic is uniform, defect free, and has good mechanical strength.

VI. VARISTOR APPLICATIONS

ZnO varistors have now been in widespread use for more than 30 years. The technology has supplanted generally less useful approaches to voltage surge protection such as silicon carbide arresters, gas discharge tubes, and to some extent semiconductor-based surge suppressors such as Zener diodes. A significant advantage of ZnO varistors as a transient protective device derives from the ceramic nature of the material. Since the material is polycrystalline with energy absorption occurring essentially at the grain boundaries distributed throughout the volume of the material, ZnO varistors are inherently able to absorb more energy than single-junction protective devices such as Zener diodes.

More than one billion ZnO varistor devices are in service. They protect circuits with voltages ranging from around 10 volts to millions of volts. The larger devices are capable of surviving transient currents greater than 100,000 A and can absorb energies greater than 10,000 J (energy absorption capability > 200 J/cm^3 is routinely available).

Consumer-purchased surge-suppressor electrical outlets based on ZnO varistor technology are commonly used to protect computers and other high-value electronic equipment. A simple example of this type of unit is shown in Figure 20. Three surge-protected power points are available. In Figure 21 we show a variety of shapes and sizes for ZnO varistors sold into the residential and industrial markets for protecting electrical equipment operating at voltages from a few tens of volts to around 1000 volts. Details of voltage and energy ratings, application notes and so forth are available from many manufacturers in the United States, Europe, and Asia.

A second useful feature of ZnO varistors deriving from the ceramic nature of the material is the ability to configure a particular device to conform to system constraints. Some applications of ZnO varistors are given to illustrate the versatil-

Figure 20   Typical consumer surge-protected electrical outlet.

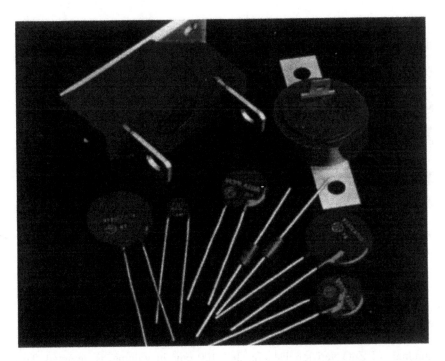

Figure 21 Various shapes and sizes for ZnO varistors sold into the residential and industrial markets for protecting electrical equipment operating at voltages from a few tens of volts to around 1000 V.

ity of the material. Figure 22 shows a miniature ZnO varistor sleeve [64] fabricated as a tube that can fit around a connector pin as small as size 22. These devices fit inside available multipin cable connectors and provide a low-inductance compact configuration for protection of sophisticated electronics against other electrical transients.

Figure 23 illustrates a miniature "chip" varistor designed to be directly mounted on a printed circuit board using surface mount technology. The absence of device leads reduces inductance and permits dense device packing on the circuit board. Devices of this shape are often made in a multilayer configuration (similar to multilayer capacitors), which facilitates the fabrication of lower voltage varistor devices.

In addition to the protection of electronic equipment, a major application of ZnO varistors is for the protection of electric power distribution and transmission systems [65]. In these applications, a surge suppressor is required to function reliably on systems with voltages ranging up to a megavolt and to absorb transient

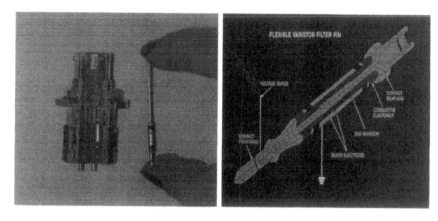

Figure 22 Tubular varistor sleeve mounted on connector pin. The pins are in turn placed inside the shell of multipin cable connectors.

energies in the megajoule range. Large volumes of varistor material are needed to meet these requirements. In Figures 24 and 25, a large station arrester containing hundreds of ZnO varistor disks, each >100 cm³ in volume, is depicted. The advantages to the power system designer of using ZnO varistor transient protection technology is described extensively elsewhere [66]. It is interesting to note, however, that ZnO varistor technology has significantly simplified the design and increased the reliability of electrical power system protection against transient surges.

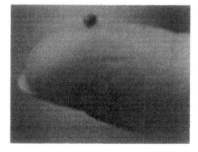

Figure 23 Surface-mount chip varistor. These leadless devices can be directly soldered onto a printed circuit board or ceramic hybrid assembly.

Figure 24   Large power station arrester containing hundreds of ZnO varistor disks each >100 cm³ in volume.

VII.   CONCLUSIONS

ZnO varistors have found acceptance as a simple, cost-effective way to protect electrical systems from transient voltage surges. Device operation is controlled by grain–boundary effects at the ZnO–ZnO interface. By virtue of their ceramic nature, they can be fabricated into a variety of sizes and shapes, and this feature

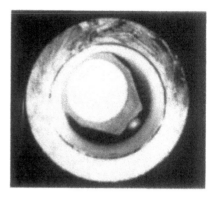

Figure 25 ZnO varistor components used in the construction of the high-voltage station arrester shown in Figure 24. The ZnO varistor station arrester disks are surrounded by a silicone rubber collar and pressed against the side of a porcelain insulating housing to enhance thermal transfer.

facilitates a high degree of user flexibility. Varistors are inherently multijunction grain–boundary devices, and any transient surge energy to be absorbed when the device acts in a protective mode is distributed among the many ZnO grain junctions. The multijunction feature of varistor behavior is the heart of the high-current and energy-absorption capability of the material.

REFERENCES

1. Castleberry DE. IEEE Trans. Electron Devices, ED-26, R1123, Castleberry DE, Levinson LM. SID International Symposium Paper 18.5, 1980.
2. Dosch RG, Tuttle BA, Brooks RA. J Matlr Res 1986; 1(1):90–99.
3. Martzloff FD, Hahn GD. IEEE Trans. Power Apparat. Syst., PAS-89 (6). July–Aug. 1970:1049–1056.
4. Fox RW. Electron. Des. 22. May 24, 1974; 52.
5. Matsuoka M. Jpn J Appl Phys 1971; 10:736.
6. Valeyev KhS, Knayazev VA, Drozdov NG. Electrichestro, 4. 1964:72.
7. Kosman MS, Pettsol'd EG. Uch. Zap. Leningr. Gos. Pedagog Inst. A. I. Gertsena, 207. 1961:191.
8. Levinson LM, Philipp HR. Am Ceram Soc Bull 1986; 65(4):639–646. See also the web site http://VaristorTechnology.com.
9. Einzinger R. Annu. Rev. Mater. Sci. 1987; 17:299–321.
10. Gupta TK. J. Am. Ceram. Soc. 1990; 73(7):1817–1840.
11. Blatter G, Greuter F. Semicond Sci Technol 1990; 5(1):111–137.

12. Pike GE. In: Swain MV, ed. Materials Science and Technology. Vol. 11. Weinheim: VCH, 1994:731–754.
13. Clarke David R. J. Am. Ceram. Soc. 1999; 82(3):485–502.
14. Philipp HR, Levinson LM. Adv Ceram 1981; 1:309.
15. Clarke DR. J. Appl. Phys. 1978; 49(4):2407–2411.
16. Clarke DR. J. Appl. Phys. 1979; 50(11):6829–6832.
17. Kingery WD, Vander Sande JB, Mitamura T. J Am Ceram Soc 1979; 62(3–4): 221–222.
18. Chiang Y-M, Lee J-R, Wang H. in Ceramic Microstructure: Control at the Atomic Level Tomsia AP, Glaeser A, eds. New York: Plenum Press, 1998:131–147.
19. Morris WG, Cahn JW. in Grain Boundaries in Engineering Materials, 1975: 223–234. Walter JL, Westbrook JH, Woodford DA, eds. LA: Claitors, Baton Rouge.
19a. J. Am. Ceram. Soc. 1992; 75(11):3127–3128.
20. Wong J. J. Appl. Phys. 1975; 46(4):1653–1659.
21. Santhanam AT, Gupta TK, Carlson WG. J. Appl. Phys. 1979; 50(2):852–859.
22. Inada M. Jpn. J. Appl. Phys. 1978; 17(1):1–10.
23. Levinson LM, Philipp HR. J Appl Phys 1976; 47:1117.
24. Levinson LM, Philipp HR. J Appl Phys 1975; 46:1332.
25. Morris WG. J. Vac. Sci. Technol. 1976; 13:926.
26. Levine JD. Crit. Rev. Solid State Sci. 1975; 5:597.
27. Bernasconi J, Klein HP, Knecht B, Strassler S. J Electron Mater 1976; 5:473.
28. Levinson LM, Philipp HR. J. Appl. Phys. 1976; 47:3116.
29. Mahan GD, Levinson LM, Philipp HR. J Appl Phys 1980; 50:2799.
30. Philipp HR, Levinson LM. J. Appl. Phys. 1979; 50:383.
31. Bernasconi J, Strassler S, Knecht B, Klein HP, Menth A. Solid State Commun. 1977; 21:867.
32. Emtage PR. J. Appl. Phys. 1977; 48:4372.
33. Hower PL, Gupta TK. J. Appl. Phys. 1979; 50:4847.
34. Pike GE. Mater. Res. Soc. Proc. 1982; 5:369.
35. Greuter F, Blatter G, Rossinelli M, Stucki F. In: Levinson LM, ed. Ceramic Transactions: Advances in Varistor Technology. Vol. 3. Columbus, Ohio: American Ceramic Society, 1989:31.
36. Greuter F, Blatter G. Semicond. Sci. Technol. 1990; 5:111.
37. Pike GE, Kurtz SR, Gourley PL, Philipp HR, Levinson LM. J. Appl. Phys. 1985; 57:5512.
38. Blatter G, Greuter F. Phys. Rev. B: Condens Matter. 1986; 34(12):8555–8572.
39. Einzinger R. Appl. Surf.Sci. 1979; 3(1):390–408.
40. Wen Q, Clarke DR. Ceramic Transactions Levinson LM, Hirano S-I, eds. Grain Boundaries and Interfacial Phenomena in Electronic Ceramics. Vol. 41. Westerville, OH: American Ceramic Society, 1994:217–230.
41. Vojta A, Wen Q, Clarke DR. Comput. Mater. Sci. 1996; 6(1):51–62.
42. Vojta A, Clarke DR. J. Appl. Phys. 1997; 81(2):985–993.
43. Nan C-W, Clarke DR. J. Am. Ceram. Soc. 1996; 79(12):3185–3192.
44. Bartkowiak M, Mahan GD. Phys. Rev. B: Condens. Matter. 1995; 51(16): 10825–10832.

45. Bartowiak M, Mahan GD, Modine FA, Alim MA. J. Appl. Phys. 1996; 79(1): 273–281.
46. Wang H, Bartowiak M, Modine FA, Dinwiddie RB, Boatner LA, Mahan GD. J. Am. Ceram. Soc. 1998; 81(8):2013–2022.
47. Eda K. in Grain Boundaries in Semiconductors. Materials Research Society Symposium Proceedings Leamy HJ eds.,, Pike GE, eds. New York: Elsevier, 1997:1982.
48. Myoshi T, Maida K, Takahashi K, Yamaryaki T. Adv. Ceram. 1981; 1:309.
49. Carlson WG, Gupta TK. J. Appl. Phys. 1982; 53:5746.
50. Philipp HR, Levinson LM. J. Appl. Phys. 1977; 48:1621.
51. Heiland G, Mollwo E, Stockman F. in Solid State Physics Seitz F eds.,, Turnbull D, eds. Vol. 8. New York: Academic Press, 1959.
52. Philipp HR, Mahan GD, Levinson LM. Report ORNL/Sub/84–17457/1, Subcontract 86X-17457C, July 1984 for DOE under Contract DE-AC05-840R 21400.
53. Eda K, Iga A, Matsuoka M. Jpn. J. Appl. Phys. 1979; 18(5):997–998.
54. Shirley CG, Paulson WM. J. Appl. Phys. 1979; 50(9):5782–5789.
55. Eda K, Iga A, Matsuoka M. J. Appl. Phys. 1980; 51(5):2678–2684.
56. Gupta TK, Carlson WG, Hower PL. Current instability phenomena in ZnO varistors. J Appl Phys. Vol. 52, 1981:4104–4111.
57. Moldenhauer W, Bather KH, Bruckner W, Hinz D, Buhling D. Phys. Status Solidi. 1981; 67:533–542.
58. Hayashi M, Haba M, Hirano S, Okamoto M, Watanbe M. J. Appl. Phys. 1982; 53(8):5754–5762.
59. Philipp HR, Levinson LM. in Advances in CeramicsEdited by Yan MF, Heuer AH, eds. Additives and Interfaces in Electronic Ceramics. Vol. 7. Columbus, OH: American Ceramic Society, 1983:1–21.
60. Sato K, Takada Y. J. Appl. Phys. 1982; 53(5):8819.
61. Gupta TK, Carlson WG. J. Mater. Sci. 1985; 20:3487–500.
62. Gupta TK, Miller AC. J. Mater. Res. 1988; 3(4):745–754.
63. Eda K. J. Appl. Phys. 1984; 56(10):2948–2955.
64. Aviation Week and Space Technology. May 20, 1985.
65. Sakshaug EC, Kresge JS, Miske SA. IEEE Trans Power Apparat Syst, PAS-96 1977:647–656.
66. Application Guide, Tranquell Station Surge Arresters, General Electric Co., GE Industrial Systems, Fort Edward, NY, 12828. See the web sites http:// www.geindustrial.com/products/brochures/arrester.pdf and http:// www.geindustrial.com/cwc/products?famid = 2.

8

Electroceramic Thin Films for Microelectronics and Microsystems

A. I. Kingon
North Carolina State University
Raleigh, North Carolina, U.S.A.

P. Muralt and N. Setter
Swiss Federal Institute of Technology
Lausanne, Switzerland

R. Waser
Aachen University of Technology
Aachen, Germany

I. INTRODUCTION

Functional ceramic thin-films have been investigated for device applications since the mid-1980s. Products based on such films have been commercialized from the mid-1990s onward. The field interfaces with various engineering disciplines such as communications engineering, microelectronics, microtechnologies, and informatics and is gaining importance in these domains.

The interest in electroceramic thin films was triggered by two events. First, the newly discovered high-temperature oxide superconductors promised to show considerably larger critical current densities if deposited as thin films. Second, an idea originating in the late 1960s of ferroelectric semiconductor memories was taken up again by utilizing the experience gained with the highly advanced Si semiconductor technology. These pioneering lines led to a broad research and development effort worldwide to combine the large variety of functions offered

465

by electroceramic thin films with the standard integrated circuit chips. The range comprises commercial "high-k" (high permittivity) dielectric ceramic films for decoupling capacitors in monolithic microwave integrated circuits for mobile telephones [1], 3D integration of high-k multilayer thin film capacitors [2], and cell dielectrics for gigabit dynamic random access memory (DRAM) devices [3]. Ferroelectric materials are employed in nonvolatile ferroelectric random access memories (FRAMs) [4] with a recent further development of the so-called ferroelectric transistor (a field-effect transistor with a ferroelectric gate oxide) [5].

The integration of sensors into microprocessors emerged in the 1980s. The excellent micromachinability of silicon added new perspectives in sensor design; micromachined silicon structures such as cantilever beams, free-standing bridges, membranes, and channels led to novel structures of miniature transducers resulting in the emergence of the new field of microelectromechanical systems (MEMS). MEMS are miniature devices or arrays of devices that combine electronics with other components, such as sensors, transducers, and actuators, and are fabricated by integrated circuit processing techniques. For a number of applications the MEMS concept facilitates better device quality (e.g., higher sensitivity, faster response, and better reproducibility), and results in lower costs with, very often, lower power consumption than discrete devices. Examples of MEMS products are the air-bag accelerometer, which reached the market at the beginning of the 1990s, and the more recent digital micromirror display (TV screen) in which a large array of electronically controlled mirrors etched onto a MEMS converts video data to high-resolution TV images. Often devices based on micromachined silicon are classified as MEMS even if the electronics is not (or not yet) integrated with the device. This convention is followed in the present chapter.

The integration of functional ceramics in the form of thin or thick films with silicon adds versatility to the MEMS and opens new possibilities. A number of microdevices based on functional ceramic thin films have been developed, some of which have been commercialized. The integration of thick films with silicon lags behind; the still-too-high temperature of densification of thick films (here thick films refers to films made using powder-based technologies) hinders their integration onto silicon electronics. The main effort in thick-film development is therefore focused on reduction of the processing temperature. In this chapter we focus on thin films.

All types of electroceramics are of interest for MEMSs and microelectronics applications: conductors (e.g., IrO_2) are used as electrodes, semiconductors are used in microsensors (e.g., gas sensors) [6], films based on $(La,Sr)MnO_3$ and other perovskite manganites and related structures show a colossal magnetoresistive behavior [7], high-temperature superconductor thin films are employed in SQUID magnetic field sensors [8] and in coatings of microwave resonators that increase the Q factor of the device.

In this chapter we center on linear and nonlinear dielectric ceramic thin films (polar ceramics, piezoelectrics, electrostrictive, high-permittivity materials and ferroelectrics). The main applications of this group of materials in microelectronics are memories and components such as filters and resonators; examples of materials are (Ba,Sr) TiO_3, $Pb(Zr,Ti)O_3$, and aluminum nitride (AlN). In MEMS, polar and piezoelectric materials are suitable for various microsensors and microactuators.

Thin-film deposition techniques such as sputtering, pulsed laser deposition (PLD), molecular beam epitaxy (MBE), chemical solution deposition (CSD), and chemical vapor deposition (CVD) are being used in the processing of ceramic thin films. During these deposition processes the electroceramic materials are synthesized on a microscopic scale without powder processing as an intermediate step, usually at temperatures much below the typical sintering temperatures of bulk ceramics. The processing of thin films is presented in the first part of this chapter.

Thin films cannot be treated in an isolated manner. They are always a part of a system that comprises a substrate and various interfacial layers, e.g., diffusion barriers, electrodes, passivation layers, layers for stress compensation and for nucleation and growth control, etc. For realization of devices, the patterning of the electroceramic films is an important issue too. Patterning of thin films is accomplished mainly by two approaches: a reductive approach of etching the annealed continuous layers, and an additive approach in which a treatment of the substrate results in obtention of patterned film prior to the annealing step. All these issues are discussed in the part of this chapter devoted to integration processes.

Thin-film properties often differ from bulk properties: Down-scaling increases the importance of interfaces with respect to volume effects. Films are often textured whereas bulk ceramics are isotropic in most cases. The lower temperature of processing in comparison to bulk ceramics changes the defect concentration, and this influences the properties as well. Stress due to thermal expansion mismatch between the various layers or lattice mismatch influences the properties as well. Various devices based on ceramic thin films have already reached the market; others are at an advanced development stage; and novel applications continue to emerge. Several important applications that have been commercialized recently and a few that are considered highly promising are outlined. Properties are discussed in this chapter in the context of device applications.

II. DEPOSITION METHODS OF CERAMIC FILMS

In general, thin-film deposition methods can be subdivided, on one hand, into chemical and physical techniques and, on the other hand, into techniques in which the solid film arises from an adjacent liquid phase and an adjacent gas phase, respectively. This section will outline those techniques that found widespread use in the deposition of thin films of complex oxides. A comparison of thin-film deposition techniques for all major classes of electronic materials is given in Ref. P.

Chemical deposition of thin oxide films can be performed through the gas phase (CVD) or the solution phase (CSD). In both cases, chemical precursors are employed that undergo chemical reactions for the formation of the oxide film. Typically, the precursor molecules contain the metal atoms, M, and organic groups, R. In most cases, such as in alkoxides, ketonates, and carboxylates, the metal atom is bound to the organic group through an oxygen atom, i.e., M-O-R. This is somewhat different to the case of precursors used for the deposition of compound semiconductors by metal organic chemical vapor deposition (MOCVD) where the precursors are conventional metal organic compounds, i.e., M-R systems.

Physical deposition of thin oxide films are exclusively vapor phase techniques (physical vapor deposition, PVD). The liquid phase epitaxy (LPE) is not employed for oxides. We will give a short introduction to the basic principles of the different physical deposition methods and some comments on their advantages and drawbacks. Emphasis will be placed on the description of the oxygen control for the various techniques, since this presents the major difference to standard PVD techniques used for the fabrication of semiconductor and metal films.

A. Chemical Solution Deposition

The use of alkoxide solutions for the coatings of simple oxides (SiO_2, TiO_2) is known since the 1930s [10]. Thin films of complex multicomponent oxides such as lead zirconate titanate (PZT) have been first applied to CSD-type processing techniques by Gurkowich and Blum [11] as well as Budd, et al. [12]. Since then, the CSD technique has been employed by numerous groups (see, for example, Refs. 13 to 20).

A generalized flow chart of the CSD of oxide thin films is shown in Figure 1. The process starts with the preparation of a suitable coating solution from precursors according to the designated film composition and the chemical route to be used. Besides mixing, the preparation may include the addition of stabilizers, partial hydrolysis, refluxing, etc. (see, for example, Refs. 21–23). The coating solution is then deposited onto the substrates by

> Spin coating where typically a photoresist spinner is employed and which
>   is suitable for semiconductor wafers,
> Dip coating, which is often used in the optics industry for large or nonplanar
>   substrates, and

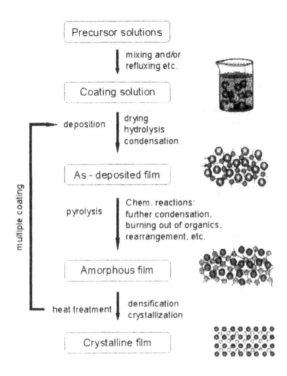

Figure 1   Illustration of the CSD technique. The bars describe the states during the CSD procedure, whereas the arrows indicate the treatment and the internal processes.

Spray coating, which is based on a misting of the coating solution and deposition of the mist exploiting gravitation or an electrostatic force [24].

The wet film may undergo drying, hydrolysis, and condensation reactions depending on the chemical route. The as-deposited film possibly represents a chemical or physical network, depending on the specific route (see below). Upon subsequent heat treatment, an additional hydrolysis and condensation and/or a pyrolysis of organic ligands may take place, again depending on the chemical route. The resulting film consists of amorphous or nanocrystalline oxides and/or carbonates. Upon further heat treatment, any carbonate will decompose and the film will crystallize through a homogeneous or a heterogeneous nucleation. Typically, the desired final film thickness is built up by multiple coatings and annealings.

The CSD method comprises a range of deposition techniques and of chemical routes, the most important of which are shown in Figure 2. Depending on the type and reactivity of the precursors, the chemistry shows a wide spectrum

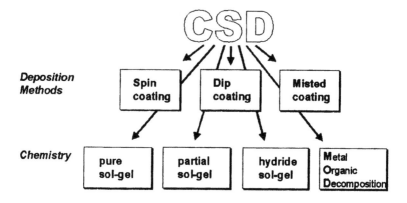

Figure 2 Variety of CSD deposition techniques and chemical route types.

of reaction types. On the one hand, there are the pure sol-gel reactions in which alkoxide precursor systems undergo hydrolysis and condensation reactions. The formation of SiO_2 coatings starting from Si alkoxides is the classical example of this type of reaction. The condensation leads to a chemical gelation in which—under appropriate reaction conditions—no pyrolysis reaction of any organic ligand occurs. At the other extreme, there is metal organic decomposition (MOD), which typically starts from carboxylates of the cations or, in special cases, from the nitrates [25]. The carboxylates do not chemically react with water. Consequently, during the heat treatment, first the solvents are evaporated, a process sometimes referred to as physical gelation. Upon further heating, the carboxylates pyrolytically decompose into amorphous or nanocrystalline oxides or carbonates. There is a wide spectrum of possible reaction routes between the pure sol-gel route and the MOD route. Depending on the type of alkoxides and a possible stabilization of the precursors, there may be partial hydrolysis and condensation, whereas some organic ligands remain in the gelated film and undergo pyrolysis upon further heat treatment. In the synthesis of multicomponent oxide films, hydride routes are often followed, i.e., some precursors may be employed that tend to follow the sol-gel or partial sol-gel route whereas others undergo typical MOD reactions.

The *advantages* of CSD are (1) an excellent control of the film composition through the stoichiometry of the coating solution, (2) a relatively low capital investment, and (3) an easy fabrication over large areas, up to multiple square meters for dip and spray-coating techniques. The *disadvantages* are (1) some specific obstacles to achieving epitaxial films, (2) the lack of an opportunity to deposit atomic layer superstructures, and (3) a poor step coverage for narrow submicrometer 3D structures.

B. Metal Organic Chemical Vapor Deposition

MOCVD of oxide thin films is based on the evaporation of the precursors that decompose on a hot substrate in a suitable reaction chamber in which the temperatures, as well as the pressure of gases such as oxygen or some other oxidizing gas, inert gases, and the precursor vapor, are controlled. For a more detailed review, the reader is referred to Ref. 26.

1. Precursor Systems

The most essential requirements for the precursors are a sufficient volatility at low temperature and a sufficiently high decomposition temperature. The range between the two temperatures defines the process window. For heavy cations, a reasonable volatility is often reached by β-diketonates with several alkyl groups, such as in tetramethylheptadionates (thd). The two keto groups chelate the cation while the outer alkyl groups effectively shield any polar region of the molecule. Typically two (thd) ligands are reacted with one cation, yielding e.g., $Ba(thd)_2$. Since the outer shell of this molecule consists entirely of alkyl groups there is only a weak van der Waals interaction between neighboring molecules resulting in a relatively low boiling temperature. For cations with a higher electronegativity (i.e., a lower tendency to form ionic bonds), alkyl compounds, such as $Pb(C_2H_5)_4$, and alkoxide compounds, such as $Ti(OC_3H_7)_4$, may give rise to sufficient volatility.

2. Precursor Delivery and Reactor Design

Over the years, different techniques have been developed to evaporate the precursors and to deliver them into the reaction chamber (Figure 3). The main difference is mixing in the vapor phase or in the condensed phase. In the first case, the liquid, dissolved, or solid precursor is brought to a designated temperature. A carrier gas flowing through or over the precursor system picks up the precursor at a designated rate. For precursor solutions, these precursor delivery systems are called *bubbler systems* [26]. For multication films, the vapor flows must be mixed with high precision. In the latter case, the precursors are mixed in a solution either as a cocktail in a batch or by mixing the single solution using liquid mass flow meters. In front of the reactor, the mixed solution is flash evaporated by contact with high-temperature metal frit or by misting it into a hot carrier gas.

The *reactor design* has to support a continuous, completely homogeneous, and highly reproducible deposition reaction. For small wafers, a reactor with a horizontal gas flow is most appropriate [26]. For higher throughput, the same horizontal gas flow principle is translated into a *multiwafer planetary reactor* [27], in which there is a central supply of the precursor and carrier gas and the wafers are grouped symmetrically around this center. The wafers are driven into

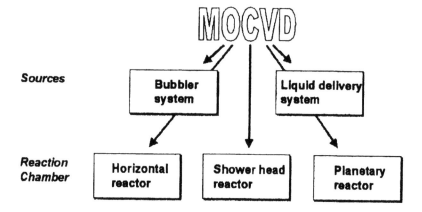

Figure 3 Overview of the typical precursor source systems and reactor design principles used for MOCVD of oxide thin films.

a slow planetary motion in order to avoid precursor depletion effects in the gas phase from the center to the outside and, thus, to guarantee completely homogeneous deposition. These systems show a very high overall precursor efficiency. Another system especially suitable for very large (e.g., 300 mm) wafers is the *showerhead* design [28]. Here the precursor vapor is distributed over a temperature-controlled plate with numerous defined holes (so-called showerhead) which supplies the precursor vapor homogeneously over a large area.

3. Deposition Mechanism and Epitaxy

The chemical mechanisms involved in the MOCVD of complex oxides comprise aspects such as the gas-phase reaction pathway, the decomposition temperatures and activation energies, the adsorption and surface migration processes, as well as the incorporation efficiencies. Based on mass spectrometric and thermodynamic data, modeling results have been reported for $(Ba,Sr)TiO_3$ (BST) deposition using β-diketonate precursors [29]. The first stage of decomposition $M(thd)_2 \rightarrow M(thd) + (thd)$, where $M = Ba$, Sr, or TiO, occurs around 250°C. The subsequent stages involving $M(thd)$ and thd radicals occur around 350–450°C and finally lead to metal oxides and the light species H_2O, CO, and CO_2. Considering in addition the solute and oxidizer fluxes, the incorporation rate for Ti and Sr in the growing BST layer could be predicted.

Some remarks on the temperature dependence of the deposition rate seem worthwhile as the reaction can quite generally be divided into two regimes: at low temperatures a regime of reaction-controlled growth and at high temperatures a regime of transport-limited growth.

1. *Reaction limited growth:* Although the reaction kinetics includes many different process steps, the growth rate, j, can often be described by an Arrhenius law with an effective activation energy, E_a: $j \propto \exp(-E_a/k_BT)$. Due to the dominating temperature dependence the deposition rate is only weakly dependent on the flow homogeneity, and this regime is best suited for conformal deposition. However, the exponential temperature dependence necessitates a very high-temperature stability and uniformity over large wafers.

2. *Transport limited regime:* At medium and higher temperatures the major limitations of the reaction are given by the transport of the precursors and reactants to the surface. For the case that the diffusivity of the gases, D, is the dominating parameter, the growth rate may be written as $j \propto D^{1/2}T$. As the diffusivity for gases is determined by the average velocity, u, and the mean free path length, λ, i.e., $D = \lambda u/3$, only a weak temperature dependence, $\propto 1/T^{1/4}$, is expected. Hence, the deposition rate is insensitive to small temperature gradients; however, control of the flow pattern becomes important.

MOCVD often leads to an *epitaxial growth* of the films if a range of basic requirements are met: (1) the ratio of precursors in the gas phase should be appropriate concerning the sticking and redesorption coefficients in order to allow a film composition close to the designated stoichiometry; (2) the total growth rate determined by precursor supply, flow conditions, and temperature should be moderate to allow incorporation of the adatoms into the correct lattice sites during growth; and (3) the temperature should be high enough to ensure a sufficient surface diffusion of the adatoms.

4. Conformal Coverage

An increasingly important aspect of the integration of complex oxides into future 3D-type, high-density devices will be the quality of the step coverage, along the route growing aspect ratios [30]. The quality includes the degree of conformity, i.e., the uniformity of the layer thickness on horizontal and vertical parts of the structure, as well as the uniformity of the stoichiometry of the film, the microstructure, the orientation, and the density. An excellent example is given in Figure 4 which shows an MOCVD processed $SrTiO_3$ thin layer in an approximately 900-nm-deep and 150-nm-wide hole in SiO_2 [31]. The bottom-to-top step coverage is 0.99 and the Sr/Ti stoichiometry variation is within $\pm 5\%$, as determined by energy dispersive spectroscopy (EDS) transmission electron microscopy (TEM).

To summarize, the *advantages* of MOCVD are (1) the opportunity to deposit epitaxial thin films relatively easily because of the molecular reaction, lay-down, and incorporation into the crystal lattice at the surface; (2) homogeneous deposi-

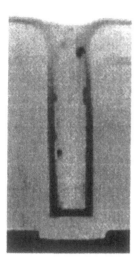

Figure 4 MOCVD of SrTiO$_3$ thin films into a test hole of an SiO$_2$ layer with an aspect ratio of 1:6 and a width of 150 nm. The deposition was performed in a dome-type reactor at a wafer temperature of 420°C. The TEM cross section shows the very high conformality of the thin film [31]. (Reproduced by permission of C. S. Hwang.)

tion over large areas for some of the reactor designs; (3) compatibility with semiconductor fabrication techniques; and (4) opportunity to achieve a high-quality step coverage even for 3D structures of high aspect ratio. This capability is unique to MOCVD compared to all other thin-film deposition techniques. The major *disadvantage* compared to CSD is the requirement for a relatively lengthy adjustment of the film stoichiometry for new film compositions.

C. Molecular Beam Epitaxy

1. System Setup and Operational Principle

Historically, the first PVD technique was the evaporation of metals in high vacuum and the deposition of the vapor onto substrates. MBE has evolved from these simple evaporation techniques by the application of ultrahigh vacuum (UHV) techniques to obtain mean free pathlengths of the atoms which are larger than the dimension of the source to substrate distances, and by including additional systems for beam control and in situ analysis [32]. A schematic view of a system is shown in Figure 5 including several different sources that allow the controlled deposition of multielement compounds. The shutters are very important to control the growth of dopant profiles or the deposition of multilayers. Due to the UHV

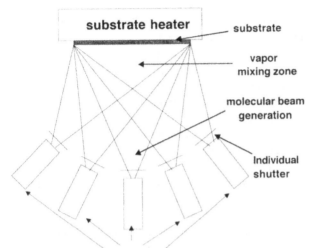

effusion cells for constituent elements and dopants

Figure 5 Schematic view of an MBE system for the growth of multielement compound films.

environment, surface analysis techniques such as reflection high-energy electron diffraction (RHEED) can be applied for in situ control of the growth process.

2. Sources and Oxygen Partial Pressure

A classical MBE source is the *Knudsen cell*. This cell is partially filled with the (melted) material to evaporize. It is heated to a suitable temperature, T, to develop an equilibrium vapor pressure, p_e, in the electron mass and k_B, the Boltzman constant. The upper part of the container has a hole that is small enough that the effusion rate does not disturb the equilibrium pressure. The evaporation rate, N_e, is described by the Hertz–Knudsen (or Langmuir) equation:

$$N_e = \frac{p_e F_e}{\sqrt{2\pi m k_B T}}$$

(1)

F_e is the area of the aperture of the hole [33]. Therefore, in principle, the source can be precisely controlled by a single parameter, the temperature. However, the technical details are very complex. Frequently, *electron beam evaporators* are employed as MBE sources. The electron beam is magnetically deflected by 270° and is centered on the source material. In this way a liquid of the source material is produced on a block of the same material, which can be held in a water-cooled cold crucible, in order to avoid contamination. Other MBE sources are classical thermal evaporators (crucibles) or controlled gas inlets.

The oxygen partial pressure for the deposition of oxide films is a major challenge for UHV technique. However, several solutions have been successfully implemented including differential pumping as indicated in Figure 6, i.e., by rotation of the substrate the film is temporarily exposed to oxygen whereas the sources and the vapor are at high vacuum condition in order to avoid oxidation and deterioration of the sources [34]. On the other hand, MBE offers the opportunity to switch between a highly oxidizing and a highly reducing atmosphere. This feature can be exploited to realize metastable phase combinations. For example, titanates cannot be directly grown on Si because they require a certain oxygen partial pressure which, however, inevitably leads to the formation of SiO_2 at the interface. In order to overcome the stability problem of oxides on Si, McKee et al. developed a two-step MBE process to deposit $SrTiO_3$ and $BaTiO_3$ epitaxially on Si(001) exploiting a layer-by-layer energy minimization at the interface [35,36] (Figure 7). The first atomic layer is grown by MBE deposition of the alkaline earth metal on the Si surface at high temperature (850°C) leading to silicide formation, followed by subsequent perovskite layers grown at much lower temperature (200°C) with oxygen being introduced at appropriate intervals. By this procedure, the interfacial formation of SiO_2 could be avoided.

D Pulsed Laser Deposition

The concept of the PLD process is shown in Figure 8. A short and energetic pulsed laser beam from an excimer laser is focused to a target. The pulse energy of 1 J/pulse leads to the immediate formation of a plasma of the target material

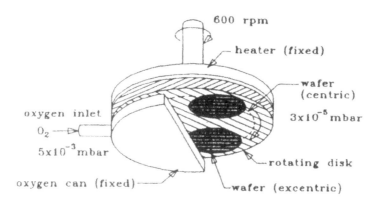

Figure 6 Control of the oxygen partial pressure by differential pumping [41].

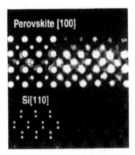

Figure 7   Z-contrast image of SrTiO₃ on (001) Si. The epitaxy that is apparent is (001) SrTiO₃ ∥ (001) Si and SrTiO₃ [100] ∥ Si [110]. On the left side there is an inset model of the perovskite/Si projection [35].

due to the high-energy density of 3–5 J/cm² at the target surface. The plasma contains energetic neutral atoms, ions, and molecules and reaches the substrate surface with a broad energy distribution of 0.1 to >10 eV. Details of the laser ablation process [37] are summarized in Figure 9. A problem of the method is that, in addition to the energetic neutral atoms, ions, and molecules, some small droplets may be deposited on the film ending up as so-called *boulders*. Different

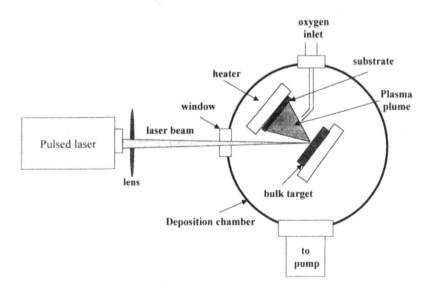

Figure 8   Setup of a PLD system for the deposition of oxide layers.

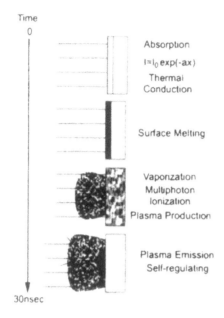

Figure 9   Time evolution of the plasma plume [37].

methods to reduce this effect have been developed, e.g., time-of-flight selection of the deposits [38], as the heavier particles are slower, or the use of an off-axis geometry. In off-axis PLD geometry the surface of the substrate is placed parallel to the expansion direction of the plasma. This geometry has the additional advantage that samples larger than the lateral extension of the plasma plume can be used. However, only if the target is rotated are homogeneous films obtained.

Based on the work on the high-temperature superconductors, the deposition of oxide films is well developed. Very satisfactory photon absorption within the oxide target is provided at ultraviolet (UV) wavelengths of 248 or 193 nm. Typically, a repetition rate of 50 Hz at a pulse length of 25 ns is used. For the growth of epitaxial oxide thin films, sufficient ion mobility is needed. Therefore, depending on the material, substrate temperatures of 700°C or more are required. The substrate heater must be oxygen resistant (e.g., an SiC heater). For the deposition of oxides, an ambient oxygen pressure of 0.3–1 mbar usually is maintained within the chamber.

E. Sputter Deposition

Sputtering of surface atoms has been known since 1852 when Sir W. R. Grove observed this effect during his investigations of plasma discharges. It was soon recognized that this sputtering could be employed for the deposition of thin films. However, the large-scale technological application was developed only during the last decades (see e.g., Refs. 39–41 for review).

1. DC Sputtering

The most simple sputtering system is the so-called DC sputtering, which is schematically shown in Figure 10. In a vacuum chamber the target material, which is eroded, is at the cathode side (negative potential), and the substrate for the film is opposed at the anode side. The potential of several 100 V between these plates leads to the ignition of a plasma discharge for the typical pressures of 10^{-1}–10^{-3} mbar and the positively charged ions are accelerated to the target. These accelerated particles sputter off the deposits, which arrive at the substrate mostly as neutral atoms.

The discharge is maintained as the accelerated electrons continuously ionize new ions by collisions with the sputter gas. The potential distribution between anode and cathode is indicated by the dotted red line: as the plasma is well conducting there is no major potential drop in the plasma region, and due to the different mobility of electrons and ions the main voltage drop is observed at the cathode (darkroom). This potential distribution is advantageous as the acceleration of the sputtering gas ions proceeds directly in front of the target and not in

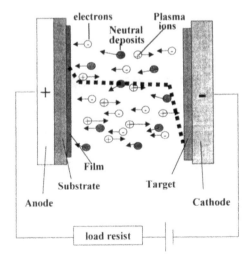

Figure 10 Schematics of a DC sputter system. The dotted line indicates the potential between anode and cathode.

a region far off, where the ions would undergo additional collisions and lose their energy on a long path to the target.

Different sputtering yields are observed for different target atoms, and this difference in principle yields the deposition of a film of different stoichiometry. However, there is generally a good self-regulating mechanism: due to the very low penetration depth of the sputtering process, the faster eroded component is denuded after a short initial time, and finally in a quasi-equilibrium the difference in yield is compensated by the enrichment. Therefore, sputtering generally allows the deposition of films with the same stoichiometry as the target.

2. Magnetron Sputtering

An ionization degree of less than 1% of the atoms is characteristic for a plasma and consequently yields a rather low sputter rate. To improve the ionization rate magnetic fields can be used that force the electron on helical paths close to the cathode and yield a much higher ionization probability. A typical arrangement using permanent magnets and the corresponding fields are shown in Figure 11 [42]. This magnetron arrangement allows additionally a lower gas pressure; however, it has the disadvantage of a more inhomogeneous target erosion than a simple planar geometry.

3. RF Sputtering

DC sputtering works fine as long as the target material has a certain electrical conductivity. However, it is problematic for insulating targets such as polar ox-

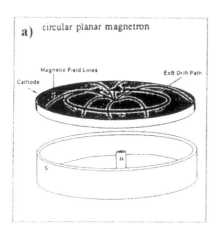
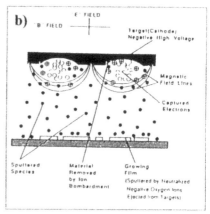

Figure 11   (a) Design of a circular planar magnetron [11] and (b) schematic of the deposition process [37].

ides. This problem can be solved by a high-frequency plasma discharge as shown in Figure 12. A typical frequency of 13.6 MHz is capacitively coupled to the target and there is only a small voltage decay across the electrode. As the electrons are much faster than the ions, a negative potential at the electrodes as compared to the plasma potential evolves during each cycle. With a symmetrical arrangement of cathode and anode we would obtain similar resputtering rates and no film growth. However, nonsymmetries may be introduced by differences in the geometry, i.e., different sizes of target and substrate, and yield some bias voltage. This is supported by the grounding of the substrate and the deposition chamber that is generally applied. Nevertheless, deposition rates are much lower than for DC sputtering.

F. Gas Pressure and Film Growth

For oxide layers the gas pressure is even more important than for metal or semiconductor films as negative oxygen ions are formed that can resputter from the film. As the resputtering yield also depends on the different elements, the stoichiometry of the growing film is changed and these changes of the film are not compensated as is the case for the sputter yield from the target discussed above. This resputtering can be reduced by the so-called high-pressure oxygen sputtering [37] where as the energy of the ions is reduced by the increased number of gas collisions connected with shorter free pathlength; however, it must be considered

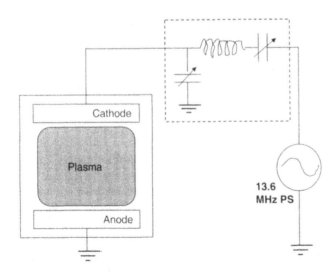

Figure 12   RF plasma sputter system with RF matching network [11].

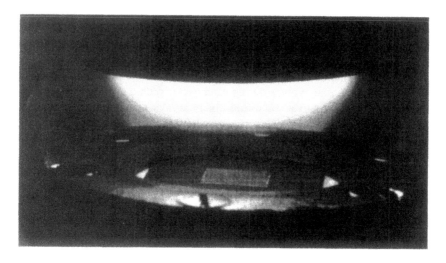

Figure 13 DC sputter deposition in *high-pressure* oxygen.

that this high gas pressure also reduces the deposition rate. This method proved
to be especially useful for the deposition of high-quality oxide thin films. Figure
13 shows a high-pressure (a few mbar) oxygen sputtering system that is operated
in DC mode for the deposition of oxide films [43]. Due to the relatively high
pressure, the ions are almost thermalized and, hence, arrive at the substrate at an
energy much lower then that for typical sputtering processes. Perfect control of
the interface and epitaxial growth can be obtained by this *low-energy* sputtering
process at rather slow growth rates. This perfect interfaces are demonstrated by
the cross-sectional high-resolution TEM view of the resulting multilayer structure
(Figure 14), which is used as a junction barrier in high-T_c devices [44].

III. INTEGRATION PROCESSES

Integration is the processing work required to add a thin-film material onto a
device. It is successful if the material provides full functionality at the end. The
thin film thus has to exhibit the correct phase and the correct composition, eventually the desired crystalline orientation; it must have the correct dimensions,
and—in case of dielectrics—to be contacted by intact electrode materials with
no second phases in between. Furthermore, integration processes should promote
good adhesion between the different layers and avoid degradation of the ferroelectric material or other parts of the device. Integration of ferroelectric materials is
extremely complex and touches the whole product development from design to

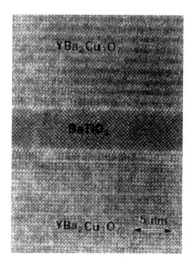

Figure 14 High-resolution TEM micrograph of a sputter-deposited trilayer system [44].

operation. There are no reviews focused on this aspect. Usually the information is scattered in articles on devices or thin-film processing. The reader is referred to a number of general reviews and books for more information and further citations than given below [45–48].

A. Stability of Electrodes

In the case of oxides grown at the relatively high temperatures of 550–700°C, oxidation of metals, interdiffusion, and alloying are the main risks. One must avoid the use of materials subjected to such risks or, if not possible, separate materials by buffer layers. High-quality $Pb(Zr_x Ti_{1-x})O_3$ (PZT) films cannot be grown directly on silicon. Buffer layers are needed to prevent interdiffusion (Pb↔Si) and oxidation reactions. For most applications, the PZT film has to be grown on an electrode. This material should not lose its good electrical conductivity during the oxide deposition process. There are three possibilities to work with: (1) the electrode material does not oxidize; (2) the electrode material is a conductive oxide; (3) the oxide of the electrode material is an electrical conductor. The choice of materials of the first group is very limited. Most ideal is the refractory noble metal platinum (Pt). If the deposition temperature is not too high, iridium (Ir) also falls into this category. Otherwise it is a member of the third category since IrO_2 is a good conductor. A further member of the third group is Ru, whose oxide RuO_2 is a good conductor as well. Gold and silver are not used, as their melting temperatures are too low resulting in high diffusivities and recrystallization phenomena at the required temperatures. The second group of materials is composed of oxide conductors. The above-mentioned rutiles IrO_2 and RuO_2 are used as are a large number of conducting perovskites, such as $SrRuO_3$, $(La,Sr)CoO_3$, $LaNiO_3$, and others [49–51]. Additional complications arise when the ferroelectric material contains lead, which is the case for many frequently used ferroelectric perovskites (e.g., $PbZr_xTi_{1-x}O_3$). Lead that is not yet incorporated in the perovskite lattice shows a high diffusivity and volatility, and in addition forms compounds with many other materials. It may form lead silicates when it is in contact with silicon dioxide, pyrochlore when in contact with tantalum, etc. So in many cases it is not only the oxygen diffusion but also the lead diffusion that is of concern.

Platinum is the standard electrode material used in MEMS applications. It was also very much investigated for FRAM applications. It was found that PZT capacitors are fatiguing too early when using Pt electrodes (at least one electrode has to be replaced by an oxide electrode). Pt can be used in combination with $SrBi_2Ta_2O_9$ (SBT) ferroelectric thin films. Most often, Pt bottom electrodes are grown on a silicon oxide film. An adhesion layer of Ti, TiO_2/Ti, Ta, or similar material is needed to avoid delamination. All pure metal adhesion layers react easily with oxygen to form oxides. Platinum does not inhibit the diffusion of Ti to the PZT side, where it reacts with oxygen and serves as nucleation centers for PZT. There is also evidence that oxygen migrates along the grain boundaries through the platinum film and reacts with the Ti layer. Lead can diffuse through the platinum layer along the grain boundaries and react with the oxide of the silicon wafer to form silicates or compounds with the adhesion layer. The lead loss near grain boundaries of the Pt electrode may lead to the local formation of pyrochlore [lead-deficient phase in the Pb-(Zr, Ti)-O system] in the PZT film. In a comparative study of stabilized Pt electrodes using Ti, Zr, and Ta adhesion layers on silicon oxide it was observed that Ti diffuses through the platinum film, whereas Zr and Ta stay mostly in place. This behavior originates from the fact that the mobile species are the Ti ions in titania and the oxygen ions in zirconia and Ta_2O_5. Oxygen thus diffuses into the Ta and Zr layer, whereas Ti migrates through the already formed oxide and through the Pt grain boundaries to react with oxygen. Pockets of TiO_2 have indeed been found between Pt grains (Figure 15) [52]. The outdiffused Ti leaves nanopores behind. In addition to oxygen, lead diffuses down into the adhesion layer. A lot of lead is found in the former Ta adhesion layer, which has reacted to pyrochlore with Pb and oxygen. Much less lead is found in preoxidized Ta and Ti layers. It is thought that oxidation of adhesion layers opens grain boundaries of the platinum. It is indeed observed that the film stress becomes very compressive at the temperature at which the adhesion layer is oxidizing. Diffusion of Pt along grain boundaries and interfaces reduces this compressive stress above 500°C (Figure 16) [53]. The art of choosing

No pre-oxidation   50 nm

Figure 15   Platinum electrode on Ti adhesion layer observed by TEM imaging. The arrow indicates a TiO_2 pocket in a grain boundary, as verified by electron energy loss spectroscopy [52].

an adhesion layer consists of keeping the balance between the chemical reactivity needed for the adhesion and the chemical passivity needed for avoiding too much oxygen diffusion.

Iridium and ruthenium oxides offer several advantages over platinum: PZT thin films with such electrodes are much less prone to fatigue; such electrodes are much better barriers against lead diffusion and oxygen diffusion; and they are mechanically more stable at higher temperatures (RuO_2 is limited to $<800°C$ due to formation of volatile RuO_4). IrO_2 is one of the key materials for stacked

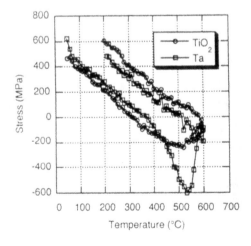

Figure 16   Stress vs. temperature curve for 100-nm-thick Pt films on oxidized silicon wafer using Ta and TiO_2 adhesion layers [53].

capacitor geometry where PZT (or SBT) has to be grown on the plug metal of
the drain contact. This metal can be a refractory but very reactive metal such as
tungsten (Figure 17) [54].

B. Nucleation and Texture

In the previous section the bottom electrode was presented under the aspect of
chemical and thermal stability. There is still another very important function of
the bottom electrode: seeding of the correct crystalline phase. The PZT (perovskite
structure) system has been well investigated in this respect. Growth of PZT thin
films is nucleation controlled, meaning that the activation energy for nucleation
of the perovskite is considerably larger than for its growth. Experimentally, these
activation energies have been determined as 4.4 and 1.1 eV/unit cell, respectively
[55]. As a consequence, heterogeneous nucleation is preferred over homogeneous
nucleation. This is very important for sol-gel deposition techniques or other post-
crystallization processes because it allows one to obtain a columnar film microstructure nucleated at the bottom electrode. Simultaneous nucleation at the top
surface and in the bulk of the film can be avoided if the pyrolyzed film is free
of particles and clusters. Nucleation phenomena can be nicely described by the
capillarity theory adapted to thin films by Pound et al. The critical free energy
for the formation ΔG^* of a nucleus of critical size is written as (see, e.g. [56]):

$$\Delta G^* = \frac{4}{3}\pi \frac{\sigma_{ev}^3}{\Delta G_{cell}^2}\left[2 - 3\frac{\sigma_{sv}-\sigma_{sc}}{\sigma_{cv}} + \left(\frac{\sigma_{sv}-\sigma_{sc}}{\sigma_{cv}}\right)^3\right]$$

(2)

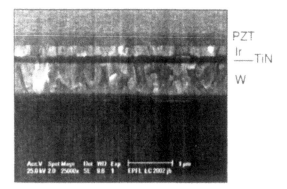

Figure 17 Layer sequence including tungsten metal as bottom electrode for PZT. The
$IrO_2/Ir/TiN$ barrier prevents oxidation of W during growth of PZT at 600°C [54].

where ΔG_{cell} stands for the volume free energy per unit cell. This energy is negative and in the case of small particles, is opposed by positive surface energies (σ), where the indices refer to the interfaces of condensate ($=$ film)–vapor (cv), substrate–vapor (sv), and substrate–condensate (sc). If we deal with homoepitaxy, the lattice of the substrate is just continued by the thin film, σ_{sv} $= \sigma_{cv}$ and $\sigma_{sc} = 0$ apply, and consequently ΔG^* is zero. No activation energy for nucleation is needed as expected. In practice we deal mostly with heterogeneous nucleation, meaning that the substrate–condensate interface σ_{sc} is larger than zero. A substrate with $\sigma_{sv} > \sigma_{sc}$ can reduce the nucleation energy as compared to nucleation inside the amorphous pyrolyzed matter (considering the latter as vapor) of pyrolyzed sol-gel films. Another consequence of nucleation-driven growth is the possibility of controlling the texture or orientation of the film by the growth substrate (most often the bottom electrode in practical applications). The bottom electrode can be prepared in such a way as to favor nucleation of one specific orientation. An evident solution is to prepare electrodes with the same unit cell structure, as, e.g., perovskite conductors for perovskite ferroelectrics. An epitaxial growth is obtained, and the orientation passes directly from the electrode to the PZT thin film on top [57]. In general we have to consider that the substrate–condensate as well as the condensate–vapor, interface depends on the crystallographic orientation of the nucleus. A further possibility for controlling nucleation consists of increasing $(\Delta G_{cell})^2$, i.e., choosing a more exothermic reaction. In the PZT system, for instance, one can start with a Ti-rich composition on the first few nanometers. The free energy of formation is lower (follows from the fact that the PbO vapor pressure is smaller for Ti-rich compositions), i.e. $(\Delta G_{cell})^2$ is larger, and thus ΔG^* is smaller. $PbTiO_3$ thus nucleates most easily of all PZT compositions. One possible approach is therefore to use a $PbTiO_3$ seeding layer—that can be grown at lower temperature and with smaller lead excess—on which PZT grows quasi-epitaxially. It has been observed that such a seed layer reduces the deposition temperature of sol-gel deposited films [58]. Moreover, it has been observed that in situ sputter-deposited $PbTiO_3$ layers grow with {100} texture on Pt (111) polycrystalline electrodes [59] when deposited with enough high-lead flux. Such seed layers are very convenient to fabricate PZT {100} films with sputtering as well as sol-gel deposition. Instead of depositing $PbTiO_3$ seeds it is also possible to influence interface chemistry by a 1- to 3-nm-thick Ti [60] or TiO_2 film. The resulting PZT texture on Pt (111) is (111). Such a behavior cannot be explained by the substrate–condensate interface energy only; it is clear that the other surface energies are important as well. Most probably a PbO-terminated (100) surface has a lower energy than a polar Ti^{4+}- terminated (111) surface. This could explain the generally observed behavior that nucleation in lead-rich conditions tends to lead to (100) orientation. Using the titania seed layer, nucleation takes place in a Ti-rich environment after enough Pb ions have diffused into the seed layer. This implies that a seed has a thickness of around

1 nm or less, i.e., about two lattice constants. The phenomenon is demonstrated in Figure 18, which shows the result of a patterned titania seed layer on which $PbTiO_3$ has been grown. The contrast of the image is due to different surface morphologies of (111) and (100) $PbTiO_3$. Chemistry alone does not explain all of the observed features. An ordered titania [presumably TiO_2 (110)] seed layer was observed to work much better. The lattice contants on the TiO_2 (110) surface match well with the PbO_3 hexagonal structure of the PZT (111) plane. The importance of the use of seed layers on Pt electrodes has been very much seen in the case of relaxor compounds that are not easy to nucleate, such as $Pb(Mg,Nb)O_3$ [61]. To date, no good films have been obtained on platinum electrodes without the use of seed layers. For such compounds, nucleation energies and growth temperatures are even higher than for PZT. Lead loss and electrode deterioration have to be prevented as much as possible. Seed layers promote nucleation at lower temperatures and in a shorter time, thus reducing the lead loss and thermal budget. Seed layers lower the nucleation energy and thus the critical nucleus radius. This manifests finally in smaller grains. This is indeed observed (Figure 19). Seeding helps to prevent pyrochlore nucleation due to lead loss and inhomogeneous sols. The latter effect was shown for a solution in which Zr precursor clustered during polymerization, leading to pyrochlore formation when the deposition was performed on bare Pt electrodes [62–64].

It has been shown that textured growth of PZT can also be achieved on rutile electrodes. Sputter deposited PZT (111) was obtained on RuO_2 (200) thin

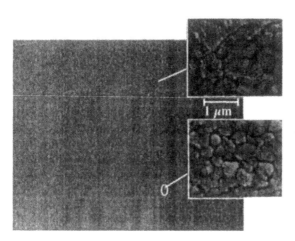

Figure 18 SEM of a $PbTiO_3$ thin film whose texture was "patterned" by a TiO_2 seed layer on a Pt (111) bottom electrode. The seed layer transforms to (111) $PbTiO_3$ during the in situ sputter deposition process at 530°C [63].

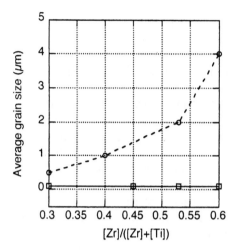

Figure 19 Average grain size of seeded and unseeded sol-gel-deposited PZT thin films. The first ones were grown on Pt (111) layers with TiO_2 or $PbTiO_3$ seed layers [64], the latter ones on sapphire.

films equipped with a titania seed layer. In another approach, the RuO_2 electrode was grown with nanosized grains that are smaller than the PZT nuclei [65]. In this case, the interface does not promote a preferential orientation. The perovskite–gel interface energy is dominating. The chosen sol-gel process conditions yielded the (111) orientation in this case as well.

C. Patterning

The inertness of electrode systems required for PZT deposition leads to difficulties in patterning. Etching of platinum and iridium is most difficult. Wet etching cannot be applied since all the known wet etchants would destroy the whole device. Dry etching including ion bombardment is required. The etching rates are in general quite low. Redeposition may lead to "fencing". For this reason low working pressures are needed. Physical etching by Ar^+ ion bombardment is a possible solution. One cannot use too large ion energies, for this would lead to a flowing of the photoresist and thus a loss of resolution. Larger ionic currents, meaning the use of high-density plasma sources, are needed in order to increase the rate. The addition of chemically attacking species, such as chlorine atoms and ions, enhances the rate further. The mixture of physical and chemical removal (reactive ion etching, RIE) made it possible to work with photoresist masks and still to achieve decent etching rates ([66] and references therein). The dry-etching tool must provide a high plasma density at relatively low pressures (Pa range or less). This requirement can be fulfilled with today's commercially available inductively coupled plasma (ICP) reactor systems, or even better with ECR ion gun systems that include an rf-powered substrate holder (Figure 20) [67].

The damage-free micrometer-scale patterning of ferroelectric thin films is one of technological difficulties in the fabrication of piezoelectric MEMSs and ferroelectric memories. Wet etching in HCl/HF solutions is only applicable when the required precision is larger than several micrometers. For micrometer precision one generally has to apply dry etching. It is difficult to achieve a high etching rate of PZT and an adequate selectivity to electrodes materials and to photoresist at the same time. In addition, the film properties usually deteriorate during dry etching, requiring a later annealing step to remove oxygen vacancies. Most of existing processes use high-density plasma RIE/ICP or ion milling methods. In general, common limitations were observed due to high-energy bombardment (typically several hundreds of volts) and relatively important pressures (1–100 Pa) leading to PZT damage and degradation of the electrical properties. Relatively good results have been obtained with gases mixing chlorine and fluorine species together with Ar at moderate ion bombardment energies of around 100 eV, yielding etch rates of up to 60 nm/min with an acceptable selectivity with respect to photoresist (<2) and Pt (<1.5). The gas mixture takes care of the fact that Ti and Zr are better removed as fluorine compounds, and lead is better removed as a chlorine compound.

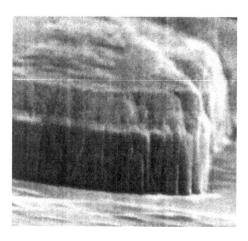

Figure 20 SEM view of patterned layer stack consisting of 300 nm PZT/100 nm Pt/ 600 nm SiO_2 as obtained by dry etching in a ECR/RF reactor [67].

IV. APPLICATIONS

A. Piezoelectric Components

Piezoelectrics are an important family of functional materials in microsystems technology, or MEMSs. They directly provide an electromechanical coupling and are thus useful for all kind of motion sensors, linear actuators, acoustic and ultrasound devices, linear actuators (see Ref. 68 for a review). Piezoelectricity has advantageous properties such as a wide range of application frequencies (mHz to 100 GHz), low power, large output forces, and good signal-to-noise ratio. For MEMS applications, the transverse piezoelectric coefficient is exploited in combination with deflective structures such as cantilevers (Figure 21) bridges, disks, and membranes. The geometrical simplicity of this planar transduction mechanism can be an advantage as well, as for instance in active cantilevers for atomic force microscopy, (AFM). ZnO was in many cases the material of choice to demonstrate piezoelectric thin-film devices (see, e.g., Refs. 69 and 70). Among the bulk materials, however, PZT ceramics is known to be a much stronger piezoelectric material. Many attempts have been undertaken during the past 10 years to integrate PZT into MEMS devices and to demonstrate applications such as ultrasonic micromotors, Lamb wave devices for pumping and particle filtering, cantilever actuators, active AFM cantilevers, accelerometers and gyroscopes, and ultrasonic transducers for medical applications. More recently there has been a growing interest to apply piezoelectric thin films also for microwave resonators as needed in telecommunication. For this purpose, bulk acoustic waves (thickness mode resonance) are excited in acoustically isolated piezoelectric thin films. The resonance frequencies are typically at 2–4 GHz for a 1-μm-thick film. Materials with good acoustic properties are required at such frequencies. To date AlN and ZnO are the only compounds that meet the quality factor requirements.

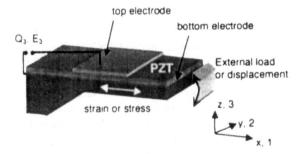

Figure 21 Transverse piezoelectric effect in deflecting structure.

1. Piezoelectric Coefficients

In most of the structures applied in MEMSs, the piezoelectric film is part of a composite structure, i.e., it is clamped to another elastic body. A rigorous treatment of this problem requires the solution of the equations of state with two piezoelectric and several clastic coefficients. The latter are, however, usually not known precisely. A more pragmatic way is to consider effective piezoelectric coefficients of films clamped to a rigid substrate. d_{33f} describes the thickness change as a function of the applied field, i.e., the longitudinal effect; e_{31f} the in-plane stress as a function of the applied field, i.e., the transverse effect. The film is clamped in the film plane (coordinates 1,2). In off-plane direction (coordinate 3), the film is free to move ($\sigma_3 = 0$). This corresponds to a mixed-boundary condition. Directly measured piezoelectric coefficients of thin films on substrates are therefore functions of standard piezoelectric coefficients and elastic constants. These effective coefficients are related to the ordinary ones by the following relations:

$$e_{31,f} = \frac{d_{31}}{s_{11}^E + s_{12}^E} = e_{31} - \frac{c_{13}^E}{c_{33}^E} e_{33} \qquad |e_{31,f}| > |e_{31}|$$

$$d_{33,f} = \frac{e_{33}}{c_{33}^E} = d_{33} - \frac{2 s_{13}^E}{s_{11}^E + s_{12}^E} d_{31} \qquad < d_{33} \tag{3}$$

e_{31f} is determined either by substrate bending (variation of x_1 and x_2 at $\sigma_3 = 0$, $E_3 = 0$) and collecting the developed charges that are related to the in-plane strains as:

$$D_3 = e_{31,f}(x_1 + x_2) \tag{4}$$

or by applying a field and measuring the deflection of the substrate which is governed by the in-plane stresses:

$$\sigma_{1,2} = e_{31,f} E_3 \tag{5}$$

Note that $|e_{31,f}|$ is always larger than the bulk coefficient $|e_{31}|$. This originates from the fact that larger piezoelectric stresses can be developed in the transverse directions if the sample is free to move in the longitudinal direction.

Most of the potential applications are based on the transverse coefficient $e_{31,f}$ (see Ref. 47 for a review of its determinations). Bending of beams and deflections of membranes are much more suitable for obtaining large responses or large excursions. For this reason, this coefficient is discussed in more detail below. In terms of piezoelectric coefficients, PZT is clearly the leader among the above materials. This translates into a superior performance in force, torque, and output power of actuators and motors, and also of sensors with current detection. However, when voltages are detected, when the dielectric noise current limits signal-to-noise ratio, and when the coupling coefficient is important (power consumption, power yield, transducer response), the dielectric constant and the dielectric losses have to be considered as well. In these cases, PZT is not anymore so brilliant because of its high dielectric constant and large dielectric losses (several %) as compared to AlN (0.2%). AlN and ZnO are more suited for voltage detection (Table 1) and AlN exhibits a larger signal-to-noise ratio. The coupling coefficient in thin-film composite structures needs to be considered in a different way than in homogeneous bulk materials. The stiffness of the structure usually depends more on the passive part, e.g., silicon, thermal oxide, silicon nitride, etc., than on PZT itself. On silicon structures, optimal coupling coefficients are obtained for a thickness of the passive layers that is somewhat larger than the PZT thickness. This means that one should consider the compliance of the substrate rather than that of PZT. In analogy with the planar coupling coefficient k_p, the following material figure of merit for the coupling factor is therefore considered:

$$k_{p,f}^2 = \frac{2e_{31,f}^2}{\varepsilon_0\varepsilon_{33,f}} \left(\frac{1-v}{Y}\right)_{Si} \tag{6}$$

The numbers given in Table 1 show that the texture of the PZT thin films is quite important for the piezoelectric properties. The properties of PZT(100) films one much superior to these of (111)-textured films and approach those of optimized, i.e., doped, PZT ceramics. The same table also shows the values for the frequently used ZnO, and the semiconductor compatible AlN.

In thin-film structures, the coupling coefficient does not only depend on material parameters. Film stresses play a role, too. Film stresses are hardly avoidable. In spite of efforts to reduce or to compensate such stresses, there will be a

Table 1 Comparison of Transverse Piezoelectric Coefficients as Measured in Thin Films

Figure of merits	ZnO	AlN	PZT (111) 45 : 55 film	PZT (100) 53 : 47 film
Force, current response: $e_{31,f}$ (C m^{-2})	−0.7	−1.0	−8.5	−12
Voltage response $(e_{31,f}/\varepsilon_0\varepsilon_{33})$ (GV/m)	−7.2	−10.3	−1.2	−1.4
Signal-to-noise ratio $\dfrac{\|e_{31,f}\|}{\sqrt{\varepsilon_0\varepsilon_{33}\tan\delta}}$ (10^5 Pa$^{1/2}$)	5	20	5.7	7.1
Coupling coefficient $(k_{p,f})^2$ on Si	0.06	0.11	0.11	0.19

residual value between 10 to 100 MPa. Such stresses give a prestrain or a precurvature to micromechanical structures. Poling of PZT thin films may lead to a change of the residual stress in PZT thin films. In some cases, this stress has to be taken into account in the design phase of the device. In very thin membranes, tensile stresses increase the resonance frequency and reduce the coupling coefficient. In thin-film diaphragms subjected to tensile stress, a transition from disk behavior (resonance frequencies depend on rigidity of plate) to membrane behavior (resonance frequencies depend on stretching forces) is observed when thinning down the diaphragm.

2. Piezoelectric Flexural Thin-Film Devices

There are many potential applications of piezoelectric thin films in the MEMS area. Yet to date (to the knowledge of the author) only one application has been commercially exploited i.e., the active AFM cantilevers. It is too soon to precisely assess the potential of other applications. The MEMS area is new, and it will take some time to develop exploitable applications. In addition, there are competing concepts for a given application. Piezoelectricity is only one actuation principle. Electrostatic, thermal, and magnetic actuation principles have their advantages as well. Piezoelectrics are rather suited for ultrasonic devices, low-voltage devices, transducers (actuator/sensor combination), and larger excursions-force products that are not suited for electrostatic actuation.

A generic, basic device is the piezoelectric-laminated membrane that has been demonstrated for a variety of applications: ultrasonic micromotors, micropumps, fluid ejectors for ink-jet printing and other droplet depositions, ultrasonic transducers for imaging and proximity sensors, and Lamb wave devices [71] for pumping and filtering. Figure 22 shows a scanning electron microscopy (SEM) image of a PZT-coated micromachined membrane [72]. The membrane is free except for four bridges where it is fixed. The admittance of a 300-μm-diameter membrane is depicted in Figure 23. The corresponding membrane was obtained by micromachining of a silicon-on-oxide wafer with a silicon membrane thickness of 3.5 μm. The PZT thin film was 2 μm thick and deposited by sol-gel deposition. In air, a coupling factor of $k^2 = 5.3\%$ with a quality factor of 140 was obtained.

3. Bulk Acoustic Wave Devices

Filters and resonators based on piezoelectric materials have an essential role in mobile telecommunications. The reason is that piezoelectric bodies at resonance best fulfill requirements with respect to quality factor, coupling factor, and size. They are used as tanks for signal filtering and channel definition. One of the most challenging applications in mobile phones is the so-called *rf filter. These separate the incoming signal from disturbing signals at frequencies outside the*

Figure 22 SEM image of PZT-coated micromachined membrane anchored by bridges. The view shows the membrane and one of the bridges [72].

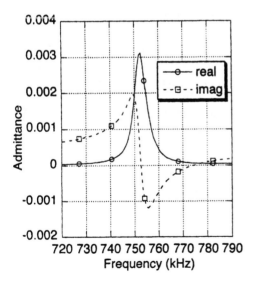

Figure 23 Admittance curve of piezoelectric-coated membrane at fundamental resonance [72].

band of a given carrier frequency (e.g., 850–900 MHz, 1.85–1.9 GHz, etc.). The required bandwidth amounts typically to a few percent of the carrier frequency (2–4%). These filters are mostly surface acoustic wave (SAW) devices on $LiNbO_3$ and $LiTaO_3$ single-crystal wafers. In the recent years a new possibility based on bulk acoustic waves (BAWs) [73] has been investigated and is currently being prepared for industrial production. Best suited is the fundamental mode of a piezoelectric plate, i.e., a thickness vibration. Neglecting loading effects by electrodes, the thickness of the plate corresponds to $D = \dfrac{\lambda}{2} = \dfrac{v_s}{2f_0}$, where f_0 is the frequency and v_s the sound velocity. The latter is typically in the order of 10,000 m/s, thus requiring plates of 2.5 μm thickness for 2 GHz. The vibrating plate must be acoustically isolated from a solid support. An ultrasonic wave excited in a piezoelectric thin film would not totally reflect at the film–substrate interface but instead would propagate into the substrate. Most of the energy would be lost and only small quality factors would be achieved. There are two means to prevent the acoustic emission into the substrate. One is the local removal of the substrate below the film, either by bulk micromachining or by surface micromachining (Figure 24). Such devices are usually called thin-film bulk acoustic resonators (TFBARs). In this way one can build bridge structures with quasi-free-standing resonators. The second technique applies an acoustic reflector to send the acoustic power back into the piezoelectric thin film. This technique is usually referred to as SMR (solidly mounted resonator). The reflector consists of a set of quarter-wave layers of alternating high and low acoustic impedance materials.

BAW filters offer several advantages. (1) They require much less space than non-BAW filters Several thousand filters can be produced on one 100-mm wafer. With increasing frequency, less space is needed. (2) Standard silicon or GaAs substrates can be used. No special piezoelectric substrates such as $LiTaO_3$ or $LiNbO_3$ are required. (3) There is a possibility of integrating transistors and filters on the same chip (especially when using materials such as AlN). (4) BAWs are more apt to deal with larger powers (transmitter filter).

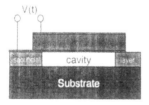

Figure 24   Schematic structure of a TFBAR-type BAW device using surface micromachining.

The most suitable materials for thin-film BAW devices tested to date AlN and zinc oxide (ZnO). These are polar, nonferroelectric materials. In the past, ZnO was used most of the time to demonstrate BAW devices because of stress and quality problems encountered with AlN thin films. However, growth of highly oriented AlN thin films with low stresses can be obtained on metal films such as platinum and aluminum by carefully choosing the level of ion bombardment [74]. Deposited at 400°C by pulsed reactive sputtering, these films exhibited nearly single-crystal values of the relevant properties. Most interesting also is the possibility of achieving a zero temperature drift of the resonance frequency thanks to a compensation of the sound velocity drift in AlN by the one in SiO_2. Figure 25 shows the admittance of an AlN resonator working at 7.7 GHz exhibiting a good coupling constant of 5.5% and a high quality factor of 400 [75]. Figure 26 shows the s12 scattering parameter for a BAW filter centered at 2.14 GHz.

B. Pyroelectric Thin Films for Infrared Detector Array

Pyroelectricity is one of the best performing principles for the detection of temperature changes. Therefore, bulk crystals and ceramics have been used for many years to fabricate thermal infrared detectors. They were and still are applied for contactless temperature measurement devices, security detectors (intruder

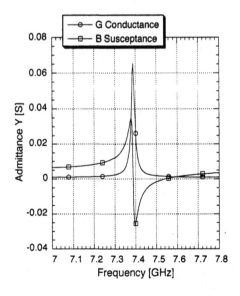

Figure 25   Admittance curve of SMR resonator based on AlN thin film [75].

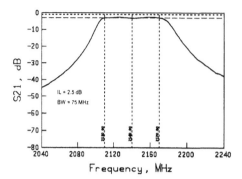

Figure 26 S12 scattering parameter of BAW filter. (Courtesy of K.M. Lakin, TFR Technologies.)

alarms), and human presence sensors. With respect to semiconductor devices, thermal detectors are competitive in the important wavelength interval from 8 to 12 μm. Their special attraction lies in the fact that they do not need cooling. Thermal detectors are too slow to be used in IR imaging with scanning mirrors. However, they are fast enough if 2D arrays can be realized. In this case the readout rate for each pixel is identical to the frame rate (30–50 Hz). Surface micromachining techniques combined with thin film deposition of pyroelectric thin films (see Ref. 76 for a review) allowed the realization of such 2D arrays in a monolithic way directly on the readout chip [77]. The downscaling of arrays and their cheap production by batch processing opens also new fields of applications. As an example, a linear arrays for an infrared spectrometer are treated in more detail.

With their small thermal capacity (H), thin-film structures have a considerable advantage over bulk pyroelectric detectors. The reason is that the response at frequencies above the inverse thermal time constant H/G, where G is the thermal conduction to the heat sink of the device, proportional to $1/H$. However, G must be small enough that the inverse thermal time constant is lower than or equal to the operation frequency (normally the chopper frequency). For operation at a few tens of Hz, the heat conductivity must already be quite small to achieve this condition. The role of silicon micromachining techniques is to provide for a good thermal insulation. Silicon is too good a thermal conductor; therefore, a ceramic membrane was chosen to carry the pyroelectric elements (Figure 27). The membrane consisted of a LPCVD Si_3N_4 film grown on a thermal oxide (SiO_2). The thicknesses were chosen to compensate for the mechanical stresses of these films. As these films also coated the backside of the double-sided polished wafer, they served also as a mask for backside etching in KOH. Bottom-electrode

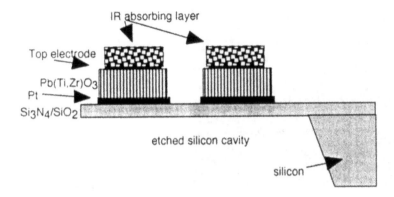

Figure 27   Typical structure of pyroelectric elements of a linear array on a thin membrane fabricated by means of micromachining. The elements are contacted to pads in the other direction than the one seen in this cut.

and pyroelectric film (PZT15/85) are deposited by sputtering and sol-gel, respectively. The top electrode is deposited and patterned by a lift-off technique. Windows to access the top electrodes are opened by a CF_4-reactive ion etching. The PZT elements on the membrane part are etched free in an HCl:F solution, leaving only narrow bridges between the elements and the bulk silicon part, as needed for separation of bottom and top conductor. The platinum bottom electrode is removed between the elements by electrochemical etching. This etching technique does not attack the membrane material. After deposition and patterning of the conductor lines, pads (Au/Cr), and absorbing layer, the silicon is removed below the elements by backside etching, as defined by a window in the backside nitride layer. Inherent to thermal detectors is the need to absorb infrared radiation. A black platinum film was utilized as aborber. Black Pt exhibits a dendritic morphology and grows at some given current densities and concentrations in an electrochemical bath (Figure 28).

A first series of arrays with 12 elements of 0.36 mm^2 save good voltage responses at 1 Hz of 800 V/W in air and 3000 V/W in vacuum (Figure 28). The much larger heat conduction for operation in air was due to heat transfer in air between membrane and device socket. At higher frequencies, current measurement is preferred. At 30 Hz, the current response amounted to 15–20 µA/W, with only small changes as a function of air pressure. The latter is due the fact that above the inverse thermal time constant the current response is determined by the inverse heat capacity and not by the thermal conductivity.

The smaller elements (0.125 mm^2) of a 50-element array (Figure 29) [78] showed a smaller voltage response at low frequencies (460 V/W). The current

Figure 28 Electrochemically deposited black platinum grown on a Cr-Au top electrode on a $PZT/Pt/Ta/Si_3N_4/SiO_2$ layer stack.

response (16 $\mu A/W$) was about the same. Current detection at 10-Hz chopper frequency was chosen as operation mode for the gas spectrometer. The infrared source was a simple hot filament. The measurement of absorption spectra for CO_2 and CO has been demonstrated (Figure 30) [79]. The low noise equivalent power of 1 $nW\ Hz^{-1/2}$ allows detection of a few ppm of CO_2, provided that the electronics does not increase the noise level.

C. Integrated Capacitors

1. Background

For several years it has been recognized that it may be useful to utilize the large values of permittivity associated with ferroelectric films, similar to the way ferroelectric ceramic materials are utilized for their permittivity in multilayer ceramic capacitors. We generally categorize this as ferroelectric films for ''integrated capacitor'' applications. However, it must be pointed out that this covers a fairly wide spectrum of applications, involving several segments of the electronics industry. In this section we will briefly summarize these applications.

Integrated ferroelectric capacitor applications may be categorized as follows:

Capacitors for DRAM memory cells
Capacitors that are integrated onto a Si chip at the ''back end'' (or interconnect wiring) to perform functions such as power–ground decoupling
Capacitors that are integrated onto the ''package'' or ''multichip module'' (MCM) on which the Si chip is located, with the package technology

Figure 29   Top view on 50-element array with 200-μm period obtained with bulk micro-machining, membrane size 2 × 11 mm. The black platinum absorbers, the Cr-Au contact lines, the membrane layers between the elements, and the SiO_2 layer for reduction of parasitic capacitance are well visible [78].

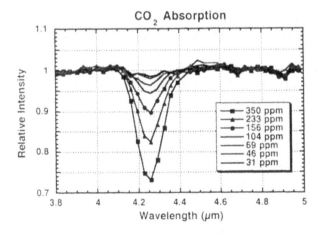

Figure 30 CO_2 absorption spectrum measured by means of thin-film pyroelectric array [79].

based on Si (MCM-D), multilayer ceramic (MCM-C), or polymer (MCM-L)

Thin-film integrated capacitors with these various technologies face distinctly different challenges, and are discussed in turn.

2. High-Permittivity Films for DRAM Capacitor Cells

This application for ferroelectric films has been recognized and investigated for a number of years. At issue are the demands imposed by the aggressive scaling of semiconductor logic and memory to smaller dimensions and high device densities. While dimensions have scaled, the required capacitance per cell has remained almost constant. The resultant scaled increase of capacitance density was accommodated for many years by decreasing the thickness of the SiO_2 capacitor dielectric. When this thickness approached a lower limit dictated by electron tunneling, an effective increase in capacitor area was needed at each new DRAM generation. Industry moved in two directions to accomplish this: either by etching trenches in the Si substrate to form the capacitor; or by building up complex, large-area structures above the DRAM transistor on which to form the dielectric (the stacked-capacitor approach). However, with even more aggressive scaling, the aspect ratios (height to diameter) of the trenches or stacked cells become problematic, and replacement of the SiO_2 or silicon oxynitride capacitor dielectric by a high-permittivity alternative becomes a more attractive alternative. The dimensions, requirements, and possible dielectrics for each technology generation are laid out in the *International Technology Roadmap for Semiconductors* (issued every 2 years) [80].

For a number of years the ferroelectric perovskite solid solution $(Ba,Sr)TiO_3$ (BST) has been investigated as a possible high-permittivity DRAM dielectric for the future generations toward the end of the roadmap. Considerable progress has been made in understanding the relationships between BST properties and DRAM performance [81–90], Examples include the nonlinear dielectric response [86,87]; the impact of strain and thickness on the dielectric function [87]; the role of dielectric relaxation processes on charge loss of the DRAM capacitor [88]; specific degradation mechanisms [89]; and accommodation of the deviation of cation stoichiometry [90]. Despite this progress, there remain significant technical obstacles to implementation, related to conformal coverage of the 3D electrode by the MOCVD process, reliable oxygen barriers, adequate control of stoichiometry, reliability in terms of the required 10-year lifetimes, and availability of MOCVD equipment suitable for the manufacturing environment. As a result, for intermediate generations of Gbit memories entering production at the present time, lower permittivity but less complex dielectrics such as Ta_2O_5 and Al_2O_3 will be incorporated.

3. On-Chip Capacitors

There are a number of capacitors integrated into typical analog and digital integrated circuits that are of sufficiently large value as to occupy considerable space. In particular, decoupling capacitors can occupy a significant fraction of the chip area and thereby contribute significantly to chip cost. Therefore, there is considerable instruction to replace these capacitors, fabricated from silicon oxynitride, with considerably smaller ones fabricated from a high-permittivity dielectric. The difficulty is that these capacitors need to be accommodated in the back-end process of the chip, which effectively limits process temperatures to less than 450°C and requires direct exposure to reducing anneal atmospheres. A number of years ago, Matsushita managed to integrate a BST capacitor into the back end of a GaAs MIMIC (optoelectronic chip), which is in volume production. At the low process temperatures, capacitor properties are poor (high loss tangents); therefore, this remains an isolated example. Efforts have been made to develop new materials specifically for low process temperatures, and the group of Trolier–McKinstry has enjoyed some success with the pyrochlore- and zirconalite-structured films [91].

4. On-Package Capacitors

There is presently a strong effort to replace the large number of discrete passive components (capacitors, resistors, and inductors) that are surface mounted on the "package" (or high-density interconnect package) along with the integrated circuits. The driving force is the potential space and cost saving. There are two general approaches to the problem: (1) group these passive components on integrated passive modules, which are generally small Si chips with thin film passive component networks formed on the surface, the passive modules then being surface mounted onto the substrate like a single discrete component; and (2) *embed* the passive components *into buried layers* of the package. This technology is being developed for both ceramic (called MCM-C or LTCC) and polymer-based substrates (called MCM-L, or high density interconnect printed wiring boards). For polymer-based substrates, used for high-volume and consumer applications, a major issue is temperature and process compatibility with the polymers (usually photodefinable epoxies). This is particularly problematic for embedded capacitors made of complex oxides where dielectric quality is closely related to the process temperatures. One approach under development is the deposition of thin-film perovskite capacitors directly onto Cu foils, which are subsequently laminated into lower layers of the multilayer substrate [92]. Upper layers containing inductors and the necessary high-density interconnect lines are built up using standard layers of photodefinable epoxy and plated copper. This approach is particularly interesting as it melds advanced thin-film processing methods with historically old technology that has evolved out of the printed wiring board industry.

D. Ferroelectric Memory Devices

1. Background

From the earliest days of ferroelectric/piezoelectric ceramic materials, there have been attempts to utilize the switchable spontaneous polarization of a ferroelectric material as the basis for solid-state memory arrays. A review of the early work in this field may be found in the 1972 book chapter by H.H. Wieder [93]. The primary approach taken at that time was the "cross-point array," a simple memory array architecture in which a single ferroelectric element was written by raising the voltage of both the appropriate line and row to $V/2$. With the ferroelectric element exposed to a resultant voltage V, where V was greater than the coercive voltage, the polarization of the bit could be switched. The major flaw was the fact that the coercive fields of all elements in the array were not tightly defined, leading to unacceptably large bit errors. Efforts to circumvent the problem at that time included photoconductive/ferroelectric devices [94] that addressed the selected bit both optically and electrically, but still did not yield commercialized devices.

A breakthrough came a several years later with the development of the ferroelectric memory as we know it now, which has one or two Si-based access transistors per bit, allowing a single bit to be uniquely addressed [95]. The configuration is known as the FERAM [96]. The access transistor allows a short pulse to be uniquely applied to the ferroelectric capacitor of the cell, writing (i.e., "poling") the ferroelectric capacitor in one of two possible net polar orientations, corresponding to bit states "0" or "1." The cell is read by again applying a voltage pulse to the capacitor and monitoring the resultant current pulse through a sense amplifier. As shown in Figure 31, the current pulse corresponds to two possible cases. If no switching occurs because the pulse is the same sign as the pulse that wrote the bit, then the current pulse corresponds to capacitor charging only. If, however, the read voltage pulse is of opposite sign to the write pulse, the resultant current pulse will contain the additional current corresponding to the polarization switching charge. Figure 30 indicates that the two cases can be discriminated, i.e., the state of the cell can be read. It should be noted that this is a "destructive" readout memory device. This is important, as both read and write cycles must be included in determining the number of switching cycles that the memory must endure to achieve reliability and lifetime goals.

The reason for the ongoing interest in ferroelectric memories lies in the unique combination of properties that it offers. A comparison of the salient features of various memory types is presented in Table 2 [97]. The DRAM is the primary chip-based solid-state memory, which is both used as a stand-alone memory and "embedded" onto the same chip as logic circuits. It can be seen from the table that the advantages of DRAM are the small cell size (implying higher density and lower cost), the short read and write times, and the large number of

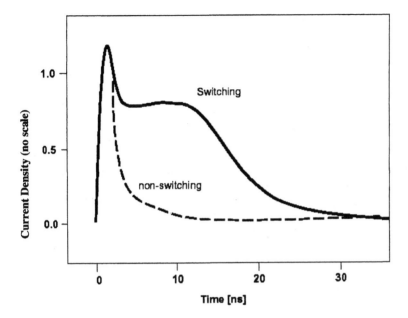

Figure 31 Schematic showing typical current pulses for the switching and nonswitching cases of polarization for the ferreoelectric capacitor.

read/write cycles. The major disadvantage is the nonvolatility, implying that the written data for every bit must be continually refreshed on a subsecond basis. This has two implications, namely, that the "standby power" requirement is relatively high and that information is lost if the memory is depowered. SRAMs are fast-access memory, but also nonvolatile, and with a very large cell size. EEPROM and Flash EEPROM (usually now just referred to as "Flash" memory) are utilized primarily for their nonvolatility but suffer from disadvantages of large write voltages and a limited number of write cycles.

As shown in Table 2, ferroelectric RAM offers several unique features. As a result, FERAM has already found application in niche markets where these features are highly desirable. However, the FERAMs currently produced (2004) are only available as low-device-density memory arrays. The result of the large cell sizes and low density is the larger cost, and the restriction to niche applications that only require low device density. A prime example is the FERAM application to smart cards, where the memory can be embedded with the processor, and the nonvolatility and low voltage operation are a desirable feature for contactless operation [98]. The topic of device density is critical for the future of ferroelectric memories and is discussed in greater detail in Section IV. D, 4.

Table 2 Comparison of the Materials, Designs, and State of Integration of FERAM Devices from Various Manufacturers [97]

Attribute	Non volatile memory	Read-only memory		Random access memory	
	FERAM	Flash EEPROM	EEPROM	DRAM	SRAM
Cell area	$1.5~\mu m^2$	$\sim 0.08~\mu m^2$	Varies	$0.06~\mu m^2$	$\sim 1~\mu m^2$
Cell area[a]	$24~F^2~(8~F^2)^b$	6–$12~F^2$	$\sim 40~F^2$	$6~F^2$	$\sim 100~F^2$
F in 2003	$250~nm^c$	115 nm	Varies	100 nm	100 nm
Nonvolatile	Yes	Yes	Yes	No	No
Programming voltage	2.5 V	~ 12 V	~ 12 V	2×0.9 V	2×0.9 V
Access time	55 ns	120 ns	150 ns	55 ns	55 ns
Programming time	~ 100 ns	10 us	10 ms	55 ns	55 ns
Endurance	$1.00~E + 14$	$1.00~E + 04$	$1.00~E + 04$	n/a	n/a
Retention	10 years	10 years	10 years	64 ms	64 ms

[a] "F" is the technology node or minimum feature size.
[b] Cell area for 2003 is estimated to be around $24~F^2$. However, technically, the cell size can be as small as $8~F^2$, or even $6~F^2$.
[c] The minimal feature size for FERAM for 2003 is several technology generations behind DRAM. This gap must be closed for FERAM to compete directly with mainstream memory such as DRAM.

We must point out that the discussion to this point has centered on FERAM. However, there is another type of memory that has been under parallel development for about 30 years and that has seen a resurgence of interest. This is the ferroelectric memory field effect transistor (FEMFET), which is a CMOS-based FET in which the SiO_2 gate dielectric is replaced by a ferroelectric material, which is used to modulate the conduction of the channel. The status of this type of ferroelectric memory is discussed in a later section.

2. Material Types

The primary ferroelectric materials which are utilized in ferroelectric memories are the $Pb(Zr,Ti)O_3$ (PZT) solid solution system, and modified $SrBi_2Ta_2O_9$ (SBT). The pseudobinary subsolidus phase diagram for the PZT system is shown elsewhere in this book. Ferroelectric films have been produced from compositions over the entire range of Zr/Ti ratios. Figure 32 shows a set of polarization versus field $(P–E)$ hysteresis loops for a set of 200-nm PZT films, all (111) oriented, on Pt-coated Si substrates [99]. The set has a range of Zr/Ti ratios, including Zr-rich rhombohedral symmetry compositions, compositions near the morphotropic phase boundary (Zr/Ti = 52:48), and Ti-rich compositions with tetragonal symmetry. There are several points worth noting with respect to Figure 32. First, the coercive fields are smaller for the Zr-rich compositions, increasing monotonically with increasing Ti content. This corresponds to the increasing c/a ratio of the unit cells. Second, unlike the case of piezoelectric ceramics discussed elsewhere in the book, the largest values of remanent polarization are not achieved at the morphotropic boundary compositions. This is simply due to the substantially larger fields that can be applied to the films without breakdown, implying that polarization saturation can be achieved. The marked difference between films and bulk ceramics is captured in Figure 33, which shows values for the remanent polarization of ceramics versus the thin films as a function of B-site cation composition. For the case of thin films, there is clearly no requirement to limit the choice of composition to those close to the morphotropic boundary. The third point is that the rhombohedral compositions display considerably less "square" loops, i.e., they are more "slanted," implying that the polarization reduces considerably after removal of the switching pulse. The most common PZT compositions utilized for FERAM are the Ti-rich compositions, primarily due to the squareness of the P–E hysteresis loops, and therefore the well-defined value of the switching voltage. In addition, the Zr-rich compositions are typically more difficult to achieve as phase-pure perovskite, and these nucleation issues are discussed in Section 3.

The phenomenon of slanted loops for the rhombohedral compositions appears to be consistently observed in polycrystalline and oriented films on Si substrates. There are a number of possible origins [100], but the effect is to significantly reduce the switchable charge. The behavior is also in contrast to that observed for epitaxial films on single-crystal perovskite substrates [101,102]. This information is also included in Figure 33. The figure indicates that for the films with tetragonal symmetry, both the oriented and epitaxial films show remanent polarizations approaching the values predicted from phenomenological theory [103]. In contrast, for the films of rhombohedral symmetry, only the epitaxial films on perovskite substrates are close to the predicted values, while the oriented films display considerably reduced values of both spontaneous and remanent polarization. There are two important implications for nonvolatile memories. First, the desirably low coercive fields of the rhombohedral compositions are of limited utility in the case of non-square loops, as an operating field that is significantly greater than the coercive field must be used to ensure complete switching. Thus, the operating voltages for rhombohedral compositions are for practical purposes not lower than those of the tetragonal compositions. Second, the origin of the lower switchable charge of the rhombohedral compositions must be identi-

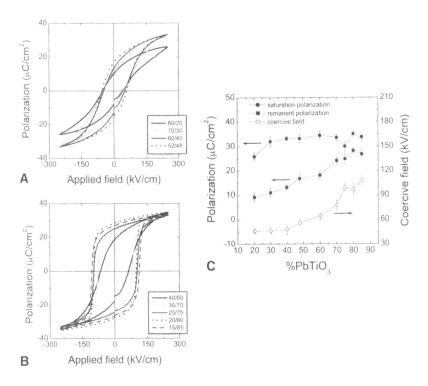

Figure 32 *P–E* hysteresis loops for (111)-oriented (fiber-textured) PZT films for various Zr/Ti ratios [99]. The films are 200 nm thick and are produced by CSD on Pt/SiO$_2$ substrates, with Pt top electrodes.

fied and rectified before these compositions become serious contenders for FERAM on Si substrates.

The PZT thin films are normally compositionally modified by cation substitution or dopants, analogous to the PZT bulk ceramics. The dopants are utilized to control the temperature of the phase transitions (e.g., Ca, Sr, and La oxide additions lower the Curie point and reduce the coercive fields), while aliovalen dopants are used to modify the electronic defects, thereby modifying resistivity and domain wall mobility. These approaches are well reviewed in the bulk ceramic literature [104,105]. However, the literature for PZT thin films is not completely consistent with the bulk ceramic picture, with some variation being reported for the effect of aliovalent dopants on resistivity, permittivity, and the degradation mechanisms of fatigue, as well as imprint (see below). There are two likely reasons for the reported inconsistencies: (1) the concentration of Pb vacancies in the films being higher than in the bulk and thus dominating the defect equilibria;

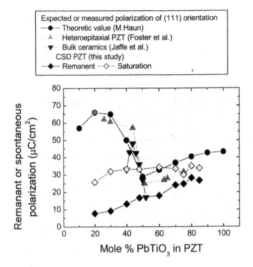

Figure 33 Remanent polarization (Pr) and spontaneous polarization (Ps) as a function of Zr/Ti ratio for oriented [99] and epitaxial [101,102] PZT thin films, compared to values derived from phenomenological theory [103]. The polarization values for all films are converted to those for (111) orientations in order to make the comparisons direct. Values for poled bulk ceramics are also shown in order to emphasize the sharp peak in the remanent polarization values for the morphotropic boundary compositions associated with ease of poling.

and/or the lack of control and thus variation of oxygen activity during processing by various thin-film techniques (e.g., CSD may be more reducing during pyrolysis than sputtering).

The orientation of the PZT films is typically controlled to be fiber textured (111) or (100/001), implying that the in-plane orientation is random, and the orientation normal to the film plane is controlled. The orientation control can be achieved by a number of methods, involving either local epitaxy on a template or control of the orientation of nuclei, in both chemical and physical deposition methods, as discussed in Section 3. The orientation impacts the domain structure and therefore also the switching of the polarization. For example, a tetragonal PZT film deposited with pseudo-cubic (100) orientation on Si substrates will typically be under biaxial tension after cooling from the processing temperature to room temperature. The tensile stress can be partly relieved by the film assuming a domain structure that consists of the tetragonal c axis aligned partly in the plane of the film. It is important to note that on writing a digital state into the capacitor, if complete switching (complete "poling," in ceramic terminology) were to occur, then this would involve the motion of 90° domain walls, and a change in the stress state of the capacitor. We refer to these domain walls as "ferroelastic." There is now considerable evidence that the ferroelastic walls are not mobile in *thin* tetragonal films on Si substrates, due to the large cost in elastic strain energy [106]. This is in marked contrast to bulk ceramics.

It should now be clear that a reason for selecting the (111)-oriented Ti-rich tetragonal films is that switching could occur without ferroelastic effects, i.e., orientation reversal *could* always occur without the movement of ferroelastic 90° domain walls, but rather through the movement of nonferroelastic 180° domain walls.

The second primary material utilized in FERAM is the layered perovskite, $SrBi_2Ta_2O_9$ (SBT). The layered perovskites of this class have been known for a considerable time [107,108] but only investigated for application in FERAM over the past decade. Advantages of this material include low coercive fields and good reliability with standard Pt electrodes (see next section), while the remanent polarization is lower than that of the tetragonal PZT compositions, as shown in Figure 34. At room temperature the structure is rhombohedral, with a relatively small deviation from cubic symmetry, in comparison with tetragonal PZT. The origin of the lower coercive fields is believed to be the fact that the spontaneous polarization lies only in the $a-b$ plane [108], and there is a negligible or zero strain associated with switching (i.e., little elastic strain energy associated with ferroelastic domain wall motion). The layered perovskite structure can be easily induced to grow fiber textured with the c axis normal to the film plane. However, this is not useful for simple planar ferroelectric capacitors, as it yields no remanent polarization in the desired field directions. As a result, films are currently grown with randomly oriented polycrystalline microstructures. This is acceptable for low-density (small-area) capacitors, but presents a problem when the capacitor dimensions are reduced down toward the grain dimensions [109]. With regard to composition, SBT is not normally deposited as the stoichiometric compound, but typically Sr-deficient and Bi-rich [110,111]. The compositional modifications are empirically determined, and direct correlations between composition, structure, and properties have not been reliably established. A complicating feature is the direct reaction between Bi from the ferroelectric layer with the Pt electrodes [112]. Niobium can also be partially substituted for Ta in the B site, with the effect of increasing both E_c and P_r.

Due to the recognition of problems associated with each of the primary materials, investigation into alternative ferroelectric films for FERAM has been considered. Among the more promising options are solid solutions based on $Bi_4Ti_3O_{12}$, in particular those with rare earth substitutions [113,114]. The attractiveness lies in the apparent lack of fatigue with noble metal electrodes (like SBT), and the large values of remanent polarization (like the PZT solid solutions).

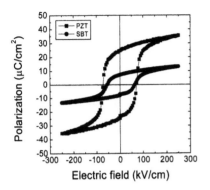

Figure 34 Comparison of typical *P–E* hysteresis loops for SBT and PZT films for FERAM applications.

Investigations of the processing and reliability of these materials are currently at an early stage.

3. Reliability and Lifetime Issues

The development of ferroelectric thin-film memories has been slowed by the complexity of the ferroelectric behavior, in particular the issues related to the reliability of the devices. The three reliability issues generally considered are polarization "fatigue", "retention," and "imprint". All three impact the reliability with which data are stored and read on a ferroelectric bit. These will be discussed in turn.

Polarization fatigue is the decrease in the polarization (both remanent and saturation) which occurs as the ferroelectric capacitor is switched through read and/or write cycles, as shown in Figure 35(a). The phenomenon is clearly observed for PZT films with Pt electrodes, but not for SBT films with Pt electrodes. An empirical solution was found for PZT films by employing oxide electrodes [115], one Pt and one oxide electrode [116], or hybrid Pt/oxide electrodes [117]. These solutions have been demonstrated to allow capacitors to be cycled to approximately 10E15 cycles, although one needs to ensure that the fatigue measurements are correctly undertaken [118]. Mechanisms of fatigue remain the subject of heated debate, but an extensive review of the correlations between theory and experimental observations may be found in Ref. 119.

The polarization is also known to reduce with time *without switching* (in contrast to fatigue) [120]. The mechanisms are a subject of debate [120] and *may* be related to the aging that is observed in bulk ferroelectric materials. The process is thermally activated, and it is estimated that retention loss should not result in

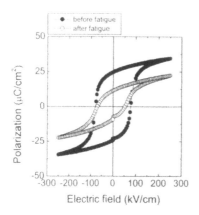

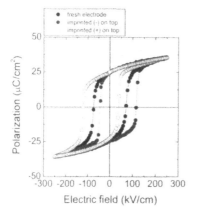

Figure 35 Schematic showing physical changes in P–E hysteresis loops, corresponding to modes of failure of ferroelectric capacitor. (a) Polarization fatigue. (b) Imprint.

device failure within the standard 10-year/125°C lifetime requirements. However, the aging or ''relaxation'' phenomenon also implies that polarization switching is dependent on the magnitude and duration of the voltage pulse, and this may become a problem for low-voltage subsequent generations of ferroelectric memories [120b].

Another degradation phenomenon, imprint, is currently considered to be a major problem. The phenomenon corresponds to the development of an internal bias in a written capacitor with time, as shown in Figure 35b. The implication is that one polarization state becomes more stable than the other. The phenomenon is observed in both SBT and PZT, and with all electrode types. It correlates (directly or indirectly) with the Zr/Ti ratio of PZT thin-film capacitors. Mechanisms are still unresolved [121–124] but it should be noted that mechanisms that lead to polarization being screened in the film rather than the electrode result in a bias to the capacitor. The origin could be a polarization discontinuity near the film–electrode interface, resulting in incomplete screening of the polarization by the charge on the electrode. This in turn leads to a depolarizing field compensated by a separation of mobile charges in the bulk, which constitute a space charge near the interface. Asymmetry of the space charge at the interfaces implies an overall bias to the capacitor. The time constants for the motion of these space charges are considerably smaller than the time constants for switching, leading to a tendency toward reversal of the polarization after removal of the switching pulse. An alternative mechanism observed in bulk ferroelectrics involves the polarization-induced alignment of defect dipoles [122] but does not appear to be the operative mechanism for the thin films that have been studied. A methodology for eliminating failures due to imprint has not yet been established. The occurrence of occasional and unpredicted imprint failures in an array of ferroelectric memory devices remains a threat to the technology, especially for low-voltage devices.

4. Device Density, Cell Geometry, and Integration

The issue of device density is very important. As stated previously, low-density FERAM devices (both embedded and stand-alone) are currently in commercial production, with densities ranging from 64 kb to 4 Mb [125]. This is considerably less than the current state of the art for mainstream memories. For example, standalone DRAM is currently 256 Mb [80,125], and low-density devices are necessarily relegated to niche markets. There is therefore a considerable effort to increase device density.

The first of the current limitations are related to cell designs and geometries. While we have pointed out that the standard FERAM consists of a single-access transistor with a single ferroelectric capacitor (1T1C), most of the devices currently (2002) in commercial production are 2T2C. This is to circumvent the problem of having a voltage reference when the properties of the capacitor are changing with time and cycling, as described in the previous section. In effect, the use of a 2T2C configuration yields an internal reference for every bit. The cell size is, however, doubled—an unacceptable penalty for most applications. The major developers are therefore concentrating in achieving the material stability required for 1T1C architecture. Most of the current designs also simplify the integration with the underlying CMOS transistors by utilizing as so-called strapped cell. In this geometry the underlying CMOS structures are completed and passivated with a dielectric isolation layer prior to depositing and defining the capacitor structures. Contact is subsequently made to both the bottom and top electrodes of the capacitor from above. This requires a larger area for the lower electrode, with a concomitant increase in cell size. The higher density alternative has the ferroelectric capacitor "stacked" above the transistor (COT), and the contact is made from transistor to bottom electrode from below via a conductive plug. This approach is clearly desirable from the perspective of cell size but brings with it a severe increase in process integration complexity. This is because the plug needs to be processed under reducing conditions, while the ferroelectric film is deposited or processed under oxidizing conditions. The lower electrode structure must therefore incorporate a conductive oxygen barrier [126]. Another configuration worth reporting is that developed by NEC, in which the capacitor is under the bit line.

E. Ferroelectric Field Effect Transistors

Another memory concept that has been discussed and developed for many years is the ferroelectric field effect transistor (FEFET or FEMFET). The concept is simple, replacing the gate dielectric of a field effect transistor by a ferroelectric material. The direction of polarization controls the flat-band voltage of the capacitor, i.e., it gives two characteristic voltages at which the transistor can be turned on. This means that the transistor can be used as a simple nonvolatile memory, with the added advantage that it has a nondestructive readout (NDRO), i.e., it does not need to be switched in order to read the bit state. The other major advantage is that it is a one transistor memory; therefore, in principal it must be much smaller than any other memory type. The early work utilized bismuth titanate films and uncovered a significant difficulty, namely, the fact that the turn-on voltage, particularly in one orientation, changed with time. The origin of the phenomenon is clear: the polarization must be screened, and there are several competing sources of screening charge other than the desired carriers from the transistor channel, namely, charge that has leaked through the ferroelectric gate, or mobile charge from within the ferroelectric gate. The result is that the screening process occurs over a long time period (days and months), with a concomitant change in the turn-on voltage (the characteristic memory window). This is known as a retention problem. The problem is usually worse for one polarization, i.e., that direction for which screening requires minority carriers from the channel.

Research by Westinghouse in the 1980s and early 1990s focused on the problem of retention by addressing dielectric and interface quality. They moved from $Bi_4Ti_3O_{12}$ to PZT and then to MBE-grown ferroelectric $BaMgF_4$, but were unsuccessful in solving the retention problem [127]. More recently, research on the device type has renewed, particularly in Japan [128]. Approaches have included the incorporation of an additional insulator layer between ferroelectric and Si channel (MFIS type), or even a floating gate (MFMIS type); the matching of charge requirements at all interfaces; and minimization of leakage currents [128]. Retention of a few days has been achieved. The 1T ferroelectric memory thus remains tantalizing: the obstacles continue to loom large, but success would promise in a revolution in the microelectronics industry.

F. Tunable and Switchable Components Based on Ceramic Thin Films

Dielectric tunability, the change in permittivity under electric field, is a useful property for a number of microwave devices. Tunability, n, is expressed usually as $n = \varepsilon(0)/\varepsilon(E)$ where $\varepsilon(0)$ is the zero dc permittivity and $\varepsilon(E)$ is the permittivity under a specified given field. The potential applicability of ferroelectrics as field-tunable dielectrics has been realized for more than 40 years (basically for military radar applications). Microwave applications for tunable devices increased in importance recently due to rapid advances in telecommunications. The significant advancements in thin and thick ferroelectric films deposition technology and the corresponding reduction in the required voltage for tunability make these materials even more attractive. References 129 and 130 give recent reviews on applications and properties of ferroelectric tunable microwave devices.

The underlying phenomena are the strong field dependence of the permittivity of paraelectric materials and the resulting modification of properties such as the velocity of electromagnetic wave propagation through the dielectric. The facts that the permittivity change can be obtained at a high speed and with a small power loss are important advantages.

The competing materials and technologies for tunable devices include ferrites, semiconductor diodes, and MEMS switches. Ferroelectrics have an advantage over semiconductors at high frequencies (> 10 GHz) where the quality factor of the semiconductors degrades. Their advantages over ferrites are in size reduction and in substantial reduction in power consumption. MEMS switches are in development at present, they are mechanical components, as the tunability is achieved by changing the distance of the gap between the metallic conductors. This is a potential disadvantage whereas the low-power consumption is an obvious advantage.

Tunable microwave components and devices for which ferroelectric thin films are suitable include basically three groups of applications: (1) varactor applications, (2) phase shifters, and (3) tunable resonators and filters.

The simplest component based on tunable dielectric material is the varactor (field-dependent capacitor). A varactor is a fast switch used in many devices such as voltage-controlled oscillator (VCO) and frequency modulators. In its parallel plate, "sandwich structure" (Figure 36a), it requires low voltage and is advantageous in low-power applications. The in-plane capacitor structure (Figure 36b) is suitable in high-power applications. The in-plane capacitor structure is somewhat

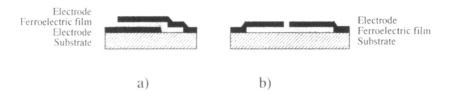

Electrode
Ferroelectric film   Electrode
Electrode   Ferroelectric film
Substrate   Substrate

a)   b)

Figure 36 Common structures of ferroelectric thin-film varactors: (a) sandwich structure in which the field is applied through the thickness of the film; (b) "flip-chip" configuration in which the field is applied across the gap between the electrodes in plane of the film.

simpler in the processing and the structuring of the devices and is also advantageous for growth of epitaxial films (substrate–film epitaxial relationship).

Tunable phase shifters are of particular interest for phased array antennas, replacing the mechanical steering system with an electronic one. A phased array antenna consists of a number of elements each with its own controllable phase. When the phase shifter is made of a tunable dielectric, it is possible to produce electronically steerable antenna arrays. The result is a possibility to shape and tune emitted and received beams, useful in military and commercial radar systems. Tunable ferroelectric thin films are attractive for these applications because of the possibility of obtaining rapid and continuous tuning at low voltages.

Tunable resonators and filters constitute the third group, with applications in wireless and satellite communications. This group is perhaps the most demanding as far as low dielectric loss in the material is concerned. Tunable filters having 4% bandwidth and 12% tunability at 19 GHz were reported recently [131]. Some applications of tunable ferroelectrics at cryogenic temperatures (e.g., space applications) may benefit from loss-less electrodes using superconductive thin ceramic films.

The most important properties of ferroelectric thin films for tunable applications are a high tunability and low dielectric losses at the operation temperature range. The combination of these two properties makes the paraelectric phases of ferroelectric materials attractive. Those with T_c not too far below the operating temperature range show the best performance. Tunability at the paraelectric phase is not sensitive to frequency change and is stronger when the permittivity is higher. However, intrinsic losses (those that originate from the lattice response to the ac field) also increase with the permittivity. This makes the search for the best tunable materials a difficult task.

A useful figure of merit [129] for materials for bi-stable tunable devices, called the commutation quality factor K, is:

$$K = \frac{(n-1)^2}{n \cdot \tan \delta(0) \cdot \tan \delta(E)} \qquad (7)$$

where tan $\delta(0)$ and tan $\delta(E)$ are loss tangents of the material at zero and at a given E field, respectively. $K = 2000$ gives a rough order of magnitude of well-performing materials. This corresponds approximately to $n = 2$ and tan $\delta = 0.01$–0.02. The applied field for reaching this performance has of course to be limited to a practical level.

The dielectric losses at microwave frequencies are a matter of great concern. The losses of ferroelectric films are higher than those of bulk materials of equivalent compositions. The origin of this phenomenon could be grain boundaries, impurities, and pores, as well as interaction of the ac field with local dipoles. Another source of increased losses, of intrinsic origin, is the quasi-Debye losses [132]; these are losses arising from the fact that under field the paraelectric material becomes polar and therefore of a symmetry that allows coupling of the ac oscillations to the phonon spectrum. While the permittivity of paraelectric phases is frequency independent, the losses of intrinsic origin increase linearly with frequency in the microwave frequency range [132].

Single crystals of the incipient ferroelectrics $SrTiO_3$ (STO) and $KTaO_3$ (KTO) exhibit, at low temperatures, very low losses and a high permittivity ($\varepsilon_1 \approx 2500$ and tan $\delta \approx 0.001$ at 80 K). Thin films of these materials show low losses as well as a high tunability (Figure 37); however, at room temperature the dielectric constant and hence the tunability are low. Recent works, both theoretical [133] and experimental [134], show the possibility of shifting T_c of STO thin film to a higher temperature by the introduction of uniform stresses (e.g., by using an appropriate substrate), resulting in an increase of its permittivity, and consequently its tunability, at room temperature.

Presently, the best performance at room temperature is obtained by barium strontium titanate ($Ba_xSr_{1-x}TiO_3$, [BST]) thin films. The compositions of interest

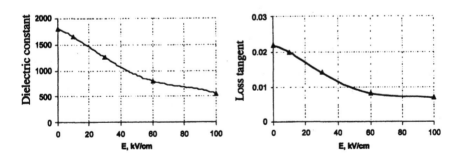

Figure 37   Dielectric constant (a) and loss factor (b) of STO thin film measured at 80 K (film thickness 0.25 μm, MgO substrate) as a function of dc field. Frequency of loss measurements: 8 GHz. [Reproduced courtesy of K. Astafiev (EPFL) and T. Rivkin (NRL).]

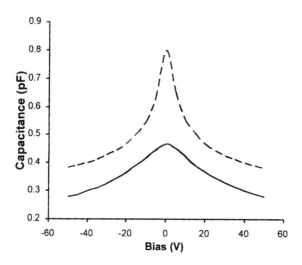

Figure 38 Capacitance as a function of applied de voltage on epitaxial BST ($x = 0.4$) films on MgO and LAO substrates at 300 K. (Adapted from Ref. 135.)

are those in which T_c is just below room temperature, namely, those of $x = 0.4$–0.6. Depending on the substrate, on the film thickness, and on the configuration of the electrodes, these films have shown at room temperature a dielectric constant of 200–5000, tunability up to $n \sim 2$, and losses down to ~0.02. To illustrate this point, Figure 38 shows the tunability of 300-nm epitaxial BST ($x = 0.4$) films grown by PLD on MgO and LaAlO$_3$ [LAO] single crystals and annealed at 1100°C in oxygen [135]. With the BST on LAO, phase shift of 45° per Debye was reported at 30 GHz [136]. Elsewhere, phase shifts up to 110° per Debye were measured in the planar plate structure [137] in X band and using BST films.

Although various issues are still open on both the fundamental level (e.g., origin of increased losses in thin films in comparison to bulk) and at the device level (e.g., cost-effective architecture), the performance that has been demonstrated to date, the moderate voltage needed, and the compatibility with monolithic fabrication techniques show great promise for implication of tunable ferroelectric thin films in radar and communication systems, in particular for Ku band and above.

V. PERSPECTIVES

This chapter has outlined the state of the art in ferroelectric thin films as an important example of the advances in the wider field of electroceramic thin films.

A wealth of knowledge and know-how exists in film deposition, in microstructure, and in the characterization of the functional and structural properties of the films. While deposition on various substrates is advanced, the patterning is still an open field and is bound to see further development in the near future.

Ferroelectric ceramic films in particular, and electroceramics thin films in general are poised for new applications. Both the needs for further miniaturization in microelectronics and the expansion expected in sensor and sensor network markets require this.

Nanoscience and nanotechnology of electroceramics is an emerging field of interest that is tied to thin films. Recently, a number of reports demonstrated the fabrication of structured films having lateral dimensions smaller than 100 nm [138–140]. Domain writing using probe microscopy techniques is another approach to obtain submicrometer-scale functional elements [141]. Although this field is still in its infancy, results in nanoanalysis techniques, in nanopowder fabrication, and in nanofabrication of devices obtained so far promise valuable and exciting developments in this area. The existing knowledge on processing and properties of electroceramic thin films is undoubtedly one of the important foundations on which these developments are likely to be bared in the coming years.

REFERENCES

1. Noma A, Ueda D. Reliability study on BST capacitors for GaAs MMIC. Integr Ferroelectr 1997; 15:69–78.
2. Yoshida M, Yabuta H, Yamagouchi S, Yamaguchi H, Sone K, Arita T, Iizuka S, Nishimoto S, Kato Y. Plasma CVD of (BaSr)TiO3 dielectrics for gigabit DRAM capacitors. J Electroceram 1999; 3:123–133.
3. Sommerfelt SR. (Ba,Sr)TiO$_3$ thin films for DRAMs. In: Ramesh R, ed. Thin Film Ferroelectric Materials and Devices. Boston: Kluwer Academic, 1997:1–42.
4. Scott JF, Paz de Araujo CA. Ferroelectric memories. Science 1989; 246:1400–1405.
5. Ishiwara H. Current status of fabrication and integration of ferroelectric gate FET's. Proc Mat Res Soc Symp 2000; 596:427–436.
6. Semanik S, Cavicchi R. Kinetically-controlled chemical sensing using micromachined structures. Acc Chem Res 1998; 31:279–287.
7. Tokura Y, ed. Colossal Magnetoresistive Oxides. New York: Gordon and Breach, 2000.
8. Tian YJ, Linzen S, Schmidl F, Dörrer L, Weidl R, Seidel R. High-T$_c$ directly coupled direct current SQUID gradiometer with flip-chip flux transformer. App Phy Lett 1999; 74:1302–1304.
9. Ehrhart P. Film Deposition Methods. Chapter 8 in Nanoelectronics and Information Technology. (Advanced Electronic Materials and Novel Devices) Waser R, ed. New York: John Wiley and Sons, 2003.
10. Geffchen W, Berger E. Deutsches Reichspatent 736 411: Jenaer Glaswerk Schott & Gen., (1939).

11. Blum JB, Gurkowich SR. Sol-gel-derived $PbTiO_3$. J Mater Sci 1985; 20:4479.
12. Budd KD, Dey SK, Payne DA. Sol-gel processing of $PbTiO_3$, $PbZrO_3$, PZT, and PLZT thin films. Br Ceram Proc 1985; 36:107.
13. Yi G, Sayer M. Sol-gel processing of complex oxide films. Ceram Bull 1991; 70: 1173.
14. Campion J-F, Payne DA, Chae HK, Xu Z. Chemical processing of barium titanate powders and thin layer dielectrics, Ceram. Trans Ceram Powder Sci 1991; IV:477.
15. Schwartz RW, Assink RA, Headley T. Spectroscopic and microstructural characterization of solution chemistry effects in PZT thin film processing. MRS Proc 1992; 243:245.
16. Klee M, Larsen PK. Ferroelectric thin films for memory applications: sol-gel processing and decomposition of organo-metallic compounds. Ferroelectrics 1992; 133:91.
17. Joshi PC, Krupanidhi SB. Structural and electrical characteristics of $SrTiO_3$ thin films for dynamic random access memory applications. J Appl Phys 1993; 73:7627.
18. Lakeman CDE, Xu Z, Payne DA. On the evolution of structure and composition in sol-gel-derived lead zirconate titanate thin layers. J Mater Res 1995; 10:2042.
19. Lange FF. Chemical solution routes to single-crystal thin films. Science 1996; 273: 903.
20. Hoffmann S, Hasenkox U, Waser R, Jia CL, Urban K. Chemical solution deposited $BaTiO_3$ and $SrTiO_3$ thin films with columnar microstructure. MRS Proc 1997; 474: 9.
21. Willems GJ, Wouters DJ, Maes HE, Nouwen R. Nucleation and orientation of sol-gel PZT-films on Pt electrodes. Integr Ferroelect 1997; 15:19.
22. Schwartz RW. Chemical solution deposition of perowskite thin films. Chem Mater 1997; 9:2325.
23. Waser R, Schneller T, Hoffmann-Eifert S, Ehrhart P. Integr Ferroelectr. 2001; 36: 3.
24. Solayappan N, McMillan LD, Paz De Araujo CA, Grant B. Integr Ferroelectr. 1997; 18:127.
25. Javoric S, Kosec M, Malic B. Integr Ferroelectr. 2000; 30:309.
26. Foster CM. In: Ramesh R, ed. Thin Film Ferroelectric Materials and Devices. Boston: Kluwer Academic, 1997:167–197.
27. Juergensen H. CVD engineering for multilayer multicomponent materials: optoelectronics. Mater Chem Phys;, Deschler M, Woelk E, Schmitz D, Strauch G, Juergensen H. Multiwafer MOCVD systems for ferroelectrics. Integr Ferroelectr 1997; 18:119–125.
28. Bilodeau SM, Carl R, Van Buskirk PC, Ward J. MOCVD $BaSrTiO_3$ for $\geq$1-Gbit DRAMs. Solid State Technology 1997; 40:235–242.
29. Mymrin VF, Smirnov SA, Komissarov AE, Karpov S Yu, Przhevalski IN, Mkarov YuN, Dauelsberg M, Schumacher M, Strzyzewski P, Strauch G, Juergensen H. Integr Ferroelectrics. 2000; 30:271.
30. Kotecki DE, Baniecki JD, Shen H, Laibowitz RB, Saenger KL, Lian JJ, Shaw TM, Athavale SD, Cabral, Jr. C, Duncombe PR, Gutsche M, Kunkel G, Park Y-J, Wang Y-Y, Wise R. $(Ba,Sr)TiO_3$ dielectrics for future stacked-capacitor DRAM. IBM J Res Dev 1999; 43:367–382.
31. Hwang CS. Invited presentation at the ISIF 2001, Colorado Springs, March 12–14, 2001.

32. Herman MA, Sitter H. Molecular beam epitaxy: fundamentals and current status, 2nd ed. Springer Series in Materials Science. Vol. 7. Berlin: Springer, 1996.

33. Bunshah RF. Handbook of Deposition Technologies for Films and Coatings. 2nd ed.. Park Ridge: Noyes Publ., 1994.

34. Chopra KL. Thin Film Phenomena. New York: McGraw-Hill, 1969.

35. McKee RA, Walker FJ, Chisholm MF. Crystalline oxides on silicon: the first five monolayers. Phys Rev Lett 1998; 81:3014–3017.

36. Droopad R, Wang J, Eisenbeiser K, Yu Z, Ramdani J, Curless JA, Overgaard CD, Finder JM, Hallmark JA, Kaushik V, Nguyen BY, Marshall DS, Ooms WJ. Epitaxial oxide films on silicon: Growth, modeling and device properties. MRS Proc 2000; 619:155–165.

37. Auciello O, Kingon AI, Krauss AR, Lichtenwalner DJ. In: Auciello O, Engemann J, eds. Multicomponent and Multilayered Thin Films for Advanced Microtechnologies: Techniques, Fundamentals and Devices. NATO ASI.. Dordrecht: Kluwer Acad., 1993:151–208.

38. Venkatesan T, Wu XD, Muenchhausen R, Pique A. Mat Res Soc Bull. 1992; 17: 54.

39. Rossnagel SM. Plasma sputter deposition systems: plasma generation, basic physics and characterisation. In Ref 37:1–20.

40. Wasa K, Hayakawa S. Handbook of sputter deposition technology. Park Ridge: Noyes, 1992.

41. Wehner GH, Anderson GS. The nature of physical sputtering. In: Maissel LI, Glang R, eds. Handbook of Thin Film Technologies. New York: McGraw-Hill, 1970.

42. Stroscio JA, Pierce DT, Dragoset RA. Homoepitaxial growth of iron and a real space view of reflection-high-energy-electron diffraction. Phys Rev Lett 1993; 70: 3615–3618.

43. Poppe U, Klein N, Dähne U, Soltner H, Jia CL, Kabius B, Urban K, Lubig A, Schmidt K, Hensen S, Orbach S, Müller G, Piel H. Low-resistivity epitaxial YBa₂Cu₃O₇ thin films with improved microstructure and reduced microwave losses. J Appl Phys 1992; 71:5572–5578.

44. Poppe U, Hojczyk R, Jia CL, Faley MI, Evers W, Bobba F, Urban K, Horstmann C, Dittmann R, Breuer U, Holzbrecher H. IEEE Trans Appl Supercond. 1999; 9: 3452.

45. de Araujo CP, Scott JF, Taylor GW. Ferroelectric thin films: synthesis and basic properties. In: Taylor GW, ed. Ferroelectricity and Related Phenomena, Vol. 10. New York: Gordon and Breach, 1996.

46. Muralt P. Ultrasonic micromotors based on PZT thin films. J Electroceram 1999; 3:143–150.

47. Muralt P. Ferroelectric thin films for microsensors and actuators: a review. Micromech Microeng 2000; 10(2):136–146.

48. Muralt P. PZT thin films for micro sensors and actuators: Where do we stand? IEEE Trans. Ultrasonics Ferroelectrics Frequency Control (UFFC) 2000; 47:903–915.

49. Goodenough JB. Transition metal oxides with metallic conductivity. Bull Soc Chim Fr 1965; 4:1200.

50. Tamura H, Yoneyama H, Matsumoto Y. Physiochemical and electrochemical properties of perovskite oxides. In Trasatti S, ed. Electrodes of Conductive Oxides. Amsterdam: Elsevier:, 1980:261–300.

51. Trasatti S, Lodi G. Properties of conductive transition metal oxides with rutile-type structure. In: Electrodes of Conductive Metallic Oxides. Trasatti S. ed. Amsterdam: 301–358.

52. Maeder T, Sagalowics L, Muralt P. Stabilized platinum electrodes for PZT thin film deposition using Ti, Zr, and Ta adhesion layers. Jpn J Appl Phys 1998; 37: 2007–2012.

53. Muralt P. unpublished.

54. Trupina L, Baborowski J, Muralt et al. P. Tungsten based electrodes for stacked capacitor ferroelectric memories. Jpn J Appl Phys 2002; 41(11B):6862–6866.

55. Kwok CK, Desu SB. Formation kinetics of PZT thin films. J Mater Res 1994; 9: 1728–1733.

56. Chopra K. Thin Film Phenomena. New York: McGraw-Hill, 1969.

57. Ramesh R, Gilchrist H, Sands T, Keramidas VG, Haakenaasen R, Fork DK. Ferroelectric LSCO/PZT/LSCO heterostructures on silicon via template growth. Appl Phys Lett 1993; 63:3592–3594.

58. Kwok CK, Desu SB. Low temperature perovskite formation of PZT thin films by a seeding process. J Mater Res 1993; 8:339–344.

59. Maeder T, Muralt P, Kohli M, Kholkin A, Setter N. $Pb(Zr,Ti)O_3$ Thin films by in-situ reactive sputtering on micromachined membranes for micromechanical applications. Br Ceram Proc 1995; 54:206–218.

60. Aoki K, Fukuda Y, Numata K, Nishimura A. Effects of titanmium buffer layer on PZT crystallization porecess in sol-gel deposition technique. Jpn J Appl Phys 1995; 34:192–195.

61. Kighelman Z, Damjanovic D, Seifert A, Sagalowicz L, Setter N. Relaxor behavior and electromechanical properties of $Pb(Mg,Nb)O_3$ thin films. Appl Phys Lett 1998; 73:2281–2283.

62. Malic B, Kosec M, Arcon I, Kodre A, Hiboux S, Muralt P. PZT thin films prepared from modified zirconium alkoxide. Integr Ferroelectr 2001; 30:81–89.

63. Muralt P, Maeder T, Sagalowicz, et al. L. Texture control of $PbTiO_3$ and PZT thin films with TiO_2 seeding. J Appl Phys 1998; 83(7):3835–3841.

64. Seifert A, Ledermann N, Hiboux S, Baborowski J, Muralt P, Setter N. Integr Ferroelectr. 2001; 35:159–166.

65. Norga GJ, Jin S, Fe L, Wouters DJ, Bender H, Maes HE. Growth of (111)-oriented $Pb(Zr,Ti)O_3$ layers on nanocrystalline RuO_2 electrodes using the sol-gel technique. J Mater Res 2001; 16:828–833.

66. Muralt P, Ledermann N, Baborowski J, Gentil S. Integration of Piezoelectric $Pb(Zr_xTi_{1-x})O_3$ (PZT) thin films into micromachined sensors and acuators. In: Materials and Process Integration for MEMS. Tay FEH. ed. Boston: Kluwer Academic, 2002.

67. Baborowski J, Muralt P, Ledermann N, Colla E, Seifert A, Gentil S, Setter N. Mechanisms of PZT thin film etching with ECR/RF reactor. Integr Ferroelectr 2001; 31:261–271.

68. Muralt P. Piezoelectric thin films for MEMS. In Encyclopedia of Materials Science and Technology. Amsterdam: Elsevier, 2001:6999–7009.

69. Moroney RM, White RM, Howe RT. Microtransport induced by ultrasonic Lamb waves. Appl Phys Lett 1991; 59:774–776.

70. Percin G, Khuri-Yakub BT. Piezoelectric droplet ejector for ink-jet printing of fluids and solid particles. Rev Sci Instrum 2003; 74:1120–1127.

71. Lugienbuhl P, Collins SD, Racine G-A, Grétillat M-A, de Rooij NF, Brooks KG, Setter N. Microfabricated Lamb wave device based on PZT sol-gel thin film for mechanical transport of solid particles and liquids. J Micromech Syst 1997; 6: 337–346.

72. Muralt P, et al. Study of PZT coated membrane structures for micromachined ultrasonic transducers. Proc IEEE Ultrasonic Transducers 2001:907–911.

73. Lakin KM, Kline GR, McCarron KT. Thin film bulk acoustic wave filter for GPS. Proc IEEE Ultrasonic Symp 1992:471–476.

74. Dubois M-A, Muralt P. Properties of AlN thin films for piezoelectric transducers and microwave filter applications. Appl Phys Lett 1999; 74:3032–3034.

75. Lanz R, Dubois M-A, Muralt P. Solidly mounted BAW filters for the 6 to 8 GHz range based on AlN thin films. Proc IEEE Ultrasonics Symp 2001; 1:843–846.

76. Muralt P. Micromachined infrared detectors based on pyroelectric thin films. Rep Progr Phys 64:1339–1388.

77. Hanson CM, Beratan HR, Belcher JF. SPI Proc. on Infrared Detectors and Focal Plan Arrays VI. San José, CA, 2000.

78. Willing B, Kohli M, Muralt P, Setter N, Oehler O. Gas spectrometry based on pyroelectric thin film arrays integrated on silicon. Sensors and Actuators A 1998; 6:109–113.

79. Willing B, Kohli M, Muralt P, Oehler O. Thin film pyroelectric array as a detector for an infrared gas spectrometer. Infrared Phys Technol 1998; 39:443–449.

80. http://public.itrs.net/International Technology Roadmap for Semiconductors, 2001 Edition. Front End Processes.

81. Kotecki DE, et al. IBM. J Res Dev 1999; 43:367.

82. Kingon AI, Streiffer SK, Basceri C, Summerfelt SR. High-permittivity perovskite thin films for dynamic random access memories. Mater Res Soc Bull 1996; 21: 46–53.

83. Summerfelt SR. $(Ba,Sr)TiO_3$ Thin Films for DRAM's. In: Ramesh R, ed. Thin Film Ferroelectric Materials and Devices. Boston: Kluwer Academic, 1997:1–42.

84. Kotecki DE. A review of high dielectric materials for DRAM applications. Integr Ferroelectr 1997; 16:1–19.

85. Kingon AI, Maria J-P, Streiffer S K. Alternative dielectrics to silicon dioxide for memory and logic devices. Nature 2000; 406:1032–1038.

86. Basceri C, Streiffer SK, Kingon AI, Waser R. The dielectric response as a function of temperature and film thickness of fiber-textured $(Ba,Sr)TiO_3$ thin films grown by chemical vapor deposition. J Appl Phys 1997; 82(5):2497–2504.

87. Streiffer SK, Basceri C, Parker CB, Lash SE, Kingon AI. Ferroelectricity in thin films: the dielectric response of fiber-textured $(Ba_xSr_{1-x}Ti_{1+y}O_{3+z})$ thin films grown by chemical vapor deposition. J Appl Phys 1999; 86(8):4565–4575.

88. Baniecki JD, Laibowitz RB, Shaw TM, Duncombe PR, Neumayer DA, Kotecki DE, Shen H, Ma QM. Dielectric relaxation of $Ba_{0.7}Sr_{0.3}TiO_3$ thin films from 1 mHz to 20 GHz. Appl Phys Lett 1998; 72:498–500.

89. Grossmann M, Hoffmann S, Gusowski S, Waser R, Basceri C, Lash SE, Parker CB, Streiffer SK, Kingon AI. Resistance degradation behavior of $Ba_{0.7}Sr_{0.3}TiO_3$

thin films compared to mechanisms found in titanate ceramics and single crystals. Integr Ferroelectr 1998; 22(1–4):603–614.

90. Stemmer S, Streiffer SK, Browning ND, Kingon AI. Accommodation of nonstoichiometry in (100) fiber-textured $(Ba_xSr_{1-x}Ti_{1+y}O_{3+z}$ thin films grown by chemical vapor deposition. Appl Phys Lett 1999; 74(17):2432–2434.

91. Ren W, Trolier-McKinstry S, Randall CA, Shrout TR. Bismuth zinc niobate pyrochlore dielectric thin films for capacitive applications. J Appl Phys 2001; 89(1): 767–774.

92. Maria J-P, Cheek K, Streiffer SK, Kim SH, Dunn G, Kingon AI. Lead zirconate titanate thin films on base-metal foils: An approach for embedded high K passive components. J Am Ceram Soc; 2001; 84(10):2436–2438. See also the proceedings of the IMAPS workshop on Integrated Passive Components (Agonquit, Maine, June 2002). Proceedings available on CDROM from the International Microelectronics and Packaging Society.

93. Wieder HH. in Large Capacity Memory Techniques for Computing Systems. Yovits M, ed. New York: Macmillan, 1962:277.

94. Chapman DW. J Appl Phys;, Chapman DW. Thin film photoconductive-ferroelectric memory device. J Vac Sci Technol 1972; 9(1):425–431.

95. Eaton SS, Butler DB, Parris M, Wilson D, McNeillie H. Proc. IEEE Int Solid State Circuits Conference, 1988.

96. The acronym FRAM is sometimes used for this type of ferroelectric memory, instead of FERAM. FRAM is a registered trademark of Ramtron Internation Corporation, an early and continuing pioneer in the field of ferroelectric memory.

97. Some of the data for this table were compiled from the International Technology Roadmap for Semiconductors [Ref. 80]. F is the minimum feature size for the given technology node or generation. For DRAMs, it corresponds to the pitch between DRAM cells.

98. Takasu H. Ferroelectric memories and their applications. Microelectron Eng 2001; 59:237–246.

99. Kim DJ. PhD thesis, Department of Materials Science and Engineering, North Carolina State University. J Appl Phys 2003; 93(9):5568–5575.

100a. Hwang SC, Arlt G. Switching in ferroelectric polycrystals. J Appl Phys; 2000; 87(2):869–875.

100b. Arlt G. A model for switching and hysteresis in ferroelectric ceramics. Integr Ferroelectrics 1997; 16(1–4):229–236.

100c. Ganpule et al. CS. Polarization relaxation kinetics and 180 domain wall dynamics in ferroelectric thin films. Phys Rev B 014101 (2001); 65.

101. Foster CM, Bai G-R, Csencsits R, Vetrone J, Jammy R, Wills LA, Carr E, Amano J. Single crystal $Pb(Zr_xTi_{1-x})O_3$ thin films prepared by metal-organic chemical vapor deposition: systematic compositional variation of electronic and optical properties. J Appl Phys 1997; 81:2349–2357.

102. Oikawa T, Aratani M, Saito K, Funakubo H. Composition dependence of ferroelectric properties of epitaxial $Pb(Zr_xTi_{1-x})O_3$ thin films grown by metalorganic chemical vapor deposition. J Crystal Growth 2002; 237–239(Pt. 1):455–458.

103. Haun MJ, Furman E, Jang SJ, Cross LE. Thermodynamic theory of the lead zirconate titanate solid solution system. V. Theoretical calculations. Ferroelectrics 1989; 99:63–86.

104. Smyth DM. Aliovalent doping in perovskite oxides. Ceram Trans 104 (Perovskite Oxides for Electronic, Energy Conversion, and Energy Efficiency Applications). 2000:109–123.

105. Smyth DM. Defect structure in perovskite titanates. Curr Opin Solid State Mat Sci 1996; 1(5):692–697.

106. Tuttle B. Presentation and extended abstract of the US–Japan Workshop on Dielectric and Piezoelectric Ceramics, Providence, RI, Sept. 27–29, 2001. See also [99].

107. Smolenskii GA, Isupov VU. Sov. Phys. Solid State. 1961; 3:661.

108. Newnham RE, Wolf RW, Horsey RS, Diaz-Colon FA. Mater Res Bull 1973; 8: 1183.

109. Gruverman A. Scaling effect on statistical behavior of switching parameters of ferroelectric capacitors. Appl Phys Lett 1999; 75:1452–1452.

110. Noguchi T, Hase T, Miyasaka Y. Analysis of the dependence of ferroelectric properties of strontium bismuth tantalate (SBT) thin films on the composition and process temperature. Jpn J Appl Phys, Part 1 1996; 35(9B):4900–4904.

111. Palanduz AC, Smyth DM. Defect chemistry of $SrBi_2Ta_2O_9$ and ferroelectric fatigue endurance. J Electroceram 2000; 5(1):21–30.

112. Thomas DT, Fujimura N, Streiffer SK, Kingon AI. Composition and electrode effects on the electrical properties of $SrBi_2Ta_2O_9$. Proc Mater Res Soc 1998; 493 (Ferroelectric Thin Films VI):153–158.

113. Park BH, Kang BS, Bu SD, Noh TW, Lee L, Joe W. Nature. 1999; 401:682.

114. Watanabe T, Kojima T, Sakai T, Funakubo H, Osada M, Noguchi Y, Miyayama M. Large remanent polarization of Bi4Ti3O12-based thin films modified by the site engineering technique. J Appl Phys 2002; 92(3):1518–1521.

115. Al-Shareef HN, Kingon AI. Electrode materials for ferroelectric thin film capacitors and their effect on the electrical properties. Chapter 7 of Ferroelectric Thin Films: Synthesis and Basic Properties. Paz de Araujo CA, Scott JF, Taylor GW, eds. New York: Gordon and Breach, 1996.

116. Stolichnov I, Tagantsev A, Setter N, Cross JS, Tsukada M. Top-interface-controlled switching and fatigue endurance of (Pb,La)(Zr,Ti)O₃ ferroelectric capacitors. Appl Phys Lett 1999; 74(23):3552–3554.

117. Al-Shareef HN, Auciello O, Kingon AI. Electrical properties of ferroelectric thin-film capacitors with hybrid (Pt,RuO_2) electrodes for nonvolatile memory applications. J Appl Phys 1995; 77(5):2146–2154.

118. Grossmann M, Bolten D, Lohse O, Boettger U, Waser R, Tiedke S. Correlation between switching and fatigue in $Pb(Zr_{0.3}Ti_{0.7})O_3$ thin films. Appl Phys Lett 2000; 77(12):1894–1896.

119. Tagantsev AK, Stolichnov I, Colla E, Setter N. Polarisation fatigue in ferroelectric films: basic experimental findings, phenomenological scenarios, and microscopic features. J Appl Phys 2001; 90(3):1387–1402.

120a. Benedetto JM, Moore RA, McLean FB. Effects of operating conditions on the fast-decay component of the retained polarization in lead zirconate titanate thin films. J Appl Phys; 1994; 75(1):460–466.

120b. Lohse O, Grossmann M, Boettger U, Bolten D, Waser R. Relaxation mechanism of ferroelectric switching in Pb(Zr, Ti)O₃ thin films. J Appl Phys 2001; 89(4): 2332–2336.

121. Series of papers from the Sandia group. See, for example: Dimos D, Warren WL, Sinclair M, Tuttle BA, Schwartz RW. J. Appl. Phys;, Al-Shareef HN, Dimos D, Warren WL, Tuttle BA. J Appl Phys 1996; 80:4573.
122. Arlt G, Neumann H. Ferroelectrics 1988; 87:109.
123. Grossmann M, Lohse O, Bolten D, Boettger U, Waser R, Hartner W, Kastner M, Schindler G. Appl. Phys. Lett. 2000; 76:363.
124. Gruverman A, Rodriguez BJ, Nemanich RJ, Kingon AI. Nanoscale observation of photoinduced domain pinning and investigation of imprint behavior in ferroelectric thin films. J Appl Phys 2002; 92:2734–2739.
125. See for example: Nikkei Electronics Asia, October 2001 issue, published by Nikkei Business Publications.
126. For information on barriers, see the DRAM reviews [82,85].
127. Sinharoy S, Buhay H, Francombe MH, Lampe DR. $BaMgF_4$ thin film development and processing for ferroelectric FETs. Integr Ferroelectrics 1993; 3(3):217–23.
128. Ishiwara H. Recent progress of FET-type ferroelectric memories. Integr Ferroelectr 2001; 34(1–4):11–20.
129. Vendik OG, Hollmann EK, Kozyrev AB, Prudan AM. Ferroelectric tuning of planar and bulk microwave devices. J Superconduct 1999; 12:325–338.
130. Lancaster MJ, Powell J, Porch A. Thin film ferroelectric microwave devices, Supercond. Sci Technol 1998; 11:1323–1334.
131. Miranda FA, Subramanyam G, Van Keuls FW, Warner KD, Mueller CH. Design and development of ferroelectric tunable microwave components for Ku- and K-band satellite communication systems. IEEE Trans MTT 2000; 48:1181–1189.
132. Tagantsev AK. Degradation and dielectric loss mechanisms in dielectric films. In. Encyclopedia of Materials: Science and Technology. Amsterdam: Elsevier Science, 2001:2027–2036.
133. Pertsev N, Zembilgotov AG, Tagantsev AK. Effect of mechanical boundary conditions on phase diagrams of epitaxial ferroelectric thin films. Phys Rev Lett 1998; 80:1988–1991.
134. Schlom D. unpublished.
135. Carlson CM, Rivkin TV, Parilla PA, Perkins JD, Ginley DS, Kozyrev AB, Oshadchy VN, Pavlov AS. Large dielectric constant BST thin films for high performance microwave phase shifters. Appl Phys Lett 2000; 76:1920–1922.
136. Van Keuls FW, Muller CH, Romanofsky RR, Warner JD, Miranda FA, Jiang H. A comparison of MOCVD with PLD $BaSrTiO_3$ thin films on $LaAlO_3$ for tunable microwave applications. Integr Ferroelectr 2001; 39:437–448.
137. Kozyrev A, et al. Integr Ferroelectr. 1999; 24:287.
138. Alexe M, Harnagea C, Hesse D, Gösele U. Patterning and switching of nanosize ferroelectric memory cells. APL 1999; 73(12):1793–1795.
139. Ganpule CS, Stanishevsky A, Su Q, Aggarwal S, Melngailis J, Williams E, Ramesh R. Scaling of ferroelectric properties in thin films. Appl Phys Lett 1999; 73:409–411.
140. Bühlmann S, Dwir B, Baborowski J, Muralt P. Size effect in mesoscopic epitaxial ferroelectric structures: increase of piezoelectric response with decreasing feature size. Appl Phys Lett 2002; 80:3195–3197.
141. Ahn CH, Tybell T, Antognazza L, Char K, Hammond RH, Beasley MR, Fischer Ø, Triscone J-M. Local, nonvolatile *electronic* writing of epitaxial $Pb(Zr_{0.52}Ti_{0.48})O_3/SrRuO_3$ heterostructures. Science 1997; 276:1100–1103.

9

Ceramic Thick-Films
Process and Materials

William Borland
DuPont Electronic Technologies
Research Triangle Park, North Carolina, U.S.A.

I. INTRODUCTION

The ceramic thick-film process can be defined as the sequential screen printing and firing at elevated temperatures of a conductor, resistor, and/or dielectric onto a suitable substrate. The thickness of each thick-film element depends on material and application but is typically between 8 and 50 microns (μm). This distinguishes thick film from thin film, in which thickness is generally less than 1 μm. The thick-film process uses simple processing equipment that requires low capital outlay. Since the manufacturing process is simple, design changes can easily be made. Tooling for a new circuit is relatively inexpensive, so that small production runs can be easily accommodated. Production rates can be high, especially when multi-up designs are manufactured. The process also allows the substrates to be inspected after each step. However, materials can be expensive and the overall process often requires many steps, although this can be partially addressed by cofiring more than one print of material.

Thick-film materials are used to manufacture an ever-widening diverse set of products. The largest application, however, is the thick-film hybrid. A thick-film hybrid can be defined as the combination of electronic circuitry manufactured by the thick-film process with discrete add-on devices in such a way that an electronic function is performed. The vast majority of thick-film hybrids are used either to modify the function of integrated circuits (ICs) by integration with passive components, or to interface ICs principally to convert analog signals to digital form and vice versa. Thick-film hybrids can also be used to interconnect ICs only. This is a pure multilayer construction of conductors separated by dielectric layers and is often called a multilayer ceramic module or ceramic printed wiring board (PWB). Thick-film hybrids can be printed individually or multi-up. Multi-up constructions have many circuits on one substrate. After printing and firing has been completed, substrates are laser scribed and broken into the individual circuits. Thick-film hybrids can be single sided or double sided, single layer or multilayer. Single-sided hybrids have circuitry on one side of the substrate only. This is the most common hybrid build. Double-sided hybrids have circuitry on both sides of the substrate with electrical connection through either holes in the substrate or edge connections. This is equivalent to a two-metal-layer multilayer circuit. Single-layer circuits are conductor traces printed and fired onto a substrate. Where traces need to cross over each other, pads of cross-over dielectric are printed and fired over the first conductor trace. Resistors and glass encapsulants are printed and fired as the final operation before trimming, component attachment, and packaging. Multilayer hybrids are constructed by alternately printing and firing conductive traces and dielectric layers. Connection between layers is achieved by vias (holes in the dielectric layers) filled with conductor material during a via-fill printing and firing process. This sequential process is repeated until the desired number of layers has been formed. Multilayers can be as few as two metal layers to several layers thick. Resistors can be printed and fired on the surface of the multilayer and can be trimmed to high tolerance. If surface real estate is at a premium and lower tolerances are acceptable, thick-film resistors and capacitors can be buried within the multilayer, sandwiched between two layers of dielectric [1].

Thick-film materials are also used extensively in the manufacture of many resistive components such as chip resistors, resistor arrays, and potentiometers. Components are processed multi-up on large substrates and eventually separated into individual units. As in the hybrid process, conductor terminations are printed and fired on alumina first followed by the resistor and finally, if required, a glass encapsulant. Chip resistors and arrays are trimmed to value at this stage. Substrates are laser scribed and separated into individual units and surface mount conductor terminations may be applied to the individual components as the final finishing step. Other chip components, such as capacitors, resonator filters, and varistors-use the thick-film process in a simple screening of a conductive electrode on to an individual prefired ceramic body.

Sensors utilize thick-film materials in many ways. The thick-film element may function as an electrode, as an encapsulant, or as the sensing material. Force, weight, and pressure sensors are often based on the piezoresistivity exhibited in strain gauge resistors. Current limiting devices and other property-limiting sensors are based on the nonlinear resistance–temperature characteristics of positive temperature coefficient (PTC) and negative temperature coefficient (NTC) thermistors. Potentiometric sensors used in fluid level gauges, flowmeters, and position sensors are based on movement of the potentiometer wiper on the resistor track due to changes in fluid level, airflow, etc. Platinum thick-film pastes are used in temperature sensors and mass flow anemometers. Combining thick-film materials with the unique properties of certain ceramic materials allows for the manufacture of a variety of other sensing devices, such as gas and humidity sensors.

Thick-film materials are also used to make heaters. Automotive rear window heaters (often called defoggers) use silver pastes of varying resistivites matched to power requirements dictated by the design of the window heater. Lines of silver terminating at bussbars at the edge of the window are printed onto a flat piece of glass. The silver paste is often partially printed on dried black enamel paste printed along the edges of the glass to hide the bussbars. The materials are then cofired and pressed into the final shape in one process to form the finished window. High-performance electric kettles often have heating element made from thick-film materials [2]. Thick-film dielectric is either printed or laminated in tape form onto one side of a stainless steel substrate that has been machined to fit in the base of the kettle. It is then fired at 850°C followed by printing and firing a spiral heating element design onto the dielectric using silver-palladium conductors. The substrate is then fitted inside the kettle, the noncoated steel in contact with the water container. Electric kettles using this heating technology boil water very quickly.

Thick-film materials are used in the manufacture of solar cells. A fine grid pattern of silver is printed and fired onto the front face of the silicon wafer. This acts as a collector agent for the electrons. Silver pastes doped with aluminum or even pure aluminum pastes designed to cut through the silicon dioxide coating cover the back face to complete the circuit.

Thick-film pastes also have a significant role in the manufacture of flat panel plasma display (PDP) televisions and monitors [3]. A PDP structure made by the thick-film process consists of two parallel glass substrates sandwiching the thick-film circuitry. The sealed panel is about 1.5 in. thick and contains a gas mixture that when energized causes emission of UV light. This impinges on colored phosphors in pixels to create visible light. Each individual pixel on the panel contains phosphors in a dielectric cavity and has associated electrodes on both sides of the panel. The design allows each pixel to be addressed individually to create the full colors and images. The pixel density requires high precision, fine conductor lines, and dielectric patterning. This is accomplished by use of thick-film photoimageable conductors and dielectric tape and pastes fired on the glass panels at temperatures between 450°C and 600°C.

II. HISTORY

Ceramic thick-film technology has its origins in pottery decoration and was originally practiced by the Chinese more than 2000 years ago. Colors and glazes based on inorganic oxides are printed on pottery and fired to permanently fuse the design into the ceramic. Thick-film technology for microelectronics dates to the early 1920s with the development of modern screen printing technology. By the 1930s, fabrication of simple electronic circuitry by screen printing was well underway. Thick-film applications grew substantially during World War II due to the military's need for electronic circuit miniaturization. Rapid growth in use of the technology came from the inventions of the transistor, diode, and IC in the late 1940s and 1950s. Since then, thick-film applications have steadily grown due to a constant flow of new needs and the ability to develop processes and materials to satisfy them.

III. PROCESS

A. Screen Printing and Leveling

The selective deposition of thick-film materials onto a substrate essentially depends on forcing ink through a patterned screen, stencil, or metal mask mounted under tension on a metal frame. The screen's circuit pattern is formed photographically on the mesh. Ultraviolet (UV)–sensitive emulsion is applied to the mesh, exposed through a positive or negative of the desired pattern, and developed in such a manner that the holes in the mesh are left open where deposition of paste on the substrate is desired. Screens can be made from nylon, polyester, or stainless steel. The majority of thick-film applications use stainless steel mesh screens. Table 1 lists typical specifications of stainless steel screens. Use of 325-mesh screens is common, allowing for 200-μm (8 mil) lines and spaces to be resolved comfortably, suitable for most applications. For finer features, sophisticated higher mesh count screens employing fine wires (14–20 μm diameter) under high tension and very thin emulsion thickness [4,5] allows for 75-μm (3-mil) lines and spaces to be resolved. Emulsion thickness is the thickness in excess of the screen mesh thickness, as shown in Figure 1, and may vary from zero to 50 μm or more for some solder paste printing. Etched metal masks [6] can achieve 50-μm (2-mil) lines and photoimageable compositions [7,8] allow for 40-μm (1.5-mil) lines and 50-μm (2-mil) vias to be resolved (9). Photoetching or photodefining [10,11], where a photoresist is applied to a previously fired thick-film conductor, imaged, and the excess conductor chemically etched away, can achieve 10μm (0.4-mil) lines and spaces.

The paste to be printed is placed on the screen in front of a polyurethane squeegee. Since thick-film paste does not flow through the screen mesh by the

Table 1 Typical Specifications for Stainless Steel Screens

Mesh	Wire diameter (μm)	Mesh aperture (μm)	Open area (%)	Cloth thickness (μm)
120	66	145	47.3	132–147
145	56	119	46.4	122–132
180	46	97	45.7	94–109
200	41	86	46.2	81–97
230	36	74	45.9	71–86
280	31	61	44.1	66–81
325	28	51	41.3	58–71
400	25	38	36.0	50–61

action of gravity alone, pressure is applied to the squeegee. As shown schematically in Figure 2, as the squeegee traverses the screen, the screen is brought into contact with the substrate. Paste is forced into the open areas of the screen and wets the substrate's surface. As the squeegee continues its traverse, the screen behind the squeegee separates (peels) from the substrate and regains its original position. Paste that has wetted and adhered to the substrate is drawn through the screen openings and remains on the substrate, creating the screened pattern. The amount of pressure developed in the paste depends on the pressure applied to the squeegee, the angle of inclination of the squeegee blade, the speed of traverse, and the viscosity and amount of paste on the screen. Correct pressure applied to the squeegee will ensure complete filling of the open areas of the screen, without spreading the paste outside the boundaries defined by the emulsion [12–18].

The driving force for the leveling process is the minimization of the surface Gibbs free energy (G_s) given by

$$\Delta G_s = \gamma_{LV} \Delta A_{LV} \tag{1}$$

where γ_{LV} is the liquid–vapor interfacial energy (surface tension of the paste) and A_{LV} is the area of the paste–vapor interface. However, forces proportional

Figure 1 Cloth (t_c) and emulsion thickness (t_e) in screen printing.

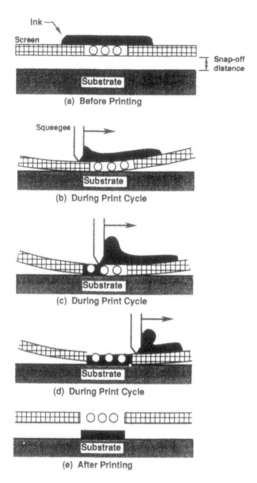

Figure 2 The screen printing process. (Reproduced from Electronic Ceramics: Properties, Devices, and Applications, edited by Lionel M. Levinson, Marcel Dekker, New York, 1987, p. 344.)

to the viscosity of the paste oppose this. Therefore, maximum leveling is achieved with a paste of low viscosity during the leveling period. For a small printed pattern, the emulsion thickness prevents the mesh from making contact with the substrate. In such a case, the initial print is a pattern of cylinders corresponding to the openings in the screen on a bed of paste the thickness of the emulsion. This will level according to the process depicted in Figure 3. If a screen with zero emulsion thickness is used, only a series of cylinders are printed and the

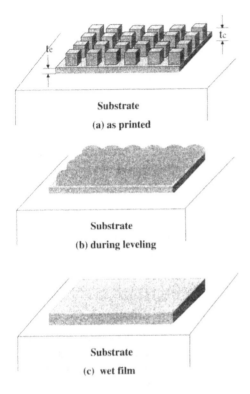

Substrate

(a) as printed

Substrate

(b) during leveling

Substrate

(c) wet film

Figure 3 The leveling process with a finite emulsion. (Reproduced from Electronic Ceramics: Properties, Devices, and Applications, edited by Lionel M. Levinson, Marcel Dekker, New York, 1987, p. 355.)

paste must spread on the substrate to achieve the situation depicted in Figure 4(b) before the leveling process can begin. This requires that the contact angle (θ) must be less than some maximal value, which can be calculated knowing the screen mesh thickness, wire diameter, and opening diameter. The equilibrium contact angle for any fluid on a planar solid surface is given by

$$\theta = \cos^{-1}\left(\frac{\gamma_{SV} - \gamma_{SL}}{\gamma_{LV}}\right) \qquad (2)$$

where γ_{SV} and γ_{SL} are the solid–vapor and solid–liquid interfacial energies, respectively. Reducing the surface tension of the paste γ_{LV} by additions of surfactants will reduce θ and improve leveling. Too small a contact angle, however, will lead to excessive spreading of the paste. If a pattern involves large open

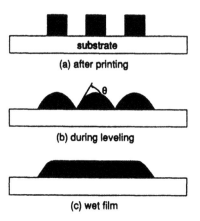

(a) after printing

(b) during leveling

(c) wet film

Figure 4 The leveling process with zero emulsion. (Reproduced from Electronic Ceramics: Properties, Devices, and Applications, edited by Lionel M. Levinson, Marcel Dekker, New York, 1987, p. 335.)

areas, the mesh will make contact with the substrate in the center of the open area. At this point, only cylinders are printed. Near the edges of the open area, the emulsion thickness prevents the mesh from making contact creating thicker prints. This lead to printed cross-sections that, as depicted in Figure 5, depart from ideal rectangles.

The wet film thickness (t_w) and fired film thickness (t_f) can be calculated from the formulas

$$t_w = \left(\frac{t_c A_o}{100}\right)(1-S) + t_e \tag{3}$$

$$t_f = \frac{t_w V_s}{1-P_f} \tag{4}$$

where t_c is the screen cloth thickness, A_o the percent open area of the screen, S the screen retention factor, t_e the emulsion thickness, V_s the volume fraction of solids in the ink, and P_f the fired film porosity. Equation (3) gives the maximum wet film thickness because it assumes the geometry shown in Figure 3(c). Figure 4(c) is the more realistic wet-film geometry. The screen retention factor, S, is a measure of how completely the paste is removed from the screen mesh during a print cycle and depends on paste properties and machine variables. A tacky (sticky) paste will give a larger S, which can be reduced to some extent by the proper organic additives. The principle machine variables that influence S are

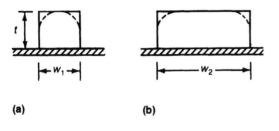

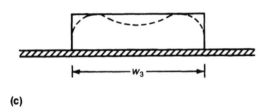

Figure 5 Departure from ideal rectangles. Courtesy of DuPont Technical Service Literature.

screen tension, snap-off distance (see Figure 2), and squeegee speed. Increasing any one of these variables will increase the velocity of the screen peel from the substrate after the squeegee has passed, which will decrease S and increase the wet film thickness [19].

The force retarding leveling is proportional to the paste viscosity (η), which is the internal resistance exerted by a fluid to the relative motion of its parts, and is given by

$$\eta = \frac{\tau}{\gamma}$$
(5)

where τ is the shear stress and γ is the shear strain rate. Figure 6 illustrates four ways in which fluids can respond to shear. A fluid is said to be Newtonian if its viscosity is independent of shear rate and is a function of temperature only. A dilatant fluid becomes more viscous as shear rate increases. Neither of these characteristics is conducive to printing. The viscosity of a pseudoplastic material decreases with increased shear rate. The viscosity of a thixotropic material increases with time after the shear stress has been removed. Most screen-printed materials are designed to have both pseudoplastic and thixotropic properties. Such materials decrease in viscosity upon shearing and recover to high viscosity upon removal of the shear stress after a short period. The ideal paste for screening

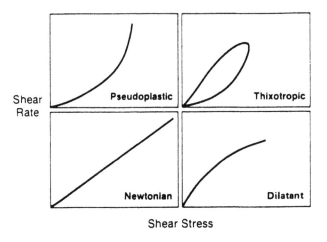

Figure 6 Fluid response to shear.

should have a relatively low viscosity (approx. 10 Pa-s) at the high shear strain rate produced when the squeegee traverses the pattern, so that paste transfer to the substrate is easily effected. The shear strain rate at this point in the print cycle has been estimated to be as high as 1000 s^{-1} [20], although more recent work suggests that in standard thick-film printing conditions, it is lower than this, probably close to 200 s^{-1} [21]. The viscosity should remain relatively low (<1000 Pa-s) for a short time after the squeegee passes and the screen snaps off of the substrate, so that leveling can occur in a reasonable period of time (<15 min). The shear strain rate during leveling driven by surface tension forces is estimated to be approximately 10^{-2} s^{-1} [22]. After leveling, the shear strain rate driven by gravity is approximately 10^{-2} s^{-1} and the viscosity should become very high (>10^4 Pas-s) in order to prevent spreading of the film. This is an impossible set of rheological requirements for any real paste. Ideal behavior is approximated by proper selection of high molecular weight polymer and solvent combination, by use of proper particle sizes of the inorganic constituents, and by addition of surfactants and thixotropes.

B. Drying and Firing

After the paste has been allowed to level, it is dried at 100°C to 150°C for 10–15 min to remove solvents. Photosensitive pastes that can be thermally polymerized may be limited to 70–80°C followed by the imaging and developing process to produce the features desired. Firing is then undertaken in a continuous belt furnace at temperatures between 500°C and 1000°C. Sometimes the drying and firing processes are combined, the drying occurring in the initial zones of the belt furnace. The total cycle time from when the parts enter the furnace until they exit (door-to-door time) varies from 15 min to 60 min. In thick-film microelectronic applications it is typically 30 and 60 min for air and nitrogen-fired parts, respectively. Peak temperatures are normally 850°C for air and 900°C for nitrogen systems, and time at peak is normally 10 min. Typical firing profiles for air and nitrogen systems are illustrated in Figures 7 and 8. For some other applications, such as plasma displays and automotive rear window heaters where the glass substrates cannot tolerate such high temperatures, peak temperature may be as low as 600°C with time at peak just a few minutes.

In the initial phase of firing (200°C to 500°C), organic resins are burnt off in the burnout zones of the furnace. Decomposition of the polymers occurs by simple oxidation in air-fired systems producing CO_2, H_2O, and light-weight fly ash by-products. However, in nitrogen firing, due to the low level of oxygen, decomposition of the organic resins occurs by pyrolysis. Cleavage of the polymer at the ethyl groups and dehydrogenation takes place resulting in a complex mixture of polymers, monomers, and gases that burn off as gas, vapors, and droplets.

A clean, dry atmosphere is necessary in the furnace. In addition, halogenated solvents must be avoided as they can have deleterious effects on resistor properties. Typically the volume of air entering the furnace per minute should be approximately equal to the total volume of the furnace muffle, with the air flow countercurrent to the direction of substrate movement. Higher production rates require higher air flows in order to keep burnout products out of the peak firing zone. Good ventilation and exhaust during the burnout phase is therefore essential. The amount of air required for burnoff of the organics in the burnout zone of an air firing furnace can be calculated from

$$V = PLAWS \tag{6}$$

where V is the volume of airflow required, P is ratio of printed paste to total substrate area, L is the ratio of substrate area to furnace belt area, A is the constant representing the burnout requirements of the thick film paste, W is the furnace belt width, and S is the furnace belt speed.

Nitrogen firing of base metal thick-film pastes requires complete burn off of organics in the burnout zone or else chemical unsaturation, cross-linking, and eventual carbonization of the resin will occur. This will lead to entrapment of carbonaceous residues in the film causing reducing conditions later in the firing process that can lead to blistering and shorting problems [23,24]. Complete burnout demands adequate nitrogen flows into and out of the furnace. Typically, half the flow of nitrogen should enter and exit the burnout zone. Good distribution of gas by use of sparge pipes is also essential to ensure that each part is continually subjected to fresh gas. The burnoff process is often assisted through doping of the burnout zone nitrogen with up to 100 ppm of oxygen. However, since copper

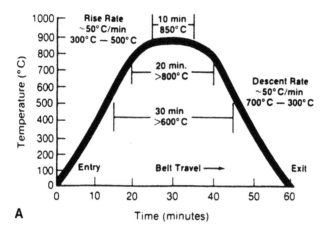

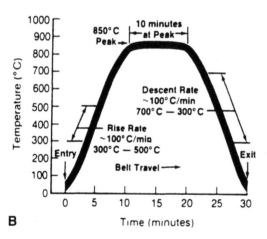

Figure 7   Air firing furnace profiles.

oxidizes readily, the high-temperature firing zone still has to be maintained at
<5 ppm oxygen. It is also important that the exit temperature be as low as possible
so that oxidation is prevented as parts move from the protection of nitrogen to
air.

As the temperature rises, the microstructure begins to develop. At temperatures between about 450–700°C, glass begins to soften and flow, metal particles
begin to sinter, and a film is formed. At this point adhesion between the thick
film and the substrate begins to develop. The material is held at the peak firing

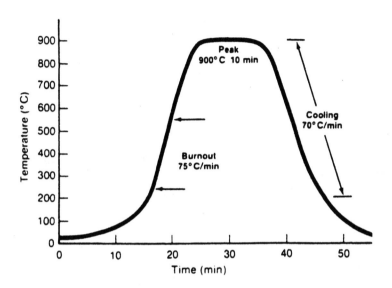

Figure 8 Nitrogen firing furnace profile.

temperature for approximately 10 min to fully develop the microstructure and properties of the film. On cooling, any difference in thermal expansion coefficients of the various thick-film components will begin to assert itself from approximately 300–400°C and down, leading to strains built into the films.

IV. SUBSTRATES

Substrates provide the mechanical base and electrical insulating material on which thick-film materials are fabricated. They may function as a simple passive carrier providing strength as in many hybrid microelectronic applications, or may be a key active component of the circuit as in silicon solar cells. All thick-film substrates should have the ability to withstand high temperatures and have high electrical resistivity, mechanical strength, dielectric breakdown voltage, and thermal shock resistance. In addition to these general requirements, other important properties that depend on the application include thermal conductivity, thermal expansion, surface smoothness, dielectric constant, and dissipation factor. Table 2 lists key physical properties of various substrates.

Ceramic substrates are the preferred substrate for most thick-film applications due to dimension stability and inertness at typical thick-film firing temperatures. Ceramics have very high electrical resistivities in the order of 10^{13} Ω-cm

Table 2 Physical Properties of Selected Substrate Materials

Property	96% Al_2O_3	99.5% Al_2O_3	99.5% BeO	AlN	SiC
Coefficient of thermal expansion ppm/°C, 25–300°C	6.4–7.2	6.6	7.2–8.0	3.8–4.4	3.7–3.8
Thermal conductivity W/m-K, 25°C	20–26	29–37	260–290	140–260	70–270
Dielectric constant 1 MHz, 25°C	8.9–10.2	9.7–10.5	6.5–7.0	8.0–9.2	40–42
Dissipation factor (%) 1 MHz, 25°C	0.03–0.1	0.01	0.02–0.04	0.03–0.1	5
Dielectric strength kv/mm, 25°C	14–24	15–36	10–43	14–27	1.0
Volume resistivity Ohm-cm, 25°C	$>10^{14}$	$>10^{14}$	$>10^{14}$	$>10^{14}$	$>10^{14}$
Flexural strength MPa, 25°C	200–400	420–500	170–270	275–500	440

and dielectric breakdown voltages in excess of 500 V/mil, making them ideal for high-voltage circuitry. However, ceramics, are inherently brittle materials that fracture without plastic deformation, giving rise to occasional breakage during handling. Still, they are two to five times stronger than porcelain or glass. The ability of a substrate to withstand thermal shock is determined by the rate of temperature change during the shock process [25], its temperature coefficient of expansion, and its elastic modulus. A coefficient of thermal endurance, F, can be used to compare materials [26] and is given by

$$F = \frac{P \ k}{\alpha E \ \rho c} \tag{7}$$

where P is tensile strength, α is the temperature coefficient of expansion, k is thermal conductivity, E is Young's modulus, ρ is density, and c is specific heat. Using this relationship, the relative shock resistance of four materials is shown in Table 3, together with their coefficients of expansion. Fused silica has a very high thermal shock resistance due to its very low thermal coefficient of expansion.

The frequent need to conduct heat away from transistor chips makes thermal conductivity an important design characteristic of substrates. High thermal conductivity is only realized in highly crystalline materials. Glass is a poor thermal conductor, which is one reason it has not found wide acceptance as a substrate.

Table 3 Relative Shock Resistance of Substrate Materials

Material	Thermal endurance factor, F	Coeff. of thermal expansion ppm/°C, 25–300°C
Fused silica	13.0	0.56
Alumina	3.7	6.8
Beryllia	3.0	7.8
Glass	0.9	9.0

When high thermal conductivity is required, beryllia or aluminum nitride may be chosen. Thermal expansion properties become important when operation at extreme temperatures or when components run hot in a part of the circuit. In such cases, stresses can be generated in the heating and cooling cycle if the thermal coeffients of expansion of the different elements of the circuit are different. In extreme cases, components can break or "pop" off the substrate.

A relatively rough substrate surface aids in paste transfer in the printing process, allows for better adhesion of the thick film, and generally costs less. However, high-frequency circuits or capacitor components benefit from smoother surfaces. Camber (a measure of flatness) becomes more important as the overall size of the substrate increases. The dielectric constant determines the capacitance associated with different film elements on the substrate, e.g., the capacitance coupling between conductor runs on the same side of the substrate. Use of both sides of the substrate can produce capacitors. For most circuits this is not an issue, but in high-frequency applications, dielectric constant becomes critical because it affects cross-talk and speed of operation. Dissipation factor is a measure of the electrical loss characteristics of the material and becomes important at higher frequency, particularly at microwave levels where very low loss dielectrics become desirable.

Alumina has become the most widely used ceramic substrate for thick-film applications because it combines electrical, mechanical, and economical advantages. It has a high elastic modulus and strength, giving it the highest fracture strength among the refractory oxides. Most substrates are made from a tape process. Alumina powder, glass powder, and other additives are mixed with a polymer solution or emulsion, generally acrylic, and cast into a tape from which the "green" substrates can be cut. These are then fired at approximately 1400°C to sinter the alumina powder. Most circuits use 96% alumina substrates, and the majority of thick-film pastes have been optimized for this chemistry. The 4% additions are designed to provide optimal electrical and mechanical properties, while maximizing densification without grain growth of alumina crystals at the sintering temperature. Typically, these additives are magnesia and silica. Magnesia is added to control the grain growth. It acts by segregating to grain boundaries of the alumina and preventing them from moving. Silica is added to promote densification at the sintering temperature. It does this by forming a glassy matrix with the alumina, known as mullite, which coalesces and binds the alumina crystals together. The sintered microstructure is composed of close-packed grains of crystalline alumina, separated by a small amount of glass with occasional voids or porosity. If ultralow-loss alumina is required for high-frequency applications, the glass component is eliminated. However, 99.5% alumina purity substrates, have to be sintered at temperature between 1800°C and 1900°C for adequate density. The firing of thick-film materials on alumina is generally accompanied by some degree of chemical interaction between the substrate and the glass contained in the paste. This may or may not be a critical element of the property development but is often the reason why pastes designed for firing on alumina do not perform equally as well on other substrates. For example, Figure 9 [27] shows the amount of 96% alumina substrate dissolved into the lead borosilicate glass of a thick-film resistor as a function of time at three different firing temperatures. A typical firing time for thick-film resistors is 10 min at peak, which means as much as 14% of the fired resistor is made up of alumina from the substrate.

Berrylia has a thermal conductivity of 260 W/m.K at room temperature, an order of magnitude higher than alumina and approximately half that of copper.

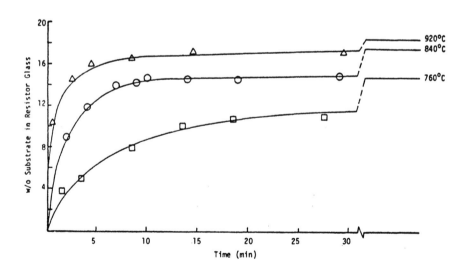

Figure 9   Dissolution of 96% Al_2O_3 substrate in fired resistors. (Reproduced from Ceramic Materials for Electronics, Materials Aspects of Thick-Film Technology, p. 445.)

The combination of high strength and thermal conductivity give berryllia a good thermal shock resistance. It is primarily used for applications that require rapid significant heat transfer through the substrate. The coefficient of thermal expansion is slightly higher than alumina and its dielectric constant is slightly lower. The disadvantages with berryllia are the high cost and potential toxicity problems associated with its use.

Aluminum nitride has excellent thermal conductivity and does not have any of the toxicity issues of beryllia. The coefficient of thermal expansion of aluminum nitride is between 3.8 and 4.4 ppm/K, close to that of silicon and gallium arsenide, making it very suitable for direct attachment of large dies. It has good thermal shock resistance and a high flexural strength, exceeding that of alumina and beryllia. It can be easily machined or scribed due to its low hardness and its chemical inertness. Substrates are manufactured by sintering submicron aluminum nitride powders in nitrogen at 1900°C. Sintering additives such as yttrium oxide (Y_2O_3), yttrium fluoride (YF_3) [28], or calcium carbide (CaC_2) [29], enhance sintering from formation of liquid phases. Carbon from the calcium carbide is also used to reduce the oxygen impurity level. Single-crystal oxygen-free aluminum nitride has a thermal conductivity of 320 W/m.K [30,31] but the thermal conductivity depends on its oxygen content [30,32]. Commercial substrates vary in thermal conductivity from about 140 to 230 W/m.K. Aluminum nitride can be fired in reducing atmospheres to 1600°C and can be fired in air to 900°C, forming only a very thin aluminum oxide layer. The process of oxidation is slow, as the oxide formed creates a protective barrier to further oxidation. If this oxide layer becomes too thick it acts as a thermal barrier. A thick oxide layer can be produced by choice of the wrong thick-film material. Aluminum nitride is reactive to ruthenium dioxide and some of the conventional glass binders used in thick-film compositions. This can lead to chemical reduction and formation of aluminum oxide at the interface during the firing process such as

$$3RuO_2 + 4AlN \rightarrow 2Al_2O_3 + 3Ru + 2N_2 \qquad (8)$$

$$3PbO(glass) + 2AlN \rightarrow Al_2O_3 + 3Pb + N_2 \qquad (9)$$

Reduction of the glass will also cause blistering and poor long-term adhesion. Compositions for aluminum nitride are formulated with nonreducible binder systems that form reactive bonds with the AlN substrate without formation of oxide layers [33–35].

Stainless steel substrates are strong, vibration shock resistant, mechanically rugged, and have good thermal properties, potentially making them ideal for harsh environments. The metal can be machined and punched to produce holes or mounting brackets before the insulating layer is applied. This can be accomplished by spraying paint containing glass powder onto the metal, laminating a dielectric tape to the surface, or printing a thick-film dielectric on the metal followed by firing. The substrates have excellent electromagnetic and electrostatic shielding properties and can provide a built-in ground plane and heatsink. Early systems based on porcelain on steel (POS) suffered from low processing temperatures as the glazes readily flowed at temperatures of 650°C, making them unsuitable for standard thick-film materials. Thick-film on 430 stainless steel overcomes this issue, allowing the use of standard thick-film firing profiles [36,37]. Except for electric kettles and some automotive applications, however, thick film on stainless steel has seen limited adoption in microelectronics. This is partly due to the difficulty in making multi-up circuits followed by separation into individual units, making them less economically attractive.

Glass substrates are used in two major applications: flat panel plasma displays and automotive rear window heaters and antennas. Soda-lime glasses are mostly used for these applications, and have excellent surface finishes, low warpage, camber, waviness, and optical transparency. However, processing temperatures, are low, between 450°C and 600°C. Its low-impact fracture resistance and poor thermal properties have limited its applications in areas where optical transparency is not a necessity.

Silicon substrates used in solar cell manufacture can be single-crystal, polycrystalline, or amorphous silicon. Single and polycrystalline silicon substrates or wafers are made from silicon ingots or blocks. These are made from molten silicon, either using a pulling technique or casting into a mold. These are then cut to form the thin wafers. Amorphous silicon substrates are made by deposition of a thin film of silicon on indium-tin oxide (ITO)–coated glass using plasma chemical vapor deposition. Polycrystalline substrates make up the vast majority of substrates used in making thick-film solar cells.

Silicon carbide, steatite, forsterite, titanate and other substrates have been used for some very limited and specific applications. Silicon carbide has a low thermal expansion coefficient and a relatively high thermal conductivity, similar in key features to aluminum nitride. It is doped with oxides such as boron oxide (B_2O_3) or beryllia (BeO), which forms a grain boundary phase and converts silicon carbide from a semiconductor to an insulator. The thermal conductivity given in Table 2 is for BeO-doped silicon carbide. Dielectric constants and dissipation factors of steatite and fosterite are relatively independent of temperature at high frequencies, useful for insulation rather than substrates. Titanate substrates can have dielectric constants between 15 and 15,000 and can be designed as an integral part of the circuit. High dielectric constant chemistry is based on barium titanate doped with materials such as niobium pentoxide (Nb_2O_5) and zirconia (ZrO_2). These position the Curie point around room temperature and control temperature coefficients of capacitance. Glass sintering additives enhance sintering at about 1400°C.

V. THICK-FILM PASTE CONSTITUENTS

Thick-film pastes consist of an active phase, an adherent or binder phase and an organic vehicle. The active phase gives the fired paste its essential electrical properties and can be metal, glass, glass-ceramic, ceramic, conducting and/or semiconducting oxide. The adherent phase or binder is necessary for adhesion to the substrate and can influence the electrical properties of the fired film. It generally consists of a glass frit and/or crystalline oxide powders. The organic vehicle contains a resin dissolved in a volatile solvent. It acts as a carrier for the inorganic phases so that the composition can be printed. In special cases, the organic vehicle may be photosensitive as in the case of photoimageable compositions for fine-line or small via resolution.

Metal powders and conducting oxides used in conductors and resistors are generally on the order of 1 μm in diameter. They are mostly prepared by reduction of metal salts in aqueous solutions, typically nitrates and chlorides. Additives are used to control the rates of nucleation and growth, flocculation, and state of agglomeration of the precipitated powders. By such means, powders with specific sizes and morphology can be produced. Fine spherical metal powders can also be prepared by spray pyrolysis [38], and metal powder flake morphologies can be produced by mechanical deformation of precipitated powders by ball milling or attritor milling.

Glass powders are prepared by batching the appropriate mix of oxides and melting the composition in platinum or alumina crucibles. A homogeneous glass melt is generally achieved at temperatures between 1200°C and 1500°C. The melt is quenched (fritted) into water or poured on to steel rolls to form highly strained pellets or ribbons of glass that can be easily milled to a desired size range, commonly in the order of 1 μm. This is usually accomplished by ball milling for initial size reduction followed by vibratory milling to achieve the final particle size. There are certain general properties that these glasses should possess. These include high values in thermal shock resistance, abrasion resistance, mechanical strength, dielectric strength, and environmental inertness as well as low dielectric loss. In addition, depending on the application, they should possess appropriate values in thermal expansion, viscosity–temperature relationship, surface tension–temperature relationship, and reactivity to the substrate and other inorganic components of the system. Conductor glasses are often high bismuth-borosilicates modified with Al_2O_3, PbO, ZnO_2, and/or BaO. Dielectric glasses and glass-ceramics are generally based on Al_2O_3, MgO, BaO, CaO, and SiO_2 compositions. Resistor glasses are often high-lead systems.

The vehicle is a key element in dispersing the thick-film inorganic phases and developing suitable printing behavior. It also determines drying rate, as well as the green strength of the dried film and the bond of the film to the substrate.

The organic vehicle system typically consists of high molecular weight polymers dissolved in a mixture of low-vapor pressure solvents. Common high molecular weight polymers used are ethyl cellulose modified by rosins to impart green strength. Suitable low vapor pressure solvents are terpineol, butyl carbitol, and butyl carbitol acetate. The low vapor pressure solvent is required so that the viscosity of the paste does not change while spread out on the screen, giving the system a long screen life. Dispersing agents and rheological additives may be added in small quantities to adjust viscosity, fine-feature printability, and shelf life. In addition to being a carrier for the inorganic phases, the organic components have to be completely removed during firing without interfering with the chemistry of the film. This requirement is often quite difficult, as will be discussed in connection with nitrogen firing systems.

Vehicles used in photoimageable pastes incorporate a photoinitiator, which is sensitive to UV light, in the thick-film paste [7,8]. An oversized area of paste is screen printed onto the substrate, generally through a relatively coarse mesh such as 120- to 200-mesh screen. After leveling, the wet film is dried at about 80°C so as not to deactivate the photosensivity associated with higher drying temperatures. The dried film is exposed with a suitable phototool and an Hg or Hg/Xe UV light source. The photoinitiator cross-links the polymer in the exposed areas, making it resistant to washing solvents. It is then developed (washed) to reveal the features exposed and then fired. Photoimageable pastes can resolve 40-μm-wide conductor tracks and 50-μm-diameter vias in dielectrics [9].

The thick-film paste must have high viscosity to maintain particles in suspension. Blending of the composition is carried out on a three-roll mill until a reproducible degree of dispersion has been attained. Care is taken when the paste contains a ductile metal, as the powder can be converted to large flakes if the pressure on the rolls is excessive. Another function of the roll mill is often to break up and disperse agglomerates of particles. This is evaluated by using a fineness-of-grind gauge, and the roll milling is continued until the agglomerate size is below some predetermined limit, e.g., 10 μm.

VI. CONDUCTORS

Thick-film conductors are used primarily to transmit signals from one circuit location to another and as pads for attachment of dies, components, and wirebonds. Other uses include low-value resistors, capacitor electrodes, shielding, heatsinks, heating elements, sealing package lids, and electroplating bases. The general requirements for thick-film conductors are low electrical resistivity and good adhesion to the substrate. The choice of a suitable conductor depends on several factors, which may include some or all of the following:

Resistivity
Solderability
Solder leach resistance
Wire bondability
Migration resistance
Thermal aged adhesion
Thermal cycled adhesion
Line resolution
Compatibility with resistors and dielectrics
Cost

No single conductor will have all properties. Therefore, several conductor metallurgies exist. Elements that are used in thick-film conductors are given in Table 4 in two groups—the noble metals and the base metals. Also given in Table 4 are the sheet resistivities normalized to 25 μm, melting points, and thermal expansion coefficients. The sheet resistivity is the resistance of a square film at a defined thickness. In Table 4, the defined thickness is 25 μm, which is the standard normally used. In practice however, conductors are printed thinner than this, typically firing out in the 10- to 12-μm range. The sheet resistivity is a materials property, independent of film geometry. The effect of sheet resistivity, number and size of squares, and aspect ratios on resistance is discussed in more detail in the section on resistors. The values in Table 4 are the sheet resistivities of the pure element. In practice, conductors do not achieve these values due to dilution by the glass binder and porosity.

Table 4 Elements Used in Thick-Film Conductors

Elements	Resistivity (μΩ-cm at 25°C)	Sheet resistivity (mΩ/□/25 μm)	Melting point (°C)	Thermal expansion (ppm/°C)
Ag	1.6	0.64	961	19.7
Au	2.3	0.92	1063	14.2
Pt	10.5	4.2	1769	9.0
Pd	10.8	4.3	1552	11.7
Cu	1.7	0.68	1083	16.5
Al	2.67	1.07	660	22.4
Mo	5.2	2.04	2610	5.1
W	5.6	2.24	3410	4.6
Ni	6.8	2.72	1453	13.3

A. Noble Metal Conductors

From Table 4, pure silver would be an obvious choice among the noble metals for a conductor because of its low resistivity and cost. Silver conductors require very little glass addition to sinter to full density. Therefore, they realize most of the benefits of silver's conductivity. Silver is used as the buried conductor in multilayer hybrids, in plasma display panels, in defoggers, and in solar cells. However, for surface conductors, its use is somewhat limited due to either its poor leach resistance or fear of migration across the surface of the substrate in the presence of an electric field and a humid atmosphere, or both. Shorting by silver migration can be avoided by eliminating large voltage differences between adjacent conductors in the circuit design. If this is not feasible, encapsulation of the silver with a high-temperature glass encapsulant protects silver from migration [39,40]. Silver has excellent solderability but dissolves very rapidly in molten solder, i.e., it has very poor leach resistance. The addition of 1% platinum (Pt) improves the leach resistance as effectively as additions of 10% palladium (Pd) [24] and has minimal effect on conductivity, as additional glass to aid sintering is not needed. Such low-glass, high-silver compositions have been found to hold up well in thermal cycling situations and have found use in automotive applications [41]. Small additions of platinum, however, do not alleviate the tendency for silver to migrate. Encapsulation is still necessary.

Silver and palladium show complete solid solubility, and the addition of palladium minimizes silver migration and improves solder leach resistance. At levels around 33% palladium, migration is effectively stopped. As a result, silver-palladium conductors and compositions with silver-palladium ratios of 2.5:1 and 3:1 have been commonly used in the manufacture of thick-film hybrids. As shown in Figure 10, however, the resistance of silver increases substantially with additions of palladium. However, good solder leach resistance can be obtained with solder containing silver. Therefore, silver palladium ratios of around 6:1 are routinely used—a compromise on silver migration, leach resistance, and conductivity.

Palladium undergoes an oxidation reaction during the firing process. The formation of palladium oxide from its elements is given by

$$Pd(s) + \frac{1}{2}O_2(g) = PdO(s) \tag{10}$$

then

$$P_{O_2}^{-1/2} = \exp\left(\frac{-\Delta G_f^{\circ}}{RT}\right) \tag{11}$$

where R is the gas constant, T is the absolute temperature, and ΔG_f° is the standard free energy of formation of 1 mol of PdO from its elements, which is given as

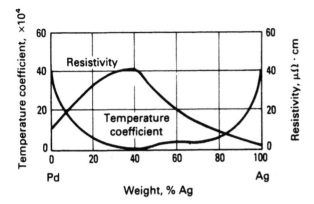

Figure 10 Silver-palladium phase diagram. Courtesy of DuPont literature.

a function of temperature in standard compilations of thermodynamic data [42]. If $Po_2 = 0.21$ atm (air) is substituted into Eq. (11), the temperature is calculated to be 802°C. This means that palladium metal is the thermodynamically stable phase in air above 802°C and that PdO is the stable phase at lower temperatures. During cooling from temperatures above 802°C, PdO will form on the surface of the metal. This calculation is strictly valid only for pure palladium. In combination with silver, this oxidation temperature is lowered, thus lowering the kinetics of oxide formation. The amount of PdO formed also depends on cooling rates. Faster rates reduce the amount of PdO formed. The presence of PdO adversely affects solderability and inhibits solid solution formation [43]. High palladium compositions, therefore, have a tendency to less than perfect solderability.

Gold-based conductors are used for applications where diffusion of silver from the conductor terminations into resistors has to be avoided or there is a need for reliable wire bonds. Gold (Au) conductors have high conductivity, high migration resistance, are readily compatible with many thick-film materials (especially resistors), and can be formulated for excellent wire bondability. Compositions can be mixed bonded (glass and oxides) or pure oxide bonded. High conductivity and migration resistance is an inherent property of gold. Optimal wire bondability is achieved by the development during firing of a high-density gold film that has the adherent phase concentrated at the substrate–conductor interface. Such gold conductors consist of mono-sized spherical gold powders, sometimes mixed with powders of flake morphology, with very small additions of oxides for adhesion. In the case of aluminum wire bonding, pure gold conductors suffer from strength degradation and resistance increase of the bond after aging at 150°C or higher [44]. This is shown in Figure 11 (curve a). During the high-temperature

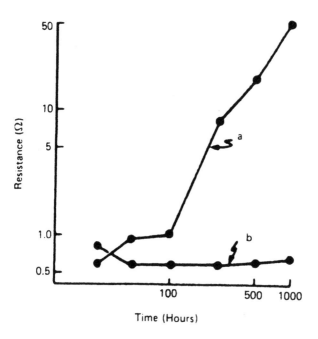

Figure 11   Resistance of aluminum wire bonds after storage at 150°C. (a) Without Pd and (b) with Pd.

aging, gold diffuses more rapidly into the aluminum than aluminum into gold-forming Al_2Au_5 and $AlAu_4$ intermetallic compounds. This leads to Kirkendall void formation within the gold metallization, thereby weakening the bonds. Alloying about 1% of palladium with the gold reduces the diffusion rate so that high bond strength and low resistance at the bond can be maintained (Figure 11, curve b) [44], making gold-palladium conductors the choice for aluminum wire bonding. For gold wires bonded to gold metallizations, the Kirkendall effect is not an issue, and bond strength remains stable with temperature. Since gold conductors are expensive, a common practice is to join the gold conductor to silver-palladium conductors a short distance from where the wire bonds are made. Gold conductors are also very suitable for resistor terminations, die attach pads, and highly reliable electrical interconnections. However, pure gold conductors are not readily solderable using conventional solders, due to poor solder leach resistance. Solder leach resistance can be significantly improved by the addition of substantial amounts of platinum or palladium at the expense of conductivity. Platinum-gold conductors are the best overall for reliability, but their cost tends to restrict their use to cost-insensitive applications or situations where their improved performance makes them essential. For example, some circuits that contain very critical thick-film resistors use platinum-gold conductors to terminate the resistors, but like the case of wire bondable gold conductors, a short distance away from the resistor, the platinum-gold conductor joins to a silver-palladium conductor to complete the circuit.

B. Base Metal Conductors

Copper has very low resistivity, is easily solderable, has excellent leach and migration resistance, and is inherently a low-cost metal, making it an ideal conductor candidate. However, to avoid oxidation, copper conductors need to be fired in low-P_{O_2} atmospheres. The atmosphere requirements for firing copper conductors can be calculated from

$$\log P_{O_2} = \frac{\Delta G^\circ}{2.3RT} \tag{12}$$

where ΔG° is the standard free energy change for conversion of copper to the lowest oxide of copper (Cu_2O), R is the gas constant, and P_{O_2} is the oxygen partial pressure at which copper metal and Cu_2O will coexist at temperature T. This can be done for any metal–metal oxide system, and plotting Eq. (12) will create a phase stability diagram. This is shown in Figure 12 for systems of interest in thick-film technology [42]. The metal is the stable phase in the T-P_{O_2} region below the line, and the oxide is the stable phase above the line. The lines of Figure 12 are strictly valid only for the equlibrium of a pure metal and a pure oxide. When the oxide is a constituent of a glass, its line on Figure 12 will be shifted downward. It is also necessary to recognize the importance of the position of the carbon curve (the C-CO-CO_2 equilibrium for 1 atm total pressure of CO + CO_2) on Figure 12 because the organic vehicle of the pastes must be removed, most commonly by oxidation. This means that thick-film firing should be carried out in the T-P_{O_2} region above the carbon curve. As can be seen in Figure 12, there is a processing window between the Cu line and the C curve where carbon will be oxidized and copper will be thermodynamically stable. However, commercial copper conductors are typically not processed in this window but rather at 900°C in an atmosphere containing 1–10 ppm O_2, which is in the Cu_2O phase field. Due to the oxidation kinetics of copper, however, metallic films of copper are produced under these conditions. In an atmosphere containing 1–10 ppm O_2 in N_2, the oxidation of copper occurs very slowly and the diffusion of oxygen into copper metal at 900°C occurs very rapidly. The oxygen that diffuses into the copper at the firing temperature precipitates as Cu_2O if it exceeds the solubility limit of 0.0016 wt % at 900°C [45]. The result is a shiny copper film even though

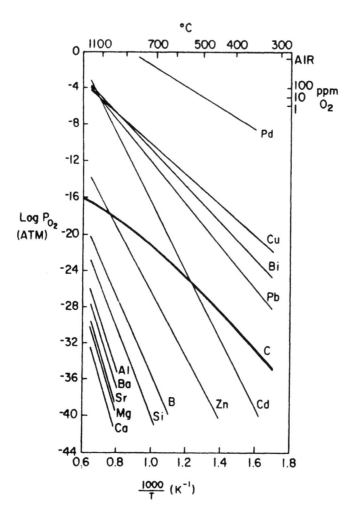

Figure 12 Phase stability diagram for selected elements of interest in thick-film technology.

the processing was under conditions where Cu_2O was the thermodynamically stable phase.

Firing copper films in low-Po_2 atmospheres, especially in heavily loaded furnaces, can lead to strongly reducing conditions in the high-temperature zone of the furnace, due to insufficient removal of organics in the burnout zone. This can lead to reduction of oxides to their metallic state whose lines in Figure 12 are above the carbon curve at the firing temperature. It is very difficult to remove organics from copper films in low-P_{O_2} atmospheres. Any temperature along the carbon curve gives a sufficiently low P_{O_2} to reduce PbO, Cu_2O, and Bi_2O_3, for example, even if they are constituents of glass rather than individual oxides. Once reduced, the kinetics of oxidation of are sufficiently slow in low partial pressures of oxygen that they are never converted back to their oxides, even if the organics are completely removed during normal processing. Low-P_{O_2} atmosphere firing of copper can produce low adhesion of the copper film due to a lack of copper oxide [46] and in the case of bismuth, the metal can evaporate and subsequently condense on cooler parts of the furnace. Metallic bismuth has been shown to be a contributory factor in opens in conductor traces and shorts in the dielectric [23,24,47]. Glass modifiers whose constituents are below the carbon curve eliminate this issue.

Other base metal conductors include aluminum paste, which is used to metallize the back face of solar cells. Aluminum is very reducing, allowing the metal to penetrate the silica on the surface of a silicon wafer and make contact with the silicon. Compositions are fired in air at 850–900°C to form an aluminum-silicon (Al-Si) eutectic alloy. This gives rise to an electric field known as a back surface field in the cell due to a highly aluminum-doped back surface of the silicon producing a p^{++} layer. The field at the back enhances photocurrent collection in conjunction with the larger field present at the p-n junction. Aluminum also enhances the penetration of previously deposited hydrogen into the silicon. Hydrogen can diffuse to a greater depth into the silicon with aluminum on the back than into uncoated silicon. The hydrogen passivates crystalline defects in the silicon, such as grain boundaries in multicrystalline silicon. Both these effects enhance the cell's efficiency. Aluminum pastes can be fired in air because of aluminum's self-limiting oxidation characteristics and the fact that it melts at 660°C. When the aluminum powder melts, it forms a molten film and its exposed surface undergoes oxidation. However, the volume of oxide formed is the same as that of the metal from which it was formed. Hence, a thin dense oxide film is formed that is stress free. As a result, it does not crack, preventing continued oxidation. The cooled film consists of an aluminum film coated with a very thin layer of alumina. Electrical contact is made either from mechanical means, such as clips, or solder attachment on silver pads printed directly onto the silicon through "windows" in the aluminum paste and cofired with the aluminum.

C. Densification and Adhesion

Sintering of metal powder is surface energy driven. Finer metal particles sinter more rapidly and to a higher final density than coarser particles of the same metallurgy. This is in agreement with sintering models that have solid-state diffusion as the rate-limiting step. However, even films made with fine powders exhibit some porosity in the fired state. Equivalent-sized particles of different metallurgy sinter at different rates and to different final densities. The final density is also influenced by the green-state packing density; the higher the green density, the higher the fired density. Silver conductors containing only silver spherical particles have porosity levels at the printed and dried stage of about 50% [48]. This is close to that expected for simple cubic packing of spheres (0.476). Addition of large agglomerates, fine irregular glass particles, or higher organic content to the paste tends to lower the green density.

Solid-state diffusion begins at relatively low temperatures by "neck" formation between adjacent metal particles. Pure gold and pure silver conductors can reach a high sintered density by solid-state diffusion alone. However, the thick-film firing process is too short for most of the alloy conductors whose melting points are higher that that of silver or gold. In such cases, solid-state diffusion is augmented by additions of glass to promote reactive liquid-phase sintering. The glass aids densification and fills any voids that may be left after sintering. More glass is typically needed with higher melting point compositions. This leads to significantly higher resistivity than the bulk metal of the same composition. This is illustrated in Figure 13 for silver-palladium alloys. Near peak firing temperatures, metal oxides can dissolve in the liquid glass phase and rapidly transport themselves. This accelerates the sintering process, allowing higher density to be achieved. This is why silver-palladium and platinum-gold alloys typically contain glass as a component of their bonding chemistry and why pure silver and gold compositions contain very little, if any.

The content and chemistry of the thick-film conductor binder determines the adhesion of the conductor to the substrate and also can affect other conductor

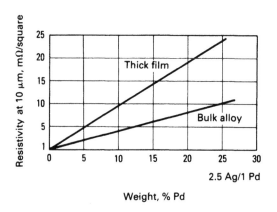

Figure 13 Effect of binder content on silver-palladium alloys. Courtesy of DuPont literature.

properties such as solderability and resistivity. Adhesion of a conductor to the substrate can be mechanical or chemical or a combination of both. Three primary methods are used to achieve adhesion. These are reactive bonding, flux bonding, and frit bonding. Reactive-bonded conductors (often called oxide-bonded conductors) on alumina substrates typically have small amounts (1%) of a reactive oxide such as CuO included in the composition. During firing, these oxides react with the alumina substrate to form compounds such as copper aluminate ($CuAlO_2$) or MAl_2O_4-type spinel phases to provide adhesion to the metal film [49–51]. An advantage of reactive bonding is the small amount of additive needed, which means that the electrical resistivity of the conductor is kept as low as possible and the surface is nearly pure metal, which enhances solderability and wire bonding. However, reactive bonding is only feasible in compositions that sinter to high density by solid-state diffusion alone, such as pure silver and gold compositions. Flux bonding involves the addition of 1–5% of an oxide that forms a liquid phase with the substrate at the firing temperature. Bismuth oxide is commonly used for films on alumina substrates because the Bi_2O_3-Al_2O_3 eutectic temperature is 820°C, and a transient liquid is formed when the conductor is fired above this temperature. Since this liquid wets both substrate and metal, good adhesion can be achieved.

The third most common method of achieving adhesion is frit bonding, which is a pure mechanical bonding. It involves the addition of 2–10% glass powder (often modified borosilicates) relative to metal in the conductor composition. During firing, the glass wets the substrate and partially penetrates the metal network, filling any voids without coating the surface of the conductor. To achieve this, it is necessary for the glass to have the proper surface tension and viscosity during the firing operation relative to the surface energies of the metal and substrate and the interfacial energies between glass and substrate and metal and glass. Frit-bonded conductors tend to be more sensitive to firing conditions than chemically bonded systems. For example, firing a pure frit-bonded conductor at too high a temperature or for too long a time can result in a continuous film of glass between the metal and the substrate. This occurs because the glass usually has a smaller contact angle on the substrate than on the metal. Such a structure can crack at lower mechanical stresses than when the glass partially penetrates the metal network, and is therefore undesirable.

Binder chemistry and content can also have a strong influence on solderability. For example, too much glass in the conductor composition can completely cover the surface of the conductor with glass. In the case of a chemically stable glass, poor solderability would result. If bismuth oxide is used, however, molten tin in the solder will reduce the surface bismuth oxide to bismuth metal, and solderability will be maintained. When used effectively, bismuth oxide can reduce the effects of high palladium content, different substrates, high glass content, and firing conditions on solderability. In practice, many commercial thick-film conductors for alumina substrates are formulated as "mixed bonded" systems and may contain glass frit, bismuth oxide, and oxides such as copper oxide. They represent an optimal approach to maximize adhesion and solderability while minimizing sensitivity to firing.

VII. DIELECTRICS

Thick-film dielectrics are used in three primary applications: encapsulants that protect circuitry from hostile environments, cross-over and multilayer insulation, and capacitors. There are some general properties that dielectrics used in any of these applications should possess. These include high electrical resistivity, breakdown voltage and mechanical strength, environmental stability, and compatibility (no detrimental chemical reaction) with other circuit elements with which it will come in contact. In addition, the printed dielectric should be free of pinholes that traverse the entire thickness of the film. These can lead to electrical shorts via environmental degradation. This requirement is sometimes satisfied by the application of a minimum of two prints for the dielectric.

An encapsulant is a protective film applied after all other components of the circuit have been printed and fired. It can also be used for solder dams to prevent wetting of certain conductor tracks by solder. Organic and inorganic encapsulants are available, but only glass systems give complete protection against the elements. Glass encapsulants were originally designed to protect early silver-palladium resistors, which underwent large resistance changes when exposed to humidity or high temperatures. As such, encapsulants were generally designed to fire at around 500–550°C so as not to affect the properties of the underlying resistor. The move to more stable ruthenium-based resistors reduced resistor encapsulation needs so that encapsulants are now primarily used for enhancing reliability of the entire thick-film hybrid. Typically applied as a single print, their key requirement is to thoroughly wet the underlying circuitry and fire to a pinhole-free dense film.

A cross-over insulator is a thick-film dielectric pad that electrically isolates two conductor tracks one of which crosses over the other. The dielectric pad is printed and fired on top of the first conductor track followed by printing and firing the second conductor. In addition to the general properties required, cross-over dielectrics must have a low dielectric constant and low dielectric loss to minimize coupling between the conductors below and above the cross-over. They must also be capable of being refired to allow the top conductor to be applied. Multilayer dielectrics are used to isolate complete layers of conductive tracks. A multilayer will cover most, if not all, of the previous conductor. Electrical connection between conductors on different layers is achieved by the use of vias. Dielectrics used for multilayers must satisfy all the requirements for cross-overs and in addition must provide a smooth surface for printing the next layer, and have good via resolution. The most critical requirement is its coefficient of thermal expansion (CTE) should be equal to or very slightly less than the substrate on which it will be applied. This ensures that the dielectric never comes under tensile stresses. Designing the CTE to be slightly less than that of the substrate maintains the dielectric under slight compression, which produces maximal strength.

Thick-film capacitors are produced by sequentially printing and firing a bottom electrode, the capacitor dielectric, and a top electrode. In addition to the general dielectric requirements, capacitor dielectrics should have low dielectric loss, high dielectric constant and strength, and be capable of being refired so that the top electrode conductor can be applied. Thick-film capacitors do not have the volumetric efficiency of discrete multilayer chip capacitors making them unusable in cases of high surface component density where high capacitance is needed. They are used in cases of relatively low-capacitance needs and where requirements demand low profiles or where soldered components are avoided.

A. Dielectric Chemistry

There are three common types of dielectric compositions used in thick-film applications: amorphous glass, ceramic-filled amorphous glass, and crystallizable glass (also called devitrifying glass). Amorphous glass compositions are generally used for encapsulants. Typically modified alumino borosilicates, they exhibit low viscosity at the firing temperature thereby achieving a thin pinhole-free glass film. In order to laser trim resistors through the glass, fired thickness is generally kept between 10 and 15 μm, although for protecting silver-bearing conductors a thicker film might be used.

Ceramic-filled amorphous glass insulation dielectrics were originally developed to have a good expansion match with the substrate and better thermal conductivity. Use of a dielectric with a CTE mismatch with the substrate can cause the substrate to bow. If the dielectric has a higher CTE than the substrate, severe tensile stresses can result in the dielectric leading to fracture when subjected to thermal or mechanical shock. Ceramic-filled glass dielectrics for insulation between conductors consist of an amorphous glass powder and a ceramic filler powder, typically alumina, silica, magnesia, or combinations thereof. Since solid-state sintering of the ceramic powder cannot take place at thick-film firing temperatures, a significant amount of glass must be used so that a continuous glass phase exists during firing to prevent formation of pinholes. Choice of the correct crystalline ceramic filler, however, allows latitude in designing composite CTEs and improving electrical and thermal properties over the matrix glass. The linear thermal expansion coefficient (α) of such a composite dielectric can be calculated from the Turner equation [52]:

$$\alpha = \frac{\sum_i \alpha_i B_i F_i / d_i}{\sum_i B_i F_i / d_i} \tag{13}$$

where α_i, B_i, F_i, and d_i are the coefficient, bulk modulus, weight fraction, and density of the ith component in the composite dielectric. Use of the Turner equation allows for the selection of ceramic phases and glass to design the CTE of the dielectric to be slightly less than that of the substrate. The ceramic phase retards the flow of glass during any subsequent refire, making these dielectrics suitable for multilayer systems. In some cases, an additional benefit can be gained by partial dissolution of the ceramic phase in the matrix glass at the firing temperature. On cooling, some of the filler that dissolves at the firing temperature will precipitate to give a fine glass-ceramic microstructure. The filler that remains dissolved converts the glass to a more refractory composition retarding flow of the matrix glass on any subsequent refire.

Crystallizable glasses were developed to allow refiring of the insulation dielectric without further softening of the glass. During the firing process, the glass softens and flows to produce a pinhole-free film and then undergoes a partial or complete crystallization in which a crystalline phase of higher melting point is precipitated. This strengthens the glass and eliminates flow during any subsequent firing. The nucleation and precipitation process gives rise to exothermic reactions, as shown in Figure 14, the rates of which are controlled by the glass composition to complete crystallization before reaching the peak temperature of the firing step. For example, barium aluminosilicate glass can be designed to precipitate

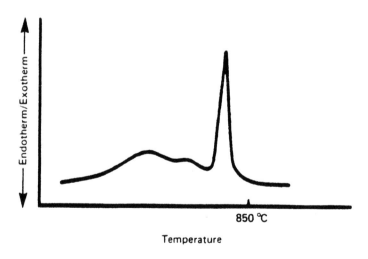

Figure 14   DTA scan of a crystallizable glass dielectric.

$BaAl_2Si_2O_8$ by 800°C [44]. This allows for cofiring conductors on top of the dielectric without danger of the conductor sinking into the dielectric. Conductors can also be fired on top of previously fired dielectric, if desired. The CTE of crystallizable glasses is determined by the amount of precipitated phase and the glass that remains. Crystallizable glasses remove some lattitude in designing for thermal and electrical properties and optimizing matched CTE, especially if the level of crystallization is high. One way around this is to design a hybrid system where only partial crystallization occurs and the residual glass can take ceramic filler [41]. In this way, the best properties of ceramic-filled glasses and crystallizable glasses can be combined.

Capacitor dielectrics with dielectric constant (ε_T) values between 10 and approximately 60, and with stable temperature characteristics are common. They are either based on crystallizable glasses where the precipitated crystalline phase is predominately barium-neodymium-titanate, or ceramic-filled glass where the ceramic phase is titania (TiO_2) or other stable relatively high-ε_T fillers. Glasses in general have low dielectric constants, and the dielectric mixing rule limits the dielectric constant values that can be achieved using glass-based approaches. The empirical logarithmic rule for the dielectric constant (ε_T) of a composite in terms of the volume fraction (V_i) and dielectric constants (ε_T) of the individual phases is given by

$$\log \varepsilon_T = \sum_i V_i \log \varepsilon_T \tag{14}$$

For a glass with a dielectric constant of 10 and a ceramic phase having a dielectric constant of 1000, Eq. (14) predicts a dielectric constant of only 159 with 60% ceramic phase and 40% glass. Since it has not proved possible to precipitate high volume fractions of high-ε_T phases in thick-film crystallizable glass dielectric compositions, most high-ε_T dielectrics are based on ceramic-filled glass. Such compositions can be made with doped barium titanate ($BaTiO_3$) fillers that achieve ε_T values of over 1000. Use of very-high-K relaxor fillers, such as lead-magnesium-niobate (PMN), can achieve ε_T values of up to 12,000 [53], but low glass binder levels are used to minimize the dilution of K. This can make the dielectric porous. For this reason, high-ε_T dielectrics are always encapsulated to protect them from the environment.

B. Dielectric Reliability

Thick-film dielectrics can electrically fail in service due to two basic causes and effects: the presence of interconnected porosity, cracks, or other void defects, with humidity producing shorts from electrochemical migration; or a dielectric that readily dissolves silver oxide, leading to shorts from solid-state migration. Both mechanisms are a direct or indirect consequence of the dielectric chemistry.

Electrochemical migration occurs when moisture condenses onto the surface of an interconnected pore or crack [54]. The moisture dissolves atmospheric CO_2, ionic contaminants, and very small quantities of the dielectric glass to form an electrically conductive electrolyte. Conductive pathways between electrodes are then formed in the porosity or crack resulting in a decreased dielectric insulation resistance. This allows a small current to flow supported by migration of the dissolved species. The current flow initiates release of metal ions at the anode and these are driven to the cathode by the electric field where they are electrodeposited. Eventually a filament of metal grows back through the porosity or crack to the anode, resulting in a short circuit.

Solid-state migration generally occurs through ionic migration combined with electronic conduction. The dielectric chemistry and the electrode metallurgy are key factors in this process. Ionic migration occurs most readily during firing of the dielectric when oxides from the electrode, such as Cu_2O or Ag_2O, can dissolve in the glass, altering the dielectric composition and its insulation resistance. The extent to which this occurs depends on the rate of dissolution of the oxides and their solubility limits in the dielectric glass. Repeat firings and higher peak firing temperatures increase the intensity of this reaction. Once oxides have been introduced into the glass, migration of the cations away from the electrodes occurs by means of exchange with other monovalent ions in the glass [54]. Ions of nearly equivalent sizes, such as Ag^+ at 0.13 nm (1.26 Å) and Na^+ at 0.1 nm (0.97 Å), exchange well due to low strain and polarization factors. At high temperatures, silver ions can migrate over the entire thickness of a dielectric film in a matter of minutes when the glass contains significant levels of sodium.

Electronic conduction of the dielectric depends on how easily the anode can release ions to the glass under an electric field and how easily the dielectric can transport them to the cathode. The release of ions depends on voltage, temperature, and metallurgy of the anode [55]. Silver can supply cations (Ag^+) quite readily at moderate electric fields and temperatures, copper and gold less readily. The electric field initially drives migration by ion exchange until the cation level in the glass can support electronic conduction. Once electronic conduction is initiated, electrons flow to the cathode at increasingly higher velocities and in greater numbers, resulting in more current flow. Ultimately, the number of electrons flowing and their velocities are such that the dielectric breaks down. Copper or gold anodes do not readily release ions into the glass. In this case, the electric field initially drives migration by ion exchange, but then a zone of dielectric next to the anode forms that is depleted of ions and the process stalls. Gold is used for high reliability applications due to this property.

Glass chemistry suitable for thick-film insulation dielectrics minimizes or eliminates those elements, such as boron, that increase the glass solubility in water. Additionally, monovalent ions such as Na^+, K^+ or Li^+ are avoided. Common glass constituents, such as lead and boron, often increase the solubility for the oxides of the electrode, and these are avoided. Incorporation of high levels of inert fillers, whether added or internally crystallized, also reduces the tendency of the electrode oxide to dissolve in the glass due to lower residual glass content. Finally, a fired dense dielectric is also important. Typical glasses that can meet the above criteria are often based on barium-aluminosilicate chemistry. Elements to modify properties, such as CTE, dissipation factor, and other electrical properties, are also added, but the base chemistry is more often than not the determining factor for the dielectric reliability.

Nitrogen firing of multilayer dielectrics presents some special challenges to the design of the dielectric. If the glass flows before all organic materials are removed, carbon becomes entrapped in the glass and reduces the dielectric insulation resistance [56]. Carbon entrapment can also lead to blistering [57] in the dielectric on refiring of subsequent layers. This is often due to formation of carbon dioxide from reaction with other glass constituents. Solutions such as making the dielectric porous and encapsulating the structure with a hermetic system have been implemented to overcome this issue but shorting has been observed [23]. These issues are why copper systems have never gained great popularity and their use is now primarily single-layer hybrids.

VIII. RESISTORS

Thick-film resistors are widely used because they have properties, such as high sheet resistance and power dissipation, that are difficult or impossible to achieve with silicon monolithic ICs. They can be fabricated directly on the thick-film circuit or surface mounted as chip resistors. The resistance of any thick-film resistor is directly proportional to its length and inversely proportional to its cross-sectional area, and is given by

$$R = \rho L/A \qquad (13)$$

where ρ is resistivity, L is length, and A is cross-sectional area. For a constant thickness, this can be reduced to

$$R = R_s L/W \qquad (14)$$

where R_s is the sheet resistivity, which is generally described in ohms/square ($\Omega/\square$), W is the width of the resistor, and L/W becomes the number of squares (N). The resistance can now be expressed by the sheet resistivity of the material and the geometry of the resistor as described by the number of squares (L/W) as shown below and illustrated in Figure 15.

$$R = R_s N \qquad (15)$$

The need for standardization led to the normalization of print thickness to 25 μm (1 mil) as this was the recommended dry print thickness of many commercial

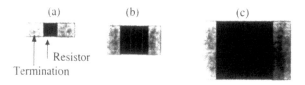

In Fig. (a), (b) and (c), the resistor length is equal to its
width. The resistance of each resistor is 1 kOhm

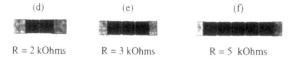

R = 2 kOhms   R = 3 kOhms   R = 5 kOhms

In Fig. (d), (e), and (f) the length is 2, 3, and 5 times the
width respectively leading to resistors if 2, 3, and 5 kOhms.

In Fig. (g) the length is half that of the width leading to a
resistance of 500 Ohms. In Fig (h), the width is five times
that of the length giving a resistor of 200 Ohms.

Figure 15   Effect of squares and aspect ratio on resistance. Courtesy of DuPont Technical
Service Literature.

resistors. This avoided the need for the resistor to always fire to the same thickness
and assumed that resistance is inversely proportional to thickness. However, physical and chemical nonhomogeneity in the resistor thickness makes this assumption
not strictly valid. As a result, resistors that are printed thinner than 25 μm can
be more accurately quoted at "printed thickness" for sheet resistivity (R).

   The general requirement for thick-film resistors is that the system covers
a wide range of resistivities. They should also be compatible with other components of the circuits with which they will come in contact, have low process
sensitivity, and show minimal permanent changes in resistance with elevated
temperatures, relative humidity, or mechanical stresses. However, the choice of
a suitable system depends on several factors, including, but not limited to:

   Sheet resistivity values
   Temperature coefficient of resistance (TCR)
   Current noise

Voltage stability
Compatibility with substrate and conductors
Process sensitivity
Stability

Thick-film resistors are commonly available with sheet resistivity values from 1 to 10^6 $\Omega/\square$. Resistors with values lower than this range are also available, often as a separate system based on different chemistries. The temperature coefficient of resistance (TCR) defines the rate of change of resistance with temperature and is given by

$$\text{TCR} = \frac{R_T - R_{25}}{R_{25}(T - 25)} \times 10^6 \, (\text{ppm}/^\circ\text{C}) \tag{16}$$

where R_T is the resistance measured at a specified temperature. Figure 16 shows a typical resistance as a function of temperature for a ruthenium-based resistor. Two TCRs are specified. The hot value is calculated between 25°C and 125°C and the cold value between −55°C and 25°C. High-quality resistors have hot and cold TCR values of 100 ppm/°C or less, so that a 100°C change in temperature will cause a 1% resistance change. However, many commercial systems are in the order of 200–250 ppm/°C. Special classes of resistors have very high TCRs, but in all other properties are identical to standard resistors.

Noise is a measure of how well the signal is carried through the conductive matrix. Thick-film resistors should have low current noise so that they function in applications involving low signal strength. Current noise is partly design re-

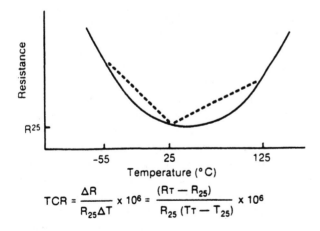

Figure 16 Resistance versus temperature in thick-film ruthenium.

lated, but the resistor composition has a big effect. Noise increases as resistor size is decreased [58] and is also generated at the conductor–resistor interface. Lower current noise in a specific design, however, is generally achieved with a resistor composition that has a higher volume fraction of conductive material and a very homogeneous fired microstructure.

When successively higher voltages are applied across a resistor the resistor will, at a certain voltage, exhibit a decrease in resistance followed by an increase in resistance as voltage is further increased. The initial decrease results from breakdown in insulation of glassy networks resulting in more conductive chains. Resistance eventually rises as the higher voltages break down conductive chains. The magnitude of change varies depending on the type of resistor material but is reproducible for specific materials, allowing these changes to be characterized by the short-time overload voltage (STOL). This is the voltage required (5 s duration) to induce a specified permanent resistance change (usually 0.25% on a 1 mm × 1 mm trimmed resistor) at 25°C. The STOL should be as high as possible since the standard working voltage is defined as 0.4 × STOL. The maximum rated power dissipation (MRPD) is the square of the standard working voltage divided by the resistance. In order to properly state the MRPD, the trimmed resistance value should be used for the calculation.

A. Materials

Most resistor pastes contain an insulating glass frit and an electrically conductive powder dispersed in an organic screening vehicle. In addition, small additions of other constituents are often added to control TCR or improve a particular property. The relative proportions of insulating and conducting phases have a large impact on properties and to some extent on the final resistivity. The conducting phase should have a relatively low resistivity, be stable in the firing atmosphere at high temperature, and be readily wet by the glass. It should also exhibit a wide range of resistivity across varying conductive phase concentrations. If firing is done in air, the choice is limited to a few highly conducting oxides and metals. Figure 17 shows resistivity versus conductive content for a variety of conductive materials that are stable in air at high temperatures. Early resistor compositions were based on silver-palladium-palladium oxide (Ag-Pd-PdO) conductive phases [59–61] combined with varying amounts of glass frit. Key to the development of resistance in these resistors was that during firing palladium oxidizes until approximately at 800°C it decomposes back to pure palladium. However, the temperature of the oxidation–reduction reaction depends on the silver-palladium ratio. As silver is added, the reversion back to palladium occurs at lower temperatures. The fired film was designed to contain mixtures of semiconducting palladium oxide and metallic silver-palladium alloy. The relative amounts of oxide, metal, and glass frit defined the resistivity. These resistors were very sensitive

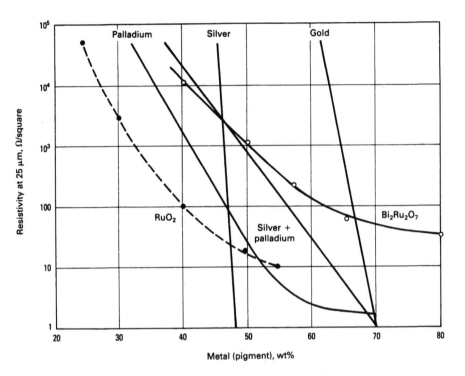

Figure 17   Resistivity versus concentration of metal powders and resistor pigments in PbO-B$_2$O$_3$-SiO$_2$ frit. Courtesy of DuPont.

to slight changes to time and temperature in the firing process and to reducing atmospheres in use, prompting the development of ruthenium-based systems.

Resistors based on ruthenium dioxide and ruthenium pyrochlores have become the standard for the industry [62–65]. Ruthenium dioxide has low resistivity and a very positive TCR. It can be used as a basis for a whole system but is more generally used in low-value resistors. Ruthenium dioxide can be partially reduced above 350°C during organic burnout resulting in a lower oxide [66], but it reoxidizes at higher temperatures. If not bound by glass, the lower oxide will evaporate and condense on to other areas of the substrate. As a result, resistors containing high concentrations of ruthenium dioxide can cause "staining" of adjacent conductors. Metal additions are often used in low-value resistors to maintain a ruthenium dioxide concentration below the level at which staining would occur.

The ruthenium pyrochlore family contains a wide variety of conducting oxides. Pyrochlores of different conductivity and temperature coefficient of resistance can be obtained by substituting suitable elements on the appropriate crystal lattice site. Ruthenates can be made with lead, bismuth, barium, strontium, and gadolinium, for example. Changing or adding conducting phases from this group provides a broad capability to adjust properties. For example, power handling, current noise, and voltage stability properties are directly related to the concentration of conductive phase. In high-resistivity resistors, good properties can be achieved by use of low-conductivity pyrochlores, rather than smaller amounts of the more highly conducting varieties.

Nitrogen firing of resistors requires conductive phases and glasses that are stable in low-oxygen environments. Ruthenium dioxide, lead, and bismuth ruthenates decompose in nitrogen at high temperatures, making them unsuitable. Lanthanum boride [67,68], tin oxide [67,69–71], indium oxide [72], strontium ruthenate [73,74], and metal silicides [75,76] are typical conducting phases used for nitrogen firing. Doping of the conductive is often employed to extend resistivity and control TCR, but most conductives do not cover the resistivity range needed for a full system. Hence, most resistor systems for nitrogen firing are composed of subsystems requiring blend breaks.

Optimizing all resistor properties simultaneously often requires the addition of small quantities of additional inorganic constituents. Additives to improve stability, modify CTE, and provide decreased sensitivity to variation in the processing conditions include oxides such as alumina, silica, zirconia, and titania. In addition, semiconducting oxides such as manganese dioxide (MnO_2) [77], cadmium oxide (CdO) [78], rhodium oxide (Rh_2O_3) [79], and vanadium oxide (V_2O_5) [80] can be added to control TCR.

B. Special Case Resistors

Thick-film thermistors used for temperature compensation are designed to have a large, well-defined change in resistance with temperature. The resistance of a NTC thick-film resistor is given by

$$R = R_0 \exp\left(\frac{\beta}{T}\right)$$

(17)

The temperature dependence is controlled by the parameter β, and the magnitude of the resistance is controlled by the pre-exponential term, R_0. The materials used in thermistor inks are generally a mixture of doped transition metal oxides and one or more glass frit compositions. Thick-film resistors with high positive TCRs are also useful as temperature sensors, and in such cases TCR values approach 2500 ppm/°C.

Strain gauge sensors utilize the piezoelectric property of resistors where resistance changes with substrate flexure. Strain gauge resistors show a reversible relative change in resistance ($\Delta R/R$) with strain (ε) defined as the fractional change in resistor length ($\Delta l/l$). The relative resistance change is directly proportional to the strain, and the proportionality constant is defined as the gauge factor (GF):

$$GF = \frac{\Delta R / R}{\varepsilon} \tag{18}$$

To be useful in a strain gauge or pressure transducer application, gauge factors should exceed 10, the higher the better. Strain gauge resistors should also have a low temperature coefficient of gauge factor (TCGF) and a low TCR.

Surge protection circuits are typically constructed from specially formulated conductors, called surge resistors. Electrical surges are high-energy, short-duration, electromagnetic pulses caused by lightning strikes, power crosses from heavy-power switching, and electrostatic discharges from humans or furniture. Surges are transient waves of voltage, current, or power. Voltage surges due to lightning strikes and switching are very short and subside within microseconds. Such pulses, however, can be very destructive to consumer and telecommunication equipment. Surge resistor designs are typically long serpentine patterns. Resistor metallurgy typically is silver-palladium [81,82], but copper-nickel [83] and other base metal compositions can be used. Surge resistors have low sheet resistivity, low TCR, and excellent stability under applied electric surge. Sheet resistivities typically range from 100 mΩ/□ to 1 Ω/□, and TCR requirements are typically less than ± 100 ppm/°C. Resistivity is controlled by the metal-to-glass ratio and TCR by the metallurgy ratio. As seen in Figure 10, the TCR of silver-palladium approaches zero at 60 wt % palladium. Similar low TCR can be achieved in the copper-nickel system between 55% and 65% copper. Other key electrical properties, such as stability under voltage pulse, depend mainly on the microstructure. Optimal values are achieved by a high degree of homogeneity and density in the fired film. Raw material particle size and glass chemistry play a key role in controlling this factor.

C. Resistance

Sheet resistivity can vary from 10 meg Ω/□ to 10 Ω/□ as the volume fraction of conducting phase relative to conducting plus insulating phase is varied from 0.02 to 0.3 [84]. This relationship has prompted various theoretical studies designed to describe the resistivity of thick-film resistors in terms of composition. Most were based on percolation theory [85,86], effective medium theory [87,88], or pertubation [89] or variational [90] approaches that were originally developed to analyze the properties of disordered materials that are statistically homogeneous and isotropic. All have a common origin in statistical continuum mechanics and strictly apply only to random and isotropic media, even though ordering parameters [91] and critical path analyses [92] are sometimes invoked. A modified bond percolation model [93] in which metallic particles were partial bonds in a lattice connecting the interstices of larger close-packed glass particles demonstrated that conductivity is generated at small-volume fractions. A model based on effective medium theory [94,95] that took into account the size difference between the glass and conductor particles led to an equation for resistivity in terms of volume fraction, particle size ratio, and other parameters. A site percolation model [96] included packing fractions for glass and conductives and fitted experimental data quite well. The theoretical models assumed some fixed microstructure, usually an idealized one, and either diverged from experimental data at some point or used empirically derived values for some parameters to enable a better fit to experimental data.

Electrical continuity at extremely low-volume fractions of the conducting phase precludes the possibility of a uniform distribution of the conducting phase in the insulating glass matrix. Networks of chains of conducting phase are observed in optical and electron micrographs of thick-film resistors [97]. Models [98–100] for microstructure development, taking into account the large difference in particle sizes, involves a series of steps after organic removal starting with segregation of the fine conducting phase particles or agglomerates to the surfaces and interstices of the larger glass particles in the film. Further heating leads to sintering of the glass particles by viscous flow. The glass infiltrates into agglomerates of conducting phase particles, and rearrangement of the particles and agglomerates occur as a result of surface tension forces. Sintering of the particles immersed in the liquid glass begins, and isolated particles or agglomerates in the liquid glass move to either join or leave preexisting chains. Modeling of the penetration of lead borosilicate glass into ruthenium dioxide agglomerates [101] and the liquid phase sintering of the ruthenium dioxide particles [102] demonstrated that the glass penetration is very rapid and sintering proceeds to an appreciable extent after 10 min at 850°C.

D. Temperature Dependence

The value of the TCR depends on the volume concentration of the conducting phase and the TCR-driving effects of semiconducting oxide additives. Low-resistivity resistors typically contain a high-volume loading of conducting oxides and the TCR is naturally positive. TCR control is achieved by semiconducting oxide additions that serve to drive the TCR negative. In high-resistivity compositions, the volume fraction of conducting oxide is generally lower and the TCR tends to be slightly negative.

If the material making up the conducting chains is the only conducting phase present, and assuming that the particles are sintered together in the chain, then the TCR of the resistor should be the same as the conducting phase to a first approximation. The TCR of the conducting phase used in resistors is typically very high (TCR of RuO_2 = 5670 ppm/°C). However, the TCR of a good resistor series is usually less than ±100 ppm/°C throughout the entire resistivity range.

This electrical behavior of resistors has prompted a number of explanations. Early work centered on the formation of boundary phases between or on the surface of sintered conducting particles that develop during the firing process [103–109]. Such phases could be the result of glass diffusion into the conducting particles, partial dissolution of the conducting oxide into the glass, or dopants forming semiconducting regions between sintered conducting particles. The semiconducting boundary phase had a negative TCR that compensated the positive TCR of the conducting phase.

A tunneling barrier model [110] assumed the conducting particles do not sinter together being separated by glass. The very small particle size of the conducting oxide produces low activation energies for electronic transport and the existence of impurities within the tunnel barrier increases the barrier transmission coefficient. An alternative model [111] assumed that the doping impurities in the glass interface form a narrow band. Models also proposed that both tunneling and hopping through the semiconducting impurities in the glass contributed to transport [112,113]. An attempt to identify the contributions of hopping, tunneling, and narrow bands produced a model based on conduction around the percolation threshold [114,115]. All nonsintered models lead to negative TCR. Combining the sintered and nonsintered approach [116] allows for different numbers of positive TCR (sintered) and negative TCR (nonsintered) contacts in the conducting chain. The addition of semiconducting oxides creates more nonsintered contacts thereby, driving the TCR more negative.

E. Material Interactions

As mentioned previously, during the firing process, alumina from the substrate diffuses into the resistor forming an aluminum concentration gradient reaching several micrometers into the resistor. At the fired thickness recommended by the manufacturer, the homogeneous portion of the resistor dominates electrical performance. If the fired thickness is reduced, the region that contains alumina has a progressively larger effect. At a fired thickness of less than 7 μm, most resistors lose their utility.

As shown in Figure 18, the resistor is thicker at the conductor–resistor interface, leading to reduced resistivity at the ends. Longer resistors reduce the effect of this factor. Chemical effects can also occur at the termination. Diffusion of metallic or glass elements from the conductor into the resistor can lead to

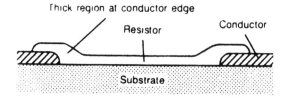

Figure 18 Longitudinal section of a resistor showing thickening at the terminations.

lowering or raising of the resistance value. Lower resistance is attributed to diffusion of the conductor metal into the resistor and is most commonly observed with silver-bearing conductors. This effect is seen less in longer resistors. Most manufacturers minimize this problem by specifying compatible conductors for specific resistors. When the problem cannot be eliminated, length effect plots such as those shown in Figure 19 are supplied for use in design. Experimental data on resistance for different lengths are plotted and the resistance extrapolated to zero squares. If the resistor is totally uniform along its length, the extrapolation will be to zero resistance. If the interaction causes higher or lower resistance, the extrapolation will either have a positive or negative value. This value should remain uniform for a given resistor–conductor combination and can be factored

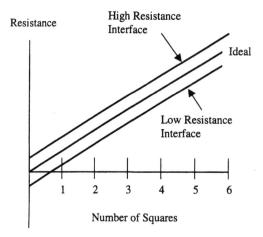

Figure 19 Design plot to correct for resistor–conductor interaction. (Reproduced from Ceramic Materials for Electronics, Materials Aspects of Thick-Film Technology, p. 482.)

into the design. Other reactions due to chemical incompatibility include micro-cracking or bubbling at the interface due to volume changes or gas evolution, or partial non-ohmic contact resulting from an insulating layer on the surface of the conductor. This is less of a problem in air-fired conductors being primarily associated with copper systems.

F. Trimming

Resistors cannot be consistently fired to a predetermined resistance value of better than ±15%. If greater precision is required, the resistors are trimmed to value by cutting away part of the resistor using a laser. This requires that the initial resistor value be sufficiently below the target resistance that the upper end of the resistance distribution is still below the target value. Figure 20 shows the resistance of a resistor after firing, one and two glass encapsulations, trimming, and packaging. As fired values normally increase slightly during glass encapsulation, trimming up to two times their initial resistance is generally recommended. Subsequent processing, such as attachment of components with solder, may marginally shift the value; and this can be taken into account in the design.

Trimming is typically accomplished using a neodymium-doped yttrium-aluminum-garnet (Nd-YAG), which operates at a wavelength of 1.06 μm. A high-intensity beam heats the resistor rapidly, causing it to vaporize. The integrity of the trimmed film and its subsequent stability are dependent on the cleanliness of the cut. The depth of cut should be through the resistor and into the substrate. If the cut is too deep, microcracks may form at the resistor–substrate interface due to excessive heating. This will eventually cause resistor value drift. If the cut is not deep enough, some of the resistor will be left in place, also causing subsequent

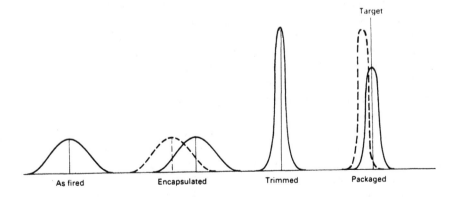

Figure 20   Processing resistors to target values. Distribution of resistance value.

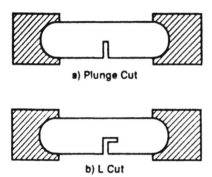

a) Plunge Cut

b) L Cut

Figure 21   Laser trim patterns.

drift. The simplest geometry for trimming is the single-plunge cut shown in Figure 21a. However, for narrowest distribution of values, the L cut shown in Figure 21b is more often used. Properly programmed laser trimmers can routinely trim resistors to <1% tolerance.

IX.   POLYMER THICK-FILM

Polymer thick-film (PTF) pastes are a special class of thick film in that they achieve all functional properties by a curing cycle at temperatures of 150°C or less. As such, they are suitable for use on low-cost plastic substrates, such as polyester, polycarbonate, paper, or FR4, as well as ceramic substrates. Since they are not sintered, polymer thick-film pastes are, strictly speaking, not ceramic thick film, but the technology is used in so many diverse applications that a short description is warranted. Uses include potentiometers and resistor networks for simple consumer applications, EMI shielding, low-temperature heating devices, current overload protection devices, electroluminescent lighting, blood glucose and other sensors [117], and iontophoretic drug delivery patches. They are also used for filling vias and embedding resistors and capacitors inside printed wiring boards [118]. However, perhaps the best known application is its use in manufacturing membrane switches.

Membrane switches operate at 1 W of power or less, at or below 30 V, and are the standard means of data input to microprocessor-controlled systems. PTF technology allows these devices to be manufactured at high volume and low cost. A basic membrane switch consists of two layers of flexible polymeric film, generally polyester, that have conductive tracks printed on adjacent sides. A spacer made from another polymer sheet or a printed dielectric film separates the two layers. When pressure is applied to the top sheet, its circuit contacts the bottom circuit through a hole in the spacer, closing the switch and making electrical contact. A graphic overlay, often embossed to give a mechanical feel to the switch, completes the system.

A. Polymer Thick-Film Materials

Resins used in polymer thick-film vehicles are critical components of the composition as they remain after curing. The solvent is chosen for drying rate and screen life, but choice of the resin depends on the desired final film properties. In general, thermoplastic resins, such as polyesters, acrylics, and vinyl copolymers, are mostly used for flexible circuitry. They dry quickly to accommodate high-speed automatic processing and the final film displays flexibility without loss of properties. Thermosetting polymers, such as epoxies and phenolics, have poor flexibility and are therefore used on rigid substrates. They often require higher curing temperatures for longer times than the thermoplastic systems, but they display excellent adhesion to epoxy-glass laminates and their abrasion resistance can meet potentiometer wiper requirements. Thermosetting resins also have good thermal stability and generally have excellent resistance to solvents, making them useful for harsher environments than thermoplastic resins.

The conductive phase for polymer thick-film conductors can be based on precious metals, base metals, or graphite or combinations thereof. The highest conductivity is achieved by the use of pure silver. Use of silver flake, 3–10 μm in diameter and 0.1–0.5 μm in thickness, can give conductivity values as low as 0.01 Ω/□. Silver is a favored conductive as besides its excellent conductivity, it's oxide is still conductive. This eliminates the issue of catastrophic loss of conductivity due to overcuring, something base metal conductors are prone to. If high conductivity is not the key priority, another material, such as silver chloride, graphite, or copper, can be mixed with silver. Silver-silver-chloride compositions have found wide use as reference electrodes in glucose blood sensors. Graphite on its own gives a conductivity in the order of 10–100 Ω/□ and is mostly used for radiofrequency interference screening and static dissipation although it is also used, sometimes doped with small amounts of platinum, as the sensing electrode in blood glucose sensors. Copper is also a common conductive. Copper compositions are mostly used in connection with printed wiring boards, such as EMI shielding and via filling.

Polymer thick-film dielectrics and encapsulants can be as simple as a polymer alone or can be filled with ceramic phases, such as barium titanate or alumina, to enhance dielectric or thermal properties. Dielectrics or encapsulants for use on rigid substrates are generally based on thermosetting materials such as epoxies, phenolics, silicones, or polyurethanes. The resins are designed to cross-link fully with curing providing a hard durable coating. Dielectrics or insulating pastes used on flexible substrates are often made from polyester systems, sometimes filled with rubber fillers. Unfilled systems can be UV curable allowing for high-speed manufacturing.

Polymer thick-film resistors are similar in composition to many PTF conductors except that resistivity and TCR need to be controlled. This can be achieved by using carbon and various metal oxides and nitrides. Silver powder is also often employed at the low resistivity ranges. Resins are usually thermosetting, such as epoxy, phenolie, or polyimide, and complete cross-linking is required to develop stable resistor properties. This means that curing temperatures are often high and long times are employed. Since all organic systems are slightly transparent to moisture, complete curing is essential or else significant drift in resistance value when subjected to humidity will result.

X. COMPLETING THE HYBRID

After the thick-film pastes have been printed and fired on the substrate, discrete components may be added and the circuit packaged. Components may include semiconductor as well as passive components. Back-bonded semiconductor chips are bonded by use of adhesives, Au-Si eutectics, or silver-glass compositions. Electrical connection is made by wire bonding to adjacent thick-film conductor pads. Flip-chips attachment and electrical connection can be done using solder reflow. The chip is designed to bond to patterned pretinned conductor pads in a face-down configuration, the geometry of the conductor land pattern matching the solder balls on the flip chip. Other semiconductor attachment approaches include beam lead, thermal compression, and tape automated bonding (TAB).

The attachment of passive components is usually accomplished by soldering. Solders that have a liquidus point below 315°C are referred to as soft solders. Those with liquidus temperatures between 315°C and 425°C are known as hard solders. Above 425°C is considered a brazing alloy. When multiple soldering operations are performed, the reflow temperature of each subsequent solder is chosen to be lower than the liquidus temperature of the previous solder.

Direct soldering to untinned conductor pads using solder paste is the most common method of soldering surface mount components. Solder pastes are formulated with relatively coarse, 10- to 75-μm diameter powder suspended in a vehicle containing any necessary fluxes. The paste is printed through a coarse mesh screen, and while the paste is still wet the components are placed in their positions using a pick-and-place machine. The paste remains tacky long enough to act as a temporary adhesive holding the components in place. The solder is then reflowed by application of a suitable heat source.

Attachment of passive components with conductive organic adhesives as an alternative to solder can be appropriate if there is a desire to avoid fluxes or if there are special considerations such as thermal cycling reliability. Typically the adhesive is silver filled epoxy, sometimes modified with plasticizers to improve resistance to stress. They can be one- or two-component systems with room temperature or elevated-temperature curing.

After final test, the hybrid is packaged. There are two principle methods of packaging: plastic and hermetic. Plastic packaging is typically accomplished by casting, potting, conformal coating, or transfer moulding plastic over the circuit. The most common packaging materials are epoxies, silicones, polyurethanes, and phenolics. Hermetic packaging consists of sealing a lid over the substrate and backfilling with an inert atmosphere. The lid is generally metal and is sealed by solder.

XI. SUMMARY

Thick-film materials are complex nonequilibrium systems. More than 50 of the elements of the periodic table are present in the various thick-film systems. Electrical, chemical, and physical properties of the ingredients, their variation with temperature, and their interactions with one another are important factors in the development of systems. Many of these factors can be modeled and where they cannot, the knowledge base in the technology has progressed to where many are known empirically.

Finally, it has to be remembered that thick-film technology is always in competition with not-in-kind technologies such as thin film, application specific ICs, low-temperature cofired ceramic, or printed wiring boards, and therefore applications are dynamic. New applications made in thick film can replaced by alternative manufacturing technologies when circumstances favor change. The popularity of thick-film technology, however, is due in large part to the simplicity of making parts, and it is this, combined with continual improvements in capability, that has allowed thick-film technology to be a viable manufacturing approach to an ever diverse set of applications.

REFERENCES

1. Kando H, Mason RC, Okabe C, Smith JD, Velasquez R. Buried resistors and capacitors for multilayer hybrids. Proc. Int. Symp. on Microelectronics, ISHM, 1995:47–52.
2. Stein SJ, Wahlers R, Heinz M, Stein MA, Tait R. Thick film heaters made from dielectric tape bonded stainless steel substrates. Proc. Int. Symp. on Microelectronics, ISHM. Los Angeles, CA, Oct. 24–26, 1995.
3. Mutoh T. Advanced photoprintable paste application for PDP, PDP consortium, 7th Section, Seminar, 1998.

4. Bacher RJ. Thick film printing technology course. Int. Symp. on Microelectronics, ISHM. Los Angeles, 1995.
5. Bacher RJ, Dobie A. Technology of screen printing. Prof. Dev. Course, IMAPS. Chicago, IL, 1999.
6. Shipton RD, Robertson CJ. 50 Micron-thick film lines and spaces with etched foil screens, 11th European Microelectronics Conf., May 14–16, Venice, Italy, 1997.
7. Cazenave JP, Suess TR. Fodel photoimageable materials—a thick film solution for high density multichip modules. Proc. Int. Symp. on Microelectronics, ISHM, 1993.
8. Cazenave JP. High density interconnections with thick film MCMs. Solid State Tech., May 1994.
9. Bacher RJ, Wang YL, Skurski MA, Crumpon JC, Nair KM. Next generation ceramic multilayer systems. Proc. Int. Symp. on Microelectronics. Boston, Sept 20–22, 2000.
10. Keusseyan R, Speck B, Chaplinsky J, Amey D, Kuty D. C Needes, S Horowitz, high Density Gold Thick Film Technology. Proc. Int. Symp. on Microelectronics, IMAPS. Philadelphia, Oct 14–16, 1997.
11. Tredinnick M, Barnwell P, Malanga D. Thick film patterning—a definitive discussion of the alternatives. Proc. Int. Symp. on Microelectronics, IMAPS. Baltimore, Oct 9–11, 2001.
12. Harper CA, Ed. Handbook of Thick Film Hybrid Microelectronics. New York: McGraw-Hill, 1974.
13. Holmes PJ, Loasby RG. Handbook of Thick Film Technology. Ayre. Scotland: Electrochemical Publishers, Ltd., 1976.
14. Riemer DE. Ink hydrodynamics of screen printing. Proc. ISHM Symp.. Anaheim, Nov. 1985:52–58.
15. Bacher RJ. High resolution thick film printing. Proc ISHM Symp Oct. 1986: 576–581.
16. Riemer DE. The shear and flow experience of ink during screen printing. Proc ISHM Symp Sept. 1987:335–340.
17. Ahmad S, Jiang T, Moden W. A simple method for optimizing screen print parameters. Proc. Int. Symp. on Microelectronics, IMAPS. Philadelphia, Oct 14–16, 1997.
18. James EJ. A practical approach to high resolution printing. Proc. Int. Symp. on Microelectronics, ISHM. Los Angeles, Oct 24–26, 1995.
19. Missele C. Screen printing primer–Part 3. Hybrid Circuit Technol May 1982; 2(5): 11.
20. Baudry H, Franconville FF. Int J Hybrid Microelectronics Dec. 1982; 5:15.
21. Bacher RJ. personal communication, 2002.
22. Trease RE, Dietz RL. Solid State Technology 1972; 15(1):39.
23. Lund RE. Circuit shorting mechanisms in copper multilayer systems. Proc. ISHM Symp.. Anaheim, 1985:463–471.
24. Borland W, Siuta VP. Materials Interactions in the Firing of Copper Thick Film Multilayer Ceramics. Mater Res Soc 1988; 108.
25. Brown R. Thin film substrates, in: Handbook of Thin Film Technology 2nd ed., Maissel LI, Glang R, eds. New York: McGraw-Hill, 1983.
26. Winklemann A, Schott O. Ann Phys Chem 1984; 51:730.

27. Palanisamy P. Influences of substrate on microstructure development in thick film resistors, PhD thesis, Purdue University, McGraw-Hill, 1979.
28. Iwase N, Anzai K, Shinozaki K, Hirao O, Thanh TD, Sugiura Y. Thick film and direct bond copper forming technologies for aluminum nitride substrates, IEEE. Trans, Compon, Hybrids Manuf Tech 1985; CHMT-8(2):253–258.
29. Kurokawa K, Utsumi K, Takamizawa K. Development of highly thermal conductive AIN ceramics by hot press techniques. Proc. 24th Sympsium of Basic Science of Ceramics, 1986.
30. Slack GA. Non-metallic crystals with high thermal conductivity. J Phys Chem Solids 1973; 34:321.
31. Werdecker W, Aldinger F. Aluminum nitride—an alternative ceramic substrate for high power applications in microelectronics, IEEE. Trans, Compon, Hybrids Manuf Tech 1984; CHMT-7(4):399–404.
32. Brown MP, Slack GA, Szymaszek JW. Thermal conductivity of commercial aluminum nitride. Am Ceram Bull 1972; 51(11).
33. Kretzschmar C, Otschik P, Griesmann H. A new paste system for AIN. Proc. Int. Symp. on Microelectronics, IMAPS. Baltimore, 2001.
34. Keusseyan RL, Parr R, Speck BS, Crumpton JC, Chaplinsky JT, Roach CJ, Valenta K, Horne GS. New gold thick film compositions for fine line printing on various substrate surfaces. Proc. Int. Symp. on Microelectronics, ISHM. Minneapolis, 1996.
35. Wang YL, Carroll AF, Smith JD, Cho Y, Bacher RJ, Anderson DK, Crumpton JC, Needes CRS. A new thick film system for AIN substrates, Ceramic Applications for Microwave and Photonic Packaging. Rhode Island: Providence, 2002.
36. Ellis ME, Richter F, Levitsky ME. Thick film on stainless steel; applications, methods and processes. Int. Symp. on Microelectronics, IMAPS. Boston, Sept. 20–22, 2000.
37. Ellis ME. Thick film on stainless steel circuit boards. Int. Symp. on Microelectronics, IMAPS. Chicago, Oct 26–28, 1999.
38. Pluym TC, Powell QH, Gurav AS, Ward TL, Kodas TT, Wang LM, Glicksman HD. Solid silver particle production by spray pyrolysis. J Aerosol Sci 1993; 24(3): 383–392.
39. Needes CRS, Nair KM, Coleman MV. The environmental reliability of series Q: a thick film materials system for the manufacture of advanced hybrid microcircuits. Proc ISHM Symp May 1988.
40. Borland W, Needes CRS, Siuta VP, Nair KM. High conductivity materials systems for advanced hybrids. Proc. Electronic Component Conference, IEEE, May 1989.
41. Keusseyan RL, LaBranche MH, Hang KW. Thick film materials systems for thermal cycle applications. Proc. Int. Symp. on Microelectronics, ISHM. Boston, 1994.
42. Turkdogan ET. Physical chemistry of high temperature technology, chapter 1. New York: Academic Press, 1980.
43. Vest RW. Materials interactions during firing of Pd/Ag conductors. Proc. ASM Conf. on Thick Films. Atlanta, 1988.
44. Larry JR, Rosenberg RM, Uhler RO. IEEE, Trans 1980; CHMT-3(2):211–225.
45. Neumann JP, Zhong T, Chang YA. Met Prog Sept. 1985; 85(4).
46. Wu HZ, Vest RW, Vest GM, Mau CS. Adhesion and densification studies of oxide free copper conductors. Proc. ISHM Syp. Atlanta, 1986:873–880.

47. Bacher RJ, Siuta VP. Proc. 36th Electronic Component Conference, IEEE, 1986.
48. Wu HZ, Vest RW. Int J Hybrid Microelectronics 1987; 10(4):20.
49. Vest RW. Ceram Bull 1986; 65(4):631–636.
50. Loasby RG, Davey N, Borlow H. Sol State Tech May 1972.
51. Taylor BE, Felten JJ, Larry JR. Progress in the technology of low cost silver containing thick film conductors, Proc. Electronic Component Conference, IEEE, 1980.
52. Turner PS. J Res NBS 1946; 37:239.
53. Yasumoto T, Iwase N, Harata M, Segawa M. High dielectric constant Y5U thick film capacitors based on relaxor. Proc. 4th US–Japan Seminar on Dielectric and Piezoelectric Ceramics. Gaithersburg, MD., 1988:748–764.
54. Kingery WD, Bowen HK, Uhlmann DR. Introduction to Ceramics. 2nd ed. New York: John Wiley and Sons, 1976.
55. Carlson DE, Hang KW, Stockdale FF. Electrode "polarization" in alkali-containing glasses. Am Ceram Soc Bull 1972; 55(7):337–341.
56. Matsuzuki T, Sato K, Suzuki T, Terashima M. High density multilayer substrate using copper thick film and thin film, Proc. Elec. Components Conf., Washington, DC, May 2, 1985.
57. Pitkanen DE, Cummings JP, Sartell JA. Int J Hybrid Microelectronics 1980; 31(1):1.
58. Kuo CY, Blank HG. The effects of resistor geometry on current noise in thick film resistors. Proc Int Soc For Hybrid Electronics. ISHM, 1968:153.
59. Hoffman LC. Precision glaze resistors. Am Ceram Soc Bull Sept. 1963; 42(9): 490–493.
60. D'Andrea JB. Ceramic composition and article, US Patent 2,924,540, Feb. 1960.
61. Hoffman LC. Resistor compositions, US Patent 3,207,706, Sept. 1965.
62. Bouchard RJ. Oxides of cubic crystal structure containing bismuth and at least one of ruthenium and indium, US Patent 3,583,931, June 1971.
63. Bouchard RJ. Composition for making electrical elements containing pyrochlore-related oxides, US Patent 3,681,262, Aug. 1972.
64. Bouchard RJ. Compositions for making resistors comprising lead-containing polynary oxide, US Patent 3,775,347, Nov. 1973.
65. Hoffman LC, Horowitz SJ. US Patent 4,302,362, Nov. 1981.
66. Bube KR. The effects of prolonged elevated temperature exposure on thick film resistors. Proc Int Soc for Hybrid Electronics 1972.
67. Donohue PC, Hormadaly J, Needes CRS, Horowitz SJ, Knaack JF. Nitrogen-fireable resistors: emerging technology for thick film hybrids, IEEE. Proc Compon Hybrids Manuf, Tech 1987; CHMT.
68. Ito O, Asai T, Ogawa T, Hasegawa M, Ikegami A. Effect of conductive particle size on LaB_6 thick film resistor, Proc. 5th Int. Microelectronics Conf., May 1988.
69. Bhide SH, Vittal N, Aiyer CR, Karekar RN, Aiyer RC. Formulation and characterization of tin based oxide resistor paste. Int J Hybrid Microelectron 1988; 11(2): 36–39.
70. Shapiro H, Merz KM. Refractory metal glazes for thick film networks, Proc. Electronic Components Conf., IEEE, 1977.
71. Wahlers RL, Merz KM. High range low cost resistor system, Proc. Electronic Components Conf., IEEE, 1978.

72. Sayers P, Petsis W. Characteristics of nitrogen fireable materials for hybrids and multilayers, Proc. Int. Microelectronics Conf. ISHM, 1986.

73. Steinberg J, Mallarga D, Cheung Y. Nitrogen fireable hybrid thick film inks. Proc. Int. Symp. on Microelectronics, ISHM, 1988.

74. Hankey DL, Panzak MX, Sutterlin R. Introduction to a novel copper compatible nitrogen firing resistor system, Proc. Int. Microelectronics Conf. ISHM, 1986.

75. Makino O, Watanabe H, Ishida T. Nitrogen fireable base metal silicide thick film resistor, Proc. 4th Int. Microelectronics Conf., May 1986.

76. Kuo CY. Thick film resistive paint and resistors made therefrom, US Patent 4,639,391, Jan., 1987.

77. Inokuma T, Taketa Y, Haradome M. Electrocomp Sci Technol 1982; 9:205.

78. Kuzel R, Broukal J. IEEE Trans 1981; CHMT-4:239.

79. Hamer DW, Biggers JV. Thick Film Hybrid Microcircuit Technology. New York: John Wiley and Sons, 1972.

80. Setty MS, Shinde RF. Active Passive. Elec Comp 1986; 12:111.

81. Bender D, Lathroy R. Novel high performance resistor design for telecommunications A.C. surge protection, Proc. Adv. Mater. Tech. Conf. ISHM, Orlando, 1989.

82. Vasudevan S. Thick film materials system for surge protection. Proc. Int. Symp. on Microelectronics, ISHM. Minneaplois, 1996.

83. Kuo CY, Martin T. Proc. Int. Symp. on Microelectronics, ISHM. San Francisco, 1992.

84. Vest RW. Conduction mechanisms in thick film circuits, Final Technical Report on ARPA Order No. 1642 (NTIS No. N76–13346), 1975:226.

85. Kirkpatrick S. Rev Mod Phys 1973; 45:574.

86. Webman I, Jortner J, Cohen MH. Phys Rev B 1975; 11:2885.

87. Kirkpatrick S. Phys Rev Lett 1971; 27:1722.

88. Springett BE. Phys Rev Lett 1973; 31:1463.

89. Nagatani T. J Phys A 1979; 12:1577.

90. Hori M. J Math Phys 1973; 14:1942.

91. Nagatani T. J Appl Phys 1980; 51:4944.

92. Ambegaokar V, Halperin BI, Langer JS. Phys Rev B 1971; 3(8):2612.

93. Pike GE, AIP Conf. Proc., No. 40 Garland CW, Tanner DB, eds, American Institute of Physics, New York, 1978.

94. Smith DPH, Anderson JC. Philos Mag B 1981; 43:797.

95. Smith DPH, Anderson JC. Thin Solid Films 1980; 71:79.

96. Ewen PJS, Robertson JM. J Phys D Appl Phys 1981; 14:2253.

97. Chitale SM, Vest RW. IEEE, Trans 1988; CHMT-11:604.

98. Inokuma T, Taketa Y, Haradome M. IEEE, Trans 1984; CHMT-7:166.

99. Vest RW. Development of conductive chains in RuO_2-glass thick film resistors. Proc. 8th Int. Symp. Reac. Solids (Gothenberg). New York: Plenum, 1977:695–670.

100. Sarma DHR, Vest RW. J Am Ceram Soc 1985; 68:249.

101. Palanisamy P, Sarma DHR, Vest RW. J Am Ceram Soc 1985; 68:C215.

102. Van Loan PR. Insulation/Circuits. Vol. 35, June 1972.

103. Hoffman LC. Bull Am Ceram Soc 1963; 42:490.

104. Melan EH, Mones AH, IEEE, Electronic Component Conf., 1964.

105. Carcia PF, Champ SE, Filppen RB, IEEE Electronic Component Conf., Washington, D.C., May 1976.
106. Sartain CC, Ryden WD, Lawson WW. J Non-Cryst Solids 1970; 5:55.
107. Cash DA, Ansell MP, Hill RM. Conduction mechanisms in thick films, Third Annual Report, MOD Contract No. Ru 7–14Chelsea College, London, Nov. 1974.
108. Iles GS. Plat Met Rev 1967; 11:126.
109. Kahan GJ. IBM J Res Dev July 1971; 15(4):313.
110. Pike GE, Seager CH. J Appl Phys 1977; 48:5152.
111. Hill RM, IERE Conf. Proc., London, 1977.
112. Halder NC. Electrocomp Sci Technol 1983; 10:21.
113. Halder NC, Snyder RJ. Electrocomp Sci Technol 1984; 11:123.
114. Licznerski BW. Conduction mechanisms identification in thick film resistors, Proc. 5th European Hybrid Microelec. Conf., Stressa, Italy, 1985.
115. Licznerski BW. Mater Sci 1987; 111(3–4):179–191.
116. Vest RW, Chitale SM, Kollipara AK. The dependence of charge transport on microstructure in thick film resistors, Proc. 5th European Hybrid Microelec. Conf., Stressa, Italy, 1985.
117. Chan MS, Kuty D, Pepin J, Parris N, Potts R, Reidy M, Tierney M, Uhegbu C, Jayalakshmi Y. Materials for fabricating biosensors for transdermal glucose monitoring. Clin Chem 1999; 45(9):1689–1690.
118. Savic J, Crosswell RT, Tungare A, Dunn G, Tang T, Lempkowski R, Zhang M, Lee T, Proc. Tech. Conf., IPC, Long Beach, CA, Mar 24–28, 2002.

10

Multilayer Ceramic Technology

Wolfram Wersing and Oliver Dernovsek
Siemens AG
Munich, Germany

I. INTRODUCTION

To match the earlier editions of *Ceramic Materials for Electronics* published in 1986 and 1992, we have retained the original title of this chapter, i.e., "Multilayer Ceramic Technology." However, to be more accurate we should have named it "Multilayer Ceramic Technology for Packaging and Modules" because today multilayer ceramic (MLC) technology not only is applied to packaging issues but covers a much wider field, including the fabrication of electronic components such as capacitors, varistors, and piezoelectric actuators.

The packaging of integrated circuits (ICs) has usually made use of a chip carrier or module to which the IC chip(s) would be joined. The module would in turn be joined to a printed circuit board, thus becoming the second level in packaging the IC. Such modules have traditionally been made using ceramic substrates as a base material. However, due to the increasing integration level of IC chips and the decreasing semiconductor prices, there was a steady demand for IC packages of lower cost and therefore there was a strong effort to increasingly replace ceramic packages with cheaper plastic packages. Only in cases where special requirements, such as high heat conduction and distribution, hermetic sealing, extremely high stability, and operation at very high frequencies, have to be fulfilled are ceramic packages still competitive. This trend applied not only to single-chip packages but to multichip carriers or modules as well. To date, depending on the type of substrate used, there are three basic kinds of multichip modules (MCM) distinguished, referred to as MCM-D, MCM-L, and MCM-C. Although the purpose of this chapter is to discuss ceramic technologies, a short overview of these technologies is presented in the next section.

In recent decades, there has also been a trend toward the appearance of microwave integrated circuits of increasing complexity. For these ICs, it is not possible to use standard packages even if they are of the ceramic type because microwave performance is considerably decreased. Due to their excellent dielectric properties, ceramic packages are desirable, but in standard MLC packaging technology wiring relies on refractory metallization (such as W); therefore, the conductor lines are inherently more resistive and relatively high microwave losses occur. In standard MLC technology, refractive electrodes must be used because of the high sintering temperature required for densification of the Al_2O_3 layers during the cofiring process. In order to overcome the conduction loss problem, i.e., to be able to use electrode materials of high conductance (e.g., Cu or even Ag), people were looking for dielectric materials that become fully dense at temperatures as low as 950°C or 900°C, respectively. It turned out that such dielectric materials could be developed on the basis of glass ceramic composites similar to those already known from ceramic hybrids [1–3]. These glass ceramic composites together with screen printing pastes based on Ag, Au, or Cu, again similar to those used for hybrids [4], formed a basic material system for MLC substrates that could be sintered at 950°C or lower. Due to the low sintering temperature, this technology was then named low-temperature cofired ceramics (LTCC). In contrast with this technology, the standard MLC technology based on Al_2O_3 tapes is known as high-temperature cofired ceramics (HTCC).

Though originally developed for microwave packages and substrates with multiple wiring layers for military systems, the LTCC technology eventually penetrated the commercial multilayer substrate market. The first applications were those in which cost played only a secondary role and/or in which the specific advantages of the technology were of great importance, as, for example, in the medical and automotive fields. Whereas the importance of the LTCC technology for multichip modules has decreased in recent years, mainly due to an increase of the IC integration level, the excellent microwave performance together with the possible integration of passive components has led to increasing application of this technology in microwave modules, used in mobile communication and information systems.

These decisive developments in MLC substrate technology related, on the one hand, to materials and processing (from HTCC to LTCC with integrated passive components) and, on the other hand, to applications ranging from multichip modules to RF and microwave modules, or even to complete microwave systems on or in a package (system in package, SIP), will be discussed in this chapter. To preserve a historical perspective in the evolution of these technologies, each processing related section starts with text on MCMs from the second edition written by W. S. Young and S. H. Knickerbocker, to which material is added incorporating new knowledge and the most recent developments in applications. This approach ensures continuity of the subject material and thus presents a more complete picture of MCM development over the last few decades.

The purpose of this chapter is to discuss laminated MLC technology, particularly the LTCC technology. Section II contains a short review of multichip module technologies including the sequential MLC process. In Section III we discuss the processing steps for the laminated MLC technology. In Section IV the different materials used in the laminated MLC are described, again with emphasis on the LTCC technology. Section V covers the design and modeling of MLC substrates and modules, ending finally with Section VI, with a discussion of applications.

II. OVERVIEW OF MULTICHIP MODULE TECHNOLOGIES

Multichip modules have features that favor smaller, lighter systems with higher performance thereby eliminating individual packages and their parasitics. In the MCM technology bare chips are attached to the substrate by using wire bonding, tape automated bonding (TAB), or flip chip. The size of the package is dictated by the size of the chips, the number of input and output connections per chip, and the additional required passive components.

Here a short review on the different multichip module technologies will be presented because these technologies are basically considered for SIP. A detailed introduction and guide to MCM technologies can be found in the *Multichip Module Technology Handbook* [1]. The principal structure of the three basic multichip module technologies is shown in Figure 1.

A. MCM-D

MCM-D is a multichip module that has interconnection structure built by deposition of thin-film metals and dielectrics on substrates such as silicon or glass wafers, ceramics, or metals (Figure 1). In principle, dielectrics for MCM-D can be of the polymeric or inorganic type. However, polymeric materials have been used to a much greater extent because of their outstanding advantages such as the capability to form thicker layers of lower stress, the speed and cost of deposition, and the better planarization and lower dielectric constant offered by polymers. In general, polyimides (PIs), benzocyclobutenes (BCBs), and cyanate esters (CEs) are the polymeric materials used commercially. Spin coating is usually used for the application of these dielectrics. In order to achieve the best layer-to-layer adhesion, a curing process in an inert atmosphere must be carried out after application of the dielectric layers.

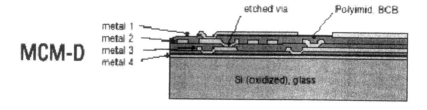

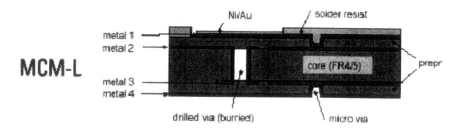

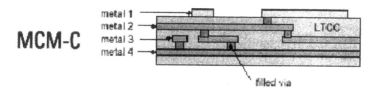

Figure 1   Basic multichip module technologies.

    To interconnect signal lines, ground, and power between different layers,
via formation in the dielectric layers is necessary. This is usually achieved by
techniques such as dry etching, wet etching, laser ablation, and use of photosensitive dielectrics.
    The most critical requirements for conductor metallization are conductivity
and reliability. Conductor metallization are deposited either by thin-film vacuum
processes such as evaporation or sputtering or by wet processes such as electroly-

tic or electroless plating. In general, electroplating is used for copper and gold, and sputtering or evaporation is used for aluminum.

The MCM-D carrier (substrate) must present a flat surface of low roughness because the subsequent thin-film fabrication requires the use of lithographic techniques. The substrate should be inert to the process chemicals, gas atmospheres, and temperatures used during fabrication. In addition, the substrate must possess certain relevant thermomechanical and electrical properties such as thermal expansion, heat conduction, and dielectric losses.

Ceramic carrier substrates are of particular interest within the scope of this chapter. Ceramics can be used in two distinct ways (1) simply as a mechanical support or (2) as an MLC substrate that contains power and ground distribution planes, some of the signal layers, and in particular embedded passives and on which thin-film signal layers and chip signal redistribution layers are deposited. It is obvious that an LTCC substrate carrier in combination with a MCM-D structure, often denoted MCM-D/C, leads to a very powerful package for an SIP. The inherent rough topography of ceramic surfaces (as fired) can be circumvented by lapping and polishing or—much cheaper—by an initial polymer planarization coat prior to the deposition of thin-film metallization [152]. As a rule of thumb, the substrate should have a roughness of less than 100 nm and should have a flatness of 10 μm or less to be suitable for thin-film lithography [151]. In the case of LTCC carrier substrates, these requirements can even be fulfilled without any postfired surface finish by certain specially developed LTCC glass ceramics.

B. MCM-L

MCM-L is a multichip module, where L represents the laminate core on which the interconnection structure is fabricated (Figure 1). The major difference between an MCM-L and a traditional printed wire board (PWB) is due to the manner of chip attach, i.e., chips are directly attached to the laminate board and are not prepackaged. MCM-L is a low-cost, high-density interconnect solution for direct chip attach. Often MCM-L is considered to include any type of PWB substrate that interconnects unpackaged, bare chips, regardless of the number of chips or the substrate density, i.e., to include all chip on board (COB) applications.

A decisive, economic advantage that MCM-L has over other MCM technologies is based on the ability to fabricate these circuits on very large panels with a multiplicity of identical patterns. A disadvantage is that MCM-L is usually more limited in interconnect density relative to other advanced MCM technologies. Furthermore, PWB laminates are poor heat conductors, making heat extraction from small, powerful chips difficult. However, as flip-chip technologies become more prevalent, an alternative thermal solution comes into play where the bulk of the heat can be extracted from the backside of the die. Another problem that had to be solved in using laminate-based MCM substrates relates to the mismatch between the inherent high coefficient of thermal expansion (CTE) of the substrate (>15 ppm/K) and the low CTE of silicon chips (<4 ppm/K). The solution to this problem was the use of a suitable underfill material placed between the die and the substrate to distribute stresses and thereby compensate for CTE mismatches. By this means, one of the major reservations to MCM-L use was eliminated. The need for higher density substrate technologies has focused the PWB industry on the development of alternative blind and buried via board constructions. Mechanically drilled through holes (most common diameter is 300 μm) and especially blind vias remain the most costly process steps in the fabrication of PWBs today, because either costly multiple drilling and lamination steps or slower throughput controlled-depth drilling processes must be used. Therefore, the development of alternative mass via generation technologies has become critical. All the various technologies considered are based on one of the three main via generation principles, namely, photo-defined vias involving photosensitive dielectric materials [153], plasma [154], and laser processing. These recent MCM-L manufacturing technology developments are substantially impacted by excimer laser processing of high-density MCMs for mainframe computer applications [155–157] that have already been used during the late 1980s.

Because the conventional PWB fabrication methods cannot support the fan-out requirements of the very dense array patterns encountered in today's ball grid arrays, chip-scale packages, and flip-chip on-board assemblies, special conductor redistribution layers, placed between the chip and the conventional PCB substrate, have been developed with the assistance of the mass via generation capabilities described above. However, as the enabling aspects of MCM-D and MCM-L are being added, this technology can be classified as MCM-D/L type.

C. MCM-C

Commercial MLC technology (Figure 1) for substrates and modules can be divided into four major categories:

> Thick-film hybrids
> High-temperature cofired ceramics (HTCC)
> Low-temperature cofired ceramics (LTCC)
> High thermal conductivity cofired aluminum nitride (AlN) ceramics

Although these four technologies are quite different regarding performance, advantages, and disadvantages, in terms of processing one has only to distinguish between two major manufacturing techniques: the sequential MLC process applied for thick-film hybrids and the laminated MLC process applied for all cofired MLCs in a very similar form. Therefore, following the text of the second edition, a short overview of these two MLC processes is given, proceeding in the next section to a detailed description of the MLC processing steps that are common to HTCC, LTCC, and AlN.

1. Sequential MLC Process of Thick-Film Hybrids

Thick-film hybrid technology has recently been described in depth [158] and therefore will be discussed only in brief in this section. In the sequential process, a fired substrate is used as a base. Alternate layers of metallurgy and dielectric are deposited on the substrate and fired. The layering and firing process is repeated, thus building a three-dimensional structure. The dielectric used was originally glass [1] but later the use of glass-ceramics [1,2] or devitrifiable glasses [1,3] became conventional. The metals used include Ag, Au, Pt, Pd, and even Cu [4]. Such a module will usually see a firing for each separate layer that is added. Many companies and organizations have been involved in building sequential MLCs [13], including GE [2,5], RCA [4], Teledyne [6], Honeywell [7], Hewlett–Packard [8], Magnavox [1], Bell Laboratories [9], and the U.S. Air Force [3]. Sequential MLCs, as large as 16.2 × 2.54 cm, have been described by Marshall and Rode [8]. Kurzweil and Loughran [5] describe paste evaluation techniques to give 100-μm lines on 200-μm centers in sequential MLCs.

To overcome the limitations of traditional screen printing higher resolution patterning techniques have been developed such as photoimageable thick-film processing and diffusion patterning. In the first case a photoactive paste is printed on the substrate, exposed through a mask to define lines and vias, developed using an aqueous process, and fired. Up to 10 layers has been demonstrated using both Cu and Au metallization [159]. Diffusion patterning involves an unpatterned dielectric layer on which the via pattern is printed using a paste that causes the dielectric layer beneath to become water soluble, so that the vias can be obtained by rinsing the dielectric layer with water [160]. Nonconventional sequential MLCs have been developed at Pactel Corp. [10] by using a polyimide dielectric on stainless steel substrates, at DuPont [11] by using porcelainized steel substrates, and at Singer Corp. [12] by using porcelainized steel or anodized aluminum substrates. Line widths as small as 25 μm on 50-μm centers may be possible [161].

In 1980 thick-film hybrids were first introduced into high-end computers by Sperry Univac. The substrates used consisted of three to five conductor layers with integral resistors [161].

2. Laminated MLC Process

In the laminated process the module is fired only once, instead of a sequence of firings, one for each layer. The process begins with the casting of unfired (green) ceramic tapes that are cut into sheets, punched with via holes, via filled and screen printed with metallurgy, laminated into a monolithic three-dimensional structure, and fired in a furnace. The ceramic and the metallurgy densify simultaneously in the same firing cycle.

A nomenclature problem arises because both sequential and laminated modules have been described in the literature as multilayer ceramic modules or substrates. The former have also been called sequential MLCs [9], multilayer substrates [8], screen-printed MLCs [1], and thick-film hybrids whereas the latter have been designated as screened multilayer ceramics [33], laminated ceramics [18], monolithic multilayers [1], and cofired MLCs. In this chapter we refer to a laminated MLC substrate simply as MLC.

Development of laminated MLC substrates has continued since Park [14] patented a tape casting technique in 1961, but the concept was developed into a process at RCA under a group headed by Bernard Schwartz. Gyurk patented the lamination and interlayer metallization concept [71], and Stetson patented the via concept [72]. These concepts were applied to Signal Corps projects for capacitors, quartz crystal packages, and early chip carriers [65,66]. RCA holds many of the basic MLC patents. Their "micromodule" capacitor was 0.3 in.^{2} and contained 11 layers. This was the beginning of the multilayer ceramic capacitor industry. Schwartz brought the concepts with him to IBM and convinced management that this was the method of the future in making chip carriers. IBM has been very successful in implementing and helping MLC technology to become an industry standard. Good reviews of the early technology development have been provided by Stetson [73], Schwartz and Wilcox [15], and Theobald et al. [16].

A number of companies have been involved in developing laminated MLC processes, including IBM [15,17–24,77,79,127,139], 3M [16,25–27], Western Electric [28], Texas Instruments [29], DuPont [30,80,81], Ferro [13], Kyocera [31,32,74,137], Hitachi [33], Ceramic Systems [34], Dielectric Systems [35], NEC [76,99,105,140], Fujitsu [85–89,102], Nippon Electric Glass, NGK, Matsushita [163], Murata [206], Westinghouse [96], Narumi [97], Honeywell [136], Hughes Aircraft [80], Northrop Grumman, National Semiconductors, Tokuyama Soda [104], Shinko Electric Industries [98], Samsung, Heraeus [103, 168], Sorep-Erulec, C-Mac, Ceram Tec, Micro Systems Engineering, Epcos [205], among others.

During the 40 years of MLC development, the sophistication of MLC technology has increased dramatically. The first commercial MLCs, which appeared about 40 years age, were one step above a planar substrate in density. A major increase in complexity occurred in November 1980 with the announcement of commercial production of a MLC for the IBM 3081 computer, shown in Figure 2. Called the thermal conduction module (TCM), it has been reviewed by Seraphim and Feinberg [67], Blodgett [68], and Patrusky [69]. It is (90-mm)^2, up to 33 layers thick (6.55 mm), has 1800 pins on the bottom, and can accept from 100 to 118 chips reflow soldered to the top. The TCM substrate has about 350,000 vias connecting various layers, contains 130 m of internal wiring, and typically

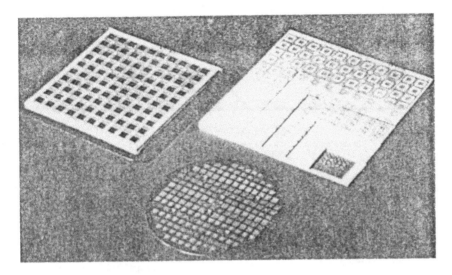

Figure 2 Ninety-millimeter MLC substrate: fired (left), mounted with 100 chips, and unfired (right), showing internal layers. (Courtesy of IBM, East Fishkill, NY)

interconnects 25,000 circuits plus 60,000 array bits [36]. This has the same number of circuits on one module as the entire IBM System 370 Model 145 has.

III. PROCESSING OF LAMINATED MULTILAYER CERAMICS

The basic process flow for building MLC substrates is illustrated in Figure 3. Different manufacturers use their own process variations and peculiarities [15,20,29,34]. Depending on the module complexity, some steps may be deleted or others added, especially after sintering.

A. Tape Casting

The science of tape casting has been discussed extensively in the literature [22,28,39,41–43,107–122]. A more recently written detailed introduction and comprehensive treatment of tape casting technology can be found in Ref. 162.

The ceramic slurry usually consists of the ceramic particles (powder), a solvent, a binder, a plasticizer, and different functional additives such as a deflocculant and a wetting agent. The different components of a ceramic slurry and

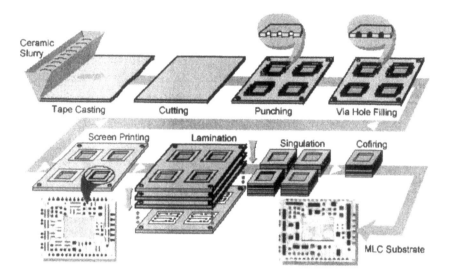

Figure 3   Multilayer ceramic process flow.

their functions are listed in Table 1. A selection of tape casting components often used is shown in Table 2. Typically, a ball mill is used to deagglomerate the ceramic and disperse it in the other components. If impact damage to the ceramic particles is to be avoided, other types of mills may be used. Complete deagglomeration and dispersion are important to ensure proper fired ceramic uniformity, particularly in shrinkage. Good dispersion is characterized by Newtonian flow behavior and low suspension viscosities in the slurry and by low total porosity and median pore radius in the cast green sheet. In the case of Al_2O_3 and polyvinyl butyral resins, the latter absorbs on the Al_2O_3 in solvents of MIBK/MeOH, creating a sterically stabilized suspension [108].

Phosphate esters have also been found to act as steric dispersants on Al_2O_3 in MEK/ethanol solvents by anchoring the long-chain molecules to the particle surfaces [115,117]. Properties of tape-cast slurries that are monitored to ensure good dispersion are typically viscosity and specific gravity. An optimal slurry viscosity is approximately 2000 CP. Impurities are another important slurry characteristic. Small quantities of impurities can result in large changes in such rheology parameters as gelation.

When choosing a slurry composition, it is important that it be compatible with the metal paste system. Dissolution of metal paste solvents into the green

Table 1 Different Components of a Ceramic Slurry and Their Functions

Component	Functions
Powders	Provide the desired material properties
Solvents	Solvate polymer binders, plasticizers, and functional additives; disperse powder particles; determine slurry viscosity
Binders	Interconnect powder particles; provide green tape strength; guarantee laminate formation; control rheological behavior
Plasticizers	Dissolve organic compounds; flexibilize polymer film
Functional additives:	
Dispersants	Disperse powder particles
	Control degree and strength of particle agglomeration
Wetting agents	Reduce surface tension and enhance wetting properties for powders and substrates
Defoamers	Prevent foam formation in slurry (especially in aqueous media)
Homogenizers	Increase mutual solubility of the components and prevent skin formation during drying
Preservatives	Suppress bacterial or fungal attacks, which occur frequently in aqueous binder systems
Flow control agents	Prevent the surface of the tape from too rapid drying and avoid crack formation
Deflocculants	Prevent dispersions from forming extremely high-density sediments

sheet can affect the integrity of the metallization. The interaction between the components in the slurry and the slurry's dependence on materials properties and their effect on the green sheet or tape are illustrated in Figure 4 [107]. Once the tape-casting slurry has been properly dispersed, it is deposited onto a moving plastic carrier sheet. This is usually done by passing under a doctor blade as in Figure 5 [120]. The solvents are driven off and the tape cures as it passes through a drying area, after which it is stripped from the plastic carrier, inspected for defects and thickness, and rolled onto spools. Smaller laboratory batch casters are also used but are poor on repeatability. The tape must be cast under the most stringent cleanliness and process controls to achieve acceptable dimensional tolerances after firing. These conditions must be maintained through lamination. The tape is now referred to as green sheet. The structure of a green tape can be defined in terms of the specific volume fractions of three different components, the ceramic powder particles V_p, the inorganic film forming polymer binders surrounding the particles V_b, and the voids V_g filled with gaseous residues of the solvents and air, with $V_p + V_b + V_g = 100\%$.

Table 2 Selection of Components Often Used in Casting Ceramic Green Sheets

Solvent	Binder	Plasticizer	Deflocculant	Wetting agent
Nonaqueous:				
Benzene	Nitrocellulose	Dibutyl phthalate	Natural fish oils	Ethyl ether of polyethylene glycol
Ethanol	Polymethacrylate (PMMA, PEMA)	Poly(ethylene glycol)		Polyoxyethylene ester
Methyl isobutyl ketone	Polyvinyl alcohol (PVA)	Polyalkylene glycol derivatives		
Toluene	Poly(vinyl butyral) (PVB)	Triethylene glycol hexoate		
Aqueous:				
Water (with defoamers, i.e., wax based,	Acrylic polymer	Butyl benzyl phthalate	Complex glassy phosphate	Nonionic octyl phenoxyethanol
silicone based, etc.)	Acrylic polymer emulsion	Dibutyl phthalate	Condensed arylsulfonic acid	
	Polyvinyl alcohol	Polyalkylene glycol		

The optimal green-sheet properties [22,28,62,107,162] include:

High tensile strength and yield strength
Flexibility for handling and via punching
Dimensional stability/defect and inclusion free
Uniform density across the sheet
An optimal range of green tape volume fractions
High compressibility for good lamination
A stable, easily volatilized binder system in reducing atmospheres
A solvent system that dries rapidly without pinholes
Sintering shrinkage that is uniform and reproducible

Taking into account the requirements considering the lamination process, the embedding of the electrode structures, and the reduction of density gradients, the optimal range of green tape volume fractions has been determined to be [162]:

$40 \text{ vol } \% < V_p < 50 \text{ vol}\%$
$20 \text{ vol } \% < V_b < 30 \text{ vol } \%$
$25 \text{ vol } \% < V_g < 35 \text{ vol}\%$

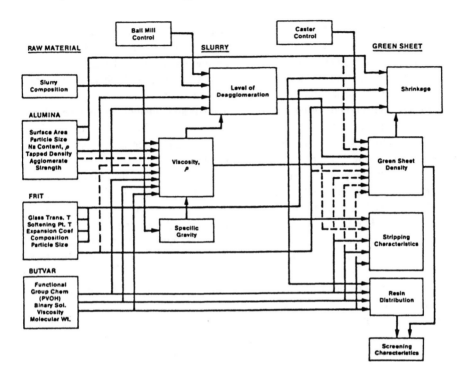

Figure 4   Green sheet materials, processes, and tool interactions. (Butvar is a trademark of Monsanto.) (From Ref. 107, with permission.)

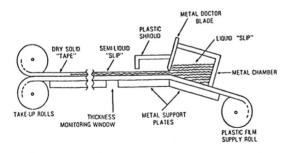

Figure 5   Schematic diagram of a casting machine. (From Ref. 120, with permission.)

The thickness of a cast sheet, nominally 60–300 μm unfired, is controlled by such factors as the doctor blade gap, temperature, carrier speed, slurry viscosity, depth of slurry behind the doctor blade, and shrinkage of the tape as it dries.

Todays continuous casters deposit 100- to 400 mm-wide sheets on a carrier foil, which is carried through several drying ovens before being inspected, stripped, and spooled. Other features on many casters include an enclosed slurry pool, multiple doctor blades, and curved beds [24,28]. Green sheet tests include compressibility, strength, bond strength, tensile strength, shrinkage, binder distribution, screening stability, and paste absorptivity [22].

Mostly the green sheets are shipped on a roll. After unrolling, the green sheet is inspected and blanked into squares using a razor, a laser, or a punch. If a laser is used its power must be controlled so as to avoid any prefiring of the sheets. Any blanks not meeting cleanliness or thickness specifications are removed. One important factor in achieving optimal sintered dimensional tolerances is the control of the green sheet thickness (i.e., ± 0.01 mm on a 0.2-mm-thick sheet).

Some tapes need to be preconditioned. That means, for example (in the case of DuPont LTCC green tape) the green sheets have to be baked for about half an hour at 120°C. The green sheets are usually punched with orientation holes—four registration holes—which will align the sheet for punching and screening, and often with additional lamination tooling holes. Because green-sheet properties can vary between the casting direction and the transverse direction [44], the layers are rotated during stacking to eliminate these effects on the fired dimensions of the substrate.

B. Via Hole Punching

Green sheet blanks begin personalization at via punching, and at this point layer-by-layer control of the blanks must be exercised through lamination [28]. Virtually every layer in a substrate is distinct from all other layers and has a unique hole pattern. Punching is typically a mechanical operation, done with a precision computer-controlled die. Punching can be done one after the other or simultaneously. Some processors use a cluster head die, which punches as many as 100 holes simultaneously [24,36]. In this process, via patterns are programmed into a computer, which in turn controls the punches. Via sizes range from 75 to 400 μm and a single green sheet may contain more than 50,000 vias. Registration of the green blanks must be held to less than ± 0.1 mm tolerance through punching, screening, and lamination [22].

As the need for smaller and smaller holes continues, limitations on mechanical punching will be reached. Laser and electron beam drilling have been demonstrated in the laboratory [143]. Recently, a UV-curable dielectric paste has been made from glass-ceramic powder and a photosensitive organic vehicle. Using photolithography technology, holes as small as 50 μm have been made on 50-μm-thick dielectric layers [144]. Other workers believe that a resolution of 10 μm on a 100-μm green sheet is possible [142]. After punching, each punched sheet is inspected for missing or plugged holes. Inspection is done optically with inspection masks and light tables. As holes become smaller, cleanliness requirements are greatly increased.

C. Line Printing and Via Hole Filling

Each punched layer is screened with the appropriate pastes, e.g., molybdenum (Mo) paste and silver (Ag) paste in case of HTCC and LTCC, respectively. Via holes can be filled using a special extrusion via filler that works with pressure of about 4 bar on a conventional thick-film screen printer equipped with a porous vacuum support. A vacuum holds the tape placed on a sheet of paper and functions as an aid for via filling. This method is somewhat limited, i.e., the via hole diameter should be larger than the tape thickness. To fill the via holes, via paste is pressed through a direct-contact mask (stencil) onto the green sheet. The stencil is usually made of 150 to 200 μm-thick stainless steel foil. Alternatively, the Mylar foil usually applied on the tape can be used as a stencil for via filling. After via holes have been filled line pattern and other conductive and dielectric patterns necessary for integrated passives are put down in several printing steps. After each printing step, the paste is dried at about 75–100°C for 5–30 min, sometimes up to 90 min [45].

The literature on thick-film screening is extensive [46–53], and some of the best sources are paste suppliers [49,50]. Several sources deal with screening techniques and problems [51–53]. See Ref. 13 for more sources. Despite the availability of references, carrying out a good screening process is an art, and the tricks of the trade are jealously guarded by developers and carefully optimized for the paste/ceramic system used. Some of the parameters that have to be optimized to achieve the best possible screening include the number of passes (back and forth), speed, paste rheology, paste solvent/green sheet interaction, side of the green sheet that is screened, tooling tolerances, and frequency of mask cleaning [45]. With good tooling and process control, line widths of 75 μm on 125-μm spacing can be achieved with high yields. Using a modern direct gravure printing method, fine conductor structures with line resolutions of 50 μm and below have been reported [204].

After screening, each layer must again be 100% inspected. This can be done using simple pattern comparators or automated using computer-controlled laser scanners. For complex patterns and large green sheets, the latter is the only reliable method. Defects in a screened pattern can be hand repaired and the sheet reinspected. In this manner, only perfect screened green sheets will be used. This minimizes the open/short defects in the fired MLC. This is one big advantage of the laminated MLC over other types.

D. Lamination and Sizing

The layers are now carefully stacked onto a lamination fixture and laminated together. During stacking, care must be taken to get every layer in its proper position and orientation in the stack, or all previous care is wasted. Precise positioning can be achieved either by using a lamination fixture with several suitable distributed position punches or by using an optically controlled positioning system that after positioning fixes each layer on the former one.

Lamination can be performed by using an uniaxial precision flat-plate press or an isostatic press. In the first case, the tapes are pressed between heated platens at a temperature of 70–80°C and at a pressure of 200 bar (20 MPa) for typically 10 mins [24]. After half the time, a 180° rotation of the laminate is often necessary. A disadvantage of this method is the flowing of the tape that results in high shrinkage tolerances, especially at the edges, and varying thickness of single parts of each layer that causes problems for microwave applications. In the case of an isostatic laminator being used, the stacked tapes are vacuum bagged and quasi-uniaxially pressed in water at a pressure of about 350 bar (35 MPa) using aluminum backplates. Again temperature is typically 70–80°C and time 10 min. In this case, cavities and windows need to have an inlay during lamination. The laminated stacked layers become an unfired substrate or "green laminate."

The individual green laminates can be sized automatically during uniaxial lamination, using lamination dies that shear the sheets as the press closes, or (in both cases) cut on special precision saws after lamination. After lamination, the edges of the green laminates are lightly sanded to eliminate sharp corners and burrs; then they are inspected for contamination, voids, or other surface damage in preparation for firing (sintering).

One aspect of the laminated MLC process that is unique is the ability to form more than one substrate per laminate, as dimensions permit. This is called "multi-up" processing. For example, using (185-mm)^2 green sheet (or green blanks), sixteen (25-mm)^2 (fired dimension), nine (35-mm)^2, or four (50-mm)^2 substrates can be obtained from each laminate [24,54]. In the example shown in Figure 2, only one (90-mm)^2 substrate is made per laminate [36]. If very large green laminates are used [e.g., (300 mm)^2] these are usually presized into smaller green laminates [e.g., (100 mm)^2] for firing. After firing, first the postsintering operations, described below, and the complete assembly of active and passive components are performed (multi-up assembly), before the final modules are sized using wafer saws. This is a common method that holds tight dimensional tolerances and allow for high-quality edges. Alternatively, an ultrasonic cutter or a laser can be used to cut the fired tapes. In both cases, the final parts show low tolerances; however, the ultrasonic process is very slow and expensive, and with the laser process the quality of the edges is very poor.

E. Binder Burnout and Sintering

1. Binder Burnout

The first phase of sintering, referred to as binder burnout, includes the removal of the plasticizer, residual solvents, dispersants, and binder from the laminate. For most MLC materials, it is important that less than 500 ppm carbon remains. As the carbon level increases, the dielectric constant and loss tangent of the substrate can increase. If levels are too high, the metal lines can even be short circuited. Other problems associated with inadequate carbon removal include cracking, blistering, warping, delamination, and anisotropic shrinkage. Binder burnout has received attention in the last few years because of its importance and complexity [37,54,55,61,123–132]. With the recent interest in lower temperature ceramics and metals that oxidize readily, it becomes increasingly difficult to remove 99.95% of the green-sheet organics at a temperature lower than the sintering temperature.

The steps in organic removal include evaporation and thermal decomposition, followed by mass transfer of the resulting volatiles out of the porous ceramic body and finally oxidation. In each case, mass/heat transport, chemical kinetics, and binder distribution all interact to determine the binder burnout characteristics. The volatiles are produced at a rate that depends on the amount of heat available. The heat contributes to the latent heat of vaporization of the polymers as well as the endothermic degradation of the organics. The binder distribution is influenced by the thickness of the ceramic, and the mass transfer resistance is determined by the morphology of the open porosity [125]. The organics near the surface of the ceramic are removed the most quickly since the diffusion distance is shorter. Vapors would be expected to be removed more rapidly than liquids because of differences in diffusivity. However, different materials often differ greatly in volatility, and transport can occur by mechanisms other than diffusion. Cima and coworkers [125] suggest that during binder burnout the binder is drawn by capillary forces from larger pores within the ceramic to smaller pores nearer the surface. They equate this to the case of drying solvents from a porous body. In order to minimize the free energy, the liquid distributes to fill the smallest pores in the body rather than distributing homogeneously throughout the structure.

The thermal degradation of one polymer is much different from that of others; therefore, each binder system would be expected to have unique binder burnout properties. The production of volatile products by thermal decomposition proceeds by several competing reactions involving radical intermediates. The first step of volatilization for materials such as acrylics is the loss of a hydrogen atom and the formation of a free radical. Depending on which polymer is used, scission, unzipping to produce a monomer, or cross-linking with another radical can occur. It is even possible that the free radical can obtain a hydrogen atom from a neighboring material and reattach to a new polymer chain. In most instances, a rapid decrease in molecular weight of the polymer occurs until it is completely volatile [126]. Another possible mechanism for thermal decomposition is side group elimination. This is the likely mechanism for polyvinyl alcohol, polyvinyl chloride, and the initial portion of polyvinyl butyral burnout. It has been observed that the surface chemistry of some ceramic oxides (acidic sites) can be important in the formation of char [126]. This char can result in strong anchoring of carbon-containing groups on the surface, which is likely to leave residual carbon after firing. Also, the intrinsic degradation reactions of polymeric and oligomeric molecules can produce nonvolatile, highly carbonaceous compounds even in the absence of adjacent ceramic surfaces [129].

In order to assist the binder burnout process, catalysts can be used. These catalysts would help pyrolyze carbon residues in nonoxidizing atmospheres. For example, Ni, Pd, Rh, Ru, Ir, and Os were found to be effective [127,128].

It is possible to reduce binder burnout problems with the use of specific organics during tape casting. Avoiding organics that readily form residues on the ceramic powder or that alter the surface chemistry of the ceramic may help. However, the impact of these altered materials on processing and green state properties may not permit their use. Thus, the binder burnout phase of the sintering cycle becomes one of the most complex and historically least understood parts of MLC processing.

The furnaces used in the binder burnout and sintering cycle may be either a continuous type or a periodic/batch type. Large substrates usually achieve better dimensional tolerances in the latter, although continuous kilns have more throughput. Figure 6 shows a stack of sintering setter tiles (kiln furniture) containing green laminates ready for sintering. Because these MLCs are alumina/Mo, the setter tiles are molybdenum, which allows for more uniform temperature control and withstands the 1500°C temperatures required for this type of MLC. The atmosphere in the sintering furnace is controlled carefully to provide enough oxygen to accomplish the binder burnout without excessive loss of SiO_2 from the ceramic [55,61]. At the same time, it is reducing enough to prevent oxidation of the metallurgy [19–21,24]. When molybdenum, tungsten, or copper is the metallurgy used, a hydrogen or hydrogen/nitrogen atmosphere is required [17,56]. The thermodynamics of the M-O-C system, where M is the metal used in the MLC, must be controlled through the P_{H_2}/P_{H_2O} ratio to control reaction of carbon oxidation [Eq. (1)] without oxidizing the metal used in the MLC. The dew point of the atmosphere is controlled by bubbling hydrogen or forming gas through a temperature-controlled water bath [21]. Thermochemical diagrams that include

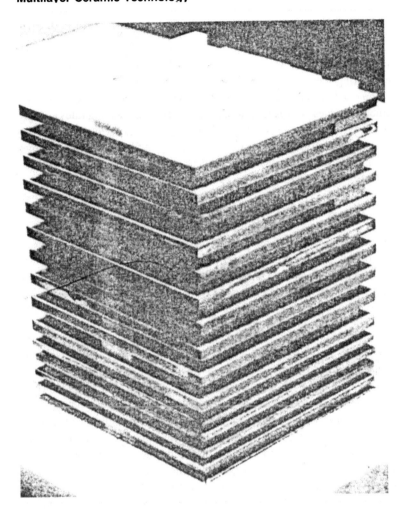

Figure 6 Load of green substrates on setter tiles ready for insertion into a sintering furnace. (Courtesy of IBM, East Fishkill, N.Y.)

M-O-C are useful in predicting dew points for the reducing atmospheres needed for cofired MLCs [21].

$$C + H_2O \rightarrow CO_2 + H_2 \tag{1}$$

An alternative approach to overcome the binder burnout problem in the case of copper metallurgy is to use copper oxide thick-film inks in place of copper inks

[163]. The green laminate is first fired in air to 500°C to completely remove the binder. It is then cooled and refired in a mixture of nitrogen and hydrogen at 300°C to reduce the copper oxide to porous copper. The porous copper completely consolidates during the subsequent cofiring process in nitrogen.

Most binder burnout cycles use a low heating rate of a few degrees per minute, although successful MLC sintering cycles have had heating rates as high as 50°C/min [149]. There is a very slight expansion of the green laminate during heat-up. The binder must give the laminate stability until densification begins but not decompose so rapidly that the laminate bulges between layers. It must be removed before sintering closes the open porosity into the laminate.

2. Free Sintering

When densification begins, the metallurgy (via) should begin shrinkage before the ceramic begins shrinking. Otherwise, cracking of the ceramic around vias occurs. The vias also shrink somewhat more than the ceramic, which means that as the cool-down begins, the via does not have the ceramic around it in tension. However, when the MLC reaches room temperature, hermiticity requires that the ceramic be tight around the via, meaning that the ceramic must be slightly in tension. In HTCC (alumina/molybdenum system) this can be accomplished by choosing materials with the thermal contraction (expansion) of the ceramic slightly higher than the metallurgy. However, this is not always possible in other systems. At the completion of the cycle, there must also be a good bond between ceramic and metallurgy if hermiticity is to be achieved [17,56,57].

All of the material selection, green sheet, and laminate parameters discussed above have a role in the success of an MLC sintering cycle. In addition, several other factors can be optimized to give the proper densification of the laminate during sintering. One factor that is especially important for LTCC dielectrics with a low glass content is particle size (or surface area of the powders). Another factor is the sintering atmosphere, which is especially important for HTCC (alumina/molybdenum system) and, in the case of LTCC, with metallurgy based on Cu. The metallurgy is more sensitive to atmosphere than the ceramic.

The alumina/molybdenum MLCs shrink about 17% during the free-sintering cycle. The hydrogen atmosphere is initially dry, with wet H_2 introduced during heat-up to facilitate the binder burnout. The sintering occurs in the range 1200–1560°C. A soak at 1560°C assures maximal densification and a good metal-ceramic bond [37]. The cooling cycle must also be done carefully, to avoid thermal shock/cracking [19], and the atmosphere is changed back to dry hydrogen. With LTCC systems, silver and copper are the preferred metallurgies because of their high conductivities. Copper metallurgy is occasionally preferred because the sintering temperature can be somewhat higher (50–100°C) than with silver metallurgy. However, the sintering of copper is even more difficult than that of molybdenum. Figures 7 and 8 show thermochemical diagrams for the Cu-O-C system, similar to those shown for the Mo-O-C system in Ref. 21. Temperatures of 500°C (binder burnout) and 900°C (densification) were chosen as examples. In this case, the Cu_2O/Cu boundary is the only one of concern. Cu_2C_2 will not form because of the presence of too much CO and O_2. CuO is inhibited by Cu_2O formation at high temperatures. Use of copper in MLCs is complicated by the presence of two oxides. Good adhesion of copper to ceramic materials, especially alumina, requires surface oxides on the copper [147]. This would be less of a problem in vitreous systems. In the Cu-O-C system, dew points can be calculated if an atmosphere is picked. Using nitrogen containing 1 ppm each of H_2 and H_2O would allow the binders to be burned out with as much as 80°C dew point (Figure 7). The dew point would then need to be reduced to below about 25°C before the sintering temperature (900°C) is reached to prevent oxide formation. Other usable atmospheres for MLCs with copper metallurgy include $N_2/CO/CO_2$ and $N_2/H_2/CO_2$, e.g., the latter with 4% H_2 in N_2 [133]. By varying the CO_2/H_2 ratios between 10 and 400, oxygen partial pressures between 10^{-11} and 10^{-3} can be achieved at 1050°C, their sintering temperature. Several MLC systems are SiO_2

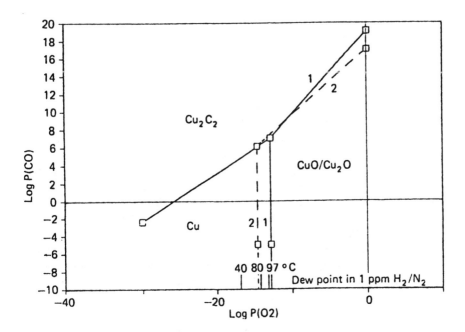

Figure 7 Cu/O/C phase diagram at 500°C.

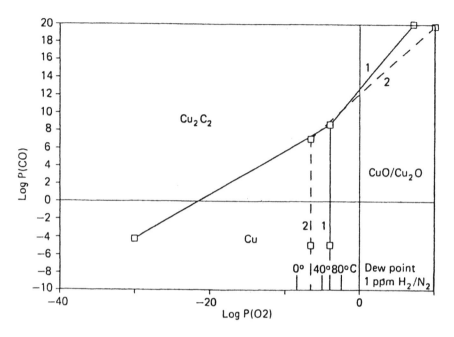

Figure 8   Cu/O/C phase diagram at 900°C.

based. In choosing atmospheres with high moisture content, care must be taken to avoid water absorption, viscosity changes, devitrification, and corrosion that could adversely affect the system [134]. For glass-ceramic with copper, Herron et al. report a sintering process using an H_2/H_2O atmosphere that changes from 10^{-6} to 10^{-4} P_{O_2} in going from below 200°C to an 800°C hold, and then changing to nitrogen for the remainder of the 950°C sintering cycle [147]. Vest reports that copper used as surface thick film is slow to oxidize and that shiny copper films can be successfully fired in atmospheres containing 1 ppm O_2 or more at 900–1000°C [148].

A second problem in sintering MLC substrates, as well as any technical ceramic, is achieving good dimensional tolerances. Complex substrates require shrinkage tolerances as low as ±0.1%. Two methods for achieving such a high degree of shrinkage control have been employed. The first is the traditional method most often used. This is done by green density control, proper substrate design, good shrinkage matching between the paste and green sheet, compatible thermal expansions between metal and ceramic (on cooling), and very low thermal gradients in the furnace. This dimensional control problem is discussed in the literature and a few examples are cited [19,41,58,59]. The other method for shrinkage control developed for LTCC is to totally constrain the lateral shrinkage and is described below. In addition to distortion in the horizontal plane of the substrates, other dimensional problems such as camber, warping, and waviness occur. It is possible to reduce these problems by balancing the metallurgy throughout the substrate and, as a last resort, by refiring the substrate with a weight on it at a slightly lower temperature. This is called flattening.

3. Constrained Sintering

In free or unconstrained sintering, densification is more or less equally distributed over the three dimensions of the MLC laminate, i.e., assuming a green density of approximately 50%, $x-y$ shrinkage varies typically between 13% and 15% and z shrinkage between 15% and 20%. In constrained or zero-shrinkage sintering the multilayer laminate is constrained during sintering in such a way that lateral shrinkage is nearly completely suppressed, i.e. $x-y$ shrinkage is 0.1–2% and z shrinkage 43–47%. These sintering processes can be classified into two groups: those where a vertical pressure is applied on the laminate during sintering (pressure assisted sintering) and those in which no pressure is applied (pressureless assisted sintering).

The pressure assisted sintering is most effective in reducing shrinkage and consequently shrinkage variance. With this method tolerances of less than 0.01% can be achieved [164]. In this process two additional tape layers have to be colaminated to the green multilayer substrate. This laminate must be stacked together with interleaved porous plates and the stack placed between two constrained dies. The dies are pressed together using a uniaxial pressure and fired in a box furnace. One of the most critical points in pressure assisted sintering is to customize the additional colaminated tape layers, which transmit pressure and constraining forces during sintering, due to the need to remove this tape cleanly at the end of the firing process.

In the case of pressure-less assisted constrained sintering, several versions have been developed. The simplest method is to fire glass-ceramic laminates of 15 layers or less on suitable prepared substrates, usually metal plates that are occasionally used in thermal management as heat spreader and heat sink. These metal plates must be thermally matched to the glass-ceramic and therefore are usually a compound of Cu and Mo [165]. Early in the sintering process, the glass-ceramic laminate adheres to the Cu-Mo-Cu metal plate, which then acts to suppress the lateral shrinkage. There is another pressure-less assisted zero shrinkage process that can be considered as the standard zero shrinkage process, as shown in Figure 9. This process also uses two additional tape layers (sacrificial layers), which have to be colminated to the laminate of LTCC tapes [166]. These two additional tape layers are fabricated of a material (e.g., Al_2O_3) with a much

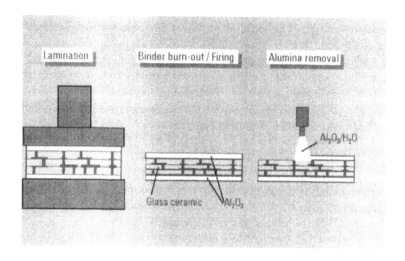

Figure 9   Zero $x-y$ shrinkage. Standard manufacturing process.

higher sintering temperature compared to LTCC. Therefore, these tapes do not shrink or sinter themselves but maintain a uniformly high frictional contact to the surface of the LTCC stack and so constrain the shrinkage of the LTCC during firing. With this method it is possible to reduce shrinkage to about 0.1–0.2% with a tolerance of less than 0.05%. The main disadvantage of this processing method is that after firing the additional sacrificial layers must be removed suffi-ciently to ensure that the surface conductors are solderable, and an additional printing and postfiring step for the outer metallizations is usually necessary.

4. Self-Constrained Sintering

Although the above-described techniques for controlling $x-y$ shrinkage are very effective, the additional processes required tend to complicate the manufacturing process and to increase fabrication cost. In the self-constrained sintering process, described in Figure 10, no additional processing steps are required; therefore, the benefits of constrained sintering techniques are realized with free sintering.

In self-constrained sintering the LTCC laminate is a compound of tapes with different sintering characteristics that can be controlled according to the type of glass-ceramic composite and the glass composition and concentration used (see Section IV). For example, in Figure 10 the laminate consists of four outer tape layers with high densification around 750°C and three inner tape layers with high densification just below 900°C. If this laminate is fired at 750°C, the

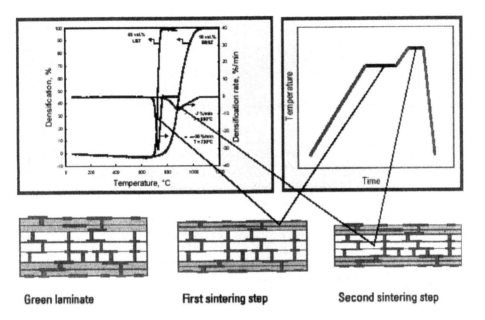

Green laminate   First sintering step   Second sintering step

Figure 10   Self-constrained zero x–y shrinkage process.

outer layers will shrink while the inner layers will not; thus, these inner layers will constrain the outer layers only to shrink in the z direction.

After densification of the outer layers is completed, the firing temperature will be increased to about 900°C so that the inner layers have a high densification rate. Now the outer layers will constrain the sintering of the inner layers, which again can only shrink in the z direction. Thus, the LTCC laminate itself will constrain its sintering, i.e., suppress its lateral shrinkage. Beside the advantage that no additional postfiring processing is required, this method has another decisive advantage, which is the decoupling of the sintering processes in the inner and outer layers. As a consequence, tape layers of materials with quite different electrical properties can be combined within one MLC substrate in order to embed passive devices of corresponding properties. This will be possible on the condition that the thermal expansion coefficients of the combined layer materials are suitably matched. As an example, Figure 11 shows an advanced LTCC module where glass-ceramic composite tapes with high dielectric constants, low dielectric losses, and excellent temperature stability have been embedded in an LTCC substrate, in order to integrate miniaturized passive microwave components such as trans-

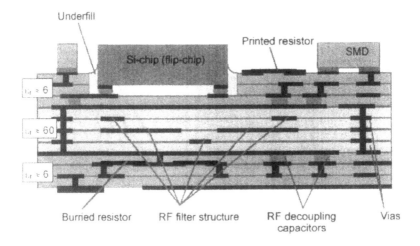

Figure 11 Advaned LTCC module with embedded high-dielectric layers for compact integration of complex microwave components.

mission line resonators, frequency filters, couplers, transformers, and baluns [167].

Another version of self-constrained sintering is obtained if the lamination of tapes with different sintering characteristics is shifted from the substrate fabrication to the tape fabrication. In this case, two thin tapes with different sintering characteristics (but identical electrical properties) are laminated in roll form as coated, after tape casting on a clear plastic backing material. The tape obtained in this way is slit to width and then delivered in roll form as a self-constrained LTCC tape that can be applied to a standard LTCC process. Advanced processing techniques that suppress the shrinkage during sintering within the substrate plane (zero xy shrinkage) to improve the geometrical precision considerably have several significant advantages especially for microwave applications. These benefits include:

Precise control of spacing in and between layers
Accurate alignment with large substrates ($> 8 \times 8$ in.)
Flip-chip bonding of bare chips
Integration of tape layers with different electrical properties
Embedded passives of corresponding properties and with decreased tolerances
Highly-solderable, camber-free conductors

In addition to these advantages, self-constrained LTCC tapes make precision cavity structures possible. Cavities cut into the laminate show minimal distortion because the walls of the cavity structures remain nearly undistorted due to the fact that each tape layer is self-constrained [168].

F. Postsintering Processing

After the substrate is fired, it must be prepared to receive surface resistors, IC chips, surface mount devices, and input/output (I/O) styles for the connections to the next layer of packaging, which is usually a printed circuit board [63].

1. Postfiring and Processing After Constrained Sintering

There are several cases where materials need to be postfired; that means the paste is applied after firing and the MLC needs to be fired again. Depending on the material used the postfiring conditions may vary. Especially postfired resistor materials need well-defined firing conditions to achieve acceptable resistor tolerances. Resistors as fired show tolerances of about 30%. Therefore, surface resistors that can be trimmed with the help of a laser should be designed to fire to about 65–85% of their nominal value; usually high-value resistors are fired to lower values (e.g., 70% of nominal value) and low-value resistors to higher values (e.g., 80% of nominal value). It should be noted that refirings may change resistivity.

In a constrained sintering process based on additional tape layers there is at least a need to remove the additional sacrificial tape layers cleanly at the end of the firing process to ensure that the surface conductors are solderable and bondable. Depending on the material system used, it also can happen that the surface conductors cannot be applied until the sacrificial tape layers have been removed. In this case at least one additional printing and postfiring step is necessary to apply the surface metallizations. If gold lines and catch pads, respectively, are to be printed on top of copper or silver-filled vias, it becomes necessary to interpose another metal such as palladium or rhodium between them to prevent the formation of voids at their interface due to vastly differing diffusion rates of one into the other (Kirkendall effect) [169].

The adhesion of thick-film components such as bonding pads and sealing bands on the substrate surface is often problematic. Although these can be cofired, it is often necessary to fire these surface patterns in postsintering steps to obtain better geometrical accuracy (especially with free sintering) or to form surface metallizations with different pastes.

With the ever-increasing need for more dense wiring, more I/O connections, and minimization of the dielectric constant around transmission lines, thin-film technology on the surface of the MLC substrate (MCM D/C; see Section II. A) has evolved. In order to prepare an MLC substrate for thin-film application, a very planar, smooth surface is desirable. Since most as-fired ceramic surfaces do not meet this requirement, a surface planarization step followed by a polishing step must be performed. The planarization and polishing processing generates debris, so the substrates need to be cleaned before proceeding to thin-film applications. Photolithographic techniques are used to generate the fine patterns of thin films. The materials most commonly used are copper conductors and polyimide dielectrics, although many others have also been tried. There are several descriptions of thin-film structures in the literature [74,105,136–141,145]. Most contain about five layers. Layer thickness is usually between 10 and 20 μm with 25-μm-wide and 5-μm-thick conductor lines.

2. Galvanic Processes and Protection Layers

In the case of HTCC and in the case of LTCC with metallurgy based on Cu there is a final plating of the metal surface features (typically with nickel and gold), and an electrical testing for opens and shorts is necessary. Nickel is used as a corrosion protector and helps gold adhesion. It is plated on and diffusion bonded to the molybdenum [37]. The gold is used to enhance wettability to solders and pin brazes. The plating process uses either an electrolytic nickel bath and electrolytic gold or an electroless nickel bath and electroless gold bath, or combinations as required by the package or module. Because the electrolytic process needs an electrical contact to each feature to be plated, electroless plating is usually preferred. It is even necessary for multichip modules because many surface features that interconnect one chip to another do not also connect to an I/O pin. For electroless plating, a bath is chosen that selectively plates all surface metal features without depositing nickel or gold on the dielectric. After depositing electroless nickel, the substrate is heat treated to improve adhesion, and thin gold is plated to prevent oxide formation and improve wettability. For processing of wire bond pads and seal bands, a thick gold layer is selectively plated to ensure robust connections.

Substrates are electrically tested by contacting top and bottom surface pads to ensure the integrity of the wiring connections. Parametric testing can ensure that net resistances and capacitances are within desired specification limits. The testing of complex MLCs requires sophisticated equipment and is speeded up with computer control and multiple-probe sets [54]. For example, on the 90-mm substrate shown in Figure 2, there are 1800 pin pads on the bottom and 100 or 118 chip sites on the top, each chip site having vias for the chip. The chip sites are surrounded by via/line/pad groups used for probing and rewiring. This makes more than 250,000 probe contacts that need to be connected and verified to adequately test this substrate [36].

3. Surface Mount and Chip Joining Techniques

The final steps of mechanical attachment and electrical interconnection, of the semiconductor devices and passive components to an MLC substrate, changes it into an MLC module. Chips can be joined to the substrate by wire bonds, beam or ribbon leads, or solder balls using wire bonding, TAB, or flip-chip technology, respectively. The latter process is preferred for high-density and microwave interconnections. It allows all the chips to be simultaneously attached to the substrate in a single reflow cycle [24,36].

The most common method of bare die assembly is attachment of the die with an adhesive and wire bonding. Several attach materials are available, including epoxy, polyimide, thermoplastics, and solder. Wire bonding does not require any special processing of the semiconductor dies. Thus, more dies are available for wire bond assembly. In the wire bonding process, the wire is brought into contact with the metallized bond pad. Thus a shearing action is produced that removes contaminants from the wire and the pad at the atomic level so that they metallurgically bond. While bonding of aluminum wires requires only ultrasonic energy and pressure (ultrasonic bonding), bonding of gold wires requires the additional application of heat (thermosonic bonding). Ultrasonic wire bonding can be used for finer bond pad pitch than thermosonic wire bonding. The limitations of wire bonding are the requirement for perimeter pads on the die, the slow sequential nature of the process, and the parasitic inductance of the wire, which can limit the performance of high-speed electronics and microwave devices.

In TAB, the die are first bonded through bumps to a copper lead pattern that is embedded in a polymer tape. Once bonded, the die can be automatically handled in a tape-and-reel format. This includes test and burn-in of the die prior to assembly, a major advantage of TAB processing. A further advantage is the sealed top surface of a bumped die. For assembly, the dies are excised from the tape and then the leads are formed for bonding in either a die face-up or face-down configuration. Because bumped wafers are not readily available, a number of alternative approaches have been developed such as bonding of gold and palladium-gold balls to the die pads using a thermosonic wire bonder [170] or by forming the bumps on the tape, e.g., plating the bumps on the ends of the copper fingers [171]. For placement of the die, epoxy and thermoplastic die attach materials (paste and preforms) are used. Soldering can be used to bond both Sn and Au plated leads to the substrate metallization. The solder is typically pulse heated with thermodes that maintain force on the leads until the solder solidifies. The Au plating thickness is important in the reliability of soldered outer lead bonds [172]. Alternatively, single-point thermocompression or thermosonic bonding can be used for bonding Au-plated TAB leads to Au substrate metallization. The disadvantages of TAB include the need for bumped die and the need to custom design a tape for each die which must be redesigned for shrinked-die versions. The limited availability of bumped die has resulted in the development of bumpless TAB processes to allow the use of standard processed die [173].

Flip-chip technology was introduced by IBM in 1964 [174]. The chips were transistors with Au/Ni-plated Cu balls embedded in Pb/Sn solder bump on the three transistor pads. A Cr/Cu/Au interface layer was deposited between the Al transistor pads and the solder bumps. The device was assembled to the substrate by inverting (flipping) the chip and reflow soldering the copper balls to the corresponding pads on the substrate. Later this copper ball technology was replaced with a pure solder bump controlled-collapse chip connection (C4) process [175]. A number of flip-chip bumping processes have been developed that use plating to form the solder bumps [176–178]. Common adhesion/barrier layers are Cr, Ti, and TiW, whereas Cu, Ni, and Pd are used for the solderable metal layers, and Au is the typical oxidation barrier to which the Pb/Sn solder is plated. After plating and etching of the background layers, the solder is reflowed to produce nearly spherical bumps. Bumping processes based on nonconductive and conductive polymers have also been developed [179, 180]. However, conductive adhesive contacts have not proven reliable to aluminum metallization. Therefore, aluminum pads first must be zincated followed by nickel and gold plating prior to screen printing the conductive adhesive. The number of I/O connections can be dramatically increased with bumps placed in an area array over the surface of the die compared to peripheral pad designs. On the other hand, peripheral designed bond pads of a chip may be on 100 μm pitch or less. Because this is too fine for flip-chip bumping and assembly on an MLC substrate, a redistribution layer on the chip is required to fan the perimeter chip pads to a wider area pad array with a pitch of 250 μm or more. Alternatively, the use of thin films on the surface of the MLC substrate (MCM D/C) can be considered, as described in Section II.A and in this section under F.1. Typically, the grid size for the thin films is one-half the size used in the MLC, so the wiring density is four times that in the ceramic. The thermal expansion mismatch between the semiconductor die and the MLC substrate results in stress on the solder joints during thermal cycling, which can lead to fatigue failure. The problem of fatigue failure increases with increasing die size, which is the trend in the semiconductor industry. In order to improve reliability, epoxy underfills are used that reduce the stress on the solder joints [181–183]. Flip-chip technology provides a high-volume, mass reflow method of assembling MCMs. As integrated circuit complexity, I/O counts, or operation frequency increases, flip-chip technology utilization will expand.

4. Module Encapsulation

Finally the module is tested and encapsulated. The encapsulation is usually accomplished by embedding the chips in a polymer or by enclosing the chips under a cap. If protection of the pin side is needed, glasses are available that can be applied before the pins are attached [64]. If enhanced cooling is needed to prevent circuit damage, liquid or gas cooling jackets can be added to the module [37]. For the IBM MLCs, the latter process is used for cooling the 90-mm substrates

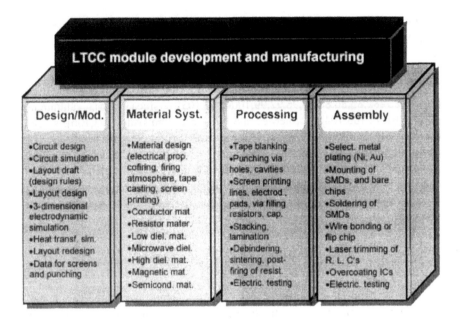

Figure 12   Overview of MLC substrate and module manufacturing.

and Kovar caps are used on the 50-mm and smaller MLCs [24]. Finally, an overview of MLC substrate and module manufacturing is shown in Figure 12.

IV. MATERIALS FOR LAMINATED MULTILAYER CERAMICS

In this section a general description of MLC materials is given, followed by a more specific discussion of new developments in the field of LTCC materials.

A. Materials Selection, Qualification, and Preparation

The raw materials used in MLC substrates include the ceramic and metal powders and any organic additives, solvents, or binder components used to make the slurry for tape casting and the various metal pastes. In the early days of MLC, aluminum oxide (92%) was the primary material of choice [79]. It was well characterized, had good strength (300 MPa), a low dielectric constant (8.5), high thermal conductivity (16.7 W/m K), and a low thermal expansion coefficient (6.5 ppm/K). Since that time, many new materials have been explored, starting with alumina-mullite [23]. Many of them have properties superior to those of alumina (see Table 2 of Ref. 75).

In the references, overviews of HTCC materials are included [74,75,79, 83,95] as well as articles dealing primarily with LTCC materials (glass-ceramics) [76–78,80–82,85,86,88,89,96,97,99,100–102,105], silica [84], aluminum nitride [91,92,103,104], cordierites and cordierite composites [93,106], silicon carbide [94], mullite [87,98], and ceramic fiber/polymer composites [90]. The permittivity or relative dielectric constant $\varepsilon_r = \varepsilon/\varepsilon_0$ of a material is of special significance because it determines the phase velocity v_{ph} of a plane wave (frequency $f = \omega/2\pi$) propagating in the material, its wavelength λ (or wave number k) in the material, or its propagation delay time T_{pd} according to Knaupp's equation [74]:

$$v_{ph} = \frac{1}{T_{pd}} = \frac{\omega}{k} = \frac{\omega}{2\pi} \cdot \lambda = f \cdot \lambda = \frac{C}{\sqrt{\varepsilon_r}} = \frac{\omega}{k_o \sqrt{\varepsilon_r}} = f \cdot \lambda_o \cdot \frac{1}{\sqrt{\varepsilon_r}} \qquad (2)$$

where ε_0, k_0, and λ_0 are dielectric constant of vacuum, vacuum wave number, and vacuum wavelength, respectively. Some of the lowest dielectric constants in Table 3 are seen for high-silica glasses. MLCs with low dielectric constants are needed for high-speed electronics and microwave devices with high operation frequencies (>10 GHz). On the other hand, materials with medium dielectric constants ($40 < \varepsilon_r < 80$) are needed for miniaturized integrated analogous devices with operation frequencies in the range 0.5–5 GHz. As semiconductor chips become larger, the thermal expansion coefficient of the substrate becomes more important. A match with the chips would avoid inducing a stress on the interconnection during heating and cooling. In the case of silicon chips, an expansion coefficient of 3 ppm/K would produce a match. In Table 3, there are several materials with similar expansion coefficients, such as borosilicate glass and SiC. On the other hand, it is also often necessary to take into consideration that SMDs usually have high thermal expansion coefficients of about 10 ppm/K. The strength of materials can be important. Borosilicate glass has a very low dielectric constant and a good expansion match to silicon; however, its strength is not high. Strength can be important in the manufacturing environment, where handling stresses can be high.

Thermal conductivity is important as chips become more and more powerful. In many advanced MLC packages, heat is extracted from above the chip instead of through the substrate, making the thermal conductivity of the substrate much less important.

The final column in Table 3 is the sintering temperature, which dictates the metals that can be cofired. Low-temperature ($<1000/900°C$) sintering materials allow the use of high-conductivity copper/silver (melting point 1083/960°C),

Table 3 Properties of Basic Dielectric Materials Used for the Development of MLCs

Material	ε_r	TCε (ppm/K)	tanδ 10^{-3}	TCE (ppm/K)	λ (W/m.K)	T_s (°C)
Alkali halides						
NaCl	5.62	+325	0.2 (1 MHz)	41.0	6.2	≈600
KCl	4.68	+295	0.1 (1 MHz)	36.6	7.0	≈600
KBr	4.78	+315	0.2 (1 MHz)	38.5	4.9	≈600
LiF	9.27	+370	0.2 (1 MHz)	34.2	5.0	≈500
Carbides, Nitrides						
SiC (BeO doped)	40.0	+40	0.5 (1 MHz)	3.7	270	≈2100
Si_3N_4	6.0			0.8	33.5	≈2000
AlN	8.8		0.5–2 (1 MHz)	4.5	320	≈1800
Oxides						
SiO_2: quartz	3.78	0	0.1 (1 MHz)	11	1.5	≈1100
SiO_2: crystobalite	3.8			50		
SiO_2: tridymite	3.8			17.5		
MgO	9.8	+110	0.3 (1 MHz)	11		
Al_2O_3	9.8	+120	0.3 (1 MHz)	6.5	30	≈1500
Ga_2O_3	9.8			7.0		
$3Al_2O_3/2SiO_2$: mullite	6.8		4 (1 MHz)	5.0	6.7	≈1500
$CaO/Al_2O_3/2SiO_2$: anorthite	6.2		0.5 (1 MHz)	4.8		
$2MgO/2Al_2O_3/5SiO_2$: cordierite	4.5–5.3		1 (1 MHz)	1.5	4.3	≈1450
$LaAlO_3$	26	+100	0.3 (1 MHz)	7.0		≈1400
$BaZrO_3$	39	−300	0.5 (1 MHz)	8.0		≈1300
$Ba(Zr_{0.3}Ti_{0.7})O_3$	3350	−35200	10–20 (1 MHz)	8.0		≈1300
$BaNd_2Ti_4O_{12}$	≈85	≈−130	≈0.3 (2 GHz)	9.0		≈1300
Glasses						
Silicate glass	3.8		0.4 (1 MHz)	0.6	1.5	≈1100
Borosilicate glass	4.6		26 (1 MHz)	3.1	1.7	≈1000
Li, K-borosilicate	4.1		0.6 (1 MHz)	3.2	1.7	≈1000
$BaO-Al_2O_3-B_2O_3-SiO_2$	10.0		1.8 (10 MHz)	7.9		≈750
$Bi_2O_3-B_2O_3-SiO_2-ZnO$	22.4	+280	4.2 (10 MHz)	11.3		T_g≈390
Glass ceramics						
$Li_2O-Al_2O_3-SiO_2$	6.6		1.8 (1 MHz)	5.0	5.4	≈1100
$Li_2O-5\%ZnO-SiO_2$	5.0		2.3 (1 MHz)	2.8		
$Li_2O-30\%ZnO-SiO_2$	5.3		6.3 (1 MHz)	8.0	2.2	≈900

whereas higher sintering temperatures (1500–1600°C) limit the metal choices to refractory metals such as Mo and W. Most metals have been tried, with tungsten or molybdenum being the most popular for alumina-cofired MLC substrates [18,19,21].

In choosing the materials for any MLC substrate, all of these factors must be weighed, as well as costs and ease of manufacture. For example, BeO looks like an attractive material until one considers toxicity. Silicon nitride is exceedingly strong, has a low dielectric constant, and has a very high thermal conductivity, but its sintering temperature (or hot pressing temperature) is very high (2000°C). Some of the most promising materials appear to be glass-ceramics and glass/ceramic composites. This includes the alumina + glass mixtures and the traditional glass-ceramics [76,77,80–83,85,86,88,95,97,99–102,106]. One example is the alumina/lead borosilicate glass-ceramic system with gold or silver-palladium metallurgy [70].

Other materials for future use that hold a great deal of promise but are not listed in Table 3 are the ceramic/polymer composites [78,90,93]. Since many polymers, such as polyimide and fluorinated polymers, have very low dielectric constants (3.5 and 2.5, respectively), they not only help increase the fracture toughness but also can help lower the overall dielectric constant. Their drawbacks include their lack of high-temperature stability and their high thermal expansion coefficients, which can range from 10 to 60 ppm/K. Ultimately, the final materials choice would depend on the specific MLC requirements and goals.

The biggest problems with raw materials include obtaining a pure, repeatable starting material with uniform properties and minimizing variations between various lots of the material [36]. Manufacturers accomplish the latter by purchasing or making very large lots of a given material and storing it for future use. Changing ceramic lots can completely alter the process control of an MLC process unless the lots have virtually identical particle shapes, particle size distributions, impurities/additives, and surface chemistry (water absorption, etc.). The subject is complex enough that entire volumes [38] and courses [39] have been written about it. Part 2 in Ref. 38 is very good for alumina characterization. The purity of the MLC starting materials is important for several reasons. One, in addition to those above, is the radiation sensitivity of the chips that will be bonded to the MLC. The number of α and β particles emitted from naturally radioactive trace elements in the starting materials looms as a serious problem as chip sophistication increases.

Numerous binder systems are in use, some of which are water soluble and some soluble only in organic solvents. Various binders, additives, and binder systems are discussed in the literature [20,28,38–41] and have already been discussed in the section on tape casting (III.A). As an example, the IBM Thermal Conduction Module will be used. It is based on an alumina/Mo MLC, and the substrate ceramic is made by using a 90:10 ratio of Bayer α-alumina and a glass frit, blended with a polyvinyl butyral–based organic binder system [24]. After firing, the ceramic is equivalent to a 94% alumina. The binder system contains a plasticizer and two solvents in addition to the polyvinyl butyral [22](Table 4).

Preparation of raw materials include milling the powders to a uniform particle size distribution and sometimes classification of the powders to remove fines or agglomerates. Ball milling [28], vibratory milling [41], and jet milling [22] are all successful techniques in use. Often two-stage milling is employed, with the binder system being added to the second stage, which becomes essentially a blending step. The product of these steps is a slurry ready for tape casting.

The processing of metallic paste is similar to that of the ceramic, although the viscosity is much greater and the quantities smaller. The process can include milling and/or classifying the powders, followed by blending with the binder system. The ultimate paste used on each layer is usually unique to that layer (ground plane versus signal layer, etc.). Paste must be carefully optimized throughout the substrate before the best shrinkage tolerances can be realized [17,19].

B. Glass-Ceramic Compositions

Three approaches have been used to obtain glass-ceramic compositions suitable for fabricating self-supporting LTCC substrates (Figure 13). In the first type, the glass-ceramics (GC), fine powder of a suitable glass composition is used that has the ability to sinter well to full density in the glassy state and simultaneously crystallize (form crystalline phases) to become a glass-ceramic. The crystalline phases make the glass-ceramic very stable against further temperature treatments such as post firing processes. A typical example of this type is the glass system $MgO\text{-}Al_2O_3\text{-}SiO_2$ having cordierite as the principal crystalline phase. During the firing process, the glass powder sinters and crystallizes to cordierite with small amounts of binary magnesium silicates such as enstatite. The sequence of these processes has been illustrated by simultaneously performing differential thermal analysis and dilatometer measurements [184]. Another example of glass-ceramics is the glass system $CaO\text{-}SiO_2\text{-}B_2O_3$ having wollastonite as the principal crystalline phase [185].

The second type, the glass-ceramic composites (GCC), consist of a mixture of a suitable glass and one or few ceramic powders, such as alumna in nearly equal proportions. Usually a volume content of more than 50% of a glass with a soften temperature T_{soft} of 20–50 °C below the sintering temperature of the composite is used. The glass (e.g. borosilicate glass) serves as a flow medium, i.e. sintering or densification is caused by a viscous flow, a dispersing and rearranging of the crystalline particles in the glass melt. Although reactive processes such as dissolution and precipitation occur at the glass ceramic interfaces, these are not essential for the densification process, i.e. densification is due to nonreac-

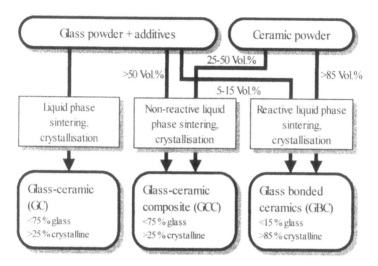

Figure 13   Manufacturing variants of glass-ceramic compositions for low temperature cofired ceramic substrates.

tive liquid phase sintering. However, the sintering kinetics can be considerably influenced by these reactive interface processes. In most cases, the reaction between the glass and the ceramic surface leads to the formation of new crystalline phases, so that the amount of glass phase in the resulting composite substrate can be considerably lower than in the original mixture of glass and ceramic powder. This usually affects the properties of the composite in the desired direction, e.g. to lower dielectric losses or to a lower temperature coefficient of the dielectric constant [186]. Because many glass and ceramic powders can be used to obtain densely sintered glass-ceramic composites, it is possible to influence or optimise composite properties such as the dielectric constant σ_r, its temperature coefficient TCε, or the coefficient of thermal expansion TCE. For example, when mixtures of alumna and lead borosilicate are fired together, the reaction product is anorthite, a calcium-alumino-silicate having low εr and low TCE. However, when a mixture of cordierite and borosilicate glass was used to obtain a low TCE composite, the reaction between the two produced a result quite opposite from what was expected (a TCE of about 17 ppm/K instead of about 3 ppm/K) due to the formation of cristobalite. This reaction can be inhibited by the addition of alumna [187-189].

In the third type, the glass bonded ceramics (GBC), only a very low volume content of 5–15% of a glass with a very low soften temperature $T_{soft} < 400\ °C$ is used to densely sinter the composite. In this case, it is necessary that the particle size of the crystalline phases be very low and the reactivity between the glass melt and the crystalline phases of the composite be particularly high, because sintering is based on a combination of rearranging and dissolving-precipitation processes of the crystalline material in the glass melt (reactive liquid phase sintering). Compared to the second case, the glass requirements which must be met to achieve full densification at 900 °C, especially the viscosity-temperature behaviour and the reactivity with the crystalline phases, are very sophisticated. Therefore, a special development of suitable glasses for this type of glass bonded ceramics is always necessary (see Section IV E).

Examples for these three types of glass-ceramic compositions are shown in Table 4.

C. Standard Dielectric Materials for LTCC

Only a few companies such as Ferro, Dupont and Heraeus offer LTCC dielectrics in the form of green tapes. Other companies, such as Kyocera, NGK and Epcos, offer LTCC material systems only in combination with their own manufacturing processes, and a third group of manufacturers, such as Murata, have developed LTCC systems only for their own produc (Table 5).

The LTCC tape is the main part of the LTCC system that defines the key properties including dielectric constant εr, its temperature coefficient $TC\varepsilon$, dissipation factor $\tan\delta$, insulation resistance, breakdown voltage, thermal coefficient of expansion $TCE \equiv \alpha$, thermal conductivity λ, and the mechanical properties such as flexural strength and fracture strength. Standard LTCC dielectrics, i.e. dielectrics which can be used for nearly all kind of applications, are usually from the glass-ceramic (GC) or from the glass-ceramic composite (GCC) type and have a dielectric constant between six and nine. However, these standard LTCC dielectrics often cannot meet all material requirements necessary for a certain application. One example are MLC substrates based on LTCC technology for wireless communication devices, one of the fastest growing segments in consumer electronics.

One additional requirement in the wireless field is very low dielectric losses or very high dielectric Q values, respectively. Having Q values exceeding 1000 seems to be adequate for most applications in the 1–5 GHz region. However, a further additional requirement is here that the dielectric exhibits a $TC\sigma$ tailored to achieve a low temperature coefficient of frequency TCf according to:

$$TCf = -\frac{1}{2} \cdot TC\varepsilon - \alpha \qquad (3)$$

This is because some of the key elements integrated within the dielectric in the high frequency circuit such as resonators and filters need to be very stable under fluctuations of the temperature. These requirements can be fulfilled, on the one

Table 4 Examples of LTCC Dielectrics and Their Main Properties

LTCC dielectric material	ε_r (1 MHz)	$TC\varepsilon$ (ppm/K)	$\alpha \equiv TCE$ (ppm/K)	Bending strength (MPa)	T_s (°C)	Conductor material	Company
Glass-ceramics							
$MgO\text{-}Al_2O_3\text{-}SiO_2$ Cordierite type	5.3–5.7		2.4–5.5	180–230	850–950	Cu	IBM
Cordierite type	4.9–5.6		2.2–3.5	170	900–1050	Au, Cu	NTK/NGK
$CaO\text{-}Al_2O_3\text{-}SiO_2\text{-}B_2O_3$	6.7		4.8	250	950	Cu, Ni	Taiyo Yuden
Glass-ceramic composites							
$SiO_2\text{-}B_2O_3 + Al_2O_3$	5.6		4.0	240	1000	Cu	Fujitsu
$Pb\text{-}SiO_2\text{-}B_2O_3 + Al_2O_3$	7.5		4.2	300	900	Au	NEC
$MgO\text{-}Al_2O_3\text{-}SiO_2$ $\text{-}B_2O_3$ + silica	4.3–5		3–8	150	850–950	Ag, Ag/Pd	Hitachi
$CaO\text{-}Al_2O_3\text{-}$ $SiO_2\text{-}B_2O_3 + Al_2O_3$	7.7	110	5.5	270	890	Ag, Au, Ag/Pd/Pt	Sumitomo
Glass + Al_2O_3 + $CaZrO_3$	8.8		7.9	210	850	Ag, Au	DuPont
$BaO\text{-}Al_2O_3\text{-}$ $SiO_2\text{-}B_2O_3$ + silica	7.0		6.7	240	850–900	Ag, Au, Ag/Pd/Pt	Heraeus
$BaO\text{-}SrO\text{-}CaO\text{-}K_2O\text{-}SiO_2$ $\text{-}B_2O_3 + Al_2O_3 + TiO_2$	9.3	−25–(−5)	8	250	850–900	Ag, Au, Ag/Pd/Pt	Motorola
Glass-bonded ceramics							
$BaNd_2Ti_4O_{12} + Bi_2O_3$ $\text{-}B_2O_3\text{-}SiO_2\text{-}ZnO$	60	−26–(−10)	9–10	300	900	Ag, Au, Ag/Pd/Pt	Siemens/Heraeus

Table 5 Companies Involved in Developing Laminated MLC Processes, Manufacturers, and Foundries (Supplier of LTCC Materials, Substrates, Modules)

Supplier of:	Mat.	Sub.	Mod.	I-Net: www.	Ref.
United States					
Ceramic Systems	✓				34
CTS		✓	✓	ctscorp.com	
Dielectric Systems	✓	✓			35
DuPont	✓			dupont.com	30,80,81
Electro-Science Laboratories	✓			electroscience.com	
Ferro	✓			ferro.com	
Honeywell		✓			136
Hughes Aircraft					80
IBM					17–24,77,127,139
Kyocera America (VisPro Div.)	✓			visprocorp.com	
National Semiconductor		✓	✓		
Northrop Grumman	✓	✓		sensor.northgrum.com	
Scrantom Engineering		✓	✓	scrantom.com	
Texas Instruments		✓			29
Westinghouse		✓			96
Japan					
Fujitsu					85–89, 102
Hitachi		✓			33
Kyocera	✓			kyocera.co.jp	31,32,74,137
Matsushita	✓	✓			163
Murata		✓		murata.com	206
Narumi					97
NEC		✓			76,99,105,140
NGK/NTK		✓		ngk.co.jp	
Nippon Electric Glass	✓			neg.co.jp	
Shinko Electric Industries					98
Tokuyama Soda					104
Korea					
Samsung	✓	✓		sem.samsung.com	
Taiwan					
ACX		✓	✓	acxc.com.tw	
Europe					
Ceram Tec		✓		ceramtec.de	
C-MAC Microtechnology		✓	✓	cmac.com/mt	
Epcos		✓	✓	epcos.de	
Heraeus	✓			4hcd.com	103
Micro Systems Engineering		✓	✓	mse-microelectronics.de	
Siemens					167,185
Thales Microel. (Sorep-Erulec)		✓	✓	sorep.com	
VIA Electronic		✓		via-electronic.de	

hand as discussed in Section 4, with advanced LTCC systems. Here the frequency defining key elements can be integrated within, especially for passive RF and microwave devices, developed as layers embedded in the substrate, together with standard dielectric layers. On the other hand, the standard dielectrics may be replaced by new dielectrics having properties similar to the standard dielectrics but with added high Q and low TCf values. If the application of these new materials grows considerably, they even might become a new standard material.

An example of such a material is the glass-ceramic composites recently developed by Motorola [190] which is also listed in Table 4. The glass used for the GCC is composed of BaO, SrO, CaO, K_2O, SiO_2, and B_2O_3 where the last three ingredients mainly determine the softening temperature and viscosity of the glass. The starting volume portion of glass in the composite is typically more than 50% as shown in Figure 13. During sintering part of the ceramic filler Al_2O_3 reacts with the glass and forms the anorthite type crystalline phases $BaSi_2Al_2O_8$, $SrSi_2Al_2O_8$ and $CaSi_2Al_2O_8$. The consumption of BaO, SrO, CaO, and SiO_2 due to this reaction greatly reduces the amount of glass in the final composite. As a result the dielectric losses of the composite are considerably reduced because the dielectric Q values of the crystalline anorthite phases are higher than 3000 and consequently much higher than the Q values of the glass phase. If the ceramic filler used consists only of Al_2O_3, then the GCC has a TCf of -78 ppm/K [190], which is mainly a result of the very positive $TC\varepsilon$ of Al_2O_3. Thus, in order to adjust the TCf to zero, a modifying filler with a negative $TC\varepsilon$ must be added. It was found that the TCf could be adjusted over a wide range, including zero, with the addition of TiO_2 which has a $TC\varepsilon$ of -750 ppm/K [190]. In this way it was possible to develop a LTCC-GCC with a dielectric constant $\varepsilon r = 9.3$, a dielectric $Q > 1000$, and a TCf within ± 5 ppm/K over the entire temperature range between -40 and $+80$ °C.

D. Conductor Pastes and Related Materials for MLC Substrates

The requirements for conductor pastes for MLC substrates are good screen print-ing quality and the ability to be co-fired with tape systems without camber or cracking problems. For cost reasons, it is desirable to have conducting material systems for internal and external requirements that may be fired in one firing operation.

1. Metal Powders and Pastes for HTCC

Metal powders cofired with aluminium oxide are generally molybdenum-and tungsten-based metals due to the high temperature (≈ 1600 °C) necessary for sintering. A thick film paste typically consists of metal powder, glass powder, organic resins and solvents. The additional glass powders are usually used to aid in densification, shrinkage matching to the dielectric, mechanical and chemical bonding to the ceramic, and for adjustment of thermal expansion. To achieve reproducible screen printing pastes, it is necessary to carefully control composition, particle size distribution, and rheological properties.

In the case of HTCC and of LTCC with metallurgy based on Cu there is a final plating of the metal surface features, usually with nickel and gold, used for processing of wire bond pads, solder and brazing pads, and seal bands.

2. Metal Powders and Pastes for LTCC

In the case of LTCC various paste products are offered by main LTCC suppliers such as DuPont, Ferro, Heraeus and Electro-Science Laboratories (ESL) to function for different electrical module connections. The paste systems can be subdivided into materials for:

cofired inner and outer conductor lines and outer via conductors
cofired metals for wire bonding (Au and Al wire bonding), soldering, and
    brazing pads
post-fired outer conductor lines and outer via conductors
post-fired metals for wire bonding (Au and Al wire bonding), soldering,
    and brazing pads
via fill conductors

LTCC metallurgy is in most cases based on Cu, Au, Ag, or on a combination of Au and Ag. As usual for thickfilm metallization pastes, LTCC metallization pastes are composed of metal powder, an inorganic glass or metal oxide binder and an organic vehicle. In the case of Cu metallurgy, copper based pastes are used for all electrical module connections. As already mentioned, a final plating of the metal surface features is usually applied according to the requirements for wire bonding, soldering or brazing. In the case of Au metallurgy, all inner and outer conductors (cofired and post fired) and via fill conductors are based on Au. Solderable conductors are based on Pt/Au or Pt/Pd/Au. In the case of Ag metallurgy, all inner and via fill conductors are based on Ag and all outer conductors on Ag/Pd or Ag/Pt. In this case, Au wire bonding is not possible. If Au wire bonding is desired, a mixed metal system can be applied where all inner and via fill conductors are based on Ag, all outer conductors (cofired and post fired) on Au, and all solderable conductors and those for heavy Al wire bonding are based on Ag/Pd.

3. Powder and Pastes for Cofired and Post-Fired Resistors

The possibility to integrate resistors, within a wide range of resistor values, was one of the main reasons for the establishment of the thickfilm hybrid technology.

622   Wersing and Dernovsek

Amongst the conductor-glass composites the system RuO_2-glass is particularly
noteworthy because of its good adherence, its continuous tunability of the resistivity with metal oxide content and its low temperature coefficient of resistance
TCR. The smooth resistivity change with RuO_2 content strongly contrasts with
the behaviour of similar noble metal based systems which display sharp metal-
insulator transitions at metal volume fractions between 12 to 36%, i.e. at the
percolation threshold [191]. Hence, the resistivity change with RuO_2 content
cannot be understood within the classical percolation model with a random distribution of the conductive particles and a switching contact between them, presumably because the latter assumption fails in the case of extremely surface-reactive
particles. Recently, Nicoloso et al. have been able to clarify the technically important anomalous resistivity behaviour of RuO_2-glass composites [192]. The correlation of the electrical transport with the microstructure has been revealed by modelling the conductivity behaviour of the composite within the framework of the
effective medium theory. The characteristic resistivity behaviour is a result of
the size and distance distribution of RuO_2 clusters and their microstructure. The
former depends on the RuO_2 content, the latter does not. The RuO_2 clusters
consist of a large number of ultrafine RuO_2 particles with an average size of
about 20 nm separated from each other by a glass film of $\leq$ 2nm thickness
which is constant even at very high metal oxide fractions. The surface chemical
interaction of the oxide with the glass gives rise to the unusual non-homogeneous
RuO_2 cluster size and distance distribution. Depending on the RuO_2 fraction,
different transport mechanisms have been identified. At high metal oxide fractions
($\geq$ 20%) metallic conduction prevails, whereas at low RuO_2 oxide contents ($\leq$
3%) ionic transport dominates. In the intermediate range where the resistance
can be adjusted continuously, variable range hopping and tunnelling contribute
considerably.

The integration of screen printed resistors in LTCC substrates is also an
important issue because it offers a considerable contribution to miniaturisation
and cost reduction of MLC modules. As described in Section III F 1, it is currently
a widely used technique to print them as postfire resistors using the resistor pastes
of the conventional thickfilm hybrid technology. Resistors embedded inside the
MLC substrate are fired together with the glass-ceramic body and undergo additional firings with each postfiring step. Screen printing pastes for cofired resistors
require a shrinkage match to the tape and must be chemically compatible to the
surrounding material. The glasses of the resistor pastes tend to interact with the
tape resulting in a change in the ratio between resistive particles and glass and
thus in a shift of square resistances and their temperature coefficients TCR. The
influence of this interaction on the properties of the final resistor decreases more
and more with increasing resistor thickness so that a printed resistor dry thickness
in the range of 20–25 μm appears to be useful for most applications.

The main issue for embedded resistors is the high tolerance of the resistance value. Depending on the maturity of the screen printing process, 3 sigma tolerances of ± 15 to 25% can be achieved. However, this must be compared with a recent survey of product requirements for wireless applications where Kapadia et al. came to the conclusion that 83% of the resistors were in the 5–10% tolerance range [193]. Therefore, efforts towards lower tolerances have been made. Lower tolerances can be achieved, on the one hand, by carefully controlling the fabrication process, primarily to obtain a uniform thickness [194, 195], and on the other hand, by laser trimming through trimming holes over the resistors [196, 197] or by the high voltage pulse trimming method [197, 198].

During the last thirty years thick film resistor materials have evolved to a class of materials with a wide resistance range and outstanding performances. However, these unique properties were achieved with conducting materials and glasses that contain cadmium and lead. In recent years, the global trend has been to restrict and eventually eliminate the use of these compounds because cadmium compounds are carcinogens and lead compounds are toxic.

The advantage of lead glasses for thick film resistors and the problems of developing lead-free thick film resistor materials have already been discussed by Vest [199]. Conventional thick film pastes for resistors are composed of conducting ruthenium compounds such as RuO_2, $Pb_2Ru_2O_{6+\delta}$, or $Bi_2Ru_2O_7$ and lead containing borosilicate glasses. Lead is added to achieve a good adhesive contact with the substrate and a dense microstructure and to hinder reaction of the resistive compound and the glass. Usually, overcoating glasses also contain lead to enable a lower firing of the overcoat.

Fukaya et al. [200] found that one major problem of lead free thickfilm resistors formed on the LTCC substrate surface is the local tensile stress which can exist in the resistor layer, although thermal expansion of the resistor material is well matched to that of the substrate. The tensile stress in the resistor layer was evaluated using FEM and it was found that it depends on the internal conductor wiring. This tensile stress causes a spreading of micro cracks generated by the laser trimming and thus increases resistance drift and decreases reliability. Based on these results they developed highly reliable Pb-free thick film resistor pastes using RuO_2 and a SiO_2-B_2O_3-K_2O glass with a low TCE. By using only a rather low amount of K_2O in the glass, the TCE of the resistor pastes was adjusted to ≤ 4.5 ppm/K so that in any case the resistors are under compression. Due to the low K_2O content in the glass, the chemical interaction of the glass with the oxide was too small to enable a low sintering temperature of 850 °C. Therefore, the RuO_2 powder was coated by K_2O to enrich the local K_2O content and to promote sintering.

Very recently, Hormadaly [201] demonstrated new Pb-free thick film resistor compositions even in the high square resistance range between 10 kΩ and 1 MΩ and with a positive TCR. Conducting phases used were ruthenium pyrochlores $M_{2-x}Cu_xRu_2O_{7-\delta}$ where M is a rare earth element, x is between 0.2 and 0.4 and δ is in the range of 0–1. Resistor pastes were prepared from these conducting phases, a zircon filler and bismuthate glasses (Bi_2O_3-SiO_2-ZnO with additions of CaO, MgO, CuO, and Al_2O_3). The dependence of the square resistance on the conductive phase behaves in a similar fashion as with standard thick film resistors. At a very low weight content ($<$ 20%) of the conductive phase the resistance became very high and TCR decreases and became negative, as expected from the used bismuthate glass.

Finally, the RF and microwave behaviour of screen printed resistors integrated in LTCC substrates is also an important issue because integration reduces parasitic effects, improves high frequency performance, and reduces electromagnetic compatibility problems by decreasing current paths to ground and power planes. Therefore, investigations related to the design, simulation, fabrication, and characterization of embedded resistors for high frequency applications are becoming increasingly important [202, 203].

E. Glass Ceramic Composites for Advanced LTCC

In order to achieve the miniaturization required by future RF devices and to overcome the limitations of the LTCC technology in its standard form the following improvements are needed:

New dielectric glass ceramic composites with medium dielectric constants around 60, fully compatible with standard LTCC materials and processing which can be co-fired to form substrates having several layers with different dielectric constants.

New fully compatible composites with high dielectric constants $\geq$ 500 which can be placed locally within the substrate to form large blocking and by-pass capacitors.

New glass ceramic composites with medium to high magnetic permeability, fully compatible with standard LTCC materials and processing which can be co-fired to form substrates having several layers with different dielectric and magnetic properties.

Advanced processing techniques as described in Section III E 4 which suppress the shrinkage during sintering within the substrate plane (zero-xy-shrinkage) to improve the geometrical precision, thereby decreasing tolerances of integrated passives and making possible flip chip bonding of bare chips.

Advanced printing techniques for fine conductor structures with a line resolution of 50 μm and below [204], as described in Section III C.

1. Medium Dielectric Materials Compatible with Standard LTCC

The opportunities provided by medium dielectric materials ($\varepsilon r \approx 40\text{--}80$) with a low temperature coefficient of permittivity (TCε) for temperature stable resonators (TCf ≈ 0) and with low dielectric losses (tanδ) can easily be seen from very compact single RF components such as frequency filters fabricated using standard ceramics [205] and multilayer techniques [206-208]. However, the medium dielectric materials used for these devices are far from being compatible with standard LTCC materials. Thus, integration of these components is not directly possible.

Therefore the main problems in the development of advanced LTCC technology are not related to the electrical material properties. The challenge is, to solve all the compatibility problems between the different materials related to chemical, thermo-mechanical and sintering aspects [209]. There have been mainly two approaches to reduce the sintering temperature of microwave ceramics so that they can be used for MLC in combination with high conductive internal Ag or Cu electrodes [210].

The first approach is based on the use of sintering aids or by modifying a microwave ceramic with a low-fireable compound. In this case often lead- or bismuth-based dielectric ceramics—known as low-fire materials—have been considered [210, 211]. However, as discussed in Section D 3, toxic Pb containing materials are no longer acceptable. In the bismuth-based Bi_2O_3-CaO-Nb_2O_5, and Bi_2O_3-CaO-ZnO-Nb_2O_5 systems sintered at 950 and 925 °C the best properties reported are $\varepsilon r = 59$, TCF $= 24$ ppm/K, Q $= 610$ (at 3.7 GHz) and $\varepsilon r = 79$, TCF $= +1$ ppm/K, Q $= 360$ (at 3.2 GHz), respectively [211. Recently, new bismuth-based dielectrics in the pyrochlore family Bi2(Ta Nb)2O7 $-$ x have been investigated for high frequency applications. Through compositional engineering and phase control, dielectric ceramics with medium permittivities $\varepsilon r = 70 -$ 150, low TCε ($<$ 30 ppm/K) and reasonably high dielectric Q values (Q·f $>$ 2000) have been realized. Processing temperatures as low as 800°C have been achieved allowing principally co-firing with LTCC substrates [212]. However, the use of bismuth-based ceramics (where Bi is the main component) together with Ag electrodes can be problematic. Usually they can only be used together with Cu electrodes, but Ag electrodes have to be preferred in LTCC technology, especially for zero shrinkage processing.

The second approach to obtain low firing ceramics is based on glass-ceramic composites (GCC) and glass bonded ceramics (GBC), as described in Section IV B. In order to obtain dielectrics with medium dielectric constants within this framework it is necessary to use ceramic powders with a dielectric constant of at least 80–100 which can be found among the high εr microwave ceramic system BaO-Re_2O_3-TiO_2 (Re: Rare Earth) [210]. This ceramic system which has been extensively studied by Kolar [213] is shown in Figure 14. Favourable properties originally found in ceramics with a molar ratio near $BaO:Nd_2O_3:TiO_2 = 1:1:5$ have been attributed to the fact that the material was single phase $BaNd_2Ti_5O_{14}$. However, subsequent work showed that in this system the exact single phase composition was $Ba_{3.75}Nd_{9.5}Ti_{18}O_{54}$. Furthermore, a solid solution described by the general formula $Ba_{6-3x}Re_{8+2x}Ti_{18}O_{54}$ extends for La containing compounds from $x = 0$ (3:2:9) through $BaO:Re_2O_3:TiO_2 = 1:1:4$ ($x = 0.5$) to $x = 1$ (3: 5:18). With decreasing rare earth ionic radius, the extension of this solid solution becomes narrower, for Nd and Pr, $0 \le x \le 0.75$, for Sm and Eu, $0 \le x \le 0.5$ and for Gd $x = 0$. These compounds exhibit a tunnel structure related to the tungsten bronzes with an infinite network of TiO_6 octahedra, linked by their corners. In the 1:1:4 family compounds one finds that with decreasing ionic size of the rare earth element εr decreases from about 100–110 for La to 76–78 for Eu together with a change in $TC\varepsilon$ from about $-(550-750)$ to $+(60-100)$ ppm/ K [213]. Since all these compounds form solid solutions, $TC\varepsilon$ can be tailored according to requirements, e.g. to achieve $TCF \approx 0$ ppm/K.

Therefore, this ceramic system is excellently suited for the development of LTCC glass ceramic composites with medium dielectric constant and a low

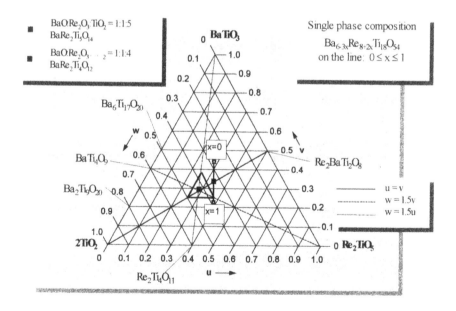

Figure 14   Phase diagram of the microwave ceramic system $BaO\text{-}Re_2O_3\text{-}TiO_2$ (Re: Rare Earth).

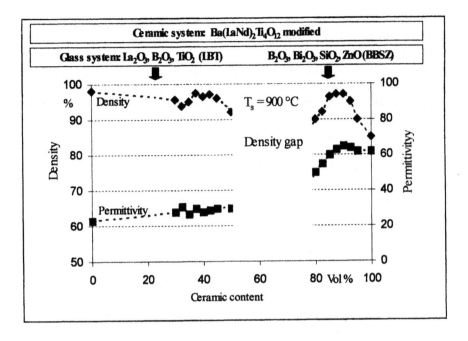

Figure 15 Glass-ceramic composite and glass bonded ceramic with low and high ceramic content, respectively.

temperature coefficient of permittivity (TCε) for temperature stable resonators. However, as shown in Figure 15 one has to take into consideration that, on the one hand, for GCC the volume amount of ceramic powder added to the composite is restricted to < 50 % and, on the other hand, for GBC the volume amount of glass powder added to the composite must be within a small range of about 5–15% in order to achieve good densification at a sintering temperature not higher than 900 °C. This is due to the fact (see Section IV B) that for GCC the sintering or densification is caused by viscous flow and a dispersing and rearranging of the crystalline particles in the glass melt. This process is hindered if the amount of filler powder becomes too high. Therefore the permittivity of a GCC, based on the ceramic system $Ba_{6-3x}Re_{8+2x}Ti_{18}O_{54}$ which can be obtained within this approach, is restricted to approximately 30 [214]. On the other hand for GBC, the densification is primarily due to the melting and dynamic crystallization behaviour of the glass during heat treatment. In this case, it is necessary that the particle size of crystalline phases be very low and the reactivity between the glass melt and the crystalline phases of the composite be particularly high. In order to fulfil the latter requirement, a new glass had to be developed. Borosilicate glass, most often found in LTCC materials, was chosen as basic substance and modified with Bi_2O_3 and ZnO to obtain a soften temperature, $T_{soft} < 400$ °C. Bi_2O_3 and ZnO were selected because they are well known to be compatible to Ba_{6-3x}-$Re_{8+2x}Ti_{18}O_{54}$ microwave ceramics [215]. Furthermore, it was found that modifying the microwave ceramic with small amounts of Bi_2O_3 or ZnO enhances the solubility of ceramic grains in molten glass and accelerates the low temperature densification process of ceramic-rich composites. Due to the very high $TC\sigma$ (280 ppm/K) of the finally chosen Bi_2O_3-B_2O_3-SiO_2-ZnO glass, the ceramic composition $BaNd_2Ti_4O_{12}$ with a negative $TC\epsilon$ was selected and modified with 1–3 wt% ZnO to obtain a composite with $TC\epsilon$ around -20 ppm/K, i.e. $TCf \approx 0$. With this approach, it is possible to realize LTCC compatible GBC with a medium dielectric constant of 60, with a low temperature coefficient of permittivity for temperature stable resonators ($TCf \approx 0$), and with low dielectric losses [214].

Tapes cast from these GBC have been blanked into squares and processed in combination with standard tapes to substrates and finally to modules, as described in Section III. Figure 16 shows a TEM bright-field image of the microstructure of a medium dielectric layer cofired in such a substrate. It can be seen that during sintering the glass phase is well distributed in a very thin layer along the grain boundaries and in the spandrels between the grains and therefore, a very dense and homogeneous microstructure is formed (Figure 17). Finally, Figure 18 shows the concentration profile of Zn, Ag, Bi (0–6%) and of Si, Ag, Ba (0–100%) as measured along a line vertical to the silver electrode. A quite sharp material profile is found at the interfaces between the medium dielectric layer, the Ag electrode, and the standard dielectric with only a very small diffusion of Ag into the dielectric layers.

2. High Dielectric Materials Compatible with Standard LTCC

As much as 90% of components used in communication packages are passive devices, which have the potential to be buried and most of these passive devices are capacitors, which are used as blocking or by-pass capacitors. Therefore, composites with high dielectric constants (e.g. $\sigma r \geq 500$) and good temperature stability (having X7R or at least Z5U characteristics) are required, which can be buried and cofired. Furthermore, in many cases it is necessary to place the buried capacitors locally within the substrate, i.e. in the immediate vicinity of the IC-connection to be blocked.

Tapes and pastes for capacitors which can be buried in standard LTCC tapes can be obtained from GBC powders which are, on the one hand, again based on a Bi_2O_3-B_2O_3-SiO_2-ZnO glass and, on the other hand, on a suitable high dielectric ceramic powder [214]. In order to achieve the required temperature characteristic, this powder can be a standard $BaTiO_3$–based capacitor powder or

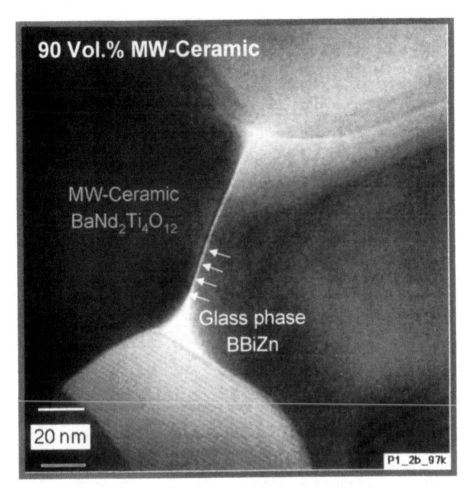

Figure 16 TEM bright-field image of the grain-boundaries and the transition phase of a glass bonded ceramic tape (ZnO doped $BaNd_2Ti_4O_{12}$ + BBSZ glass) sintered at 900 °C.

a mixture of different $(Ba_{1-u}Sr_u)TiO_3$ powders (with different Curie temperatures). The compositions (u-values) of the different $(Ba_{1-u}Sr_u)TiO_3$ powders can be chosen so that their permittivity-temperature characteristics superpose to give the required characteristic. This is possible because due to the low sintering temperature the different powders cannot significantly inter-diffuse.

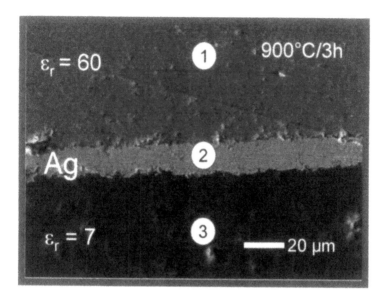

Figure 17 SEM image of the interface microstructure of an advanced LTCC substrate with medium ε_r layers integrated with standard LTCC layers.

3. Magnetic Materials Compatible with Standard LTCC

The current trend in electronics is to achieve a significant size reduction and to lower costs with more integration. This trend can also be observed for components based on ferrites. Ferrimagnetic materials are not only useful and even necessary for inductors and transformers which are used at rather low frequencies (< 10 MHz) for signal and power applications but are also very interesting for RF and microwave devices. The introduction of ferrites into the microwave field has brought with it a new class of components, which as yet has no counterpart at lower frequencies. The effect of ferrites at these frequencies is that a wave passing through the component can be influenced and directed by an external magnetic bias field. In this way it is possible to obtain useful reciprocal and non-reciprocal devices such as zirculators and isolators, modulators and fast microwave switches, tuneable attenuators and phase shifters, and tuneable resonators and filters. Although components based on bulk ceramic ferrites have been known for about five decades [216], up to now these were only used for special civil and military microwave equipment. In the future, however, a breakthrough for mass applications in the wireless systems field seems possible, due to an enormous size and cost reduction achievable with integration of ferrites into LTCC substrates. In

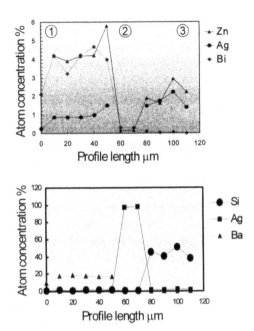

Figure 18 Concentration profile from the sample of Figure 17.

the field of low frequency applications, quite encouraging results, e.g. for low profile transformers have already been achieved [217].

Recent results have shown that glass bonded ceramics (GBC) based on NiZnCu-ferrites and Ba-, Sr-, or Ca-Hexaferrites, useful for low frequency and microwave applications respectively can be easily prepared by using a suitable Bi_2O_3-B_2O_3-SiO_2-ZnO glass. Tapes and pastes fabricated from these GBC powders can be buried in standard LTCC tapes and cofired at temperatures not higher than 900 °C.

V. DESIGN AND MODELLING FOR LAMINATED MULTILAYER CERAMICS

The proper design of an MLC substrate is the first of a number of critical steps in fabrication of the product. The tolerances achievable and the processing problems encountered are a direct result of the substrate as well as other factors, such as tooling and process parameters. Several authors have published data and ground rules on successful MLC design [16,18,19,23,25,27,30,33]. Improper design might include not balancing the metallurgy from top to bottom in the substrate, which causes camber during firing, or not balancing the patterns across each layer, which leads to fired-dimensional distortion and cracking. Electrical opens in the internal metallurgy can be caused by improper overlap in line/via connections, and so on. Shorts can arise from designing lines and vias with insufficient spacing, and so on, almost ad infinitum.

The successful design of an MLC substrate also depends on complete understanding of the green and fired properties of the ceramic and the metallurgy used and careful matching of shrinkage curves between the metal and ceramic during firing. The size and thickness of the substrates that can be successfully built rest largely on those parameters, as well as the mechanical properties of the finished metal/ceramic composite.

One factor that makes design so critical is that it is very difficult to separate design faults from process problems during the development of a given MLC substrate configuration. There is a Murphy/Miranda type of law for MLCs: Anything and everything that is done to the MLC process can and will affect dimensional control.

Therefore, all substrate foundries claim that their design rules must be strictly fulfilled by the customers. Today the design rules of the different suppliers of materials and substrates can be easily downloaded from the internet (Table 5).

A. Generation of Artwork

One of the most critical and important process steps is the generation of the artwork, which is a direct result of the design phase of the module itself. As Figure 11 shows, an MLC substrate is composed of a number of distinct layers which normally include top surface metallurgy (for chip attach and wire bonding); redistribution or fan-out layers; signal distribution layers (x and y); power distribution layers (ground planes); and bottom surface pads (for pin attachment) [29, 37]. Each of these layers require a unique punching tape for via punching and possibly an inspection mask for checking the via pattern. Each layer also requires a set of screening masks (stencils) for filling the vias and screening masks for personalizing the pattern required on that layer. The quality and integrity of the tapes, masks, screens, and so on governs the tolerance and quality seen in the substrate. Screening masks are fabricated by using photoresist processes and artwork generated by computers.

B. Design and Modelling of LTCC RF and Microwave Modules

For the successful application of the LTCC technology for integrated RF and microwave devices, it is important that the material and processing specific properties are considered from the first circuit design to the final circuit layout. Integrated passive components must be designed in such a way that the impact of materials and processing tolerances typical for this technology can be minimized or even eliminated. Therefore, well suited coupling structures for components based on coupled lines and ground plane structures such as multiple gridded ground planes must be applied [218]. Furthermore, an extensive modelling of the integrated devices is necessary. This can be performed using commercially available 2.5D or 3D electromagnetic simulation tools.

As an example, an innovative microwave filter design based on a high dielectric LTCC-material has been applied [219] to demonstrate performance and efficiency of an advanced LTCC technology. This band-pass filter has been realised using this new technology (Figure 19). Compared with a filter made up in standard LTCC technique an explicit increase in the performance and at the same

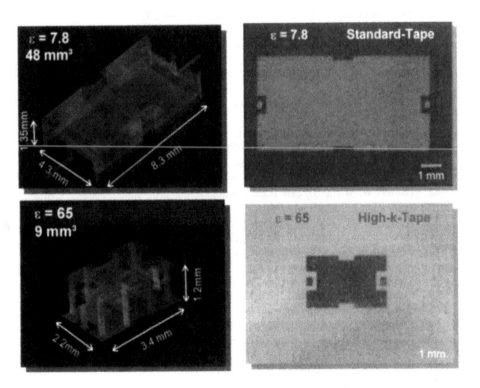

Figure 19 Design and footprint of LTCC band-pass-filters. Compared to the conventional LTCC technology, a volume up to 80% can be saved with an advanced LTCC technology.

Wersing and Dernovsek

time a volume saving of 80 % could be achieved. Such integrated components will become increasingly important for mobile communication devices. Future success for these innovative module or system-in-package technologies will decisively depend on the question of whether an efficient and precise modelling technique is available that allows short development times without many re-designs and whether it will be possible to transfer this technology into a reliable mass production.

REFERENCES

1. Keyes LK, et al. Int. Hybrid Microelectron. Symp., Beverly Hills, Calif, Nov. 1970: 641.
2. Gioia JC. Solid State Technol, May 1971:43.
3. McCormick JE, Calabrese DW. NEPCON Proc, 1969:500.
4. Kolc RF. Int. Microelectron. Conf. Proc. Anaheim, Calif, 1978:86.
5. Kruzweil K, Loughran J. IEEE Electron. Components Conf. Washington, DC, 1973: 212.
6. Redemske RF. 14th IEEE Comput. Soc. Int. Conf. San Francisco, Calif, Feb. 1977: 253.
7. Umbaugh CW. 14th IEEE Comput. Soc. Int. Conf. San Francisco, Calif, Feb. 1977: 263.
8. Marshall J, Rode F. Int. Microelectron. Sym. Proc. Minneapolis, Minn, 1978:26.
9. Close AD. Solid State Technol, May 1971:38.
10. Lebow S. 30th IEEE Electron. Components Conf, 1980:307.
11. Steinberg J. Mach. Des, Nov. 23, 1978:10.
12. Hatfield W, Wicher D. 28th IEEE Electron. Components Conf. Anaheim, Calif., 1978, Circuits Manuf, July 1978:7.
13. Agajanian AH. Solid State Technol. Oct. 1975, 56. Sept. 1976, 87.
14. Park JL. Manufacture of Ceramics, U.S. Patent 2,966,719, 1961.
15. Schwartz B, Wilcox DL. Proc. Electron. Components Conf. Washington, DC, 1967: 17.
16. Theobald PR, Davis MP, Bailey JT. Hybrid Microelectron. Symp. 1969:447.
17. Chance DA. Metall. Trans. 1970; 1:685.
18. Wilcox DL. Solid State Technol., Jan. 1971, 40. Feb. 1971, 55.
19. Chance DA, Wilcox DL. IEEE Proc. 1971; 59:1455.
20. Kaiser HD, Pakulski FJ, Schmeckenbecher AF. Solid State Technol. May 1972: 35.
21. Young WS. J. Less-Common Met. 1973; 32:321.
22. Swiss WR. Circuits Manuf. Aug. 1979:43.
23. Kumar AH, Niklewski JB. Am. Ceram. Soc. Bull. 1979; 58:1179.
24. Blodgett AJ. Proc. Electron. Components Conf. San Francisco, Calif, 1980:283.
25. Hargis BM. Solid State Technol. May 1971:47.
26. Burch ML, Margis BM. Ceramic Chip Carrier 3M Corp, 1977.

27. Guidelines for Designing Multilayer Substrates 3M Corp, 1977.
28. Mistler RE, Shanefield DJ, Runk RB. in Ceramic Processing Before Firing Onoda G, Jr, Hench L, eds. New York: Wiley, 1978.
29. Williams CE. Am. Ceram. Soc. 81st Annu. Meet. Cincinnati, Ohio, #4-E-79, 1979 See Am. Ceram. Soc. Bull, 1979:363.
30. Keller WR, et al. Proc. Electron. Components Conf., San Diego, 1969:52.
31. Rogers EL. Am. Ceram. Soc. Pacific Coast Reg. Meet., San Francisco, Calif., 20-E-80P, 1980 See Am. Ceram. Soc. Bull], 1980:830.
32. Fine Ceramics (bulletin), Kyocera International, San Diego, Calif.
33. Ihochi T. IEEE Electron. Component Conf., Washington, D.C, 1973:204.
34. Welterlen JD, Anderson SD. Int. Microelectron. Symp. Orlando, Fla. 1975:358.
35. Fagersten EG. Hybrid Microelectron. Symp. 1968:401.
36. IBM News Special Edition, Nov. 1980.
37. Blodgett AJ, Jr, Barbour DR. IBM J. Res. Dev. 1982; 26:30.
38. Onoda G, Jr, Hench L, eds. Ceramic Processing Before Firing. New York: Wiley, 1978.
39. Onoda G, Shanefield DJ. "Organic Additives and Ceramic Processing Principles," a course for the Center for Professional Advancement II, Chicago, April 30, 1980.
40. Pincus AG, Shipley LE. Ceram. Ind. Magr., Apr. 1969:106.
41. Shanefield DJ, Mistler RE. Am. Ceram. Soc. Bull. 1974; 53:416, 564.
42. Runk RB, Andrejco MJ. Am. Ceram. Soc. Bull. 1975; 54:199.
43. Williams JC. in Treatise on Materials Science and Technology Wang FY, ed. New York: Academic Press, 1976.
44. DiMarcello FV, Key PL, Williams JC. J. Am. Ceram. Soc. 1972; 55:509.
45. Desai KS. private communication.
46. Urfer EN. Am. Ceram. Soc. Bull. 1973; 52:713.
47. Kruzweil K, Loughran J. IEEE Trans. Parts, Hybrids, Packag., PHP-9, (4)/216, 1973.
48. Miller LF. Solid State Technol. 1974; 17:54.
49. Thick Film Products Catalogue Engelhard Ind., East Newark, NJ.
50. Thick Film Materials E. I. DuPont, Wilmington, Del.
51. Bradigan J. Insul./Circuits. Jan. 1980:33.
52. Technical Notes 2.1.2 (9/73) E. I. DuPont, Wilmington, Del.
53. Lipka K. Today's Screen Printing Technology. Proc ISHM 1993:475.
54. The Circuit IBM, East Fishkill, N. Y., Fib, 1979:5.
55. Gardner RA. J. Solid State Chem. 1974; 9:336.
56. Bean LW. Br. Ceram. Soc. Trans. 1971:121.
57. Pincus AG. Ceram. Age. March 1954:16.
58. Fenstermacher JE. Am. Ceram. Soc. Bull. 1969; 48:775.
59. Williams JC. Am. Ceram. Soc. Bull. 1977; 56:580.
60. Thick Film Technology 2.4.1 (3/72) E. I. DuPont, Wilmington, Del.
61. Swiss WR. private communication.
62. Gardner RA, Nufer RW. Solid State Technol. 1974; 17:38.
63. Werbizky GG, Winkler P, Haining FW. Electronics. Aug. 2, 1979.
64. Kumar AH, Tummula RR. Am. Ceram. Soc. Bull. 1978; 57:738.

65. Stetson HW, Schwartz B, Liderback WH. two papers presented at Electron. Div. Meet., Am. Ceram. Soc. San Francisco, Calif., Oct. 26, 1961 See Am. Ceram. Soc. Bull], 1961:584.

66. "Micromodule crystal units," U.S. Signal Corps Contract DA-36-039-SC-85046, March 31, 1961.

67. Seraphim DP, Feinberg I. IBM J. Res. Dev. 1981; 25:617.

68. Blodgett AJ, Jr. Sci. Am. July 1983; 249:86.

69. Patrusky B. Think, Jan./Feb. 1983, 23.

70. Shimada Y, Utsumi K, Suzuki M, Takamizawa H, Nitta M, Yano S. IEEE Electron. Comput. Conf. Orlando, Florida, May 1983:314.

71. Gyurk WJ. Methods for manufacturing multilayered monolithic ceramic bodies, U.S. Patent 3,192,086, 1965.

72. Stetson HW. Method of Making Multilayer Circuits, U.S. Patent 3,192,978, 1965.

73. Stetson HW. in Ceramics and Civilization Kingery DW, ed. Vol. Vol. III. Westerville, Ohio, 1987.

74. Kraft EH. Materials and Designs for Advanced MLC Packages. American Institute of Physics, New York, 1986:255–266.

75. Yan MF, Rhodes WW. Mater. Res. Soc. Sym. Proc. 1988; 108:439–453.

76. Shimada Y, Utsumi K, Suzuki M, Takamizawa H, Nittu M, Yano S. IEEE Trans. Components, Hybrids and Manuf. Tech., CHMT-6. 1983(4):382–388.

77. Kumar AH, McMillan PW, Tummala RR. Glass–Ceramic Structures and Sintered Substrates Thereof with Circuit Patterns of Gold, Silver or Copper, U.S. Patent 4,301,324, 1981.

78. Gerhardt R. in Electronic Packaging Material Science III Jacodine R, Jackson K, Sundahl R, eds. Pittsburgh: Materials Research Society, 1988:108, 101–105..

79. Schwartz B. Mater. Res. Soc. Sym. Proc. 1985; 40:49–59.

80. Vitriol WA, Steinberg JI. ISHM Proc, 1983:593–598.

81. Steinberg JI, Horowitz SJ, Bacher RJ. 5th European Hybrid Microelectronics Conf, 1985:302–316.

82. Whatley T. Hybrid Circuits. 1989; 19:38–41.

83. Balde JW. J. Electron. Mater. 1989; 18:221–227.

84. Sanchez LE. Int. Sym. on Ceramic Substrates and Packages. Am. Ceram. Soc., Denver, Oct. 18–21, 1987.

85. Imanaka Y, Aoki S, Kamehara N, Niwa K. Yogyo-Kyokai-Shi. 1987; 95: 1119–1121.

86. Niwa K. Ceramics. 1986; 21:188–194.

87. Hashimoto K, Niwa K, Murakawa K. Ceramic Substrate with Integrated Circuit Bonded Thereon, U.S. Patent 4,460,916, 1984.

88. Kamehara N, Kurihara K, Niwa K. Method for Producing Multilayered Glass–Ceramic Structure with Copper-Based Conductors Therein, U.S. Patent 4,504,339, 1985.

89. Kurihara K, Kamehara N, Yokoyama H, Ogawa H, Yokouchi K, Imanaka Y, Niwa K. Method for Producing a Multilayer Ceramic Circuit Board, U.S. Patent 4,642,148, 1987.

90. Bolt JD, Button DP, Yost BA. Mater. Sci. Eng., A109. 1989:207–211.

91. Kurokawa Y, Hamaguchi H, Shimada Y, Utsumi K, Takamizawa H, Kamata T, Noguchi S. 36th Elect. Com. Conf. Proc, 1984:412–418.
92. Iwase N, Tsage A, Suguira Y. Proc. Int. Microelectronic Conf, 1984:180–185.
93. Iwata Y, Saito S, Satoh Y, Okamura F. Int. Microelectronic Conf. Tokyo, 1986: 65–70.
94. Ikegami A, Yasuda T. 5th European Hybrid Microelectronics Conf, 1985:465–471.
95. Sawhill HT. Ceram. Eng. Sci. Proc. 1988; 9:1603–1617.
96. Mattox DM, Gurikovich SR, Olenick JA, Mason KM. Ceram. Eng. Sci. Proc. 1988; 9:1567–1578.
97. Nishigaki S, Yano S, Fukutu J, Fukaya M, Fuwa T. Proc. Int. Sym. on Microelectronics. 1985:225–234.
98. Horiuchi M, Mizushima K, Takeuchi Y, Wakabayashi S. IEEE Trans. Components, Hybrids and Manuf. Tech. 1988; 11:439–446.
99. Shimada Y, Yamashita Y, Takamizawa H. IEEE Trans. Components, Hybrids and Manuf. Tech. 1988; 11:163–170.
100. Tosaki H, Suzuki S. Low Temperature Co-Fired Glass Ceramic Multilayer Substrate, Inst. of Electronics and Info. Eng. of Japan, Technical Report CPM 86–60, 1986:9–14.
101. Mandai H, Sugoh K, Tsukamoto K, Tani H, Murate M. A Low Temperature Co-Fired Multilayer Ceramic Substrate Containing Copper Conductors. IMC Proc Kobe 1986:61–64.
102. Kamehara N, Niwa K, Murakawa K. Proc. 1982 Int. Microelectronics Conf, 1982: 388–400.
103. Werdecker W, Aldinger F. 35th Elect. Components Conf., IEEE, 1985:26–31.
104. Kuramoto N, Taniguchi H, Aso I. 36th Elect. Components Conf., IEEE, 1986: 424–429.
105. Watari T, Murano H. IEEE Trans. Components, Hybrids and Manuf. Tech., CHMT-8. 1985(8):462–467.
106. Thomson RG, Shyu J, Poret JC, Buckhalt C, Shealy DL, Tohver HT. Micro-Opto-electronic Mater. 1988; 877:103–110.
107. Tummala RR. in Microelectronic Packaging Handbook Tummala RR, Rymaszewshi EJ, eds. New York: Van Nostrand Reinhold, 1989:455–522.
108. Sacks MD, Scheiffele GW. Adv. Ceram. 1986; 19:175–184.
109. Cawley JD. Tape Casting as an Approach to an All-Ceramic Turbine Shroud Seal, NASA Tech. Memo. 87078. July 1985.
110. Tormey ES. Adv. Ceram. 1984; 9:140–149.
111. Hyatt EP. Ceram. Bull. 1989; 68:869–870.
112. Khadilkar CS, Sacks MD. Ceram. Trans., 1(A). 1988:397–409.
113. Sacks MD, Khadilkar CS, Scheiffele GW, Shenoy AV, Dow JH, Sheu RS. Adv. Ceram. 1987; 21:495–515.
114. TA Ring. In Advances in Ceramics MF Yan, K Niwa, HM O'Bryan, WS Young, eds. Westerville Am. Ceram. Soc.. Vol. 26, 1989:569–576.
115. Mikeska K, Cannon WR. Adv. Ceram. 1984; 9:164–183.
116. Boch P, Chartier T, Hutterpain M. J. Am. Ceram. Soc. 1986; 69:C191–C192.
117. Charter T, Streicher E, Boch P. Ceram. Bull. 1987; 66:1653–1655.

118. Gurak NR, Josty PL, Thomson RJ. Ceram. Bull. 1987; 66:1495–1497.
119. Landham RR, Nahass P, Leung DK, Ungureit M, Rhine WE, Bowen HK, Calvert PD. Ceram. Bull. 1987; 66:1513–1516.
120. Shanefield DJ. Mater. Res. Soc. Sym. Proc. 1985; 40:69–76.
121. Blum JB, Cannon WR. Mater. Res. Soc. Sym. Proc. 1985; 40:77–82.
122. Chou YT, Ko YT, Yan MF. Am. Ceram. Soc. Commun. 1987; 70:C280–C282.
123. Cima MJ, Dudziak M, Lewis JA. J. Am. Ceram. Soc. 1989; 72:1087–1090.
124. Dong C, Bowen HK. J. Am. Ceram. Soc. 1989; 72:1082–1087.
125. Cima MJ, Lewis JA, Devoe AD. J. Am. Ceram. Soc. 1989; 72:1192–1199.
126. Higgins RJ. The Chemistry of Carbon Formation During Binder Burnout in Ceramics, Report No. 88, Ceramics Processing Research Laboratory, MIT, Cambridge, Mass, September 1978.
127. Brownlow JM, Plaskett TS. Process for the Removal of Carbon Residue During the Sintering of Ceramics, U.S. Patent 4,474,731, 1984.
128. Cowen JH, Noll FE, Young LG. Catalysts for Accelerating Burnout of Organic Materials, U.S. Patent 4,778,549. October 18, 1988.
129. Higgins RJ, Rhine WE, Cima MJ, Bowen HK, Farneth WE. Chemical influences on carbon retention during non-oxidative binder removal from ceramic greenware, Report No. R17, Ceramic Processing Research Laboratory Research Summary, MIT, Cambridge, Mass., March, 1989:166–178.
130. Calvert P, Cima M. Theoretical Models for Binder Burnout, Report No. 91, Ceramics Processing Research Laboratory, MIT, Cambridge, Mass, April 1988.
131. Higgins RJ. The Chemistry of Carbon Retention During Non-Oxidative Binder Removal from Ceramic Green Bodies, Ph.D Thesis submitted to Dept. of Material Science and Engineering, MIT, Cambridge, Mass, January 1990.
132. Burn I. 1st Int. Ceram. Sci. Technol. Congress, Anaheim, Calif. Nov. 2, 1989.
133. Wynn Herron, private communication.
134. Burgess JF, Newgebauer CA, Flanagan G. J. Electrochem. Soc. 1975; 172:688.
135. Jensen RJ. Mater. Res. Soc. Sym. Proc. 1988; 108:73–79.
136. Terasawa M, Minami S, Rubin J. ISHM Proc. 1983:607–615.
137. Takusago H, Adachi K, Takada M. J. Elect. Mater. 1989; 18:319–326.
138. Ho CW, Chance DA, Bajorek CH, Acosta AE. IBM J. Res. Dev. 1982; 26:286–296.
139. Tamura T, Dohya A, Inoue T. Proc. Int. Microelectronics Symp. 1980:308–312.
140. Blade JW. J. Electron. Mater. 1989; 18:221–227.
141. Lee LD, Pober RL, Calvert PD, Bowen HK. J. Mater. Sci. Lett. 1986; 5:81–83.
142. Schwartz B. Phys. Chem. Solids. 1984; 45:1051–1068.
143. Takamizawa H, Utsumi K, Sukuhi M. NEC Res. Dev. 1987; 84.
144. Tummala RR, Ramaszewski EJ. Microelectronics Packaging Handbook. New York: Van Nostrand Reinhold, 1989:673–725.
145. Schwartz B. In Electronic Ceramics: Properties, Devices, and Applications Levinson LM, ed. New York: Marcer Dekker, 1988:1–44.
146. Hammer RB, Powell DO, Mukherjee S, Tummala R, Raj R. In: Principles of Electronic Packaging Seraphim DP, Lasky RC, Li C-Y, eds. New York: McGraw-Hill, 1989:282–332.
147. Herron LW, Master RN, Tummala RR. Method of Making Multilayered Glass-Ceramic Structures Having an Internal Distribution of Copper-Based conductors. U.S. Patent 4, 234,367, 1980.

148. Vest RW. Materials science of thick film technology. Am Ceram Soc Bull 1986; 65:631.
149. Swiss WR, Young WS. Accelerated sintering for a green ceramic sheet. U.S. Patent 4,039,338, 1977.
150. Garrou PE, Turlik I. Multichip Module Technology Handbook. New York: McGraw-Hill, 1998.
151. Leung GB, Sands SA. A Thin Film on MLC Application. Proc. ECTC, 1991:10.
152. R Himmel, J Licari. Fabrication of large-area thin film multiplayer substrates. Proc ISHM 1989:454.
153. Tsukada Y. Low Cost Thin Film Multilayer Substrate Fill Gap with LSI. Nikkei Microdevices 1991:61–67.
154. Schmidt W. A Revolutionary Answer to Today's and Future Interconnect Challenges. Proc. Printed Circuit World Convention VI, San Francisco, 1994:T12–1.
155. Bachmann F. Excimer Laser Drill for Multilayer Printed Circuit Boards: From Advanced Development to Factory Floor. MRS Bulletin 1989:49–53.
156. Redmond TF, Lankard JR, Bailz JG, Proto GR, Wassick TA. The Application of Laser Process Technology to Thin Film Packaging. Proc. 42nd ECTC 1992:763.
157. Morrison JM, Tessier TG, Gu B. A Large Format Modified TEA-CO_2 Laser Based Process for Cost Effective Small Via Generation. Proc. Int. MCM Conference, Denver, 1994:369.
158. Hankey D, Shaikh A, Vasudevan S, Khadilkar C. Thick Film Material and Processes. In: Sergent J, Harper C, eds. Hybrid Microelectronic Handbook. New York: McGraw-Hill, 1995 Chapter 3.
159. Hagen N, Henderson J, Nebe W, Osborne J. High Resolution Thick Film Material System. Int J Hybrid Microelectronics 1989; 12:175.
160. Felten J, Collins J, Wang C. High Resolution Via Patterning by Aqueous Diffusion Patterning. Proc. ISHM 1991:545.
161. Amey D. A Thick Film Hybrid Logic Module for Commercial Computer Applications. Proc. Int. Microelectronics Conf, 1981.
162a. Hellebrand H. Tape casting. In: Brook RJ, ed. Processing of Ceramics, Part 1. In: Chapter 7. Weinheim: VCH, 1996.
162b. Cahn RW, Haasen P, Kramer EJ, eds. Materials Science and Technology. Vol. 17A.
163. Baba et al Y. Co-fireable Copper Multilayer Ceramic Substrates. Proc ISHM Seattle 1988:405.
164. Mikeska KR, Jensen RH. Pressure-Assisted Sintering of Multilayer Packages. Ceramic Transactions 1990; 15:629–650.
165. Kumar AH, Prabhu AN, Thaler B. Versatile, low-cost, multiplayer ceramic board on metal core. Proc. Int. Conf. on Multichip Modules, Denver, 1995:100–107.
166. Mikeska KR, Mason RC. Dimensional Control in Cofired Glass-Ceramic Multilayers. Proc. 6th SAMPE Electronics Conf, 1992:699–712.
167. Wersing W, Gohlke S, Matz R, Eurskens W, Wannenmacher V. Integrated Passive Components Using Low Temperature Cofired Ceramics. Proc. Int. Sym. on Microelectronics IMAPS, 1998:193–199.
168. Lautzenhiser F, Amaya E, Barnwell P, Wood J. Microwave Module Design with HeraLock HL2000 LTCC. Proc. Int. Symp. on Microelectronics IMAPS 2002: 285–291.

169. Page J, Webster D. Transitioning from Gold to Copper in a Multilayer System. Proc. 30th Electronics Components Conf. San Francisco, 1980:294.

170. Larson EN, Brock MJ. Development of a Single Point Gold Bump Process for TAB Applications. Proc. Int. Conf. on Multichip Modules, Denver, 1993:391–397.

171. Sallo JS. Bumped Beam Tape for Automatic Gang Bonding. Proc. Int. Microelectronics Symp. Minneapolis, 1978:153–156.

172. Zakel E, Villain J, Reichl H. Reliability and Au Concentration in OLB Solder Fillets of TAB Devices Having a 75 μm pitch. Proc. 44th Electronics Components Conf Washington, DC, 1994:883–893.

173. Matta F, Heflinger B, Kaw R, Rastogi V. High Performance Packaging at Competitive Cost Using Demountable TAB. Int J Microcircuits Electronic Packaging 1992; 15:61–73.

174. Totta PA, Sopher RP. SLT Device Metallurgy and Its Monolithic Extension. IBM Journal of Research and Development May 1969:226–238.

175. Tummala RR, Tummala RR, Ramaszewski, eds EJ. Microelectronics Packaging Handbook. New York: Van Nostrand Reinhold, 1989:376–380.

176. Yung EK, Turlik I. Electroplated Solder Joints for Flip Chip Applications. IEEE Transactions on Components, Hybrids, and Manufacturing Technology, 1991: 549–559.

177. Chmiel G, Wolf J, Reichl H. Lead/Tin-Bumping for Flip Chip Applications. Proc Int Symp Microelectronics 1992:336–341.

178. Yu KK, Tung F. Solder Bump Fabrication by Electroplating for Flip Chip Applications. Proc. 15th IEEE/CHMT Int Electronics Manufacturing Technology (IEMT) Symp. Santa Clara, 1993:277–281.

179. Keswick K, German R, Breen M, Nolan R. Compliant Bumps for Adhesive Flip Chip Assembly. Proc. 44th Electronics Components Conf. Washington, 1994:7–15.

180. Estes RH. Fabrication and Assembly Processes for Solderless Flip Chip Assembly, 1992:322–335.

181. Suryanarayana D, Hsiao R, Gall TP, McCreart JM. Enhancement of Flip Chip Fatigue Life by Encapsulation, 1991:218–223.

182. Suryanarayana D, Wu TY, Varcoe JA. Encapsulants Used in Flip Chip Packages, 1993:858–862.

183. Clementi J, McCreary J, Niu TM, Palomaki J, Varcoe J. Flip Chip Encapsulation on Ceramic Substrates. Proc. 43th Electronics Components Conf. Orlando, 1993: 175–181.

184. Knickerbocker J, Kumar AH, Tummala RR. Sinterable Cordierite Glass-ceramics. Bull Am Ceram Soc 1993; 72:91.

185. Muralidhar SK, Roberts GJ, Shaikh AS, Hankey DL, Vlach TJ. Low Dielectric, Low Temperature Fired Glass Ceramics. U. S. Patent 5, 258, 335, 1993.

186. Eberstein M, Schiller WA, Dernovsek O, Wersing W. Adjustment of Dielectric Properties of Glass Ceramic Composites Via Crystallization. Glass Sci Technol 2000; 73:371–373.

187. Shimada Y, Kobayashi Y, Kata K, Takamizawa H, Kurano M. Large Multilayer Glass-Ceramic Substrate for Supercomputer. Proc. 40th Electronics Components Conf. Las Vegas, 1990:76.

188. Shimada Y. High Speed Computer Packaging with Multilayer Glass-Ceramic Systems. Proc ISHM 1991:176.
189. Gupta TK, Jean JH. Principles of Development of a Silica Dielectric for Microelectronic Packaging. J Mater Sci 1996; 11:243.
190. Dai SX, Huang R-F, Wilcox D. Use of titanates to achieve a temperature stable, low-temperature cofired ceramic dielectric for wireless applications. J Am Ceram Soc 2002; 85:828.
191. Stauffer D. Introduction to Percolation Theory. London and Philadelphia: Taylor and Francis, 1985.
192. Nicoloso N, LeCorre-Frisch A, Maier J, Brook RJ. Conduction mechanisms in RuO_2-Glass Composites. Silicates Industriels. Vol. 3–4, 1994:135–139.
193. Kapadia H, Cole H, Saia R, Durocher K. Evaluating the Needs for Integrated Passive Substrates. Advanced Microelectronics January/February, 1999:14–17.
194. Rellick JR, Ritter AP. Non-trimmed Buried Resistors in Green Tape Circuits, 1999: 420–424.
195. Ehlert MR, Munoz PS. Effect of Processing Variables on Cofired Resistors in LTCCs. HDI Magazine, July 1999:64–69.
196. Denlaney K, Barrett J, Barton J, Doyle R. Characterization and Performance Prediction for Integral Resistors in Low Temperature Cofired Ceramic Technology. IEEE Transactions on Advanced Packaging, 1999:78–85.
197. Thust H, Drue K-H, Thelemann T, Polzer EK, Müller J. Performance of Buried Resistors in Green Tape 951. Proc. 1997 Int. Symp. on Microelectronics, ISHM 97:48–53.
198. Thust H, Drue K-H, Thelemann T, Polzer EK, Müller J. Is Buried Better?-Evaluating the Performance of Buried Resistors in LTCCs. Advanced Packaging, March/April, 1998:40–46.
199. Vest RW. Materials Science of Thick Film Technology. Am Ceram Soc Bull 1986; 65:631.
200. Fukaya M, Higuchi C. Pb-Free Resistor for Pb-Free LTCC System:636–641.
201. Hormadaly J. New Lead-Free Thick Film Resistors. Proc. Int. Symp. on Microelectronics IMAPS 2002:543–547.
202. Wang G, Rajagopalan V, Barlow FD, Elshabini A, Brown W, Ang SS. Design and Fabrication of Embedded Resistors in LTCC for High Frequency Applications. Proc Int Symp on Microelectronics IMAPS 2001:325–331.
203. Wang G, Barlow FD, Elshabini A. Simulation, Characterization and Design of Embedded Resistors in LTCC for High Frequency Applications. Proc Int Symp on Microelectronics IMAPS 2002:851–855.
204. Hagberg J, Kittilä M, Jakku E, Leppävuori S. Fine Line LTCC-Structures by Direct Gravure Printing (DGB) Method. Proc Int Symp on Microelectronics IMAPS 2002: 51–57.
205. Block C. Keramische Streifenleitungsfilter senken Handy-Kosten. Components, 1997:39–40.
206. Okamoto A. Miniaturization and Integration of Passive Components by Multilayer Ceramic Technology. Proc. Electroceramics IV, International Conference on Electronic Ceramics and Application, Aachen, Germany, 1994:1035–1044.

207. Ender U, Helms et al. BE. Improvement of Multilayer Chip LC Filters Utilizing a New Ceramic Material for High Frequency, Low-Loss Functional Devices. Proc. Electroceramics IV, International Conference on Electronic Ceramics and Application, Aachen, Germany, 1994:1075–1082.

208. Inoue T, Kagata H, Kato J, Kameyama I. Multilayer Microwave Devices Employing Bi-Based Dielectric Ceramics with Copper Internal Conductors. Trans Mater Res Soc Jpn 1994; 14 B:1731–1734.

209. Knaijer G, Dechant K, Apté P. Low Loss, Low Temperature Cofired Ceramics with Medium Dielectric Constants. The International Journal of Microcircuits Electronic Packaging 1997; 20:246–252.

210. Wersing W. Microwave Ceramics for Resonators and Filters. Solids State & Material Science 1996; 1:715–731.

211. Inoue T, Kagata H, Kato J, Kameyama I. Multilayer microwave device employing Bi-based dielectric ceramics with copper internal conductors. Trans Mater Res Soc Jpn 1994; 14B:1731–1734.

212. MT Lanagan, D Anderson, A Baker, J Nino, S Perini, CA Randall, TR Shrout, H Sogabe, H-J Youn. High dielectric constant materials development for LTCC. Proc Int Sym on Microelectronics IMASPS 2001:155–160.

213. Kolar D. Chemistry and Properties of Perovskite-type Rare Earth Titanantes for Microwave Applications. Proc. Electroceramics IV, International Conference on Electronic Ceramics and Application, Aachen, Germany, 1994:3–10.

214. Dernovsek O, Naeini A, Preu G, Wersing W, Eberstein M, Schiller WA. LTCC Glass–Ceramic Composites for Microwave Applications. J Eur Ceram Soc 2001; 21:1693–1697.

215. W Wersing. High frequency ceramic dielectrics and their application for microwave components. In: Steele BCH, ed. Electronic Ceramics. New York: Elesvier, 1991: 67–119.

216. JH Rowen. Ferrites in microwave applications. Bell Syst T J 1953; 32:1333.

217. RL Wahlers, CYD Huang, MR Heinz, AH Feingold, J Bielawski, G Slama. Low profile LTCC transformers. Proc Int Sym on Microelectronics IMAPS 2002:76–80.

218. G Passiopoulos, K Lamacraft. The RF impact of coupled components tolerances and gridded ground planes in LTCC technology and their design countermeasures. Proc Int Sym on Microelectronics IMAPS 2002:279–284.

219. A Naeini, O Dernovsek, M Fritz, R Waser, W Wersing. Innovative microwave filter design with a new high dielectric LTCC-material. Proc. 11th MIOP Conference 2001:298–302.

Index

643

ynes UK
tent Group UK Ltd.
03071024
00031B/2631